Cholesterol Metabolism, LDL, and the LDL Receptor

N.B. Myant

Medical Research Council
Royal Postgraduate Medical School
Hammersmith Hospital
London, England

with a foreword by

M.S. Brown and J.L. Goldstein

Department of Molecular Genetics
The University of Texas Southwestern Medical Center

Cholesterol Metabolism, LDL, and the LDL Receptor

Academic Press, Inc.
Harcourt Brace Jovanovich, Publishers
San Diego New York Boston
London Sydney Tokyo Toronto

This book is printed on acid-free paper. 

Academic Press, Inc.
San Diego, California 92101

United Kingdom Edition published by
Academic Press Limited
24–28 Oval Road, London NW1 7DX

Library of Congress Cataloging-in-Publication Data

Myant, N. B.
Cholesterol metabolism, LDL, and the LDL receptor / N.B. Myant.
p. cm.
Includes index.
ISBN 0-12-512300-0 (alk. paper)
1. Low density lipoproteins. 2. Low density lipoproteins--Receptors. 3. Cholesterol--Metabolism. I. Title
[DNLM: 1. Lipoproteins, LDL Cholesterol--metabolism.
2. Receptors, LDL--physiology. QU 95 M995c]
QP99.3.L68M83 1990
599'.0192454--dc20
DNLM/DLC
for Library of Congress 89-15099
CIP

Printed in the United States of America
90 91 92 93 9 8 7 6 5 4 3 2 1

Contents

Foreword: Simplifying Cholesterol Metabolism

"Everything should be made as simple as possible, but not simpler."
Albert Einstein

At the height of its fury, the debate over cholesterol and heart attacks usually boils down to conflicting interpretations of statistical epidemiology. Ignored is the substantial body of experimental science, accumulated over more than a half century, that elucidated the mechanism by which cholesterol is absorbed and synthesized, described its physiological roles, and provided a plausible hypothesis for the harmful effects of cholesterol-carrying lipoproteins upon artery walls. Remarkably, these dispassionate biological studies came to the same conclusion that was reached by the epidemiologists—namely, that the usual concentration of cholesterol-carrying low density lipoprotein (LDL) among people in industrialized societies is entirely too high.

The breakthroughs in the understanding of cholesterol transport and metabolism grew out of the productive union of two scientific fields. Before 1970, a group of biochemists had gathered a huge body of information on the chemistry of cholesterol and its beautifully orchestrated assembly from a simple two-carbon precursor. Simultaneously, other scientists had begun to study the physiology of cholesterol-carrying lipoproteins in plasma. The lipoproteins were divided into classes, which varied in amount among individuals, often correlating positively or negatively with coronary artery disease. An early classification system apportioned quantitative variants into distinct diseases, and the beginnings of a genetic analysis emerged.

What was missing in 1970 was the integration of these two fields. How does cholesterol transport in plasma relate to cholesterol biosynthesis within cells? How does the body determine how much cholesterol needs to be secreted into blood, and how are particles as complex as cholesterol-carrying lipoproteins removed from plasma? How do genes influence these processes? Answers emerged when the sterol biochemists and the lipoprotein physiologists were welded together in the crucible of cell biology. The bonding agent was genetics and the catalyst was a human disease, familial hypercholesterolemia (FH).

Unraveling the defect in FH may be compared to the events in an Agatha Christie mystery. The prime suspect surfaced in 1973 when the cultured human fibroblast was introduced as a model for the study of genetic defects in lipoprotein metabolism. Normal human fibroblasts were shown to use cholesterol from plasma LDL and thereby to keep their own cholesterol synthesis suppressed. Cells from patients with the homozygous form of FH were unable to use LDL-cholesterol. They survived in tissue culture because they activated their endogenous cholesterol biosynthetic pathways. In tracking down the cause for this defect, it was found that normal cells possess a surface receptor that specifically binds LDL and initiates a series of events by which the LDL is internalized into the cell and degraded in lysosomes, liberating its cholesterol for metabolic purposes. Cells from FH homozygotes turned out to lack functional LDL receptors and therein lay the culprit. These initial observations stimulated a long series of studies that defined the properties of the LDL receptor and ultimately led to its purification and to the cloning of its gene. Twelve years later, in 1985, it was possible to pinpoint a specific defect in the DNA of the LDL receptor gene from an FH homozygote, thereby completing this chapter of the mystery. At the same time that these *in vitro* studies were being performed, a variety of *in vivo* studies defined the role of the LDL receptor in organs of intact animals and humans.

The past two decades of fast-moving research in LDL metabolism are admirably chronicled in this volume by Nicholas B. Myant, who himself has been an important contributor to the solution of the FH mystery. In a detailed and scholarly fashion, Myant guides us through the vast literature, pointing out many of the false clues, apparent contradictions, and unanswered questions, as well as the facts that seem to be fairly well resolved.

This book can be viewed as a companion to Myant's earlier volume on *The Biology of Cholesterol and Related Steroids*. Together, the two volumes unite the two disciplines that are fundamental to an understanding of this field—cholesterol biochemistry and lipoprotein metabolism. Inasmuch as Myant's career spans both of these disciplines, he is the ideal individual to have written these two volumes. The current book will be of great interest to anyone who wishes to read a scholarly (and not oversimplistic) evaluation of an exceedingly complex, yet ultimately important, problem in human biology. This book also provides an indispensable guide and a firm theoretical foundation to those who must continue to do battle in the epidemiological arena of cholesterol and heart attacks.

Michael S. Brown
Joseph L. Goldstein

Preface

In this book I have focused primarily on the way in which the cell content of free cholesterol is kept at an optimal level by the coordinated actions of three regulated processes: the intracellular synthesis of cholesterol, its esterification by ACAT, and the receptor-mediated uptake of LDL. I have made no attempt to survey cholesterol metabolism as a whole. The reader will find little or no mention of, for example, cholesterol absorption from the intestine, reverse cholesterol transport, or the conversion of cholesterol into biologically essential steroids. However, the approach I have taken has enabled me to deal in depth with a circumscribed aspect of cholesterol metabolism in which there have been spectacular advances at the subcellular and molecular levels within the past few years.

Knowledge of the molecular biology of lipoproteins and their cell-surface receptors is advancing so rapidly that any judgment on a controversial question is likely to be overtaken by events before it has time to appear in print. Nevertheless, where there is uncertainty, as over the question of a separate receptor for chylomicron remnants, I have tried to assess the balance of the evidence on both sides. In this regard, it is useful to bear in mind a memorandum on intelligence gathering issued by the Indian Army during World War II. "The reliability of information should be graded as follows: A, Completely reliable; B, High grade observer on matter not seen before; C, Average observer on matter seen before; D, Average observer on matter not seen before."

Many of the points discussed in the text illustrate principles whose significance extends far beyond that of the biology of cholesterol. Examples are the interaction of ligands with their cell-surface receptors, the role of coated pits in the endocytosis of receptor-bound ligands and the recycling of receptors through the interior of the cell, the intracellular transport of membrane-bound proteins, the regulation of expression of genes encoding inducible proteins, the use of natural and synthetic mutations in studies of the functions of the separate domains of a multifunctional protein, and the probable evolutionary history of genes that encode proteins with multiple domains. In most of these cases I have provided background information for the nonspecialist (not enough for some; too much for others). Also for the convenience of the nonspecialist who may not wish to read through the complete text, I have made each chapter to some extent self-contained, inevitably at the cost of some repetition.

ApoB, the protein component of LDL, could have been included in the chapter on the physical and chemical characteristics of LDL (Chapter 5). However, giving it a separate chapter (Chapter 6) has provided scope for a full discussion of recent work on the cloning of the apoB gene, on the receptor-binding domain of apoB-100, and on the unusual mode of derivation of apoB-48. In Chapter 6 I have also included some general remarks on genetic polymorphism before discussing polymorphism in the apoB gene itself.

As, I believe, most readers would expect, a substantial proportion of this book is devoted to the LDL receptor (much of Chapters 8, 9, and 10, and many references to the LDL-receptor pathway in other chapters). I hope I have succeeded in conveying something of the fascination of this remarkable protein and of its equally remarkable gene. Table 9.2 shows a partial list of the known spontaneous mutations in the receptor gene. More than 20 have already been fully characterized and it is certain that more will be discovered. Presumably, there is a limit to the number of mutations in the receptor gene that are capable of producing detectable changes in receptor function. However, we cannot at present say what this limit is likely to be.

The clinical consequences of genetic dysfunction of the LDL receptor are dealt with in the last chapter. Here I deal only with approaches to the diagnosis and treatment of familial hypercholesterolemia (FH) that are based wholly or in part on current knowledge of the LDL receptor or its gene. The practicability of gene therapy for monogenic diseases in which the mutant gene has been identified is now a topic for serious consideration. Accordingly, in Chapter 10 I have speculated on the routes that may eventually lead to successful gene therapy for receptor-negative homozygous FH.

In conclusion, I must declare a personal interest in FH. I began to study FH in 1963. Ever since then I have watched the story unfold to what, in a sense, is its conclusion. It is hardly possible to go deeper than the random genetic events underlying the whole gamut of clinical and biochemical changes that characterize FH. First, there was the proof that FH is a monogenically inherited disease. Then came the demonstration that it is primarily a disorder of LDL metabolism. This clue led to the discovery of the LDL receptor by Brown and Goldstein and the recognition, as a result of work in many laboratories, that all the manifestations of classical FH can be explained by defective LDL-receptor function. The final stages, involving isolation of the LDL receptor, the cloning of its gene, and the analysis of many of its mutant forms, are outlined in Chapter 9. The reader who has come to grips with this chapter will be in a position to appreciate how much imaginative work has gone into these investigations.

Acknowledgments

I thank G. F. Gibbons (Oxford), S. E. Humphries (Charing Cross Hospital), and D. W. Russell (Dallas) for reading and commenting on one or more chapters in draft. I also thank S. E. Humphries, M. Jones, B. L. Knight, A. K. Soutar, and J. Scott for discussion of some of the topics covered in this book. I am grateful to F. E. Baralle for allowing me to work in his laboratory under the guidance of C. C. Shoulders. This enabled me to acquire a practical knowledge of molecular genetics after I had retired from working for the Medical Research Council in 1983. Jean De Luca typed innumerable drafts of each chapter with her usual speed and accuracy. I pay tribute to the patience with which she accepted my frequent requests for deletions, insertions, or transpositions. I also wish to express my appreciation to my wife for her forbearance while I was writing at home during evenings and weekends. It would have been much harder to complete this book without help from the staff at the Wellcome Library, Royal Postgraduate Medical School (Lindsay Curtis, Liz Davis, Judith Stone, Farideh Moussavi, John Matthews, and Paul Stokes). I am also grateful to Nancy Olsen (Academic Press) for handling the production of my manuscript with exemplary care and understanding. Finally, I thank J. L. Goldstein for persuading me, on our way back from a visit to Darwin's house at Down, to write a text on the molecular biology of cholesterol metabolism.

N. B. MYANT
London
August, 1989

Chapter 1

The LDL Receptor in Perspective

All cells of vertebrates require cholesterol for their general activities of growth, replication, and maintenance. In carrying out these basic functions the typical animal cell maintains homeostasis with respect to cholesterol, the turnover and content of cell cholesterol remaining constant within quite narrow limits. Only in exceptional circumstances, such as occur during developmental differentiation or tissue regeneration after injury, is homeostasis interrupted. Homeostasis is achieved mainly by balancing input with output of cell cholesterol. However, the continual interconversion of intracellular free and esterified cholesterol provides the cell with an additional safeguard against all but the smallest fluctuations in its content of free cholesterol. As well as their general functions, in which cells are working for themselves, certain cells require cholesterol for specialized activities that serve the interests of the organism as a whole. These activities include the synthesis of bile acids by the liver and the secretion of lipoproteins by liver and intestine. In these cells, cholesterol homeostasis is less evident because the cell has to respond to the variable needs of the organism. For example, the response to sudden stress requires a rapid and massive increase in free cholesterol as substrate for steroid-hormone synthesis in the adrenal cortex. Likewise, during the absorption of fat from the intestine there are very marked changes in the cholesterol content of the intestine and liver.

In all animal cells the principal sources of cholesterol are synthesis *in situ* and uptake of cholesterol-rich lipoproteins present in the extracellular medium. Cholesterol synthesis is adjusted in accordance with the amount of sterol present in a regulatory pool within the cell, though the details of this are controversial (see Chapter 3). The lipoproteins that carry extracellular cholesterol into the cell are part of a complex system that has evolved to facilitate the transport of lipids, essentially insoluble in water, between the different tissues of the body. Most of the cellular uptake of cholesterol from the external medium is mediated by a sequence of processes that begins with the binding of lipoprotein particles to receptors on the plasma membrane. These receptors, of which there are several classes, have high affinity for a specific protein or combination of proteins in lipoproteins. Hence, the tissue distribution of lipoprotein receptors helps to determine the relative contributions of different tissues to the total uptake and degradation of each class of plasma lipoprotein.

In addition to receptor-mediated uptake, cells also take up lipoproteins by receptor-independent processes, though these processes make only a small contribution to the total supply of lipoprotein cholesterol to most tissues under physiological conditions. If there is an effective concentration gradient of free cholesterol between the polar shell of lipoproteins and the plasma membranes of cells, there may also be a net flux of free cholesterol from lipoproteins to cells. But this source of cell cholesterol is minimal except in the presence of certain abnormal lipoproteins enriched with free cholesterol. Finally, the cells of certain tissues, notably the adrenal cortex, take up esterified cholesterol from surface-bound high-density lipoproteins (HDL) without internalization of the lipoprotein particle.

In summary, then, cells obtain cholesterol from the external medium by three routes: (1) receptor-dependent and receptor-independent uptake of whole lipoprotein particles, resulting mainly in the uptake of esterified cholesterol, (2) selective uptake of esterified cholesterol from surface-bound HDL, and (3) the flow of free cholesterol from lipoproteins to cells without uptake of lipoprotein particles.

The relative importance of different lipoproteins in supplying cells with cholesterol from the external medium varies between species and between different tissues in a given species. The lipoprotein receptor responsible for the bulk of the receptor-mediated uptake of lipoproteins by extrahepatic cells in man is the LDL receptor. The activity of the LDL-receptor pathway expressed by a particular cell is controlled by the amount of sterol present in the regulatory pool that controls the rate of synthesis of cholesterol within the cell. Thus, the two major sources of supply of cholesterol to the typical extrahepatic cell—synthesis *in situ* and receptor-mediated uptake of LDL—are regulated mutually through interaction with a common regulatory pool of sterol. In this way, homeostasis at the level of the cell is maintained by reciprocal changes in cholesterol synthesis and uptake of LDL.

In cells that serve functions in relation to cholesterol over and above those of satisfying their own requirements, more complex regulatory mechanisms come into play. In these cases, the LDL-receptor pathway is not necessarily the major route for entry of extracellular cholesterol into the cell. For example, the liver, in its capacity as an organ responsible for taking up and redistributing exogenous cholesterol that enters the plasma during fat absorption, appears to use a receptor-mediated system other than the LDL-receptor pathway. Moreover, the activity of this additional receptor pathway is not controlled by the amount of cholesterol present in the regulatory pool that controls the activity of the LDL-receptor pathway.

The above survey shows how the LDL-receptor pathway fits into the general framework of cholesterol metabolism in the whole body. In Chapter 2, the general and specialized functions of cholesterol are dealt with briefly, followed by a discussion of the relative importance of synthesis *in situ* and uptake of lipoproteins to the total supply of cholesterol in different tissues.

Chapter 2

Cholesterol in Animal Tissues

I. Functions and Distribution

This section deals with the functions and distribution of cholesterol only in sufficient detail to provide a background to the text. For a more detailed discussion, see Myant (1981, Chapters 3, 6, and 7).

Cholesterol is required by all animal cells for growth and maintenance, its general function being to serve as a stabilizing constituent of the bilaminar membranes of cells. It is also a constituent of the myelin sheath of nerves, a modified plasma membrane, and is present in all plasma lipoproteins. Cholesterol is required by the liver and intestine as an essential component of lipoproteins. It is also a precursor of the bile acids formed in the liver and of steroid hormones formed in the cells of the gonads, placenta, and adrenal cortex. In each case, the requirement for cholesterol is much greater in cells that are dividing or growing rapidly than in those that are in a resting state, and the requirement by specialized cells for cholesterol as biochemical precursor varies according to the needs of the organism as a whole for bile acids or steroid hormone.

The cholesterol in tissues and plasma may be in free (unesterified) form, in which the polar 3β-OH group is exposed, or the 3β-OH group may be esterified with long-chain fatty acids. About 90% of the total cholesterol in the animal body is present as free cholesterol in cell membranes, myelin, and the polar shell of plasma lipoprotein particles. The cholesterol used as substrate for the formation of bile acids and steroid hormones is also unesterified.

All animal cells contain some esterified cholesterol, although in most tissues this accounts for less than 10% of the total. In plasma, about two-thirds of the cholesterol is esterified, predominantly with long-chain fatty acids containing two or more double bonds. In certain specialized tissues such as the adrenal cortex, an even higher proportion of the total cholesterol is esterified. The high proportion of esterified cholesterol in plasma reflects the need to transport large quantities of cholesterol through the plasma in a form that cannot interact harmfully with plasma membranes. Similarly, the esterified cholesterol in cells of the adrenal cortex and gonads is present in lipid storage droplets that provide these cells with a reservoir of cholesterol in a physiologically inactive form capable of rapidly releasing free cholesterol when this is required. In some pathological states the cells of certain tissues, particularly the reticuloendothelial system, become loaded with cholesteryl-ester-filled droplets. Owing to their histological appearance after defatting, these are known as foam cells.

II. Sources of Cholesterol in Different Tissues

All animal cells are capable of synthesizing cholesterol, though the rate of synthesis varies widely from one tissue to another and in a given tissue under different physiological conditions. In general, synthesis is relatively rapid in cells that are growing or dividing and in those that require cholesterol for the formation of bile acids, steroid hormones, or plasma lipoproteins; under most other conditions cholesterol synthesis is comparatively quiescent. For example, synthesis is very active in cells of the developing nervous system and in those of the adrenal cortex responding to maximal stress. On the other hand, cholesterol is synthesized at a low rate in skeletal muscle and mature brain, and in cells that have reached confluence in culture. In these cases, cholesterol is required only in quantities sufficient to maintain membrane turnover. However, as noted below, tissues such as skeletal muscle and skin, in which the rate of synthesis per gram is low, may make a substantial contribution to whole-body synthesis by virtue of their large mass.

Knowledge of the rate of synthesis of cholesterol in particular tissues is necessary if we are to deduce the relative contributions of synthesis *in situ* and uptake of lipoprotein cholesterol to the total supply of cholesterol to the tissues *in vivo*. Dietschy and his colleagues have developed a variety of methods for measuring rates of cholesterol synthesis in the tissues of intact animals. The most reliable are those based on measurement of the incorporation of ^3H from ^3H_2O into digitonin-precipitable sterols in tissues removed from the animal 1 hour after an intravenous injection of ^3H_2O. Using this method in combination with measurements of the incorporation of other sterol precursors into the sterols of isolated tissues *in vitro*, Turley *et al.* (1981) and Spady and Dietschy (1983) have built up a detailed picture of cholesterol synthesis in the tissues of several animal species.

In rats fed a normal diet the highest rates of cholesterol synthesis per gram of tissue occur in the liver, small intestine, and ovary. The lowest rates occur in skeletal muscle, bone marrow, adipose tissue, spleen, and kidney, with intermediate rates in skin, adrenals, lung, and the brains of growing animals. In terms of whole organs, the liver contributes about 50% of the total cholesterol synthesized in the body; the small intestine and carcass (whole body without liver and small intestine) each contribute about 25% of the total. Skin, by virtue of its large mass, makes the largest single contribution from the carcass. These values apply only to rats maintained under standard physiological conditions. Owing to the responsiveness of hepatic cholesterol synthesis to dietary cholesterol and other factors, the actual contribution of the rat's liver to whole-body synthesis may range from 10% during cholesterol feeding to 65% during cholestyramine feeding.

Comparable measurements of cholesterol synthesis in other species have shown that there are substantial species differences in the relative contributions of the liver and other tissues to whole-body synthesis of cholesterol. In particular, the liver makes a smaller contribution to total synthesis in squirrel monkeys (40%), hamsters (37%), rabbits (18%), and guinea pigs (16%) than in rats. In these species, correspondingly larger relative contributions are made by the small intestine, skin, and total mass of muscle. Another species difference lies in the rate of cholesterol synthesis in the adrenals. In the resting state, synthesis in the adrenals of rats and squirrel monkeys is low, whereas in hamsters, rabbits, and guinea pigs the rate of synthesis per gram of tissue is higher in the adrenals than in any other organ.

The methods used for determining the rate of synthesis of cholesterol in the tissues of laboratory animals *in vivo* are not applicable to man. However, indirect evidence, based on measurement of the turnover of cholesterol in the whole body and of synthesis in a few isolated tissues *in vitro,* indicates that the liver makes a much smaller contribution, and the extrahepatic tissues a much larger contribution, to total synthesis in man than in the rat (Spady and Dietschy, 1983).

In light of the above discussion we can now ask the question, "What proportion of the total supply of cholesterol to the tissues of the living animal is contributed by local synthesis and what proportion is derived from lipoproteins in the external medium?" To answer this, we need to know the rate of cholesterol synthesis and of lipoprotein uptake in the various organs *in vivo*. Using an extension of their ^3H_2O method, Turley *et al.* (1981) have measured rates of uptake of lipoprotein cholesterol in several tissues of the rat and have related these measurements to the corresponding rates of synthesis *in situ*.

In rats kept under standard laboratory conditions about 75% of the cholesterol supplied to the extrahepatic tissues as a whole is derived from local synthesis. The remaining 25% is derived from plasma lipoproteins originating predominantly in the liver. In the liver itself, the relative contributions of synthesis *in situ* and uptake of lipoproteins must vary widely according to the cholesterol content of the diet and the time in relation to absorption of fat from the intestine. In the long term, synthesis probably accounts for up to 70% of the total cholesterol supplied to the liver. The relative importance of local synthesis varies widely between different extrahepatic tissues in the rat. At one extreme the spleen and the adrenals in the resting state satisfy more than 90% of their cholesterol requirement by uptake of lipoproteins. At the other extreme, muscle and mature brain synthesize nearly all the cholesterol they require. In other organs, including heart, lung, and kidney, 70–80% of the total requirement for cholesterol is met by uptake of lipoproteins. In keeping with these differences in the degree of dependence upon lipoprotein uptake for the supply of cholesterol, drastic reduction of the plasma lipoprotein concentration by administration of 4-APP leads to increased cholesterol synthesis in adrenals, lung, and kidney but has no effect on cholesterol synthesis in skeletal muscle or brain.

Comparable estimates cannot be made for human tissues. However, as noted above, the human liver probably makes a relatively small contribution to whole-body cholesterol synthesis. This suggests that lipoprotein cholesterol originating in the liver provides a smaller proportion of the cholesterol requirement of extrahepatic tissues as a whole in man than in rats.

The quantitative importance of the various lipoproteins and of receptor-mediated and receptor-independent uptake as sources of cell cholesterol is dealt with in Chapter 7, Section II.

III. Measurement of Whole-Body Synthesis of Cholesterol

Measurements of the rate of synthesis of cholesterol in the whole body are essential in investigations of cholesterol homeostasis in the intact organism and in studies of the mechanisms by which diet and drugs alter the plasma cholesterol concentration.

Methods for measuring the absolute rate of whole-body synthesis in man (mg of cholesterol synthesized/kg body weight/day) are tedious and time consuming. If the subject is in a steady state with respect to cholesterol, the average daily rate of synthesis may be calculated from measurements of sterol balance made over a minimum of 3 weeks under carefully controlled conditions (see Myant, 1981, Chapter 10 for details). If the subject is not in a steady state throughout the study, e.g., because tissue cholesterol is being deposited or mobilized, the sterol balance method is invalid. Methods based on measurement of the turnover of radioactive cholesterol in the plasma are even more time consuming and, again, are applicable only under steady-state conditions.

These limitations have led to the development of semiquantitative methods for determining whole-body cholesterol synthesis over much shorter intervals. The most promising of these methods is based on the observation that the plasma mevalonate concentration is directly related to the rate at which cholesterol is synthesized in the whole body in rats (Popják *et al.*, 1979) and human subjects (Parker *et al.*, 1982). Presumably this relationship holds because small quantities of intracellular mevalonate leak into the plasma at a rate proportional to the rate at which this sterol precursor is formed from HMG-CoA (see Chapter 3, Section

I). The plasma mevalonate concentration responds sufficiently rapidly to reflect diurnal changes in the rate of whole-body synthesis of cholesterol in man. Parker *et al.* (1984) have also shown that a fall in plasma mevalonate concentration can be detected within 6 hours of a single oral dose of mevinolin (a powerful inhibitor of HMG-CoA reductase).

References

Myant, N. B. (1981). "The Biology of Cholesterol and Related Steroids." Heinemann, London.

Parker, T. S., McNamara, D. J., Brown, C., Garrigan, O., Kolb, R., Batwin, H., and Ahrens, E. H. (1982). Mevalonic acid in human plasma: relationship of concentration and circadian rhythm to cholesterol synthesis rates in man. *Proc. Natl. Acad. Sci. U.S.A.* **79,** 3037–3041.

Parker, T. S., McNamara, D. J., Brown, C. D., Kolb, R., Ahrens, E. H., Alberts, A. W., Tobert, J., Chen, J., and De Schepper, P. J. (1984). Plasma mevalonate as a measure of cholesterol synthesis in man. *J. Clin. Invest.* **74,** 795–804.

Popják, G., Boehm, G., Parker, T. S., Edmond, J., Edwards, P. A., and Fogelman, A. M. (1979). Determination of mevalonate in blood plasma in man and rat. Mevalonate "tolerance" tests in man. *J. Lipid Res.* **201,** 716–728.

Spady, D. K., and Dietschy, J. M. (1983). Sterol synthesis *in vivo* in 18 tissues of the squirrel monkey, guinea pig, rabbit, hamster, and rat. *J. Lipid Res.* **24,** 303–315.

Turley, S. D., Andersen, J. M., and Dietschy, J. M. (1981). Rates of sterol synthesis and uptake in the major organs of the rat *in vivo*. *J. Lipid Res.* **22,** 551–569.

Chapter 3

HMG-CoA Reductase

I. The Rate-Limiting Step in Cholesterol Biosynthesis

The initial steps in the biosynthesis of cholesterol consist in the conversion of acetyl-CoA, the normal primary precursor of sterols, into the CoA thioester of 3-hydroxy-3-methylglutaric acid (HMG-CoA). HMG-CoA is then reduced to mevalonic acid by the enzyme HMG-CoA reductase with NADPH as the hydrogen donor. This reaction takes place in a sequence of steps beginning with the formation of an HMG-CoA enzyme complex. In the presence of NADPH a histidyl residue in the active site of the enzyme is protonated. The esterified carboxyl carbon of HMG-CoA is then reduced to give the thiohemiacetal. A second reduction with NADPH gives mevalonic acid and CoA, which are released from the complex (see Rogers et al., 1983, for details). Mevalonic acid is converted into the branched C_5 isoprenoid unit, isopentenyl pyrophosphate (IPP), by phosphorylation–decarboxylation. IPP is the building unit from which several polyisoprene compounds are formed in animal cells. These include cholesterol, dolichols (acyclic polyisoprenoid alcohols containing up to 22 isoprenoid units), and the hydrophobic tail of coenzyme Q (ubiquinone); IPP also supplies the isopentenyl unit required for the synthesis of isopentenyl tRNAs (see Section V,G).

The last intermediate common to the pathways from IPP to ubiquinone, sterols, and dolichol is farnesyl pyrophosphate (FPP), a C_{15} compound formed by the sequential condensation of three isoprenoid units. In the biosynthesis of sterols two molecules of FPP condense to form squalene, a reaction catalyzed by squalene synthetase. In the presence of molecular oxygen, squalene is oxidized to 2,3-oxidosqualene by squalene epoxidase. Oxidosqualene cyclizes to form lanosterol, which is then converted into a sterol by a sequence of at least eight steps. One of these steps involves a cytochrome *P*-450-dependent hydroxylation, also requiring free O_2 (see Beytia and Porter, 1976; Myant, 1981).

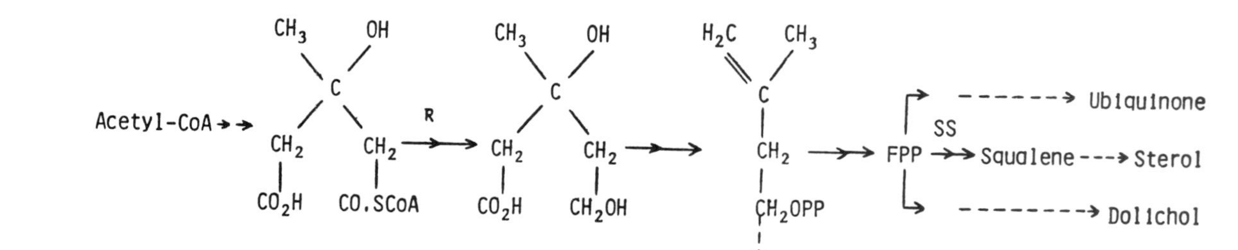

Figure 3.1. The formation of mevalonic acid and the products of its metabolism in animal cells. IPP, isopentenyl pyrophosphate; FPP, farnesyl pyrophosphate; R, reductase; SS, squalene synthetase.

The essential features of these pathways are shown in Fig. 3.1.

Under most conditions *in vivo* and in isolated cells the rate at which acetyl-CoA is converted into cholesterol is determined by the rate at which HMG-CoA is converted into mevalonic acid and this, in turn, is determined by the activity of HMG-CoA reductase. Thus, regulation of the rate of synthesis of cholesterol is, in general, mediated by regulation of the activity of HMG-CoA reductase (see Myant, 1981). There are, however, exceptions to this. For example, in some conditions the supply of acetyl-CoA or the activity of HMG-CoA synthase may become rate limiting (Clinkenbeard *et al.*, 1975). The steps that are rate limiting for cholesterol biosynthesis when mevalonic acid is the primary precursor are discussed in Section V,G.

II. Composition, Structure, and Orientation

A. Molecular Weight

HMG-CoA reductase in animal tissues is an integral membrane glycoprotein located in the smooth endoplasmic reticulum (ER). The binding of the enzyme to the ER and the way in which it is oriented in relation to the membrane explain why there has been long-standing controversy over its true molecular weight, a controversy now settled by the identification of cDNAs spanning the whole coding region of the message for the enzyme (Chin *et al.*, 1984).

Much of the early work on the enzyme present in rat liver was carried out with partially purified preparations of catalytically active protein solubilized by repeated freezing and thawing of microsomes (Heller and Gould, 1973). The enzyme prepared by this procedure has a molecular weight of 50,000–55,000 or multiples thereof. Hence, it was concluded that the native enzyme exists as a dimer or tetramer of unit M_r 50–55K. However, Ness *et al.* (1981) showed that HMG-CoA reductase cannot be solubilized by freeze-thawing in the presence of leupeptin, an inhibitor of the lysosomal proteinase cathepsin T. This, together with other evidence, indicated that the 50 to 55-K enzyme is a fragment cleaved from the cytoplasmic extension of the native membrane-bound enzyme by proteolytic enzymes released from ruptured lysosomes during the freezing and thawing. This fragment contains the active site, and its catalytic properties are identical with those of the native enzyme (Chang, 1983). It is also capable of generating antibodies that react with the native enzyme.

Chin *et al.* (1982b) and Liscum *et al.* (1983a), using detergent solubilization in the presence of proteinase inhibitors, isolated what was clearly the native enzyme from UT-1 cells, a strain of Chinese hamster ovary (CHO) cells. These cells, referred to in Section IV,D, have up to 500 times the normal content of HMG-CoA reductase and a greatly increased content of reductase mRNA. The soluble enzyme, labeled by preincubating the cells with [^{35}S]methionine, was immunoprecipitated with an antibody raised against the 50 to 55-kDa enzyme prepared from rat liver. The apparent M_r of the detergent-solubilized enzyme, determined by sodium dodecyl sulfate (SDS)–gel electrophoresis, was 90–94 kDa. This is in adequate agreement with the value of 97,092 for the molecular weight of the protein component of the enzyme, based on the amino acid sequence deduced from the nucleotide sequence of the full-length cDNA described by Chin *et al.* (1984). The agreement between the molecular weight of the solubilized native protein and that determined from the nucleotide sequence of its mRNA indicates that the enzyme is present in the cell as a monomer.

B. Glycosylation

General Remarks

The importance of the glycosylation of HMG-CoA reductase in relation to its orientation and biogenesis is such that a word or two about glycosylation of membrane-associated proteins in general may be helpful. Proteins may be glycosylated by linkage of one or more preformed oligosaccharide units to the free NH_2 groups of asparagine residues forming part of the consensus sequence -Asn-X-$^{\text{Ser}}_{\text{Thr}}$, where X is any amino acid (N-linked oligosaccharides), or they may be glycosylated by linkage of oligosaccharide to the OH of a serine, threonine, or hydroxylysine residue (O-linked oligosaccharides). The N-linkage takes place in the lumen of the ER as soon as a segment of the growing peptide chain containing an appropriate asparagine residue has entered the lumen. The oligosaccharide unit, containing *N*-acetylglucosamine, glucose, and up to nine mannose residues, is transferred enzymically to the protein from a carrier molecule of dolichol bound to the ER membrane and projecting into the lumen. The N-linked oligosaccharides are subsequently modified by removal of the terminal glucose and of one or more of the mannose residues to give ''high-mannose'' oligosaccharides, or by removal of most of the mannose residues and the addition of sialic acid and other carbohydrate residues to give ''complex'' oligosaccharides. The formation of complex and of O-linked oligosaccharides is thought to take place in the Golgi apparatus. Thus, N-linked glycosylation of a protein indicates that, during its biogenesis, the N-terminal segment of the protein has entered the ER lumen at least as far as the first asparagine residue capable of accepting an oligosaccharide unit from dolichol. Likewise, the presence of O-linked or of complex N-linked oligosaccharide indicates that the protein has, at some stage in its life, traveled from the ER to the Golgi apparatus.

Liscum *et al.* (1983a) showed that the native detergent-solubilized reductase from UT-1 cells ($M_r \cong 97$ kDa) is adsorbed specifically by concanavalin A (a plant lectin that binds to glucose and mannose residues), indicating that HMG-CoA reductase is a glycoprotein. This was confirmed by showing that [^{3}H]glucosamine is incorporated into the enzyme when the cells are incubated in the presence of the radioactive precursor. Analysis of the radioactive fragments released by digesting the ^{3}H-labeled 97-kDa enzyme with an acetylglucosaminidase showed that the carbohydrate was an N-linked oligosaccharide with "high-mannose" chains and that O-linked oligosaccharide was not present. Proteolytic cleavage of the enzyme while it was still bound to intact microsomes led to the release of catalytically active 53-kDa and 62-kDa fragments that contained no carbohydrate; all the carbohydrate present in the 97-kDa solubilized enzyme was found in a 30 to 35-kDa fragment that could be separated form the 53-kDa portion by proteolysis. Brown and Simoni (1984), using a different line of hamster cells (C100), have also reported the incorporation of [^{3}H]mannose into a solubilized 92-kDa preparation of HMG-CoA reductase and have shown that incorporation is prevented by tunicamycin, a selective inhibitor of N-glycosylation of proteins. For reasons considered in the next section, each molecule of enzyme is thought to contain one N-linked oligosaccharide unit. The implications of these findings are discussed below in relation to the orientation of the enzyme with respect to the ER membrane.

C. Structure and Orientation

The complete amino acid sequence of hamster HMG-CoA reductase has been deduced by Chin *et al.* (1984) from a cDNA spanning the whole of the coding sequence of reductase mRNA. Their first step (Luskey *et al.*, 1982) was to prepare a library of recombinant plasmids containing cDNA inserts complementary to mRNA from UT-1 cells. Colonies containing inserts encoding amino acid sequences in the enzyme were identified by differential colony hybridization, using as probes the ^{32}P-labeled cDNAs synthesized from RNA of UT-1 cells and of CHO cells in which transcription of the HMG-CoA reductase gene was suppressed by LDL (and which therefore contained no mRNA for the enzyme). Five colonies that hybridized with cDNA prepared from UT-1 cells but not with that from suppressed CHO cells were selected. From these five, three colonies were selected on the basis of their ability to hybridize with an mRNA which, when translated *in vitro*, gave rise to a protein that was precipitable with anti-reductase antibody. The plasmid from one of these colonies (pRed10) was used to generate longer reductase cDNAs by standard molecular cloning techniques. The longest of these, an insert in plasmid pRed227, was 4.5 kb long. This cDNA spanned the complete coding region of reductase mRNA and extended for 163 nucleotides into the 5′-untranslated region and for 1.65 kb into the 3′-untranslated region. The sequence of amino acids encoded by pRed227 is shown in Fig. 3.2.

The coding sequence of cDNA pRed227 specifies a protein containing 887 amino acids. The methionine encoded by the initiator codon in the message is retained in the mature enzyme protein (Brown and Simoni, 1984; Chin *et al.*, 1984). Hence, in contrast to most glycosylated membrane-associated proteins,

1 10	20	[1] 30	40	50
MLSRLFRMHG	LFVASHPWEV	IVGTVTLTIC	MMSMNMFTGN	NKICGWNYEC
60	[2] 70	80	90	[3] 100
PKFEEDVLSS	DIIILTITRC	IAILYIYFQF	QNLRQLGSKY	ILGIAGLFTI
110	120	130	[4] 140	150
FSSFVFSTVV	IHFLDKELTG	LNEALPFFLL	LIDLSRASAL	AKFALSSNSQ
160	170	[5] 180	190	[6] 200
DEVRENIARG	MAILGPTFTL	DALVECLVIG	VGTMSGVRLQ	EIMCCFGCMS
210	220	230	240	250
VLANYFVFMT	FFPACVSLVL	ELSRESREGR	PIWQLSHFAR	VLEEEENKPN
260	270	280	290	300
PVTQRVKMIM	SLGLVLVHAH	SRWIADPSPQ	[NST]TEHSKVS	LGLDEDVSKR
310	320	[7] 330	340	350
IEPSVSLWQF	YLSKMISMDI	EQVVTLSLAF	LLAVKYIFFE	QAETESTLSL
360	370	380	390	400
KNPITSPVVT	PKKAPDNCCR	REPLLVRRSE	KLSSVEEEPG	VSQDRKVEVI
410	420	430	440	450
KPLVVETESA	SRATFVLGAS	GTSPPVAART	QELEIELPSE	PRPNEECLQI

Figure 3.2. The amino acid sequence of hamster HMG-CoA reductase (N-terminal = 1; C-terminal = 887). Lines beneath letters denote postulated membrane-spanning regions; boxed letters denote potential sites for N-glycosylation. A, Ala; C, Cys; D, Asp; E, Glu; F, Phe; G, Gly; H, His; I, Ileu; K, Lys; L, Leu; M, Met; N, Asn; P, Pro; Q, Gln; R, Arg; S, Ser; T, Thr; V, Val; W, Trp; Y, Tyr. (Modified from Liscum *et al.*, 1985.)

the immediate product of translation of HMG-CoA reductase mRNA does not contain a cleavable signal sequence at its N-terminal end. Chin *et al.* (1984) have also shown that the amino acid sequence of a segment of the mature protein that includes the C-terminal alanine is the same as that predicted from the nucleotide sequence of the full-length cDNA. Thus, the amino acid sequence of the mature protein, from the N terminus to the C terminus, must be the same as that predicted from pRed227 cDNA. Points to note are that three of the asparagine

460	470	480	490	500
LESAEKGAKF	LSDAEIIQLV	NAKHIPAYKL	ETLMETHERG	VSIRRQLLST
510	520	530	540	550
KLPEPSSLQY	LPYRDY[NYS]L	VMGACCENVI	GYMPIPVGVA	GPLCLDGKEY
560	570	580	590	600
QVPMATTEGC	LVASTNRGCR	AIGLGGGASS	RVLADGMTRG	PVVRLPRACD
610	620	630	640	650
SAEVKAWLET	PEGFAVIKDA	FDSTSRFARL	QKLHVTMAGR	NLYIRFQSKT
660	670	680	690	700
GDAMGMNMIS	KGTEKALLKL	QEFFPEMQIL	AVSGNYCTDK	KPAAINWIEG
710	720	730	740	750
RGKTVVCEAV	IPAKVVREVL	KTTTEAMIDV	NINKNLVGSA	MAGSIGGYNA
760	770	780	790	800
HAANIVTAIY	IACGQDAAQN	VGSSNCITLM	EASGPTNEDL	YISCTMPSIE
810	820	830	840	850
IGTVGGGTNL	LPQQACLQML	GVQGACKDNP	GENARQLARI	VCGTVMAGEL
860	870	880	887	
SLMAALAAGH	LVRSHMVH[NR	S]KINLQDLQG	TCTKKSA	

Figure 3.2. (Cont.)

residues in the protein are at potential glycosylation sites and that the N-terminal third of the protein is hydrophobic whereas the C-terminal two-thirds are hydrophilic. There are no significant repetitions in the amino acid sequence, but plots of hydrophobicity versus amino acid residue number (Kyte and Doolittle, 1982) reveal the presence of seven nonrepetitive hydrophobic regions in the N-terminal half of the protein, each one long enough to span the membrane bilayer. Various features of the secondary structure of the enzyme are considered below.

As already mentioned, HMG-CoA reductase is a membrane-bound protein. In cell-free homogenates submitted to isopycnic centrifugation on density gradients, the enzyme sediments with microsomal vesicles derived from the smooth (ribosome-poor) ER of the intact cells (Mitropoulos *et al.*, 1978b). Moreover, the native enzyme can only be released from microsomal vesicles by dissolving their membranes with detergent (Chin *et al.*, 1982b; Hardeman *et al.*, 1983). This shows that at least a part of the enzyme is embedded within the ER membrane, i.e., HMG-CoA reductase is an *integral membrane protein.*

Two other observations, already referred to, provide clues to the orientation of the enzyme in relation to the ER membrane. First, fragments containing the catalytically active site can be released from microsomes by proteolytic enzymes. This shows that the active site is outside the microsomal vesicle, and since the orientation of the ER membrane is maintained when it forms vesicles (the cytoplasmic face of the ER becoming the outer face of the vesicle), the segment of enzyme protein bearing the active site must project into the cytoplasm in the intact cell. Second, the enzyme is glycosylated with at least one N-linked oligosaccharide. As discussed above, N-linked glycosylation involves the addition of an oligosaccharide unit to an asparagine residue in a segment of the growing peptide chain that has passed through the ER membrane and entered the lumen. It follows, therefore, that a segment of the N-terminal region of the enzyme bearing an asparagine residue projects into the ER lumen.

Taken together, these observations suggest that the enzyme is oriented in the following manner. The protein spans the membrane of the smooth ER, the hydrophilic C-terminal two-thirds projecting into the cytoplasm and containing the active site of the enzyme. The hydrophobic N-terminal third is embedded in the membrane and has an intraluminal projection bearing at least one N-linked oligosaccharide. This suggestion has been confirmed by the more detailed investigation of Liscum *et al.* (1985), using the method of Lerner (1982) to raise antibodies against synthetic peptides containing specific amino acid sequences in a protein. In particular, they have shown that the sequence of 14 amino acids at the C-terminal end of the 97-kDa enzyme protein (residues 874–887) is present in the segment released from microsomes by proteolysis and that the active site of the enzyme is on the C-terminal side of residue 470.

On the basis of the above experimental observations, combined with results of a computer-based analysis of the secondary structure of the complete amino acid sequence, Liscum *et al.* (1985) have proposed the model shown in Fig. 3.3 for the orientation of the enzyme in relation to the endoplasmic reticulum. The salient features of this model are listed as follows.

1. There are seven membrane-spanning regions, each with 23–30 amino acids in α-helical configuration, separated by three luminal and three cytoplasmic loops.

2. The N terminus projects into the ER lumen for 9 amino acids (residues 1–9) and the C-terminal end extends into the cytoplasm for 548 amino acids (residues 340–887).

3. The loop between the sixth and seventh membrane-spanning regions (residues 221–314) projects into the ER lumen and contains the only potential site for

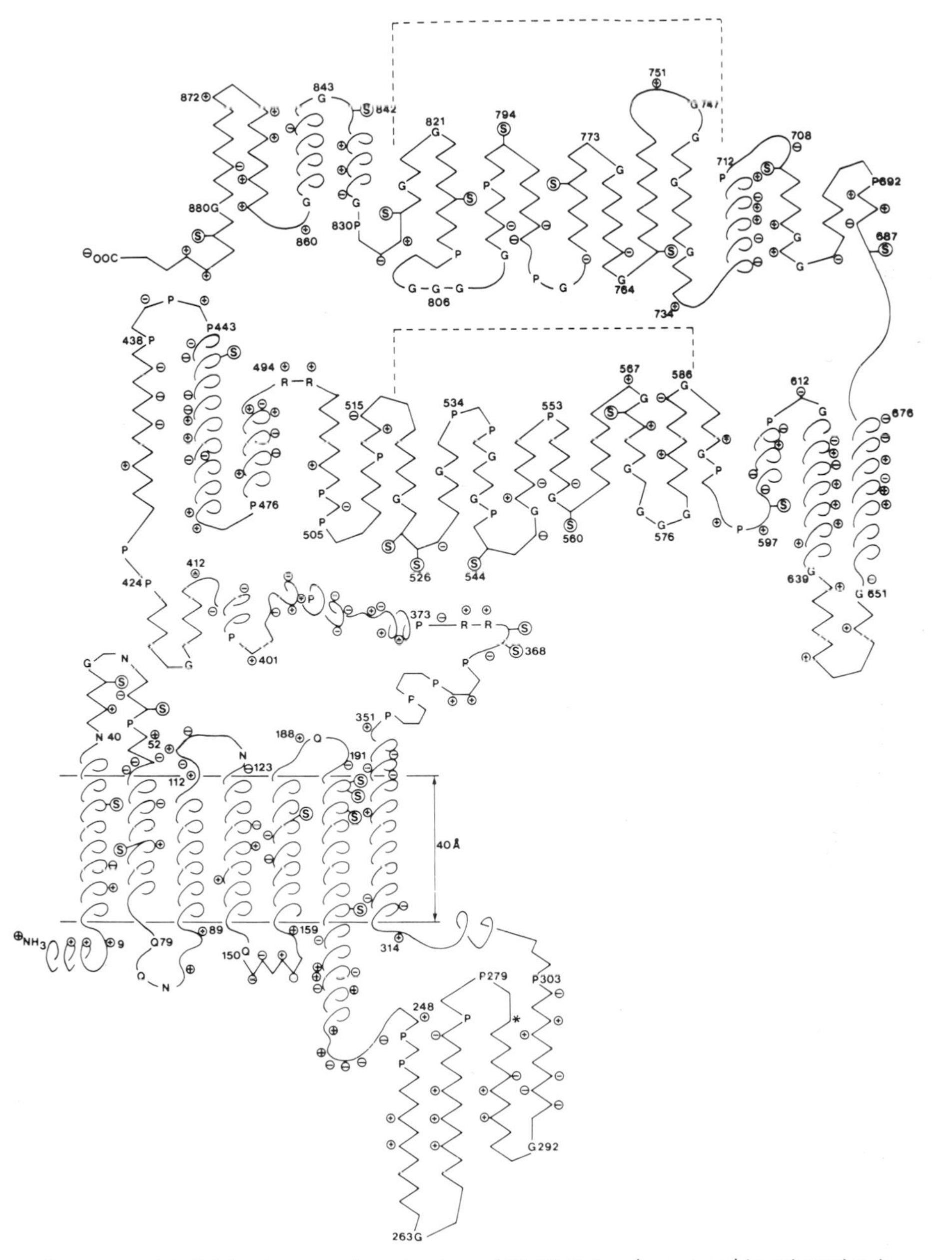

Figure 3.3. Model for the secondary structure of HMG-CoA reductase and its orientation in the ER membrane. The upper surface of the membrane shown in the model faces the cytoplasm; the lower surface faces the ER lumen. The model shows the 7 membrane-spanning α-helical segments in the N-terminal third of the protein and the cytoplasmic C-terminal segment extending from amino acid residue 340 to residue 887. The probable glycosylation site is shown as an asterisk at residue 281 (Arg). G, glycine; N, Asn; P, Pro; Q, Gln; R, Arg; (S), cysteine. (From Liscum *et al.*, 1985, with the permission of the authors.)

N-linked glycosylation ($-\underset{281}{Asn}-\underset{282}{Ser}-\underset{283}{Thr}-$) in the segment of protein known to contain carbohydrate. The two other potential sites are in the cytoplasmic domain, which does not contain carbohydrate. Thus, the model predicts that there is only one oligasaccharide unit per molecule of enzyme protein.

4. The C-terminal cytoplasmic domain contains two cysteine-rich extended β-structures flanked by amphipathic helices (residues 495–595 and 735–825). Liscum *et al.* (1985) suggest that parts of these β-domains contribute to the active site. In keeping with this, there are seven histidyl residues in the peripheral regions of the β-domains. These residues could participate in the catalytic reaction (see p. 9). Some of the seryl residues in or near these regions could also take part in the reversible phosphorylation–dephosphorylation that mediates the rapid inactivation–activation of the enzyme (Kennelly and Rodwell, 1985).

D. Insertion into the RER Membrane

1. General Remarks

Before dealing with the mechanism by which HMG-CoA reductase arrives at its orientation in the smooth ER, we shall first consider certain features of the biogenesis of proteins that associate with membranes in eukaryotic cells, either transiently or permanently. These remarks are also relevant to the question of how the LDL receptor moves from its site of synthesis in the ER to take up its functional position in the plasma membrane (see Chapter 9).

Most proteins destined for secretion by the cell or for insertion into the plasma membrane are synthesized in polysomes bound to the cytoplasmic surface of the rough (ribosome-rich) ER. Before a newly synthesized protein can reach the outside of the cell it must pass through the ER membrane into the lumen. This translocation, referred to as *vectorial discharge,* is facilitated by the presence of a *signal sequence* of 19–30 amino acids at the N-terminal end of the nascent peptide chain. Before translation of the message has been completed, the signal sequence is inserted into the membrane (co-translational insertion). Insertion is followed by discharge of the remainder of the growing peptide chain, including the C-terminal end, into the lumen. While synthesis is still in progress the signal sequence is cleaved from the protein by a protease in the ER lumen (co-translational cleavage) and N-linked oligosaccharides are added, as described above. As a consequence of these and other events, the mature secreted protein usually differs from the immediate product of translation of the message in several respects, including the absence of a signal sequence and the presence of N-linked oligosaccharide. Having entered the ER lumen, nascent proteins that will eventually be secreted are carried from the ER to the Golgi apparatus and then to the cytoplasmic face of the plasma membrane, from whence they are discharged into the external medium. Transport from the ER to the Golgi apparatus and from the Golgi apparatus to the plasma membrane is thought to be mediated by clathrin-coated vesicles (see Chapter 8 for details).

Many proteins that are synthesized in membrane-bound polysomes are not completely discharged into the ER lumen, but remain as integral membrane proteins with one or more segments in the lipid bilayer and with hydrophilic domains in the cytoplasm, the ER lumen, or both. Integral membrane proteins in the ER may be orientated in any one of several different ways. For example, cytochrome *P*-450 spans the smooth ER membrane once and has a C-terminal domain in the cytoplasm and an uncleaved N-terminal domain in the ER lumen. This orientation may be explained by co-translational insertion of the growing peptide chain and its discharge through the membrane, as described above, but with the addition of a sequence of amino acids that interrupts discharge of the nascent protein (a halt transfer signal) and the absence of cleavage of the insertion sequence (Fig. 3.4A). Transport from the rough to the smooth ER could be by lateral diffusion of the intramembrane domain. Other orientations, including those in which the protein spans the membrane more than once, may be explained by various combinations of N-terminal or nonterminal (''internal'') signal sequences and halt transfer signals, with or without cleavage of the N-terminal segment in the ER lumen. Two examples are shown in Fig. 3.4 (see also Sabatini *et al.*, 1982 for a full discussion).

Although it is generally agreed that plasma-membrane proteins are transported from their site of synthesis in the rough ER to the plasma membrane by clathrin-coated vesicles, the details of this process are largely a matter for speculation. It is possible that proteins destined for insertion into the plasma membrane are discharged completely into the ER lumen and are then transported to the Golgi apparatus, and thence to the plasma membrane, in the fluid interior of vesicles. Another possibility is that discharge across the ER membrane is interrupted, as in Fig. 3.4A, the protein remaining bound to the membrane, and that it then undergoes a sequence of membrane-to-membrane transfers (to vesicle membrane, to Golgi membrane, to vesicle membrane, to plasma membrane) by alternating budding and fusion of membranes, without ever ceasing to be membrane bound. The advantage to the cell of such a process would be that the orientation of the protein established in the ER membrane (C-terminal domain in the cytoplasm and N-glycosylated N-terminal domain in the lumen) would be conserved throughout its journey from ER to plasma membrane (see Fig. 3.5). This mechanism would also explain why most transmembrane proteins in the plasma membrane have the N-terminal glycosylated segment (which cannot cross a cell membrane) facing outside the cell.

Little is known about how proteins synthesized in the ER are directed to remain in the ER or to be transported either to the outside of the cell or to other membrane systems. There is some evidence that the presence of an N-linked oligosaccharide, which is itself determined by the presence of an acceptor

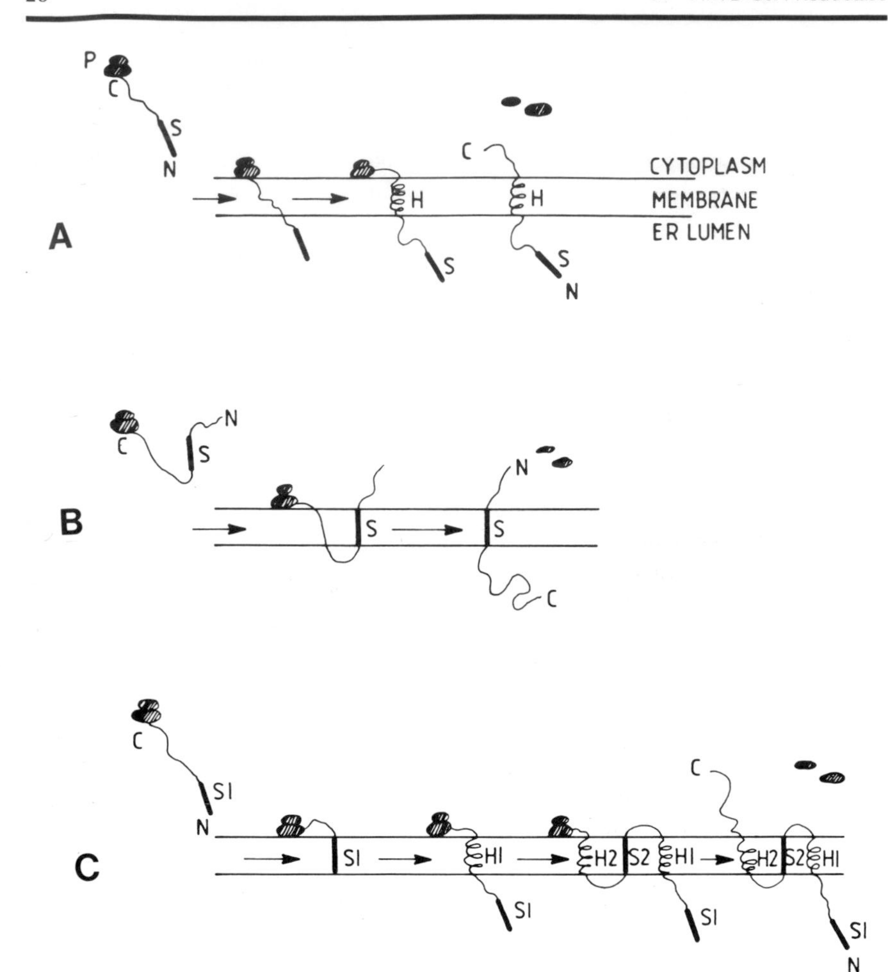

Figure 3.4. Possible mechanisms by which some integral membrane proteins take up their orientation in the ER membrane (based on Blobel, 1980; Sabatini *et al.*, 1982). P, polysome; S, signal sequence; H, halt transfer signal. (A) Topology; N, out; C, in. A terminal signal sequence emerges from the polysome, leading to binding of the polysome to the membrane and co-translational insertion of the signal sequence. Discharge of the growing protein chain across the membrane continues until it is interrupted by a halt transfer signal. The signal sequence is not cleaved. Example: cytochrome *P*-450. (B) Topology; N, in; C, out. An internal signal sequence leads to binding of the polysome to the membrane and co-translational insertion of the signal sequence in loop conformation. This is followed by uninterrupted discharge of the C-terminal segment into the ER lumen, the signal sequence remaining anchored in the membrane. No cleavage. Example (by inference): sucrase isomaltase. (C) Topology; N, out; C, in with three membrane-spanning segments. A terminal signal sequence (S1) leads to binding of the polysome to the membrane and co-translational insertion and discharge of the signal sequence into the ER lumen. Discharge across the membrane is interrupted by a halt transfer signal (H1) until a second signal sequence emerges and is inserted in loop conformation (S2). This is followed by another halt transfer signal (H2), the C-terminal segment remaining in the cytoplasm. No cleavage. Note that an odd number of membrane spans gives C and N termini on opposite sides of the membrane. (Hypothetical example).

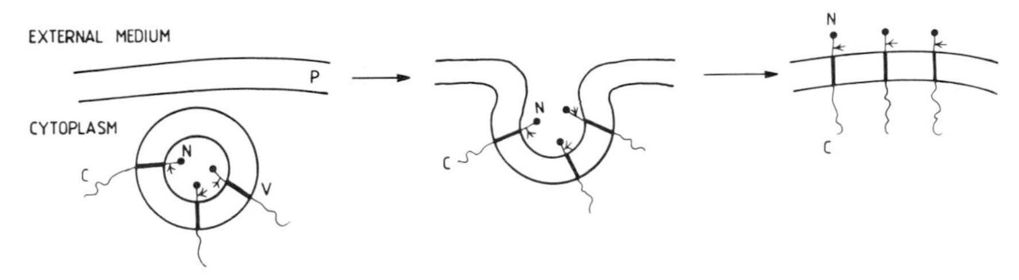

Figure 3.5. Diagram to illustrate conservation of the orientation of a transmembrane protein after membrane–membrane fusion. V is a vesicle derived by budding from another membrane containing three transmembrane protein molecules with the C-terminal domain in the cytoplasm and a glycosylated N-terminal domain in the interior of the vesicle. After fusion with the plasma membrane the C-terminal domain remains in the cytoplasm and the N-terminal domain projects into the extracellular medium. C, C terminus; N, N terminus; V, vesicle; P, plasma membrane; ᛘ, oligosaccharide.

asparagine residue in an appropriate region of the protein, acts as part of a signal for the protein to leave the ER. However, N-linked glycosylation cannot be either a sufficient or a necessary signal for the transport of a protein out of the ER membrane, since HMG-CoA reductase (which has an N-linked oligosaccharide) remains in the ER, while albumin (which is not glycosylated) is secreted by the cell.

2. Topogenesis of Reductase

With regard to the mechanisms responsible for bringing about the orientation of HMG-CoA reductase in the smooth ER, we have to explain how the enzyme comes to have seven membrane-spanning segments, a cytoplasmic C-terminal domain, and a short uncleaved N-terminal domain in the ER lumen. This orientation could result from an extension of the mechanism illustrated in Fig. 3.4C. On this model, the first step would be the co-translational insertion and translocation into the ER lumen of a short uncleaved hydrophobic signal sequence of nine amino acids, followed by a halt transfer signal. This would be followed by a sequence of three internal signal sequences inserted into the membrane in loop conformation, alternating with three halt transfer signals, the intramembrane segments being separated from each other by hydrophilic loops of variable length. The last hydrophilic loop, bearing the asparagine residue to which the oligosaccharide becomes attached, would project into the ER lumen. Interruption of discharge of the nascent protein through the membrane by the last halt transfer signal would leave a long C-terminal domain projecting into the cytoplasm. After release of the polysome from the rough ER, the membrane-bound protein would be free to diffuse laterally to the smooth ER membrane. In agreement with this scheme, Brown and Simoni (1984) have shown that synthesis of hamster HMG-CoA reductase takes place in polysomes bound to ER membranes and that insertion into the membrane is co-translational; if the enzyme is synthesized *in vitro* in a system not containing membranes and if membranes are then added to the system, no insertion takes place.

At present there is no basis for speculation as to how the enzyme is prevented from moving on to the Golgi apparatus in vesicles formed from the smooth ER. Note that if the signal sequence were inserted in loop conformation and remained within the membrane (Fig. 3.4B), as postulated by Sabatini *et al.* (1982) for the topogenesis of some integral membrane proteins, the N terminus of the completed protein would not project into the ER lumen unless there were cleavage of the hydrophilic loop between the signal sequence and the first halt transfer signal.

III. The HMG-CoA Reductase Gene and Its mRNA

The HMG-CoA reductase gene has now been characterized in considerable detail by Reynolds *et al.* (1984). Their first step was to isolate a genomic DNA clone from a phage library (prepared from hamster UT-1 cells) by hybridization with a cDNA probe complementary to part of the reductase message. A fragment of this clone was then used to isolate a second genomic clone extending further in the 5′ direction (“upstream”). These two clones contained overlapping DNA inserts covering the whole of the gene, including the 5′-untranslated region of ≅670 bp. The exon–intron structure of the gene was explored by hybridizing the cDNA clone pRed227 (see p. 13) to fragments of hamster genomic DNA produced by digesting DNA from UT-1 cells with restriction endonucleases, followed by necleotide sequencing of subcloned genomic fragments containing exon–intron junctions and other sequences of interest.

The hamster reductase gene is 25 kb long and has 20 exons separated by 19 introns (Fig. 3.6). The first intron interrupts the 5′-untranslated region and the last exon includes the 3′ end of the coding sequence plus the whole of the 3′-untranslated region. There are four potential polyadenylation signals (AATAAA) in the 3′-untranslated region of the gene (see p. 136). S1 nuclease mapping of hamster-liver and UT-1 cell mRNA, hybridized to genomic DNA extending downstream from the last of these signals, has shown that three different sites spanning 671 nucleotides can be used to terminate transcription of the gene. This could give rise to multiple reductase mRNAs of different lengths and would help to explain the presence of multiple reductase mRNAs in UT-1 cells, as revealed by Northern blotting of UT-1 RNA hybridized with a reductase cDNA probe (Chin *et al.*, 1982b).

Reynolds *et al.* (1984, 1985) have shown that the 5′-untranslated and 5′-flanking regions of the hamster reductase gene have several unusual features (Fig. 3.7). The 5′-untranslated region contains eight startpoints for transcription, grouped in two clusters extending over ≅100 bases, and up to eight ATG codons

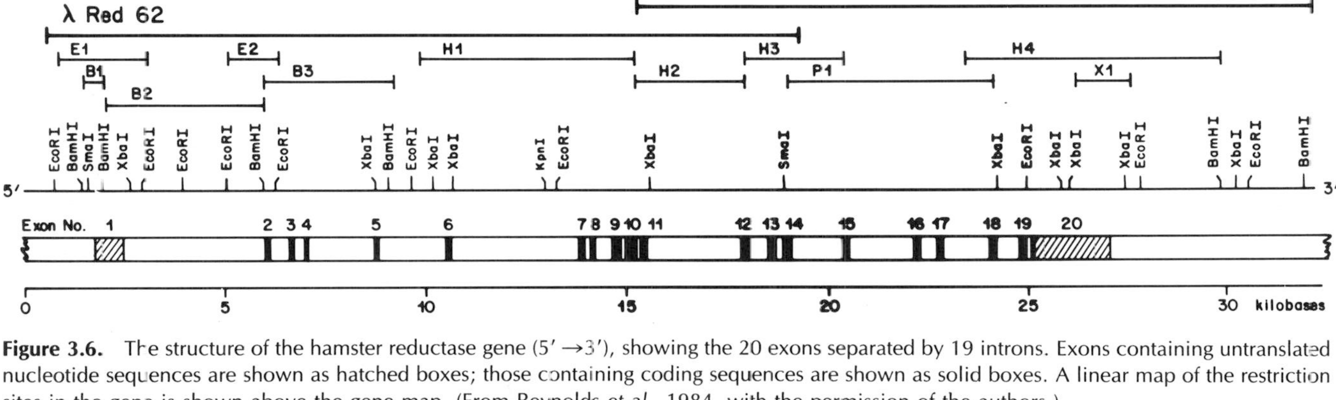

Figure 3.6. The structure of the hamster reductase gene (5′ →3′), showing the 20 exons separated by 19 introns. Exons containing untranslated nucleotide sequences are shown as hatched boxes; those containing coding sequences are shown as solid boxes. A linear map of the restriction sites in the gene is shown above the gene map. (From Reynolds *et al.*, 1984, with the permission of the authors.)

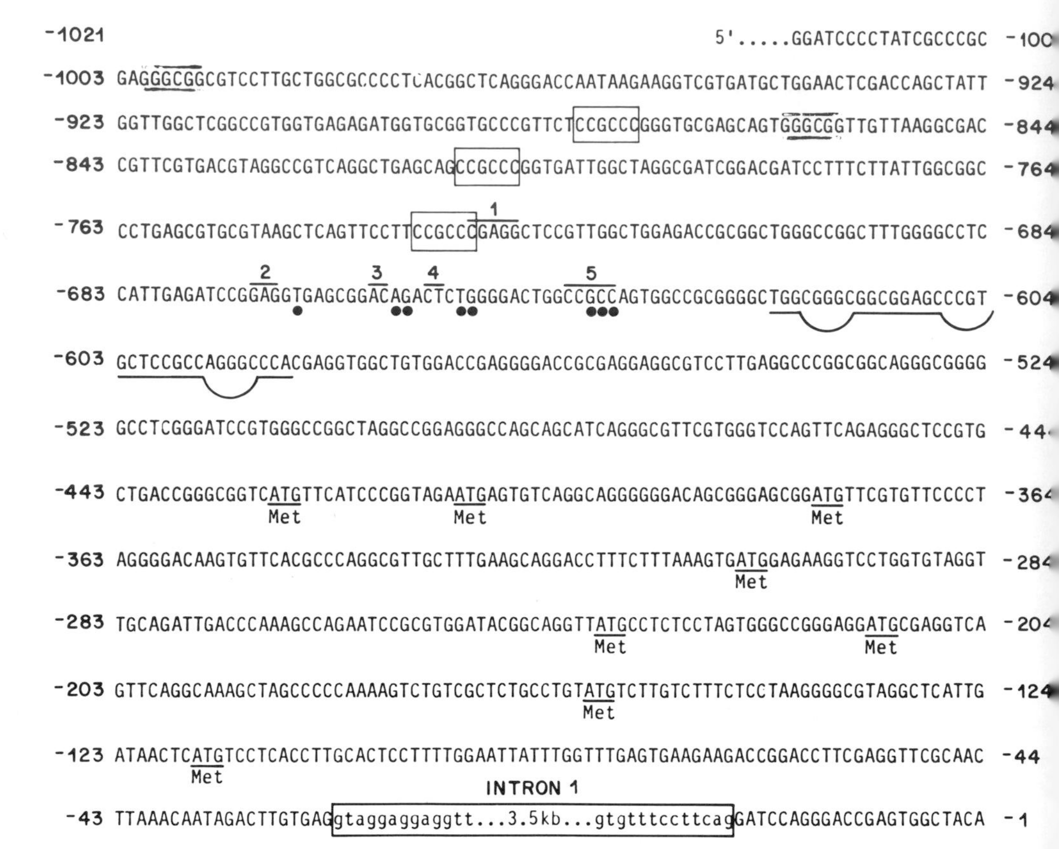

Figure 3.7. Nucleotide sequence of the 5′-untranslated region of the hamster reductase gene. Nucleotide +1 denotes the A of ATG used to encode the initiator methionine codon of reductase; residues preceding it are denoted by negative numbers. The multiple transcription initiation sites predicted by S1 nuclease analysis are indicated by numbered bars above the nucleotide; those predicted by primer extension are indicated by dots below the nucleotide. Repeated sequences CCGCCC are enclosed by boxes; the sequence GGGCGG is shown by lines above and below the sequence. The positions of eight ATG codons 5′ to the initiator ATG codon are indicated by Met below the sequence. The position of the first intron is shown by a box enclosing lower-case letters. (From Reynolds *et al.*, 1984, with the permission of the authors.)

upstream from the ATG corresponding to the AUG codon used to initiate translation of the message (the initiator codon). Moreover, Reynolds *et al.* (1985) have shown that four different 5′-splicing sites[1] can be used when the primary tran-

[1]The 5′ (left) splicing site, also called the donor site, is the exon–intron junction at the 5′ end of the intron. The 3′ (right) splicing site is at the 3′ end of the intron.

script is converted into the finished mRNA. (The 3′-splicing site is always at the same position, 23 nucleotides upstream from the A of the initiator codon.) Four of the eight startpoints use the same 5′-splicing site, but each of the remaining four startpoints can use three different splicing sites. This gives a total of 16 different 5′-untranslated regions varying in length form 68 to 670 nucleotides and containing 3, 6, or 8 ATGs upstream from the initiator codon. Thus, the existence of multiple reductase mRNAs is due to variability in the lengths of both untranslated regions and to the presence of multiple 5′-splicing sites.

Reynolds *et al.* (1985) have pointed out that one of the four splicing sites would give rise to an AUG codon in the message that would be 30 bases upstream from the initiator codon involved in the production of the 887-amino-acid enzyme shown in Fig. 3.2. It would also be in the same reading frame as the initiator codon. If this AUG could initiate translation it would result in the production of an enzyme with an additional 10 amino acids at the N-terminal end.

The 5′-flanking region of most eukaryotic genes that have been analyzed has an AT-rich sequence (the TATA box) about 20 nucleotides upstream from the startpoint for transcription and a CAAT box about 70 nucleotides upstream. Together, these constitute the *promoter* that initiates transcription of the gene. No such sequences are present in the 5′-flanking region of the reductase gene. However, the sequence TTATT occurs about 39 bp in the 5′ direction from the most upstream of the multiple initiation sites for transcription (Fig. 3.7). This atypical TATA box is required for initiation of transcription at a single site when the gene is transcribed in a cell-free system, but it has little effect on transcription in the intact cell (Osborne *et al.*, 1987, 1988). Upstream from the usual position of the TATA box there is a GC-rich region of 265 nucleotides containing three copies of the sequence CCGCCC and two copies of the reverse of its complement (GGGCGG) (see Fig. 3.7). The sequence CCGCCC is also repeated twice in each of three 21-bp tandem repeats in the early promoter of simian virus 40 (SV40), which also contains a TATA box.

Benoist and Chambon (1981) have investigated the role of the TATA box and the CCGCCC repeat region in SV40, using *deletion mutants*—recombinant plasmids containing inserts of SV40 DNA from which various segments have been excised artificially with restriction endonucleases. They have shown that selective removal of the TATA box leads to the generation of multiple new startpoints for transcription of the SV40 genome but has no effect on the rate of transcription.

On the other hand, removal of the CCGCCC repeat region 150–200 nucleotides upstream from the startpoint abolishes transcription (see also Fromm and Berg, 1982, and Hansen and Sharp, 1983, for reviews). Benoist and Chambon conclude that in SV40 the TATA box is required only for accurate localization of the region from which transcription starts and that the CCGCCC repeat region controls the rate of transcription.

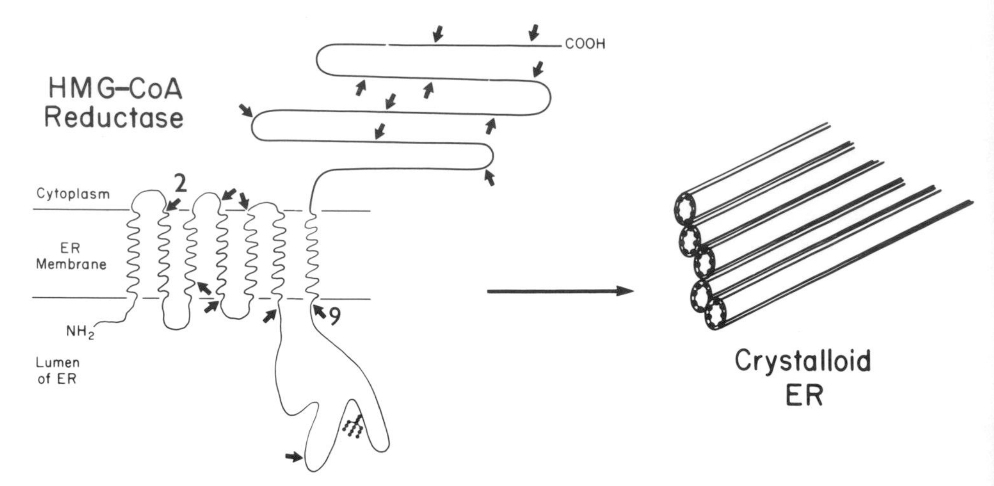

Figure 3.8. Exon–intron organization of the reductase gene in relation to domains in the reductase protein. The diagram shows the positions where introns 2–19 interrupt the protein sequence (the first intron is in the 5′-noncoding region of the gene). Introns 2, 3, 4, 5, 6, 7, and 9 interrupt the sequence at the junction of a membrane-spanning region with an external loop. Intron 8 interrupts the loop between the sixth and seventh membrane-spanning regions at amino acid residue 260. Introns 10–19 interrupt the C-terminal cytoplasmic segment of the protein. The oligosaccharide at residue 281 is shown as a trident. (From Liscum *et al.*, 1985, with the permission of the authors.)

The homology between SV40 and the hamster reductase gene suggests that the promoter for the reductase gene is in a segment of the 5′-flanking region containing CCGCCC repeats and that the presence of multiple startpoints for transcription is due to absence of a TATA box. More direct evidence concerning the position of the promoter for the reductase gene is considered in Section V,D.

Reynolds *et al.* (1984) and Osborne *et al.* (1985) suggest that some of the unusual features of the reductase gene, especially the very long 5′-untranslated region (longer than that of any other gene examined so far) and the presence of a GC-rich promoter, are related to the fact that transcription is normally maintained at a low rate by end-product inhibition. In keeping with this suggestion, Osborne *et al.* (1985) have pointed out that the genes for several other mammalian enzymes that are expressed at low activity due to end-product inhibition have GC-rich promoter regions.

As shown in Fig. 3.6, the coding region of the reductase gene is interrupted by 18 introns (introns 2–19). Seven of these (introns 2, 3, 4, 5, 6, 7, and 9) interrupt the coding sequence at a position corresponding to the junction of a membrane-spanning domain and a hydrophilic loop outside the membrane (Fig. 3.8). This exon–intron organization illustrates the general tendency for exons to code for specific structural or functional units in the protein product of a eukaryotic gene.

The relevance of this to the way in which genes coding for complex proteins have been built up over evolutionary time is discussed in Chapter 9. There are no similarities between the amino acid sequences of the seven membrane-spanning domains. Hence, these sequences could not have evolved by multiplication of a single ancestral segment of DNA in the reductase gene.

IV. Somatic-Cell Mutants in the Study of Cholesterol Metabolism

A. Random and Directed Mutation

Mutations occurring spontaneously in the germ cells have been invaluable in the study of cholesterol metabolism. The most informative of these are mutations in the gene encoding the LDL receptor; these are considered in Chapter 9. Others that have provided useful information include those that affect the structure or plasma concentration of apolipoproteins. However, the number of such mutations is very limited and none has so far been discovered that affects the structure or rate of production of HMG-CoA reductase. Random mutations can now be produced experimentally in somatic cells in culture by irradiation or treatment with a chemical mutagen. A stable line of cells, exhibiting a specific inherited abnormality, can then be established by growing the cells under conditions that favor survival of mutated cells with a particular phenotype. These cell lines can be used in the analysis of events occurring at the cellular or subcellular level. In this Section I discuss some of the ways in which somatic-cell mutants have been exploited in the study of HMG-CoA reductase.

One type of *directed* mutation (SV40 deletion mutations), as distinct from *random* mutations, has already been referred to (p. 25). Oligonucleotide-directed mutagenesis, in which specific mutations are introduced at predetermined sites in a segment of DNA generated *in vitro*, is considered below (p. 44).

B. Cholesterol and Mevalonic Acid Auxotrophs

Workers in several laboratories have isolated mutant lines of CHO cells that require cholesterol (cholesterol auxotrophs) or a sterol precursor for normal growth in culture (see Sinensky, 1985, for review). From the point of view of HMG-CoA reductase, one of the most interesting of these is the mevalonic acid auxotroph, Mev-1 (Schnitzer-Polokoff *et al.*, 1982). This mutant cell line has an absolute requirement for mevalonic acid, which cannot be replaced by cholesterol or by any other known product of the metabolism of mevalonic acid, including ubiquinone, dolichol, and isopentenyl adenosine. The genetic defect in Mev-1 cells is the absence of activity of HMG-CoA synthase, the enzyme that catalyzes the formation of HMG-CoA from acetyl-CoA and acetoacetyl-CoA. Under normal growth conditions the activity of HMG-CoA reductase is several times higher in Mev-1 cells than in wild-type CHO cells. Furthermore, reductase activity in Mev-1 cells is not completely suppressed by the reductase inhibitor 25-hydroxycholesterol, but is completely suppressed by mevalonic acid at high concentration (13.5 m*M*) or by low concentrations of mevalonic acid together with 25-hydroxycholesterol or LDL. These properties of Mev-1 cells are qualitatively similar to those of compactin-treated cells and are in line with the conclusion, discussed below, that normal growth of cells and full suppression of HMG-CoA reductase activity require the presence of one or more nonsterol products of the metabolism of mevalonic acid.

Another mevalonate auxotroph, lacking HMG-CoA reductase, has been isolated by Mosley *et al.* (1983) from CHO cells treated with a chemical mutagen. Cells from a single clone, called UT-2, were selected from the mutagen-treated culture by their ability to survive UV irradiation following exposure to 5′-bromodeoxyuridine (BrdUrd) after incubation for 24 hours in a medium lacking mevalonic acid. Under these conditions the UT-2 cells failed to grow in the mevalonate-deficient medium and were therefore unable to incorporate BrdUrd into their DNA. The normal cells incorporated BrdUrd into their DNA during growth in the absence of mevalonic acid and were killed by UV irradiation. As discussed below, UT-2 cells have been used as the starting point for the development of another line of cells capable of synthesizing reductase but in which reductase synthesis is not regulated by sterols.

C. 25-Hydroxycholesterol-Resistant Mutants

Sinensky and co-workers (Sinensky, 1985) have isolated several mutant lines of CHO cells that are resistant to the normal repression of synthesis of HMG-CoA reductase by 25-hydroxycholesterol. These cell lines were selected by growing mutagen-treated CHO cells in a medium containing 25-hydroxycholesterol and no source of cholesterol. Under these conditions the wild-type cells die because they cannot obtain sterol from the medium or by endogenous synthesis, but the mutant cells survive by making their own sterol despite the presence of 25-hydroxycholesterol in the medium. These mutants are also resistant to suppression of HMG-CoA reductase by cholesterol added to the medium in LDL. Consequently, when grown in the presence of an external source of cholesterol, the mutant cells acquire abnormally high concentrations of cholesterol in their membranes.

Complementation analysis[2] of cells hybridized by virus-induced fusion

[2] In *complementation analysis* we ask whether or not two mutations are in the same gene. For the analysis of animal cells,the two mutant chromosomes to be tested are placed in the same cell. If the

showed that recessive 25-hydroxycholesterol-resistant mutants fall into more than one complementation group. Therefore they are not all due to a mutation in the same gene. In one of these complementation groups (crB), studied by Sinensky and Mueller (1981), the mutation is associated with loss of activity of a specific 25-hydroxycholesterol-binding protein. As noted in the next section, this genetic abnormality in crB cells is relevant to the hypothesis that oxysterols mediate the inhibitory effect of LDL cholesterol on cholesterol synthesis.

D. Compactin-Resistant Cell Lines

The cell lines that have been of greatest help in the study of HMG-CoA reductase are undoubtedly those that are resistant to *compactin,* a competitive inhibitor of the enzyme. Stable lines resistant to competitive inhibition of various enzymes necessary for survival of cells in culture have been developed in several laboratories by adapting cells to stepwise increases in the concentration of the inhibitor, without pretreatment with a mutagen. At each step, the small number of surviving cells is selected and grown in the presence of a higher concentration of inhibitor. Eventually, a cell line is established that is resistant to the inhibitor at many times the minimal lethal concentration for wild-type cells. The heritable resistance is due primarily to multiplication of the gene coding for the inhibited enzyme, so that resistant cells produce greatly increased amounts of enzyme. Schimkc *et al.* (1978) suggest that gene duplication occurs spontaneously at a finite but very low rate in wild-type cells, and that the survivors at each step in adaptation are those in which gene duplication has occurred.

Compactin (Fig. 3.9) was isolated independently from *Penicillium brevicompactum* by Brown *et al.* (1976) and from *P. citrinum* by Endo *et al.* (1976) (who called it ML-236B). The drug is a potent inhibitor of HMG-CoA reductase in rat-liver microsomes (Endo *et al.*, 1976) and in cell free extracts of human fibroblasts (Brown *et al.*, 1978), its affinity for the enzyme being ≅10,000 times that of HMG-CoA. The formation of mevalonic acid from HMG-CoA in fibroblasts in culture is completely inhibited when compactin is present in the medium. Under these conditions there is a cycloheximide-sensitive increase in the amount of enzyme present in the cells, revealed by increased enzyme activity when the compactin is removed by dilution. This increase is suppressed completely by

---

mutations are in different genes, each chromosome will provide the cell with the product of the gene which is mutant in the other and the wild-type condition will be restored, i.e., the two chromosomes will complement each other. If the two mutations are allelic (at the same gene locus) no complementation will occur. In classical genetics the two mutant chromosomes are brought together by the fusion of gametes, as in human pedigree analysis or animal breeding experiments; in complementation analysis of somatic-cell mutants, cell fusion is brought about by treatment with animal viruses. Virus-fused cells are also used in "dominant–recessive" analysis of somatic-cell mutants.

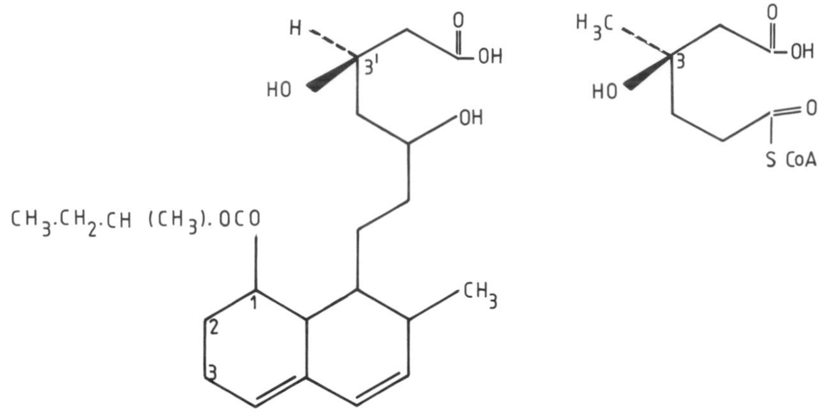

Figure 3.9. The structure of compactin (ML-236B). The structure of HMG-CoA is shown for comparison. Only the configurations of the asymmetric C-3′ of compactin and C-3 of HMG-CoA are shown. When the two molecules are orientated as shown here, the OH group at C-3′ and C-3 projects above the plane of the paper. Note that the 3-methyl group of HMG-CoA is replaced by H in compactin. In mevinolin, an analog of compactin, there is an additional α-methyl group at C-3.

adding mevalonic acid to the culture medium and is suppressed partially by adding LDL (Brown *et al.*, 1978).

Ryan *et al.* (1981) have isolated two variant lines of compactin-resistant Chinese hamster ovary (CHO-K1) cells by stepwise adaptation to growth in the presence of increasing concentrations of compactin. C100 cells grow normally in 225 μM compactin and have 40 times the normal content of HMG-CoA reductase. (Normal cells die when the concentration of compactin is 1 μM). Using a similar approach, Chin *et al.* (1982a) have isolated a line of compactin-resistant CHO cells called UT-1 and have studied them in considerable depth. As already mentioned, UT-1 cells were used by Chin *et al.* (1982b, 1984) and Reynolds *et al.* (1984) as the starting point for their study of the HMG-CoA reductase gene and mRNA, and by Liscum *et al.* (1983a, 1985) in their analysis of the structure and orientation of the enzyme. UT-1 cells have about 500 times as much reductase as wild-type CHO cells, the enzyme accounting for up to 2% of the total cell protein. The increase in enzyme content is brought about by increased enzyme synthesis due to a 15-fold multiplication in the number of copies of the reductase gene per cell, combined with an increase in the rate of transcription of each gene (Luskey *et al.*, 1983). The multiplication of reductase genes in UT-1 cells is stable to the extent that no decrease in the number of genes per cell occurs when the cells are grown in the absence of compactin for 3–4 months. When compactin is removed from the growth medium, the reductase mRNA content of the cells falls toward the normal level for CHO cells, increasing again when compactin is added back to the medium.

Chin *et al.* (1982a) have shown that the very large increase in the amount of reductase in UT-1 cells is associated with a remarkable change in their appearance. Electron microscopy of UT-1 cells revealed the presence of perinuclear inclusions consisting of numerous tubules of smooth ER membrane (≅86 nm in diameter), packed tightly in a hexagonal array or "bundled together like a handful of pencils" (Orci *et al.*, 1984) (Fig. 3.10). Fluorescence microscopy, with rabbit anti-reductase IgG followed by fluorescent anti-rabbit IgG, showed that the smooth ER tubules were filled with HGM-CoA reductase. Chin *et al.* (1982a) conclude that in UT-1 cells the smooth ER proliferates and assumes a crystalline arrangement in order to accommodate the increased mass of reductase. If LDL is added to a lipoprotein-free culture medium in which UT-1 cells are growing in the presence of compactin, the cholesterol content of the crystalloid ER membrane system increases within 2 hours and this increase is

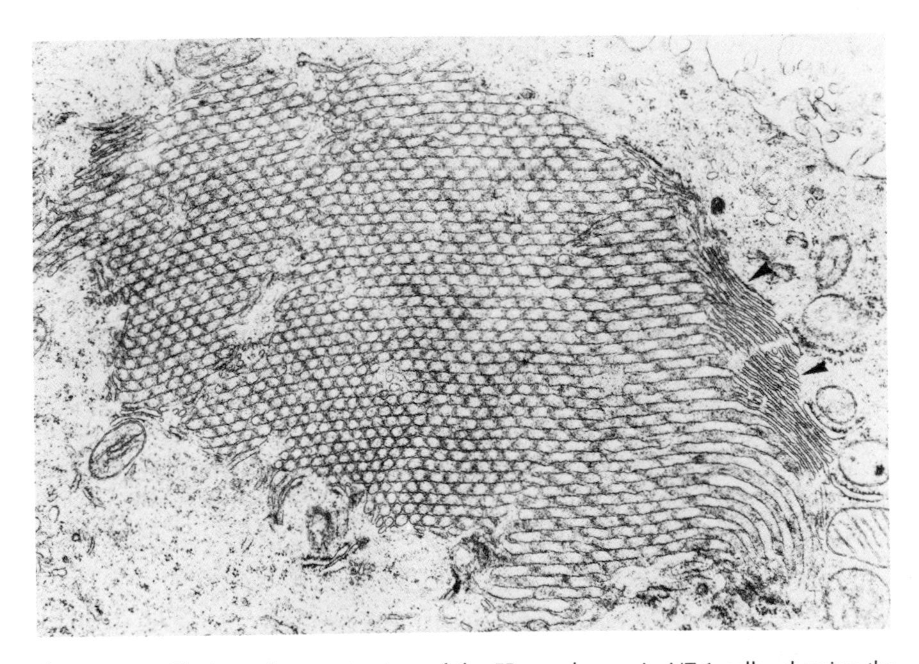

Figure 3.10. Electron-microscopic view of the ER membranes in UT-1 cells, showing the crystalloid ER and the arrangement of cisternal ER (shown by arrowheads). × 20,680. (From Orci *et al.*, 1984, with the permission of the authors.)

followed by a rapid decline in HMG-CoA reductase activity coupled with the disappearance of the crystalloid membranes (Faust *et al.*, 1982; Orci *et al.*, 1984). These observations provide a possible clue to the way in which LDL cholesterol increases the rate of degradation of HMG-CoA reductase in normal cells (see the following section).

E. An LDL-Receptor-Dependent Line of Human Cells

Human lymphoma cells of the U-937 line, first described by Sundström and Nilsson (1976), cannot synthesize their own cholesterol and are therefore unable to grow in the absence of an available source of cholesterol in the medium (Esfahani *et al.*, 1984). Frostegård *et al.* (1989) have shown that these mutant cells will grow normally if LDL is added to a serum-free growth medium and that uptake of LDL by the cells depends entirely upon binding and internalization via the LDL-receptor pathway. Making use of this property of U-937 cells, Frostegård *et al.* (1989) have developed a method for determining the relative abilities of normal and defective LDLs to bind to the LDL receptor. The rate of growth of cells in the presence of the LDL to be tested is compared with that observed in the presence of an equimolar amount of normal LDL. This method was used to demonstrate defective LDL-receptor binding and degradation of LDL from the familial defective apoB-100 (FDB) heterozygotes shown in Fig. 6.17.

V. The Regulation of HMG-CoA Reductase

A. General Considerations

HGM-CoA reductase in animal cells is controlled by a system of considerable complexity, of which the LDL-receptor pathway is an integral component. This complexity is partly a reflection of the fact that mevalonic acid, the immediate product of the enzyme, is the precursor of sterols and nonsterols that are essential for three basic activities common to most cells—maintenance, growth, and replication. If the cell is to function efficiently, these activities need to be controlled independently. As we shall see, reductase activity in the animal cell is regulated by a multivalent mechanism in which feedback suppression of the enzyme is exerted independently by cholesterol and nonsterol metabolites of mevalonic acid.

It is worth noting that in view of its presence in all plant and animal tissues and in some bacteria (Fimognari and Rodwell, 1965) HMG-CoA reductase must have a very long evolutionary history. In keeping with this, steroidal compounds, almost certainly of organic origin, have been identified in shales estimated to be at least 3000 million years old (Calvin, 1969). Some of the nonsterol metabolites of mevalonic acid mentioned above may be at least as old as sterols, in an evolutionary sense. Indeed, it is possible that their biosynthesis had already evolved before the earth's atmosphere contained enough O_2 to permit the formation of oxidosqualene from squalene, an essential step in the biosynthesis of contemporary sterols (see Gibbons *et al.*, 1982).

In the earliest organisms in which reductase was present, its activity was presumably controlled by negative feedback by sterol or nonsterol end-products formed within the cell. This mode of regulation is responsible for the control of reductase activity in present-day single-celled organisms, as in the suppression of reductase by ergosterol in yeasts (Kawaguchi, 1970). Regulation of reductase activity by end-products formed within the cell must also occur in the cells of animal tissues. However, in animal cells these ''primitive'' regulatory mechanisms are modulated by extracellular substances, including hormones, mitogens, and lipoproteins (e.g., LDL internalized by the LDL receptor).

Current evidence suggests that suppression of reductase activity by cholesterol is mediated by oxysterols produced within the cell by the oxidation of cholesterol present in a regulatory pool of free cholesterol. Regulation of reductase activity by sterols involves changes in the rates both of synthesis and of degradation of the enzyme. The state of activation of HMG-CoA reductase also changes in different conditions, though it is not clear how far these changes in activation are brought about by oxysterols generated from cholesterol. In the following sections, each of these aspects of the regulation of HMG-CoA reductase is considered, beginning with a suggested model for regulation by sterols and finishing with a discussion of multivalent suppression of reductase activity mediated by sterols and nonsterols derived from mevalonic acid.

B. A Model for Regulation of HMG-CoA Reductase by Sterols

Much of what is known about the regulation of reductase activity by sterols in mature animal cells can be fitted into the scheme shown in Fig. 3.11. In this model the amount of free cholesterol in the cell is determined by the balance between *synthesis, uptake* from the external medium, and *utilization*. Utilization includes all those processes that remove free cholesterol from the cell, including metabolism, secretion in lipoproteins, and continual leakage of free cholesterol from the plasma membrane to extracellular sterol acceptors. Within the total mass of free cholesterol there is a *regulatory pool* of membrane cholesterol. This pool monitors the free-cholesterol content of the cell and signals the cell's requirement for cholesterol to sites which control the activity of HMG-CoA reductase and the LDL-receptor pathway. These sites include the genes for reductase and the LDL receptor. The regulatory pool of sterol is introduced into the model to take into account the fact (referred to below) that, in most types of cell, free cholesterol derived from lipoproteins taken up from the external medium has no

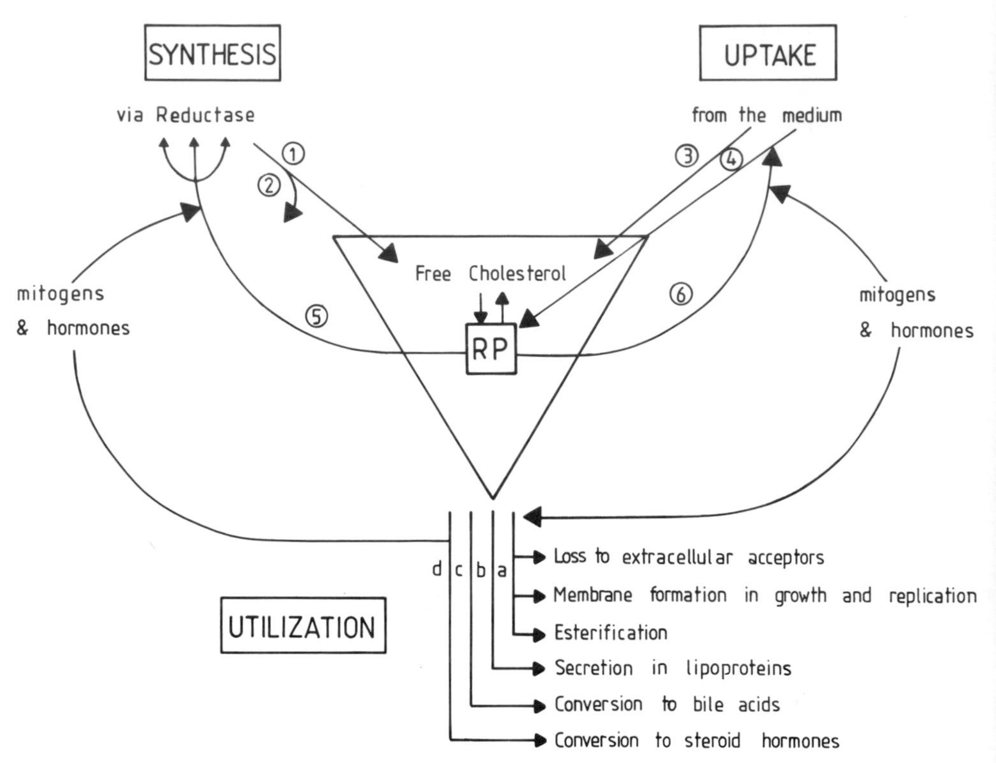

Figure 3.11. Diagram of a model for the sterol-mediated regulation of HMG-CoA reductase and the LDL-receptor pathway in animal cells.

The model is applicable to *nonspecialized cells* that use cholesterol only for replacement of free cholesterol leaked to the external medium, for membrane formation, and for esterification with fatty acids (a), and *specialized cells* that, in addition, use cholesterol for the formation of lipoproteins (b), conversion to bile acids (c), or conversion to steroid hormones (d).

The central triangle shows the mass of cellular free cholesterol, with a regulatory pool of sterol to which free cholesterol derived from receptor-mediated uptake of lipoproteins has rapid access; the regulatory pool has no known morphological counterpart. Cholesterol enters the system by synthesis *in situ* (1), by receptor-independent uptake of lipoproteins (3), and by receptor-mediated uptake of lipoproteins (4). (Entry of cholesterol by selective uptake of esterified cholesterol in HDL and by net transfer of free cholesterol from cholesterol-enriched lipoproteins is omitted from the model.) Reductase activity is regulated by short-term activation–deactivation, by induction–repression, and by changes in the rate of degradation of the enzyme (5), in response to fluctuations in the sterol content of the regulatory pool. The activity of the LDL-receptor pathway is regulated, via the regulatory pool, probably by induction–repression of the synthesis of LDL receptors (6). The activity of HMG-CoA reductase is also subject to feedback inhibition by nonsterol products of the metabolism of mevalonic acid (2). Mitogens and some hormones modulate the activities of HMG-CoA reductase and the LDL-receptor pathway. RP, regulatory pool of sterol.

short-term effect on reductase activity unless the uptake is receptor mediated. Gibson and Parker (1987) have discussed how membrane cholesterol could send information to regulatory sites at a distance. One possibility, considered in the next section, is that cholesterol is converted into a more polar compound, such as 25-hydroxycholesterol, that can be transported to the nucleus by a specific binding protein.

In most cells the input of sterol into the regulatory pool by cholesterol taken up via the LDL-receptor pathway plays a predominant role in the regulation of reductase activity. However, this is not the only mechanism by which the enzyme is regulated, as is shown by the essentially normal reductase activity expressed *in vivo* in cells that are genetically deficient in LDL receptors. The cells of some tissues may acquire cholesterol by receptor-mediated uptake of lipoproteins other than LDL. In at least one instance—the uptake of chylomicron remnants by the liver—this leads to rapid suppression of HMG-CoA reductase activity. LDL cholesterol may also enter cells by receptor-independent pathways under both normal and abnormal conditions. Cholesterol entering cells by receptor-independent pathways must be capable of inhibiting HMG-CoA reductase in the long term, but this is not usually demonstrable in short-term experiments with cells in culture. Feedback inhibition of reductase by cholesterol synthesized within the cell must also contribute to the overall control of reductase activity, particularly in those cells that do not take up cholesterol from the medium. The increase in reductase activity that occurs in the hepatocytes of cholestyramine-treated animals and in compactin-treated cells presumably reflects release from feedback inhibition by cholesterol synthesized *in situ*. Likewise, part of the inhibitory effect of large doses of mevalonic acid on reductase activity in hepatocytes *in vivo* and *in vitro* is likely to be mediated by cholesterol synthesized within the cells.

Fluctuations in the utilization of cell cholesterol are reflected in changes in the amount of sterol in the regulatory pool. These changes, in turn, cause reciprocal changes in the activity of the LDL-receptor pathway and of HMG-CoA reductase.

The activities of HMG-CoA reductase and the LDL-receptor pathway also respond reciprocally to changes in the sterol content of the regulatory pool brought about by fluctuations in receptor-mediated uptake of cholesterol. Many examples of these responses will be found in the following text.

Mitogens and certain hormones also influence the activity of reductase and the LDL-receptor pathway. How these effects are brought about is not fully understood. As shown in Fig. 3.11, they could be mediated by changes in the rate of utilization of cholesterol or in other ways that do not involve the regulatory pool of sterol. Thus, one of the effects of mitogens and of a hormone such as insulin may be to alter the threshold level of sterol in the regulatory pool at which reductase and the LDL-receptor pathway are down-regulated. Reductase activity is also regulated by nonsterol products of the metabolism of mevalonic acid. The integration of this control system with that involving sterols derived from mevalonate is discussed at the end of this chapter. Short term alteration of the state of activation of reductase by phosphorylation–dephosphorylation is discussed in Section V,F below.

The above model is fully applicable only to mature cells. The marked increase in HMG-CoA reductase activity that occurs in the cells of many developing tissues may be explained partly by a response to increased utilization of free cholesterol for membrane formation, as in developing nerve tissue undergoing myelinization. However, in most cases a developmental increase in reductase activity is best seen as a reflection of the switching on and off of genes that underlies the development of all differentiated organisms.

C. Oxysterols and Intermediates in Regulation by Cholesterol

Gould (1951) showed that cholesterol synthesis in the liver is inhibited by dietary cholesterol, an effect that was later shown to be due to suppression of hepatic HMG-CoA reductase (Linn, 1967). This led to the proposal that cholesterol synthesis in the liver is regulated by feedback inhibition of HMG-CoA reductase by cholesterol, the end-product of the pathway in which the enzyme acts. Subsequent work in several laboratories showed that reductase activity could be modulated *in vitro* in many types of cell by incubating the cells under conditions in which their cholesterol content was increased or decreased (for details see Myant, 1981, Chapter 9, and Section V,D below). These observations led to the view, now generally accepted, that in all animal cells *in vivo,* reductase activity is controlled mainly by cholesterol entering the cell by uptake of lipoprotein or by cholesterol synthesized *in situ*. However, Kandutsch and his colleagues (Kandutsch *et al.*, 1978; Kandutsch, 1982) believe that the intracellular sterol molecule directly responsible for regulating reductase activity is not cholesterol itself, but an oxysterol related metabolically to cholesterol (for a critical discussion of the oxysterol hypothesis, see Gibbons, 1983).

Kandutsch and Chen (1973, 1974, 1977) began by showing that reductase activity in mouse liver cells and L-cell fibroblasts in culture is suppressed by impure cholesterol emulsified with serum albumin but is not suppressed by purified cholesterol when an antioxidant is present in the culture medium. In the absence of antioxidant, purified cholesterol suppresses the enzyme after prolonged incubation for 12 or more hours. Investigation of the contaminants responsible for the inhibitory effect of impure preparations of cholesterol showed that reductase activity in liver cells or L-cells in culture is markedly and specifically suppressed by various C_{27} sterols containing an additional hydroxyl or ketone group in the nucleus or side chain. Most of the inhibitory sterols that have been identified have, in addition to the 3-hydroxyl group (in α or β configuration), one OH or one ketone group at any of the positions 6, 7, 15, 20, 22, 24, 25, and 32 (the methyl carbon at C-14), the most potent being 25-hydroxycholesterol (cholest-5-ene-3β,25-diol) and 7-ketocholesterol (3β-hydroxycholest-5-en-7-one). Addition of these compounds at concentrations of 1 μg/ml or less to a culture medium in which cells are growing without an external source of cholesterol leads to a fall in reductase activity to less than 5% of the control value.

The delayed inhibitory effect of purified cholesterol after incubations without antioxidant in the medium was attributed by Kandutsch and Chen (1973) to autoxidation of cholesterol in the incubation medium, leading to the formation of highly inhibitory oxysterols. The failure of purified cholesterol to inhibit HMG-CoA reductase in short-term incubations cannot be explained by inability of emulsified cholesterol to enter cells, since the cholesterol content of L-cells increases significantly within 2 hours of the addition of cholesterol to the medium (Kandutsch and Chen, 1973). Kandutsch and Chen (1977) showed that inhibition of reductase in cultured mouse L-cells by 25-hydroxycholesterol is such that the cells cease to grow and eventually die unless cholesterol, or an intermediate beyond HMG-CoA in the biosynthesis of sterols, is present in the medium.

Inhibition of reductase by oxysterols was confirmed by Brown and Goldstein (1974) and by Krieger *et al.* (1978) who, with their co-workers, have made extensive use of the inhibitory effect of 25-hydroxycholesterol in their studies of the cellular metabolism of cholesterol. Krieger *et al.* (1978) have also shown that heptane-extracted LDL particles reconstituted with 25-hydroxycholesteryl oleate (25HC-LDL) are about four times as effective as LDL reconstituted with cholesteryl oleate (C-LDL) in suppressing reductase activity in cultured normal human fibroblasts when equal quantities of 25-hydroxycholesterol and cholesterol are internalized by the cells. When normal fibroblasts are grown in the presence of 25HC-LDL with no external source of cholesterol, they cease growing and die within a few days. When fibroblasts from FH homozygotes, which lack LDL receptors, are cultured under the same conditions, they grow normally because the 25HC-LDL in the medium fails to enter the cells and thus does not inhibit endogenous synthesis of the cholesterol they need for growth and multiplication. As mentioned above, the resistance of cells lacking LDL receptors to the toxic effects of 25HC-LDL has been exploited by Krieger *et al.* (1981) in the isolation of somatic-cell mutants with mutations affecting the LDL receptor.

On the basis of their observations on the inhibitory effect of oxygenated sterols *in vivo* and *in vitro*, Kandutsch *et al.* (1978) suggested that

1. Suppression of reductase in the livers of cholesterol-fed animals, and in cells incubated in the presence of cholesterol added as an emulsion or in lipoproteins, is due entirely to traces of oxysterol present in the cholesterol added to the diet or incubation medium or generated from the added cholesterol by spontaneous autoxidation at some later stage; and

2. The endogenous sterol molecule that regulates reductase activity is not cholesterol itself but an oxysterol produced enzymically within the cell either by oxidation of cholesterol or during the conversion of lanosterol into cholesterol.

Thus, the oxysterol hypothesis has two parts that need not stand or fall together. One concerns the question of whether or not cholesterol in the external medium inhibits intracellular reductase; the other concerns the nature of the intracellular sterol that regulates reductase in accordance with the cell's requirement for cholesterol (the regulatory sterol molecule).

With regard to the first point, several groups of workers (Brown and Goldstein, 1974; Breslow *et al.*, 1975; Bell *et al.*, 1976) have confirmed the much greater inhibitory potency *in vitro* of oxysterols than of cholesterol when the sterols are added to the incubation medium in various types of emulsion. On the other hand, Bell *et al.* (1976) have shown that human LDL in which the cholesterol is chromatographically pure does inhibit HMG-CoA reductase in hepatoma cells in culture, and Chang (1983) has pointed out that the oxysterol content of LDL in normal human plasma is not nearly high enough to account for the inhibitory effect of human LDL on the reductase activity of cells in culture.

Much of the observed difference between the inhibitory effect of oxysterols and cholesterol in cells *in vitro* could be due to differences in the ability of various sterols, added as emulsions to the external medium, to cross the plasma membrane. Moreover, the relation between the concentration of a sterol in the bulk phase of an incubation medium and its concentration in the unstirred layer at the cell surface may not be the same for all sterols. In keeping with these suggestions, Bell *et al.* (1976) have shown that 25-hydroxycholesterol is taken up by hepatoma cells in culture much more rapidly than is cholesterol when both are added to the medium as albumin complexes. It should also be noted that although emulsified 25-hydroxycholesterol is at least 100 times more inhibitory than emulsified cholesterol, Krieger *et al.* (1978) have shown that, for a given amount of sterol internalized, 25-hydroxycholesterol is only four times as inhibitory as cholesterol when both sterols are incorporated in LDL particles as their fatty-acid ester (see above).

Thus, the balance of evidence favors the conclusion that, although certain oxygenated homologs of cholesterol are intrinsically more inhibitory than cholesterol itself, suppression of HMG-CoA reductase by LDL, both *in vitro* and *in vivo,* is mediated by uptake of cholesterol in the LDL particles. In any case, it is difficult to see how the undoubted regulatory effect of LDL on reductase activity of cells *in vivo* could be mediated by a component that is generated by nonregulated autoxidation. This does not, of course, exclude the possibility that oxysterol contaminants are responsible for some of the inhibitory effect of preparations of LDL that have been stored in an atmosphere containing O_2.

Kandutsch *et al.* (1978) suggest that 7α-hydroxycholesterol, formed from cholesterol in the liver, and 20α-hydroxycholesterol, formed from cholesterol in steroid-hormone-forming tissues, act as physiological regulators of reductase activity in the tissue in which they are produced. They also suggest that in the cells of other tissues the regulatory sterol molecule is a C_{30} or C_{27} precursor of cholesterol containing an additional oxygen function in the C-14 methyl group or at C-15. Several such compounds have been shown to inhibit reductase in cells in culture (see Gibbons, 1983 for references). However, this suggestion has not been tested experimentally by comparing the intracellular concentrations of these intermediates under different physiological conditions with the concentrations required for inhibition *in vitro*.

The idea that intermediates in the biosynthesis of cholesterol can act as regulators of reductase activity has been elaborated by Rudney and co-workers (Panini *et al.*, 1983). These workers have described experiments which suggest that LDL inhibits reductase in cells in culture by partial inhibition of oxidosqualene cyclase, leading to the accumulation of squalene dioxide which, in turn, is cyclized to form inhibitory oxysterols.

While it has not been established that oxygenated precursors of cholesterol are physiological regulators of HMG-CoA reductase, two lines of evidence strongly suggest that inhibition of reductase by endogenous and exogenous cholesterol is mediated by oxysterol generated enzymically from cholesterol within cells.

As mentioned above, Sinensky and co-workers (see Sinensky, 1985) have isolated mutant lines of CHO cells in which HMG-CoA reductase is resistant to inhibition by 25-hydroxycholesterol and LDL cholesterol. In one of these lines (crB), resistance appears to be due to loss of a specific oxysterol-binding protein previously described by Kandutsch and Thompson (1980) and considered by them to be essential for the inhibitory effect of oxysterols on cells in culture. The finding that crB cells are resistant to inhibition by both cholesterol and 25-hydroxycholesterol cannot be explained by the two sterols using the same protein carrier to reach their target within the cell, since Sinensky and Mueller (1981) have shown that the binding protein that is deleted in the mutant cells has high and specific affinity for the oxysterol but binds cholesterol "poorly or not at all." The defect in crB cells is difficult to explain other than by supposing that inhibition of reductase by LDL cholesterol in normal cells is mediated by some as yet unidentified oxysterol that is transported through the cytosol by the 25-hydroxycholesterol-binding protein.

The second line of evidence for an oxysterol intermediate is provided by the experiments of Gupta *et al.* (1986), who have shown that suppression of reductase in cultured intestinal cells by LDL is prevented by specific inhibitors of cytochrome *P*-450, a mixed-function oxidase. These findings suggest that inhibition of reductase by LDL is mediated by oxysterol formed from internalized cholesterol by *P*-450-dependent oxidation.

In conclusion, in the light of all the evidence discussed above, there seems little reason to doubt the orthodox view that HMG-CoA reductase is suppressed by lipoprotein cholesterol entering cells from the extracellular medium. Induction of enzyme synthesis by experimental procedures that draw cholesterol out of cells *in vitro* (see the following section) also indicates that endogenous cholesterol is one of the links in the chain by which reductase is regulated. However, there is increasing evidence that the inhibitory effect of intracellular cholesterol is mediated by an oxysterol, produced within the cell by enzymic oxidation and transported by a specific protein carrier. The molecular mechanism by which oxysterols inhibit reductase is discussed in the next section.

D. Sterol-Mediated Regulation of Enzyme Synthesis

1. At the Level of Enzyme Protein

Regulation of reductase activity by changes in the rate of enzyme synthesis was first revealed by studies of the diurnal variation in hepatic reductase activity that occurs in rats kept under conditions of controlled lighting and feeding. When feeding is restricted to the dark period of the 24-hour cycle, hepatic reductase activity rises to a maximum at about midnight and then falls to the basal level within 6–8 hours. Although LDL plays no part in this cyclic variation in reductase activity, the methods used to elucidate its basis are worth considering because they illustrate the approaches used to study changes in the rate of reductase synthesis in many other conditions.

Early observations on the effect of inhibitors of protein synthesis on hepatic reductase activity showed that the increase during the early part of the dark phase is due largely to increased enzyme synthesis (see Rodwell *et al.*, 1976, for references). More direct evidence for this was obtained by Higgins *et al.* (1971), using the incorporation of intraperitoneally injected [^{3}H]leucine into the partially purified liver enzyme as an index of its rate of synthesis. Higgins *et al.* (1971) showed that the rise in activity is accompanied by increased synthesis of enzyme protein and that synthesis ceases completely for about 6 hours after the peak of activity at midnight. They concluded that the diurnal rise and fall in enzyme activity is due entirely to changes in the rate of enzyme synthesis and that the rate of degradation and the state of activation of the enzyme remain constant throughout the 24-hour cycle.

Later work in other laboratories has provided additional evidence for the increase in enzyme synthesis in the period before the peak of activity although, as discussed below, there may also be some increase in the state of activation of the enzyme. In support of the conclusion that there is no increase in the rate of degradation of reductase during the descending limb of the activity–time curve, Gould and his co-workers (see Gould, 1977) showed that the rate of decline of enzyme activity after the peak at midnight is similar to the rate at which activity declines when cycloheximide (an inhibitor of protein synthesis) is administered at any time during the first half of the dark phase; in both cases reductase activity falls with a half-life of 2–3 hours.

Clarke *et al.* (1984) have shown that the reductase mRNA level in rat liver rises and falls during the diurnal cycle in parallel with enzyme activity, indicating that the cyclical variation in the rate of enzyme synthesis is mediated by variation in the rate of transcription of the reductase gene. The cause of the diurnal rise and fall in the rate of transcription is not known. One possibility is that it is due to diurnal variation in the secretion of hormones that affect transcription of the gene.

Brown *et al.* (1974) also made use of the response to cycloheximide in a study of the effect of LDL on reductase activity in cultured human fibroblasts. They concluded that the rate of synthesis of enzyme, estimated from the steady-state level of activity and the fractional rate of decline of activity after administration of cycloheximide, increased more than 20-fold when the cells were transferred from a medium containing serum to a lipoprotein-free medium. They also showed that when maximal doses of LDL are added to a culture medium in which cells are growing in the absence of lipoprotein, enzyme activity declines with a half-life (3.0 hours) similar to that observed after addition of cycloheximide to the medium (2.9 hours). They concluded that LDL suppresses reductase activity in fibroblasts in culture by repression of enzyme synthesis rather than by inactivating preexisting enzyme molecules or increasing their rate of degradation.

More recent studies with improved methods for estimating the turnover of reductase in cells *in vitro* have shown that modulation of the rate of synthesis of enzyme occurs under a variety of conditions. In particular, changes in the rate of synthesis of the enzyme have been investigated in several laboratories by measuring incorporation of a radioactive amino acid into immunoprecipitated reductase after incubating cells with the label during a short "pulse" (usually 1 hour or less).

Analysis by radioimmune precipitation has shown that reductase synthesis *in vitro* is suppressed in normal CHO cells by 25-hydroxycholesterol (Chang and Limanek, 1980; Sinensky *et al.*, 1981), in UT-1 cells by LDL, 25-hydroxycholesterol, and mevalonic acid (Faust *et al.*, 1982), and in rat hepatocytes by mevalonic acid and, to a small extent, by human LDL (Edwards *et al.*, 1983a, 1984).

Induction of the synthesis of reductase has been demonstrated by the radioimmune precipitation method in hepatocytes obtained from rats fed cholestyramine or mevinolin (an analog of compactin) and in normal rat hepatocytes incubated in the presence of mevinolin, lecithin dispersions, or human HDL (Edwards *et al.*, 1983b, 1984). Induction by cholestyramine, a resin that binds bile salts in the lumen of the intestine, is due to release of reductase synthesis from repression by bile salts. Physiological repression of reductase by reabsorbed bile salts may be mediated by repression of 7α-hydroxylase, the enzyme that catalyzes the conversion of free cholesterol into 7α-hydroxycholesterol. Stimulation of this enzyme would thus increase the rate at which free cholesterol is removed from the hepatocyte (see Fig. 3.11). Induction by mevinolin is due to competitive inhibition of HMG-CoA reductase, leading to a fall in the concentration of sterol and nonsterol metabolites of mevalonic acid that normally repress reductase synthesis, as discussed below in relation to compactin. The increase in enzyme synthesis that occurs in cells incubated with lecithin or HDL is probably a response to withdrawal of cholesterol from the cells by these two acceptors for free cholesterol (see Rothblat, 1969).

Cohen *et al.* (1982), on the basis of their studies of vascular endothelial cells in culture, conclude that HDL and compactin induce HMG-CoA reductase by different mechanisms. LDL reverses induction by HDL but does not reverse that brought about by compactin. Mevalonic acid, on the other hand, reverses induction of reductase by compactin. These differences may be explained in terms of the multivalent control of reductase discussed as follows. HDL, by withdrawing cholesterol from cells, releases reductase synthesis from the normal repression exerted by cholesterol or other sterol present in the regulatory pool (Fig. 3.11), whereas compactin reverses repression exerted by a nonsterol derivative of mevalonic acid as well as that due to cholesterol synthesized by the cell.

In most cases, changes in reductase synthesis brought about by sterols are mediated by changes in the rate of transcription of the reductase gene. Thus, Luskey *et al.* (1982, 1983) have shown that reductase mRNA is expressed at very high levels in UT-1 cells, in which the rate of synthesis of the enzyme is greatly increased, and that the reductase mRNA level falls in parallel with the rate of synthesis of enzyme when the cells are incubated in the presence of LDL, 25-hydroxycholesterol, or 10 m*M* mevalonic acid. As would be expected, the fall in mRNA level is accompanied by decreased incorporation of [^{3}H]uridine into the reductase message. Induction of HMG-CoA reductase in the livers of cholestyramine-treated rats is also accompanied by an increase in the level of functional reductase mRNA in the liver (Clarke *et al.*, 1983); the rise and fall in hepatic reductase mRNA in parallel with the diurnal rhythm in reductase activity has already been referred to (p. 40).

2. At the Level of Gene Transcription

Osborne *et al.* (1985) have investigated the molecular basis of induction of HMG-CoA reductase and its repression by sterols. They began by constructing recombinant plasmids in which the promoter for expression of the gene for chloramphenicol acetyltransferase (CAT) was derived from the hamster reductase gene. Various segments of the 5′-flanking and 5′-untranslated regions of the reductase gene were ligated to the 5′ end of the coding sequence of the *CAT* gene in the plasmid vector *pSVO-CAT*, producing nine different recombinants (Fig. 3.12). Mouse L-cells were transfected with these reductase–*CAT* fusion genes and the transfected cells were tested for their ability to synthesize CAT when grown in medium containing lipoprotein-deficient serum with or without added sterols.

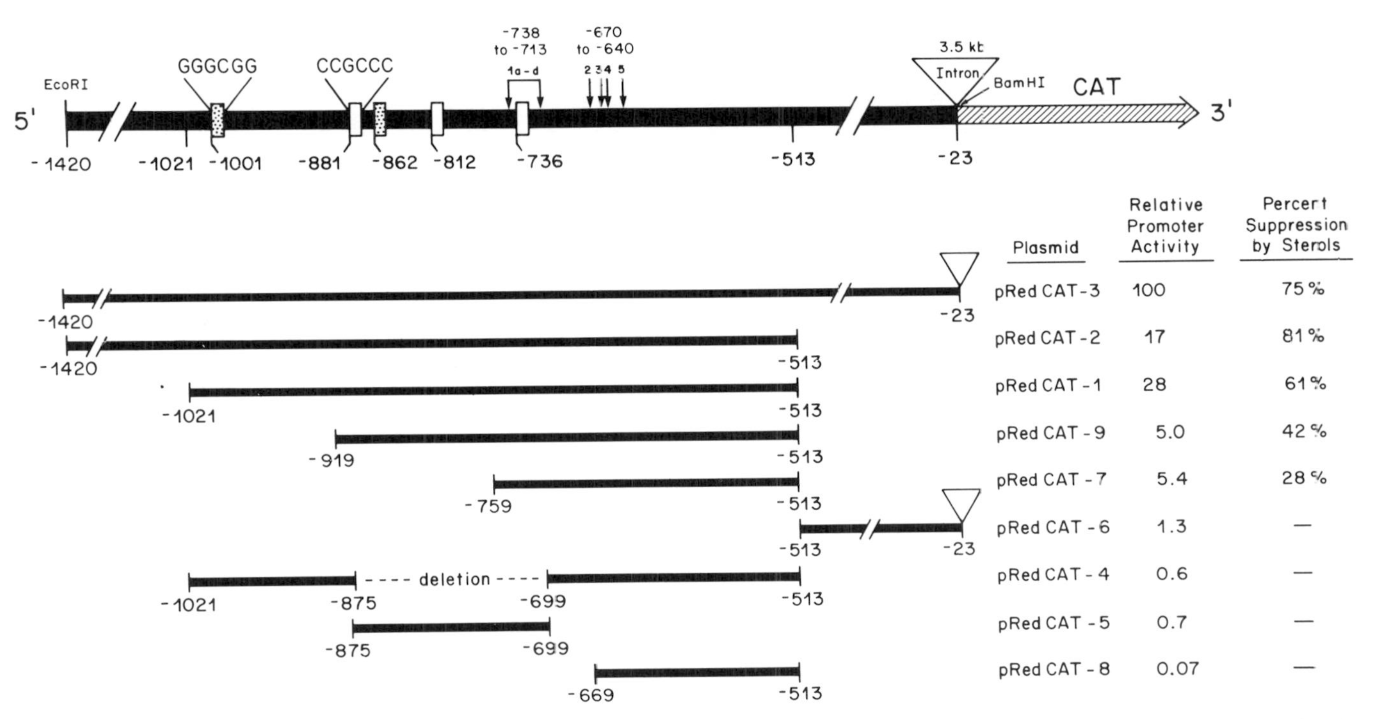

Figure 3.12. Structure and activity of hybrid genes constructed from various portions of the promoter region of the reductase gene joined to the coding region of the chloramphenicol acetyltransferase (CAT) gene. The upper bar shows details of the 5′ end of the reductase gene used to construct the nine hybrid genes (pRed *CAT-1* to -9) shown in the lower part of the figure. Nucleotides −1420 to −23 of the reductase promoter region are numbered in relation to the A of the methionine initiator codon (+1). The 3.5-kb intron that interrupts the 5′ untranslated sequence of the reductase gene between nucleotides -24 and -23 is shown as a triangle. The hexanucleotide sequences homologous to the SV40 promoter are shown as open (CCGCCC) or dotted (GGGCGG) boxes. Arrows show the positions of the multiple transcription initiation sites for reductase mRNA. The coding sequence for CAT protein is denoted by a cross-hatched arrow. The columns on the right give (1) the average CAT enzymic activity obtained with each plasmid, expressed as the percentage of the value obtained with pRed *CAT-3* (the plasmid with the longest reductase sequence), and (2) the percent suppression by 25-hydroxycholesterol plus cholesterol. (From Osborne *et al.*, 1985, with the permission of the authors.)

Since plasmid *pSVO-CAT* has no eukaryotic promoter sequences, expression of CAT in transfected cells provided an index of promoter activity in the reductase-gene sequences present in the constructs. The results of these experiments showed that all the sequences required for the promotion and sterol-dependent repression of transcription are contained within the 227 bp upstream from the most 5′ of the multiple initiation sites for transcription in the reductase gene. Osborne *et al.* (1985) concluded that efficient transcription of the reductase gene requires integrity of the whole of this flanking sequence and that promoter sites and sites that mediate inhibition of transcription by sterols are closely associated.

A comparison of the promoter region of the reductase gene with that of SV40 is relevant to the question of how transcription of the reductase gene is regulated. As we have seen (p. 25), the CCGCCC sequence (the GC box) is repeated six times in the early promoter region of SV40 and is essential for transcription from the early promoter. Dynan and Tjian (1983a,b) have shown that early transcription of SV40 by RNA polymerase II requires the presence of a promoter-specific transcription factor (Sp1) present in the cytosol of uninfected HeLa cells. Initiation of transcription of SV40 by Sp1 is dependent upon binding of Sp1 to the noncoding strand of viral DNA at the sites of the GC boxes in the early promoter (Dynan and Tjian, 1983b; Gidoni *et al.*, 1984). Dynan and Tjian (1983b) suggest that the binding of Sp1 to the promoter in some way directs or stabilizes the binding of RNA polymerase II to the region of viral DNA between the promoter and the startpoint for early transcription.

In the light of more recent evidence it seems likely that transcription of many genes involves the binding of Sp1, or of other protein transcription factors, to GC-rich regions in the promoter. Gidoni *et al.* (1984) have identified what they regard as promoter regions in the monkey genome that contain GC boxes to which Sp1 binds strongly. GC boxes have also been demonstrated in the promoter regions of the thymidine kinase gene of herpes simplex and the genes of several animal enzymes that are normally repressed under physiological conditions (for references, see Kadonaga *et al.*, 1986). However, it has yet to be shown that Sp1 is a positive transcription factor for the reductase gene.

Osborne *et al.* (1987) have shown that sterol-regulated expression of the hamster reductase gene also requires the binding of transcription factors to specific sequences in the promoter region. Using the method of DNase I footprinting (see Chapter 9, Section IV), they identified six regions (FP1 to FP6) in the 227-bp promoter that were protected from DNase I digestion by the binding of proteins present in hamster-liver nuclei (Fig. 3.13). The role of these regions in transcription of the reductase gene was investigated by constructing plasmids containing reductase–CAT fusion genes in which the nucleotide sequences in the protected regions were altered by oligonucleotide-directed mutagenesis (see Section V,E below). The mutant plasmids were then tested for their ability to express mRNA for the fusion gene in a whole extract of HeLa cells. When a

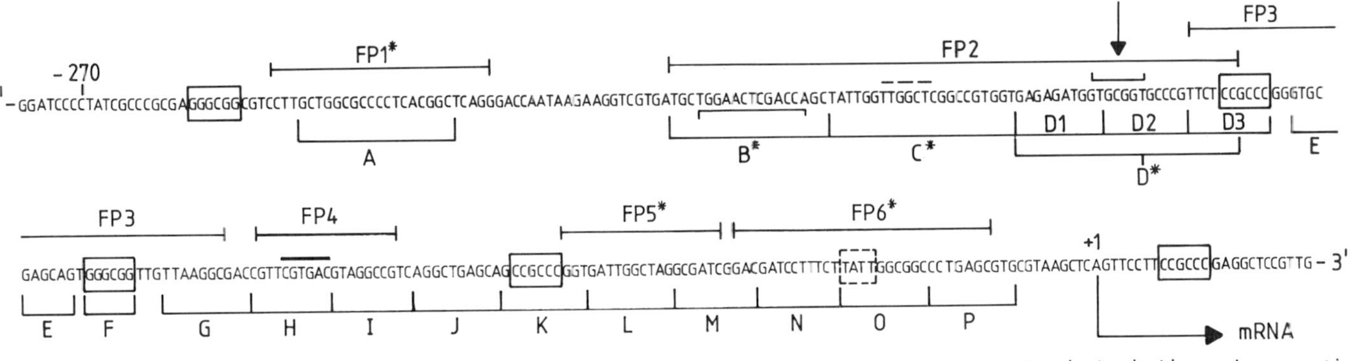

Figure 3.13. Sequence of the hamster HMG-CoA reductase promoter, showing the positions of footprints (FP1-FP6) obtained with proteins present in extracts of hamster-liver nuclei. The sequences marked A–P were subjected to substitution mutagenesis and the mutant promoters were tested for promoter activity in intact cells, after ligating them to the *CAT* gene (see text for explanation). The five GC-rich sequences are boxed. The octanucleotide GTGCGGTG in segment D is shown by a vertical arrow. The sequence TG(N_7)CCA in segment B is underlined. The atypical TATA box is enclosed by broken lines. Footprint regions that bind transcription factor RPF-1 are shown by asterisks. Nucleotide +1 is the most upstream of the multiple initiation sites for transcription of the reductase gene. Note that the numbering of nucleotides in this figure differs from that in Figs. 3.7 and 3.12. (From Osborne *et al.*, 1985, and Gil *et al.*, 1988, with the permission of the authors.)

short sequence at the 3′ end of FP2 (Fig. 3.13) was replaced by a random sequence, transcription was increased. On the other hand, "scrambling" of FP4 or of a region of FP6 containing the atypical TATA box almost abolished transcription. These results suggested that FP4 and FP6 contain positive transcriptional elements and that a sequence near the 3′ end of FP2 participates in the sterol-mediated repression of HMG-CoA reductase.

Osborne *et al.* (1988) have extended these observations by measuring the ability of mutant reductase–CAT fusion genes to express mRNA in intact hamster fibroblasts grown in the presence or absence of sterol. These experiments revealed the presence of three regions containing positive transcriptional elements (C, F, and H + I in Fig. 3.13) and one region (D1 + D2) that was required for sterol-mediated repression of transcription. When this region was altered by mutation *in vitro,* the fusion gene was expressed at a high level in the absence of sterol, and when sterol was added to the medium, expression of the gene was unaffected. Scrambling of the sequences in D1 or D2 specifically abolished the binding of hamster-liver nuclear proteins to segment D of the reductase gene, as shown in DNase I analysis.

The 20-bp D1 + D2 segment contains the octanucleotide GTGCGGTG, which has 7/8 homology with an octanucleotide in the noncoding strand of repeat 2 of the promoter of the human LDL-receptor gene (Fig. 3.14, and see Fig. 9.14). The consensus for the octanucleotide in the hamster reductase and human receptor promoters is GTG$\frac{G}{C}$GGTG. As we shall see in Chapter 9 (Section IV), the human LDL-receptor promoter contains a set of three 16-bp imperfect repeats that together mediate sterol-regulated expression of the gene. Dawson *et al.*

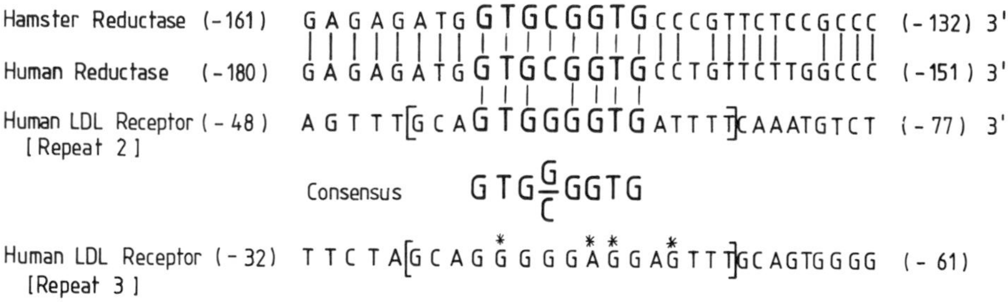

Hamster Reductase (−161) GAGAGATG**GTGCGGTG**CCCGTTCTCCGCCC (−132) 3′

Human Reductase (−180) GAGAGATG**GTGCGGTG**CCTGTTCTTGGCCC (−151) 3′

Human LDL Receptor (−48) AGTTT[GCA**GTGGGGTG**ATTTT]CAAATGTCT (−77) 3′
[Repeat 2]

Consensus GTG$\frac{G}{C}$GGTG

Human LDL Receptor (−32) TTCTA[GCAGGGGGAGGAGTTT]GCAGTGGGG (−61)
[Repeat 3]

Figure 3.14. Promoter sequences of the hamster and human HMG-CoA reductase genes (coding strands) and the human LDL-receptor gene (noncoding strand) aligned so as to show the 8-bp region of homology (bold type). The 16-bp repeat 2 of the human receptor promoter (see Figure 9.14) is enclosed in brackets. Vertical lines show perfect matches. Asterisks show the positions in repeat 3 of the human receptor promoter at which the nucleotide differs from that in repeat 2. Nucleotide +1 is the initiation site for transcription. Note that in Figure 9.14 the receptor-promoter sequences are shown in the coding strand. (From Osborne *et al.*, 1988, with the permission of the authors.)

(1988) have shown that repeat 3 is a positive transcriptional element that binds Sp1 and that, in the presence of sterol, repeat 2 suppresses the activity of repeat 3. Dawson *et al.* (1988) suggest that in the presence of sterol, repeat 2 binds a repressor protein that inhibits the binding of Sp1 to repeat 3.

The above findings suggest that regulation of the reductase and LDL-receptor genes has features in common with the negative regulation of many bacterial genes by end-product repression, exemplified by the tryptophan operon (see Lewin, 1987, Chapter 10). Transcription of the structural gene for tryptophan synthesis is controlled by a protein *repressor* that is activated by tryptophan (the *co-repressor*). The activated repressor binds to a specific sequence of ≅20 bp (the *operator*) adjacent to the promoter. Binding of the repressor prevents transcription of the tryptophan genes, probably by blocking part of the DNA site that binds RNA polymerase. In the absence of tryptophan, transcription of the tryptophan gene is maximal because the repressor does not bind to the operator. Thus, the tryptophan operon is under negative control, in the sense that the structural genes are transcribed unless transcription is prevented by the repressor protein. Mutants in which the nucleotide sequence of the operator is altered are expressed maximally whether or not the co-repressor is present. Such mutants are called *operator constitutive*.

The HMG-CoA reductase gene is also under negative control, but the control system is more complex than that of the tryptophan operon. It seems likely that the reductase-gene operator includes the octanucleotide in segment D of the promoter and that the repressor is a sterol-activated protein that binds to this region in FP2 (Fig. 3.13). On this interpretation, reductase-gene mutants in which the base sequence in segment D1 or D2 is altered are operator constitutive.

The nature of the repressor protein that binds to the octanucleotide in the reductase promoter is clearly a question of much interest. Rajavashisth *et al.* (1989) have isolated, from a HepG2-cell library, a full-length cDNA that encodes a 177-amino-acid protein. This protein [cellular nucleic acid binding protein (CNBP)] binds specifically to double- or single-stranded octanucleotide sequences that confer sterol responsiveness on the LDL-receptor promoter. CNBP contains seven imperfect 14-residue repeats homologous to "zinc finger" motifs present in other DNA-binding proteins. Rajavashisth *et al.* suggest that CNBP participates in sterol-mediated repression of HMG-CoA reductase by binding to the octanucleotide sequence and thereby disrupting the binding of NF-1-like positive transcription factors to adjacent sites in the reductase promoter (see below). In keeping with a role for CNBP in the sterol-mediated regulation of reductase, mRNA for this protein was detected in all tissues examined. Moreover, the amount of CNBP in HepG2 cells was increased by incubation in the presence of 25-hydroxycholesterol. If CNBP turns out to be the physiological sterol-activated repressor protein, it will be of interest to examine its structural and functional relationship to the oxysterol-binding protein described by

Kandutsch and co-workers (see Section V, C above). Dawson *et al.* (1989) have cloned a cDNA encoding an oxysterol-binding protein with 809 amino acids. This protein contains a sequence motif present in several other proteins that regulate gene transcription by binding specifically to DNA.

The reductase gene differs from the relatively simple control system in the tryptophan operon in that it contains three positive promoter elements, defined by mutations in segments C, F, and H + I. These positive elements are activated by the binding of hamster-liver transcription factors that leave footprints 2, 3, and 4 in the DNase I protection assay. Osborne *et al.* (1988) suggest that binding of the sterol-activated repressor to the D1/D2 region prevents binding of a protein transcription factor to segment C.

For reasons discussed by Osborne *et al.* (1988), it is unlikely that Sp1 is the transcription factor that binds to any of the segments C, F, or H + I. This is in contrast to the receptor gene, for which Sp1 is a positive transcription factor (see Chapter 9). Gil *et al.* (1988), using affinity chromatography with specific sequences of the reductase promoter coupled to sepharose, have isolated a protein with M_r 33–35 kDa from hamster-liver nuclei. This protein, called reductase promoter factor (RPF), binds to the six footprint regions shown by asterisks in Fig. 3.13. The role, if any, of RPF in transcription of the reductase gene is not known. However, Gil *et al.* (1988) note that all six regions contain the sequence TGG and that the region with highest affinity for RPF (segment B) has the 13-bp sequence TGG (N_7) CCA in the noncoding strand. This is the consensus sequence for the binding of nuclear factor NF-1, a protein that acts as a transcription factor for several eukaryotic and viral genes and is an initiation factor for adenovirus DNA replication (Jones *et al.*, 1987).

As noted above, the octanucleotide sequence in the reductase-gene operator is also present in repeat 2 of the LDL-receptor promoter. This suggests how the two genes may be regulated coordinately by a common sterol-activated repressor.

E. Sterol-Mediated Regulation of Enzyme Degradation

1. Conditions in Which Degradation Is Altered

Changes in the rate of enzyme degradation have been demonstrated in several conditions in which there is a change in reductase activity. In the first attempts to measure the turnover of reductase *in vivo* and in isolated cells *in vitro,* the rate of degradation of the enzyme was estimated from the rate at which catalytic activity declined after protein synthesis was inhibited by cycloheximide (see Gould, 1977). In these early studies the half-life of HMG-CoA reductase in rat liver was estimated to be about 3 hours at all phases of the diurnal cycle. Similar values based on the cycloheximide method were obtained for the half-life of the enzyme in cultured human fibroblasts incubated in the absence or in the presence of LDL (Brown *et al.*, 1974) and in the HTC line of rat hepatoma cells in culture (Kirsten and Watson, 1974).

Chang *et al.* (1981), on the other hand, found that in CHO cells grown in the absence of lipoproteins, reductase activity declined with a half-life of 13 hours when protein synthesis was inhibited by cycloheximide. When LDL or 25-hydroxycholesterol was added to the medium, reductase activity declined at a rate much greater than that observed in the presence of cycloheximide. This stimulatory effect on the decline of enzyme activity was almost completely abolished by adding cycloheximide with the LDL or 25-hydroxycholesterol. Chang *et al.* (1981) concluded that LDL and 25-hydroxycholesterol stimulate degradation of HMG-CoA reductase in CHO cells by a mechanism that is mediated by a protein whose half-life is so short that it disappears from the cells within hours when its synthesis is inhibited.

Shapiro and Rodwell (1969), on the basis of experiments on the effect of cycloheximide on reductase activity in rat liver, also concluded that reductase is normally degraded by a mechanism that requires a protein with a very short half-life. They suggested that the labile mediator protein is a reductase-specific proteolytic enzyme. A similar conclusion was reached by Cavenee *et al.* (1981), who showed that reductase activity remains constant for at least 8 hours in enucleated CHO cells. Their interpretation of this finding was that HMG-CoA reductase in CHO cells is degraded by a proteolytic enzyme whose intracellular concentration falls rapidly in the absence of synthesis directed by the nucleus. It should be noted, however, that in experiments in which only the catalytic activity of the enzyme is measured, it is not possible to distinguish between a fall in activity due to allosteric inactivation of the enzyme and a fall due to a decrease in the number of enzyme molecules.

The use of cycloheximide may give rise to errors in estimates of the rate of degradation of reductase, especially if a labile protein is required for normal degradation (Shapiro and Rodwell, 1969; Chang *et al.*, 1981). More direct estimates have now been obtained by the radioimmune precipitation method used to study reductase synthesis. With this method it is possible to measure the rate at which immunoreactive enzyme molecules are removed from the pool of intracellular reductase. The approach is well illustrated by the studies, already referred to, of Higgins *et al.* (1971) on the diurnal cycle in reductase activity in rat liver. After labeling of the enzyme *in vivo* by a pulse of [^{3}H]leucine, *total radioactivity* in the immunoprecipitated enzyme isolated from a standard mass of tissue declines at a constant fractional rate, from which the half-life of the labeled enzyme molecules may be calculated (subject only to a small error due to reutilization of radioactive amino acid released from the degraded enzyme).

Using this approach, Higgins *et al.* (1971) concluded that the half-life of the enzyme in rat liver is about 3 hours at all phases of the diurnal cycle, a value that in this instance is close to that obtained with the use of cycloheximide by Gould and co-workers (Gould, 1977).

Analysis by radioimmune precipitation has since been used extensively to study changes in the rate of degradation of HMG-CoA reductase in cells *in vitro* under a variety of experimental conditions; with cells *in vitro* the error due to reutilization of radioactive label is eliminated by swamping the cells with unlabeled amino acid after a short incubation in the presence of labeled amino acid (the "pulse-chase" procedure). In the investigation already referred to (p. 41), Faust *et al.* (1982) showed that the half-life of immunoprecipitable reductase in UT-1 cells grown in the presence of compactin, but in the absence of LDL, is 10 to 13 hours. When LDL or 25-hydroxycholesterol is added to the medium there is a threefold increase in the rate of degradation of the enzyme. This effect, in conjunction with a fall in the rate of synthesis of reductase to less than 5% of the baseline value, leads to a profound fall in reductase activity. An increase in the rate of degradation of reductase has also been demonstrated in compactin-resistant C100 cells after prolonged incubation in the absence of compactin (Hardeman *et al.*, 1984) and in rat hepatocytes incubated in the presence of mevalonic acid (Edwards *et al.*, 1983a).

A marked decrease in the rate of degradation of reductase was observed by Sinensky and Logel (1983) in wild-type CHO cells grown in the presence of mevinolin. A decrease in the rate of degradation of reductase also occurs when hepatocytes from the livers of rats treated with a combination of cholestyramine and mevinolin are incubated in the presence of mevinolin, and in normal hepatocytes incubated in the presence of HDL and a dispersion of lecithin (Edwards *et al.*, 1983b, 1984).

All the above changes in degradation of the enzyme can be explained on the assumption that degradation is enhanced by a product of the metabolism of mevalonic acid, which could be a sterol or a nonsterol metabolite, or both. The stabilization of the enzyme that occurs in hepatocytes incubated in the presence of HDL and lecithin suggests that the metabolite normally responsible for stimulating degradation is intracellular cholesterol, since HDL and lecithin deplete hepatocytes of free cholesterol. However, Sinensky and Logel (1983) found that mevinolin inhibits degradation of reductase in a mutant line of CHO cells (Mev-1) that are unable to synthesize mevalonic acid (see p. 27). They concluded that stabilization of reductase by mevinolin cannot be due to inhibition of the formation of a normal metabolite of mevalonic acid. They suggested that mevinolin stabilizes reductase in CHO cells by interacting directly with the enzyme.

2. Degradation of Reductase at the Molecular Level

The molecular basis of the intracellular degradation of HMG-CoA reductase is now beginning to be understood. As we have seen, most of the available evidence is consistent with the view that under physiological conditions the rate of enzyme degradation varies in parallel with the free cholesterol content of the cell. The UT-1 line of cultured cells, a genetic variant of the CHO-k1 line, provides a striking example of this. These cells have adapted partially to growth in a medium containing compactin and no cholesterol. Under these abnormal conditions, the rate of synthesis of reductase is increased more than 100-fold and the rate of degradation is reduced to less than a third of the normal. As a result of these changes the cells are able to grow, but with about half the normal cholesterol content and with a massive increase in the content of reductase, which may account for 2% of the total protein in the cell. The reductase in UT-1 cells is accommodated in a specialized extension of the smooth ER (the crystalloid endoplasmic reticulum, described on p. 31), which Anderson *et al.* (1983) have shown to be almost devoid of cholesterol.

Orci *et al.* (1984) have shown that when LDL is added to the growth medium, the cholesterol content of the crystalloid ER increases rapidly to a maximum within 2 hours and that this increase is followed by the disappearance of reductase from the ER membrane due to a combination of repression of synthesis and stimulation of degradation. These changes are followed by the complete disappearance of the crystalloid ER within 24 hours of the addition of LDL to the medium.

Orci *et al.* (1984) suggest that a similar sequence of events underlies the increased rate of degradation of HMG-CoA reductase that occurs in normal cells in culture when LDL is added to the incubation medium, with the difference that in normal cells the enzyme is embedded in normal, cholesterol-poor, smooth ER membrane. Orci *et al.* (1984) propose that free cholesterol released within lysosomes by hydrolysis of LDL cholesteryl esters is transported to the ER in membrane vesicles or in association with a protein carrier. Incorporation of free cholesterol into the ER membrane then in some way stimulates breakdown of HMG-CoA reductase and, subsequently, removal of the crystalloid ER membrane. In this scheme, stimulation of the degradation of reductase by mevalonic acid in UT-1 and other cells in culture could be mediated by the formation of endogenous cholesterol; stimulation of degradation by oxysterols might be due to incorporation of oxysterol into the ER membrane, followed by an effect on the degradation of reductase similar to that proposed for membrane cholesterol. It should also be noted that cholesterol, newly arrived in the ER membrane, might be oxidized by membrane-bound cytochrome *P*-450 to an inhibitory oxysterol, which could then be transported to the nucleus by the 25-hydroxycholesterol-binding protein.

Chin *et al.* (1985) have shown that the effect of sterols on the degradation of reductase in hamster ovary cells is independent of their effect on reductase synthesis. For this study, Chin *et al.* (1985) developed a line of hamster cells (TR-36) derived from UT-2 cells, a mutant line that lacks HMG-CoA reductase and therefore requires LDL-derived cholesterol and mevalonate for growth (see p. 28). The TR-36 cells were obtained by transfecting UT-2 cells with plasmid pRed227. This plasmid (see p. 13) contains the whole of the coding sequence of the reductase gene located 3′ to the early promoter of SV40. When multiple copies of this plasmid were integrated into the genome of UT-2 cells, the transfected cells (TR-36) expressed high levels of reductase mRNA and high reductase activity. They were therefore able to grow in the absence of LDL and mevalonic acid.

When LDL or cholesterol and 25-hydroxycholesterol were added to the growth medium, the rate of synthesis of reductase mRNA and of reductase protein remained unchanged because the SV40 promoter is not regulated by sterols. However, enzyme activity fell to half the baseline value owing to increased degradation of preformed reductase molecules. When compactin was added to the medium, the half-life of reductase increased from about 2 hours to more than 5 hours. Thus, under conditions in which synthesis of reductase is not regulated by sterols, degradation of the enzyme is affected by exogenous or endogenous sterols in a manner similar to that seen in UT-1 cells.

Chin *et al.* (1985) suggested that stimulation of the degradation of reductase by sterols is initiated by interaction of the sterol with elements of the membrane-bound domain of the enzyme. To test this hypothesis, Gil *et al.* (1985) developed a variant line of CHO cells in which the only catalytically active HMG-CoA reductase synthesized by the cells was a protein lacking the N-terminal portion that anchors the normal enzyme to the ER membrane. To accomplish this, Gil *et al.* used oligonucleotide-directed mutagenesis to construct a plasmid containing a promoter from SV40 and the base sequence coding for the whole reductase protein minus amino acids 10–341 (see Fig. 3.3).

A brief description of the main points in the strategy adopted by Gil *et al.* (1985) will illustrate the use of this method for producing defined mutants *in vitro*. The first step in the construction of the mutant reductase plasmid was the production of a single circular strand of phage M13 DNA containing a single-stranded insert derived from pRed227. The insert in this recombinant contained the SV40 early promoter and a segment from the 5′ end of the sense strand of reductase cDNA. The next step was the chemical synthesis of a 40-base oligonucleotide comprising 20 bases on each side of the sequence of 996 bases (28–1023) encoding amino acids 10–341 (see Fig. 3.15). This oligonucleotide hybridized to bases 8–27 and 1024–1043, ignoring the intervening segment of 996 bases encoding the amino acids to be deleted in the mutant reductase. The synthetic oligonucleotide was used as primer for the synthesis of a complementary copy of the recombinant phage DNA lacking the sequence of bases encoding amino acids 10–341. Replication of this mutated single strand produced double-stranded DNAs which could be used in the construction of a new plasmid. This plasmid, called pRed227Δ (Gly_{10}-Gln_{341}), encoded the nine hydrophilic N-terminal amino acids of HMG-CoA reductase joined directly to the cytoplasmic extension at the C-terminal end (amino acids 342–887). The mutant plasmid was introduced into cells of the reductase-deficient UT-2 line to produce a variant line designated TR-70.

Cells of the TR-70 line, like TR-36 cells but unlike UT-2 cells, were able to grow in a medium lacking LDL and mevalonic acid. When incubated in LDL-

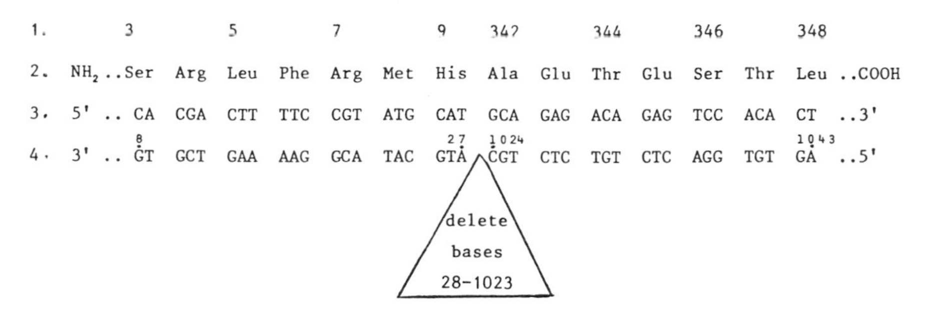

Figure 3.15. Oligonucleotide primer used for the synthesis of double-stranded circular DNA containing the nucleotide sequence encoding a segment of reductase protein from which amino acids 10–341 have been deleted. Line 1, amino acid number in reductase protein (N-terminal = 1); line 2, amino acid sequence required; line 3, mutagenic oligonucleotide primer (the sequence is in the sense mode); line 4, bases 8–27 and 1024–1043 of the template (antisense) strand in the single-stranded circular DNA. (From Gil *et al.*, 1985, with slight modification.)

deficient medium, TR-70 cells synthesized a soluble 60-kDa protein with reductase activity, representing the truncated enzyme encoded by the reductase nucleotide sequence in pRed227Δ (Gly_{10}-Gln_{341}). Like TR-36 cells, cells of the TR-70 line synthesized reductase mRNA and reductase protein at a high rate that was unaffected by the addition of sterol or mevalonic acid to the medium. This is to be expected, since in both cell lines reductase mRNA synthesis is driven by the SV40 promoter, which is not controlled by sterols. However, the rate of degradation of reductase in TR-70 cells was about five times slower than that in TR-36 cells. Moreover, whereas the rate of degradation of reductase in TR-36 cells was accelerated by adding LDL or sterol to the growth medium, neither agent had any effect on reductase degradation in TR-70 cells. In a more detailed analysis, Jingami *et al.* (1987) have shown that mutant reductase lacking two of the membrane-spanning regions is inserted into the ER membrane. However, a crystalloid ER is not formed when the mutant reductase is overexpressed. Moreover, the enzyme is degraded at a high rate ($T^{1/2} \simeq 2.5$ hours) both in the absence and in the presence of sterol. Jingami *et al.* conclude that a complete membrane-spanning domain is required for regulation of degradation by sterols and for stabilization of the enzyme when the ER membrane is depleted of cholesterol.

The properties of the truncated reductase in TR-70 cells show that normal degradation of the enzyme requires binding of its N-terminal hydrophobic domains to the smooth ER. When this cannot take place, degradation is abnormally slow and is no longer subject to regulation by sterols. This is consistent with the hypothesis that the regulatory effect of sterols and oxysterols on degradation is mediated by an interaction with membrane-bound segments of the enzyme.

How might such an interaction influence the rate of degradation of HMG-CoA reductase? One possibility is that incorporation of sterol or oxysterol into a region of the ER membrane at or close to a reductase molecule increases its susceptibility to digestion by a cytoplasmic protease, perhaps by inducing a conformational change in the enzyme. Proteolysis of membrane-bound reductase might result only in the cleavage, from the bound segment, of its cytoplasmic extension, which might then be digested in the cytoplasm or in lysosomes. Against this, however, is the fact that the soluble, truncated 60-kDa enzyme in TR-70 cells, which is almost equivalent to a cleaved hydrophilic domain of the normal membrane-bound enzyme, is degraded very slowly.

Gil *et al.* (1985) have put forward an alternative hypothesis. They suggest that under normal conditions reductase molecules cluster in specialized regions of the ER that continually bud off to form vesicles which fuse with lysosomes, where the enzyme is digested by lysosomal proteases. Incorporation of sterols into the membrane might increase the rate at which reductase-bearing vesicles are formed. Likewise, the very slow rate of degradation of reductase in UT-1 cells could be related to the virtual absence of cholesterol in the crystalloid ER, though the idea that reductase and ER membrane are removed as a unit fails to explain the observation that reductase disappears from the crystalloid ER several hours before the dissolution of the crystalloid membrane when UT-1 cells are treated with LDL (Orci *et al.*, 1984). Digestion of the enzyme by lysosomes would fit with the observation of Tanaka *et al.* (1986) that the unstimulated degradation of reductase in mouse mammary cells is inhibited when lysosomal proteolysis is inhibited. It would also fit with the finding of Parker *et al.* (1989) that monensin (an inhibitor of lysosomal activity) blocks the mevalonic acid-stimulated degradation of reductase in rat hepatocytes *in vitro*. However, nonspecific digestion by lysosomal enzymes is difficult to reconcile with the observation of Chang *et al.* (1981), suggesting that reductase is degraded by a specific protease with a very short half-life (see p. 48).

F. Changes in the State of Activation of Reductase

1. Evidence for Rapid Modulation of Enzyme Activity

Changes in the catalytic activity of existing reductase molecules, without change in their number, have been demonstrated in several conditions *in vivo* and *in vitro*. Usually, such changes occur very rapidly in response to the modulating factor and are followed by a prolonged alteration in the number of enzyme molecules.

Modulation of the specific activity of HMG-CoA reductase (catalytic activity per unit pass of enzyme protein) was first demonstrated directly by Higgins and Rudney (1973). They found that when cholesterol is added to the food of rats at the beginning of the dark period of a diurnal cycle, the usual rise and fall in the amount of immunotitratable reductase in the liver continues throughout the cycle, exactly as in control animals. However, there is a marked fall in enzyme activity, beginning about 4 hours after the cholesterol feeding. If the feeding is continued into the next 24-hour cycle, synthesis of reductase (determined from the incorporation of [^{3}H]leucine into the enzyme) and immunotitratable enzyme mass fall virtually to zero. Thus, the earliest effect of a cholesterol-containing meal on hepatic reductase is a decrease in its specific enzyme activity with no change in the number of enzyme molecules. This is then followed by a profound fall in the mass of enzyme protein.

Edwards *et al.* (1980), using a similar approach, have investigated the effect of compactin, lecithin dispersions, and mevalonate on reductase activity in isolated rat hepatocytes incubated in the presence of one or other of these agents for 3 hours. These workers concluded that the rise in reductase activity after exposure to compactin or lecithin is due mainly to increased synthesis of enzyme protein, but that there is also a small increase in the catalytic activity of existing enzyme molecules, as shown by an increase in the number of units of activity neutralized by a given mass of antireductase antibody (the antibody equivalence point). On the other hand, the decrease in enzyme activity after exposure to mevalonic acid for 3 hours appeared to be due mainly to a fall in specific enzyme activity with little change in the mass of enzyme.

Examples of the rapid modulation of the specific enzyme activity of reductase in other types of cell are mentioned below. The molecular basis of this mode of regulation has been widely discussed in terms of two mechanisms: (1) reversible phosphorylation of the enzyme, and (2) allosteric modification of the enzyme due to changes in the physical state of the ER membrane to which it is bound. Since the evidence for the physiological relevance of both mechanisms has been reviewed in detail (Gibson, 1985; Gibson and Parker, 1987; Kennelly and Rodwell, 1985; Mitropoulos and Venkatesan, 1985), the present discussion will be limited to the briefest of summaries.

2. Phosphorylation–Dephosphorylation

The idea that HMG-CoA reductase might be regulated by reversible phosphorylation was first put forward by Beg *et al.* (1973). These workers showed that reductase activity in washed liver microsomes decreases markedly when the microsomes are incubated in the presence of ATP, Mg^{2+}, and a cytosolic fraction containing protein kinase activity. They also showed that inactivated microsomal reductase could be reactivated by incubation with another cytosolic fraction in the absence of ATP. The molecular events underlying reversible inactivation–activation of reductase have since been elucidated by the combined efforts of several groups of workers. For references, see Kennelly and Rodwell (1985) and Gibson and Parker (1987). See also the reviews by Gibson and Parker (1987) and Hardie *et al.* (1989) for schemes suggesting how reversible phosphorylation of HMG-CoA reductase might be coordinated with that of other enzymes involved in the metabolism of cholesterol and fatty acids.

The ATP-dependent inactivation of reductase observed by Beg *et al.* (1973) is due to covalent phosphorylation of one or more of the 70 seryl residues in the reductase molecule by a reaction in which the γ-phosphate of ATP is transferred to the OH of serine. Analysis of the fragments obtained by proteolysis of rat reductase phosphorylated by [^{32}P]ATP suggests that a major phosphorylation site is in the linker region just outside the membrane-spanning domain (Parker *et al.*, 1989). Reactivation of the phosphorylated enzyme is due to removal of the phosphate group(s) by a fluoride-sensitive phosphoprotein phosphatase present in the cytosol of liver cells. Phosphorylation of the enzyme is catalyzed by an AMP-activated cytosolic protein kinase (AMP-PK), which itself undergoes reversible phosphorylation. (In the earlier literature, this enzyme, now known to use many different substrates, is referred to as reductase kinase.) AMP-PK is phosphorylated by a cAMP-independent cytosolic kinase (kinase kinase) and is dephosphorylated by protein phosphatase. AMP-PK, unlike reductase, is active in the phosphorylated form. Protein phosphatase is inhibited by a cytosolic protein, possibly inhibitor-1, which may be activated by a cAMP-dependent kinase. Thus, liver cells contain all the elements required for modulation of reductase by the two linked inactivation–activation cycles shown in Fig. 3.16.

The protein kinase-phosphatase system has been reported to be present in a variety of tissues, including developing brain, small intestine, mouse L-cells, and human fibroblasts, in addition to liver. In some cases, it has been shown that HMG-CoA reductase in broken-cell preparations is inactivated specifically by incubation in the presence of ATP-Mg^{2+} before assay of enzyme activity and

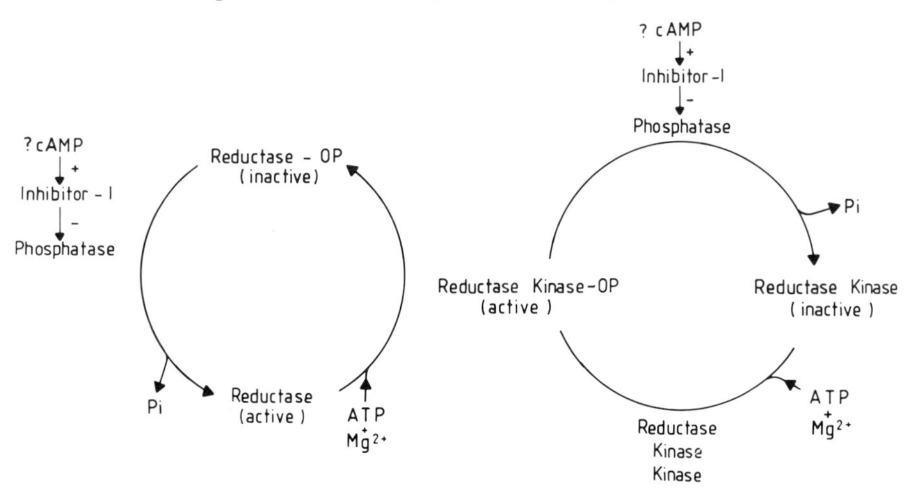

Figure 3.16. The two linked phosphorylation–dephosphorylation cycles responsible for reversible inactivation of HMG-CoA reductase. Phosphorylated reductase is dephosphorylated by phosphoprotein phosphatase and is phosphorylated by reductase kinase in its phosphorylated (activated) form. Reductase kinase is phosphorylated to its active form by reductase kinase kinase. Phosphatase is inhibited by a protein inhibitor which may exist in a phosphorylated (active) or dephosphorylated (inactive) form. cAMP-dependent activation of inhibitor-1 has not been established experimentally. Note that reductase kinase is also referred to as AMP-activated protein kinase (AMP-PK). cAMP, cyclic AMP; Pi, inorganic phosphate.

that inactivation is completely reversed by a fluoride-sensitive phosphatase (see Saucier and Kandutsch, 1979). In other cases the evidence is less direct. For example, the evidence adduced by Maltese and Volpe (1979) for reversible phosphorylation–dephosphorylation of reductase in developing rat brain consisted only in the demonstration that reductase activity in whole-brain homogenates increases during incubation, and that activation is inhibited by NaF. This was interpreted as showing that reductase in developing brain is present in a partially inactivated state and that incubation of homogenates leads to activation of reductase by an endogenous fluoride-sensitive factor.

That reductase may indeed exist *in vivo* in an inactive or latent form was first demonstrated clearly by Nordstrom *et al.* (1977), who showed that reductase activity assayed in liver microsomes prepared in the absence of NaF and EDTA is about five times that in liver microsomes prepared in the presence of both inhibitors. Nordstrom *et al.* concluded that reductase activity assayed in liver microsomes, prepared under conditions in which endogenous phosphatase and protein kinase are inhibited, represents the amount of enzyme activity expressed in the liver at the time of its removal from the animal.

These observations have provided the experimental basis for numerous attempts to test the hypothesis that rapid modulation of reductase activity *in vivo* and in intact cells *in vitro* may be mediated by changes in the proportion of the total enzyme that is in the active, dephosphorylated form. In principle, the method used in all such studies has been to measure (1) *total* reductase content of the tissue (defined as enzyme activity assayed in microsomes preincubated with phosphatase), and (2) *expressed* activity (defined as activity assayed in microsomes isolated in the presence of NaF and EDTA). It is assumed that under these conditions a change in the "phosphorylation state" of the enzyme produces a corresponding change in the ratio of expressed to total activity. Summaries of the evidence for a role of reversible phosphorylation in the regulation of reductase activity under physiological conditions will be found in reviews by Gibson and Parker (1987) and Kennelly and Rodwell (1985).

There is now general agreement that changes in the phosphorylation state of reductase make at least some contribution to the rapid changes in enzyme activity that occur in the liver and in isolated hepatocytes in response to mevalonic acid, glucagon, or insulin. In each case it has been shown that the initial change in reductase activity (a decrease with mevalonic acid or glucagon, an increase with insulin) is accompanied by a parallel change, in the same direction, in the ratio of expressed to total activity without any change in total activity. It has also been shown that, in all three cases, the short-term change in the phosphorylation state of the enzyme is followed within 1–2 hours by a change in the number of enzyme molecules due to modulation of the rate of enzyme sythesis, with or without modulation of the rate of degradation.

Gibson and Parker (1987) have discussed the possibility that phosphorylation of HMG-CoA reductase accelerates the subsequent degradation of the enzyme by increasing its susceptibility to proteolytic degradation. In support of this suggestion, Parker *et al.* (1989) have shown that phosphorylation of rat-liver reductase increases its susceptibility to proteolysis by the cytoplasmic enzyme calpain-2.

The most detailed evidence in support of the view that rapid reversible modulation of reductase occurs *in vivo,* and that a change in the ratio of expressed to total reductase activity is, in fact, due to a change in the phosphorylation state of the enzyme, has been obtained from experiments with mevalonic acid. Erickson *et al.* (1980) were the first to show that reversible modulation of hepatic reductase may occur *in vivo*. These workers showed that administration of large doses of mevalonic acid to rats by stomach tube leads to a rapid fall in hepatic reductase activity, microsomal activity falling with a half-life of 5.3 minutes. For about 30 minutes after the dose of mevalonic acid this inhibition can be reversed by incubating the liver microsomes with a cytosolic fraction of rat liver before assaying the enzyme; reversal of inhibition is prevented by NaF. At longer intervals after the dose, the diminished enzyme activity cannot be restored by the cytosolic fraction.

Extending these observations, Beg and Brewer (1981) showed that the administration of mevalonate to rats causes a reversible fall in the ratio of expressed to total hepatic reductase activity within 20 minutes of the dose, with no change in total activity, and that the fall in expressed activity is accompanied by increased incorporation of ^{32}P from [^{32}P]ATP into the enzyme. Moreover, when the activity of the inhibited enzyme was restored by incubation with protein phosphatase, there was a concomitant loss of ^{32}P from the enzyme. As in the experiments of Erickson *et al.* (1980), Beg and Brewer observed that the initial, reversible fall in expressed activity was followed 60 minutes after the dose of mevalonate by a profound fall in total enzyme activity that could not be reversed by phosphatase. Beg and Brewer (1981) also noted that at 60 minutes after the dose, when total activity had fallen to 20% of its initial value, the ratio of expressed to total activity had returned to the value observed before mevalonate was given. This is of interest in relation to the interpretation of other observations on reversible inactivation of reductase discussed below.

Beg *et al.* (1984) have shown that the initial changes in expressed activity of hepatic reductase after a dose of mevalonate are associated with changes in the activities of three enzymes that are thought to regulate the phosphorylation state of reductase. Within 20 minutes of the dose of mevalonate there is an increase in the activity of kinase kinase and in the activity and phosphorylation state of AMP-PK and a decrease in the activity of phosphoprotein phosphatase. These changes would be expected to lead to an increase in the proportion of reductase in the phosphorylated, inactive form. Beg *et al.* suggest that administration of mevalonate leads to activation of kinase kinase and inhibition of phosphatase, and that the activation of kinase kinase leads to increased phosphorylation and enzymic activity of AMP-PK. Increased activity of AMP-PK would lead, in turn, to inactivation of reductase. Since mevalonate has no direct effect on kinase kinase or phosphoprotein phosphatase, it is likely that the short-term effects observed by Beg *et al.* (1984) were mediated by one or more products of the metabolism of mevalonate. Gibson (1985) has discussed some of the metabolites of mevalonate that might mediate its effects on the phosphorylation state of the enzyme.

There is less agreement about the role of reversible phosphorylation in the modulation of reductase activity in other conditions. Brown *et al.* (1979) could find no evidence of a change in the ratio of expressed to total hepatic reductase activity in rats fed cholesterol for 12 hours. Arebalo *et al.* (1981), on the other hand, observed a fall in this ratio after 60 minutes of cholesterol feeding, with no significant change in total activity. The fall was fully reversible by phosphatase. If the feeding was continued for a further 1 hour, the fall in reductase activity became more marked and the inhibition could no longer be reversed by incubating the microsome with phosphatase. As in the experiments of Beg and Brewer (1981) with mevalonate, Arebalo *et al.* (1981) noted that after cholesterol feeding for 2 hours, when total reductase activity had fallen to half the control value, the ratio of expressed to total activity had returned to the initial value observed before cholesterol feeding.

However, these findings are not necessarily in conflict. Kennelly and Rodwell (1985) suggest that the initial response of cells to a change in demand for mevalonate or sterol is a rapid modulation of the phosphorylation state of reductase. They suggest that if the stimulus persists, the mass of reductase protein changes in the appropriate direction and the phosphorylation state returns to the baseline level, so that the enzyme is again ready for a rapid and immediate response. If this is what happens when rats are fed cholesterol one would not expect to find a change in the phosphorylation state of hepatic reductase after 12 hours of cholesterol feeding, when total activity is markedly depressed.

A mechanism that enables the cell to increase the rate of cholesterol synthesis very rapidly would be of obvious advantage to tissues, such as liver and steroid-hormone-forming organs, in which there may be a sudden demand for more cholesterol. It is not easy to see that this would also be the case in tissues that use cholesterol predominantly as a structural element. In keeping with this, Brown *et al.* (1979) found that HMG-CoA reductase in cultured fibroblasts is normally in the fully activated state, i.e., it is not "poised" for a sudden increase in activity. However, it should be noted that even in cells in which there is unlikely to be a sudden increase in the need for cholesterol, there might be rapid fluctuation in the demand for nonsterol products of the metabolism of mevalonic acid.

Higgins *et al.* (1971) (see p. 40) concluded that the diurnal variation in hepatic reductase activity observed in rats is due solely to cyclical variation in the mass of enzyme protein and that there is no variation in specific enzyme activity. In agreement with this, Brown *et al.* (1979) could find no evidence of a change in the ratio of expressed to total reductase activity in the livers of rats at any time during the feeding cycle. However, Easom and Zammit (1984) have arrived at a different conclusion. These workers have developed a method for isolating liver microsomes that is designed to minimize changes in the phosphorylation state of phosphorylatable enzymes during the isolation procedure. Using this method, they have found that the ratio of expressed to total hepatic reductase activity increases to a peak value at the fourth hour of the dark phase of the diurnal cycle and then falls to a minimum at the fourth hour of the light phase. Easom and Zammit (1984) conclude that the diurnal rhythm in hepatic reductase activity is the net result of two asynchronous cycles, one involving the phosphorylation state of the enzyme and the other involving its total activity.

7-Ketocholesterol and 25-hydroxycholesterol suppress HMG-CoA reductase in fibroblasts (Brown and Goldstein, 1974) and hepatoma cells (Bell *et al.*, 1976) more rapidly than would be expected if the only effect of oxysterols on reductase is to inhibit its synthesis. When first reported, these observations raised the possibility that oxysterols, as well as inhibiting reductase synthesis in intact cells, also bring about rapid modulation of the catalytic activity of existing enzyme molecules. In the light of the later work discussed above it now seems more likely that suppression of reductase activity by oxysterols is due to a combination of decreased synthesis and accelerated degradation of enzyme protein. If, in addition, oxysterols do exert an effect on the specific catalytic activity of reductase, this is probably not mediated by modulation of the phosphorylation state of the enzyme. This conclusion follows from the observation of Erickson *et al.* (1980) that the inhibition of hepatic reductase that occurs within 30 minutes of the addition of 25-hydroxycholesterol to a perfused rat liver cannot be reversed by incubating the liver microsomes with partially purified phosphoprotein phosphatase. Cavenee *et al.* (1981) also concluded that inhibition of reductase by oxysterols is not due to an increase in the phosphorylation state of the enzyme. They showed that 25-hydroxycholesterol has no effect on reductase activity in enucleated CHO cells. Since the components of the reductase phosphorylation–dephosphorylation system are all cytosolic, a short-term effect on the phosphorylation state of the enzyme should not require an intact cell nucleus.

Does LDL-derived cholesterol influence the phosphorylation state of HMG-CoA reductase? This question is of obvious importance in relation to the molecular mechanisms by which LDL entering cells via the LDL-receptor pathway regulates intracellular reductase activity. Faust *et al.* (1982) found that the phosphorylation state of reductase in UT-1 cells was similar when the cells were grown in the presence or absence of LDL. However, these experiments were not designed to test for the presence of a rapid change in the degree of phosphorylation of reductase in response to LDL. Beg *et al.* (1986) have shown that LDL rapidly inactivates reductase in fibroblasts preincubated in a lipoprotein-free medium. In cells exposed to LDL for up to 30 minutes, inactivation can be completely reversed by incubating the microsomes with phosphoprotein phosphatase before assaying the enzyme. After exposure to LDL for longer periods, inactivation of reductase becomes irreversible and is accompanied by a decline in *total* enzyme activity. In view of the extensive use of cultured fibroblasts in the study of the regulation of HMG-CoA reductase, the molecular basis of rapid enzyme inactivation by LDL would be worth investigating in more detail. In this regard, it is of interest that Beg *et al.* (1986) have reported the presence of typical AMP-PK and kinase activities in cultured fibroblasts.

The fact that HMG-CoA reductase exists in a partially inactivated state in intact cells of some tissues raises the question of the validity of observations on reductase activity made under conditions in which the enzyme may become activated at some stage of the assay. The extent to which activation of the enzyme takes place *in vitro* in the absence of fluoride probably varies from one laboratory to another. Factors likely to influence dephosphorylation of phosphorylated enzyme *in vitro* are the procedures used for removing the tissue from the animal and for isolating microsomes, the composition of the fluoride-free buffers used during the preparation of microsomes, the presence of cytosolic inhibitors of endogenous phosphatase, and other variables not yet defined.

In the study of Nordstrom *et al.* (1977) referred to above, activation of hepatic reductase *in vitro* in the absence of fluoride appears to have been virtually complete. Likewise, in the experiments of Faust *et al.* (1982) on UT-1 cells, reductase activity in microsomes prepared in the absence of fluoride underwent no further activation when the microsomes were incubated with exogenous phosphatase. On the other hand, in the study of Erickson *et al.* (1980) on the rapid inactivation of hepatic reductase by mevalonic acid *in vivo,* reactivation during the assay procedure cannot have been complete (despite the absence of fluoride), since activation occurred when the microsomes were treated with exogenous phosphatase. Higgins and Rudney (1973), in the work referred to on p. 54, observed a rapid decrease in the specific enzyme activity of reductase in the livers of cholesterol-fed rats. In the light of the later observations of Arebalo *et al.* (1981), this was probably due to reversible phosphorylation of the enzyme. Yet Higgins and Rudney did not use fluoride during isolation of the liver microsomes in which reductase activity was assayed.

It is difficult to reconcile all the observations so far reported on phosphorylation–dephosphorylation of HMG-CoA reductase. However, the following tentative conclusions are justified by the present evidence. In liver, and possibly in some other tissues, reductase exists in a partially phosphorylated state which is modulated *in vivo* by a variety of physiological and nonphysiological stimuli. Unless special precautions are taken to inhibit dephosphorylation during the preparation of microsomes, reductase activity assayed *in vitro* will not accurately reflect activity expressed *in vivo.* Depending upon the procedures used, the measured activity may represent any value from the activity expressed *in vivo* to that of the fully activated enzyme. Usually, though not invariably, increased phosphorylation of reductase *in vivo* will not be detected *in vitro* unless the microsomes in which activity is assayed are prepared in the presence of fluoride. Observations on changes in the mass of reductase based on immunoprecipitation with a specific antibody (including those on rates of synthesis and degradation) are valid whether or not activation takes place during the isolation procedure. This is so because the phosphorylated and dephosphorylated forms of reductase react equally with antireductase antibodies.

3. Modulation of the Fluidity of the ER Membrane

Mitropoulos and co-workers (see Mitropoulos and Venkatesan, 1985, for review) have suggested that unesterified cholesterol may modulate reductase activity *in vivo* without changing enzyme concentration by modifying the fluidity of the ER membrane in which the enzyme is embedded. They postulate that a change in the cholesterol content of a region of the membrane in the immediate vicinity of a reductase molecule alters its conformation and that this leads to a change in its catalytic activity. Mitropoulos and co-workers propose that the above mechanism is responsible for some of the change in specific catalytic activity of hepatic reductase that occurs *in vivo* soon after the feeding of cholesterol, mevalonic acid, or cholestyramine. There is a substantial body of evidence consistent with the ''membrane fluidity'' hypothesis. In particular, as mentioned below, it has been shown that reductase activity in isolated microsomes can be modified reversibly by changing the free cholesterol content of the microsomal membranes. However, in these and other experiments designed to test the hypothesis it has proved difficult to distinguish between an effect of cholesterol on reductase activity that is mediated by a change in membrane fluidity from one that is due to a direct interaction of cholesterol with the membrane-bound enzyme itself.

The effects of cholesterol on the physical state of biological membranes are discussed in the monograph of Gibbons *et al.* (1982). Only the salient facts need to be mentioned here. The element common to all natural membranes is a phospholipid bilayer in which the fatty acyl chains lie parallel to one another and face inward into the hydrophobic interior of the membrane. The conformation and mobility of the acyl chains under various experimental conditions have been studied extensively in artificial bilayers of one or more defined species of phospholipid and in natural membranes comprising a complex mixture of different phospholipids together with other lipid and nonlipid components.

In artificial bilayers of a single species of phospholipid, below a temperature characteristic for each phospholipid (T_c, the phase transition temperature) the acyl chains are in a rigid crystalline state. When the temperature is raised progressively from below the transition temperature there is an abrupt change in the physical state of the acyl chains at the T_c. The chains ''melt'' and become freely mobile, though the long-range order of the bilayer is maintained because the polar head groups remain anchored at the lipid/water interface. This state is known as the liquid-crystalline phase. Proteins incorporated into the bilayer are immobile at temperatures below the T_c. When the acyl chains melt, proteins are free to rotate and to diffuse laterally and may undergo reversible conformational changes.

When unesterified cholesterol is added to the bilayer, cholesterol molecules enter the spaces between the acyl chains, causing marked changes in the physical state of the membrane. As the cholesterol : phospholipid molar ratio is increased, the phase transition becomes progressively less abrupt until at molar ratios of about 1 : 2 a temperature-induced phase transition can no longer be detected. In the presence of cholesterol at these concentrations, over a wide range of temperatures above and below the T_c the acyl chains are maintained in a state known as the intermediate gel phase. In this state the mobility of the acyl carbon atoms adjacent to the cholesterol ring system is decreased, while the proximal and distal segments of the acyl chains remain fluid (Stoffel *et al.*, 1974). Thus, one effect of the addition of cholesterol to a phospholipid bilayer is to "rigidify" the acyl chains, and hence to increase their viscosity, above the transition temperature.

In certain circumstances, a phospholipid bilayer may contain regions that are in different states of fluidity at a given temperature. This may occur, for example, in artificial bilayers composed of mixtures of different phospholipids or in those containing cholesterol at a concentration below that at which every phospholipid molecule can interact with a cholesterol molecule. In the latter case the cholesterol in the bilayer may be distributed nonrandomly, leading to the formation of separate domains, some of which are in a crystalline state while others are in the intermediate gel or liquid crystalline state. This phenomenon, known as lateral phase separation, may also occur in complex membranes in which the cholesterol : phospholipid molar ratio is above 1 : 2, a point of some importance in relation to the effects of cholesterol on biological membranes.

Clear-cut phase transitions induced by temperature are usually difficult to detect in biological membranes. This is due to the complexity of their lipid composition and to the presence of cholesterol at concentrations high enough to maintain the bulk of the phospholipids in the intermediate gel phase over a wide range of temperature. However, comparatively sharp temperature-induced transitions have been observed in a variety of natural membranes. Such changes may occur in the membranes of microorganisms and of eukaryotic cells within the physiological temperature range, especially in those membranes in which there is lateral phase separation due to nonrandom distribution of cholesterol, as in mitochondria and other cell membranes with low cholesterol : phospholipid ratios (see DeKruyff *et al.*, 1974).

In view of the marked influence of conformational changes on the catalytic activity of many enzymes, one might expect the activity of a membrane-bound enzyme to change in response to a change in the viscosity of its microenvironment. There is now convincing evidence that this does, in fact, occur with some membrane-bound enzymes, though not with all. The most striking examples are those in which an abrupt change in membrane fluidity induced by temperature is associated with an equally abrupt change in the activation energy of an enzyme bound to the membrane.

The relation between temperature and the activity of an enzyme is most conveniently expressed in the form of a curve relating the log of the reaction rate to the reciprocal of the absolute temperature (an Arrhenius plot). Most soluble enzymes and many membrane-bound enzymes give linear Arrhenius plots over a wide range of temperature, the activation energy being related directly to the slope of the straight line. With some membrane-bound enzymes, however, the Arrhenius plot shows an abrupt change in slope (see Fig 3.17). Below the break the activation energy is higher than at temperatures above the break. This shows that when the temperature is reduced progressively there is an abrupt increase, at a particular temperature, in the energy required to activate the enzyme. In several cases the break in the Arrhenius plot occurs at the temperature at which the membrane lipids begin to undergo a phase transition (see Gibbons *et al.*, 1982). Moreover, DeKruyff *et al.* (1973) have shown that incorporation of cholesterol into the membranes of *Acholeplasma laidlawii* leads to concomitant changes in the temperature at which there is a phase transition in the membrane and in the temperature at which a break occurs in the Arrhenius plot for membrane-bound ATPase. Clearly, cholesterol is capable of modifying the activity of some membrane-bound enzymes by changing the fluidity of the bulk lipids in the membrane.

As shown in Fig. 3.17, Arrhenius plots of HMG-CoA reductase activity in microsomes from the livers of rats fed a normal diet show a break at about 28°C (Sabine and James, 1976). Mitropoulos and Venkatesan (1977) suggest that this discontinuity is due to the effect of an abrupt phase transition in the microsomal ER membrane on the activity of bound reductase. They propose that when the temperature is reduced progressively there is a sharp decrease at 28°C in the fluidity of the phospholipid bilayer surrounding the enzyme and that this modifies the conformation of the enzyme, resulting in a fall in its catalytic activity.

It has not been possible to correlate the Arrhenius discontinuity with a phase transition in the bulk lipids of hepatic ER membranes. To explain this in terms of their hypothesis, Mitropoulos and Venkatesan (1985) postulate the presence of lateral phase separation in the ER membrane such that each reductase molecule is surrounded by a region of cholesterol-poor phospholipid that is normally in a liquid-crystalline state at temperatures above about 28°C. In addition, they postulate that when the free cholesterol content of hepatic microsomes is increased, cholesterol molecules are incorporated selectively into the phospholipid domains surrounding reductase molecules. If cholesterol is capable of modifying reductase activity in this way, a cholesterol-induced change in the activity of membrane-bound enzyme should be accompanied by a change in the Arrhenius plot. In accordance with this expectation, Mitropoulos and Venkatesan (1977) and Mitropoulos *et al.* (1978a) have shown that when the free cholesterol content of

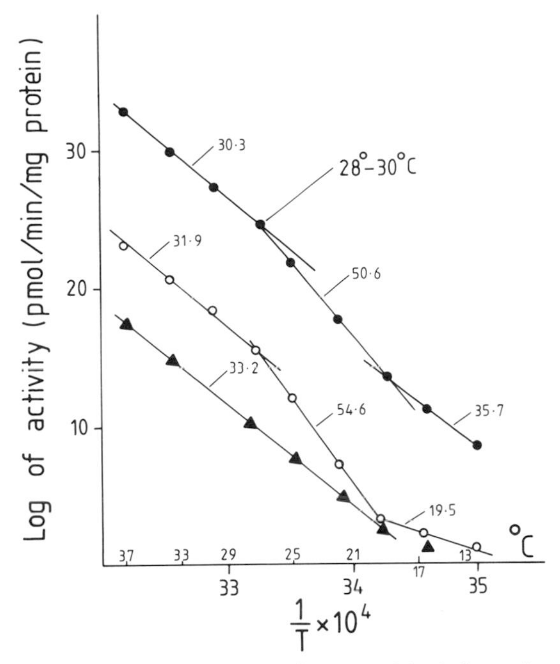

Figure 3.17. Arrhenius plots of HMG-CoA reductase activity in liver microsome from rats fed a normal chow diet (○), a cholesterol-supplemented diet (▲), or a cholestyramine-supplemented diet (●). Activation energies, calculated from the slopes of the line, are indicated beside each line (kcal/mol per °K). Numbers above the horizontal axis are degrees on the centigrade scale. T is the temperature on the Kelvin scale. (From Mitropoulos and Venkatesan (1977) with the permission of the authors.)

rat liver microsomes is increased by feeding the rats cholesterol or by giving them an injection of mevalonate, the fall in microsomal reductase activity is accompanied by loss of the break in the Arrhenius plot. Venkatesan and Mitropoulos (1982) have also shown that when the free cholesterol content of liver microsomes is increased by incubating them with human serum, reductase activity falls and the Arrhenius discontinuity disappears. Moreover, the state of phosphorylation of the enzyme has no effect on the Arrhenius discontinuity in untreated microsomes or on its disappearance in microsomes incubated with serum.

These very suggestive results are consistent with the hypothesis that cholesterol can modulate reductase activity via an effect on membrane fluidity. However, they do not exclude the possibility that changes in enzyme activity induced by temperature or by modification of the free-cholesterol content of the ER membrane are due to direct effects on the enzyme itself. In this regard it is worth noting that Heller and Gould (1974) observed a sharp break at 28°C in the Arrhenius plot for rat-liver reductase solubilized by the freeze–thaw method. A phase transition restricted to the microenvironment of microsomal reductase, whether induced by temperature or by cholesterol, has not been demonstrated. Nor is there any direct evidence that the conformation of the enzyme changes at the break point in the Arrhenius plot or when enzyme activity is suppressed by increasing the cholesterol content of the membrane. Some of these difficulties might be resolved if it were possible to study the motion of reductase molecules in the ER membrane by the methods that have been used to investigate the mobility of ATPase in sarcoplasmic reticulum (Thomas and Hidalgo, 1978).

G. Multivalent Feedback Regulation

1. Nonsterol Metabolites of Mevalonate Are Required for Growth and Division of Cells

Two independent lines of investigation have recently converged to show that animal cells require nonsterol products of mevalonic acid, in addition to cholesterol, for cell division and for normal regulation of HMG-CoA reductase. In this section the role of nonsterol metabolites of mevalonic acid in DNA synthesis will be mentioned briefly, but we shall be concerned mainly with the complex mechanism by which reductase is regulated, as revealed by the study of cells in which the production of mevalonic acid is inhibited by compactin.

As shown in Fig. 3.1, IPP, the first isoprene[3] compound derived from mevalonic acid, is the precursor of the isopentenyl residue of isopentenyl tRNAs and of at least three polyisoprenoids. In the cells of vertebrates these include cholesterol, ubiquinone, and dolichol.

In the formation of an isopentenyl tRNA the isopentenyl unit is transferred from dimethylallyl PP (formed by the isomerization of IPP) to an adenosine near the 3′ end of the anticodon of a preformed tRNA molecule (see Hall, 1970). Thus, neither isopentenyladenosine nor isopentenyladenine is a precursor of isopentenyl tRNA. As we saw in Section I previously, cholesterol is derived from squalene, the C_{30} acyclic polyisoprenoid formed by the condensation of two molecules of FPP. This reaction is catalyzed by squalene synthetase. Ubiquinone is formed essentially by the successive addition of isoprene units to FPP, followed by transfer of the polyisoprene chain to a C_6 ring derived from tyrosine, with the addition of three methyl groups derived from methionine. The final stages of assembly take place in the mitochondria. In FPP, squalene, and ubiq-

[3]The isoprene unit (C, C>C—C—C) is a branched C_5 structure with the carbon skeleton of isoprene. Many biological compounds, referred to as polyisoprenoids, are formed by the polymerization of isoprene units, with or without cyclization of the initial product of condensation.

uinone all the double bonds are *trans*. Dolichols are formed by the *cis* addition of up to 19 isoprene units to FPP.

It will be apparent from this summary that FPP is the last common intermediate in the biosynthetic pathways leading from mevalonic acid to cholesterol, ubiquinone, and dolichol. A molecule of FPP may condense with another FPP molecule to form squalene, or with a succession of isoprene units in *trans* configuration to form the polyisoprenoid chain of ubiquinone, or with isoprene units in *cis* configuration to form dolichols.

A supply of either endogenous or exogenous cholesterol is needed by animal cells in culture for sustained growth and multiplication. This became apparent when it was shown that cells in culture do not survive indefinitely in the presence of oxysterol inhibitors of HMG-CoA reductase unless an available source of cholesterol is present in the medium (see p. 37).

The first hint that HMG-CoA reductase also supplies dividing cells with an essential metabolite other than cholesterol came from the study of DNA synthesis in cells undergoing synchronous mitosis *in vitro*. Chen *et al.* (1975) showed that when lymphocytes incubated in the presence of LDL are stimulated by a mitogen, the increase in DNA synthesis that follows within 24 hours is preceded by a rise in sterol synthesis, reflecting increased reductase activity. Chen *et al.* concluded that increased sterol synthesis is required for the rise in DNA synthesis, as well as for the increased rate of membrane formation that occurs during cell division. It may well be that dividing lymphocytes, because of their low LDL-receptor activity, require an increased supply of endogenous cholesterol for the formation of new membrane, even when LDL is available to the cells. However, in the light of subsequent work on dividing cells it is likely that the rise in reductase activity observed by Chen *et al.* (1975) was partly a response of the cells to a requirement for increased production of nonsterol metabolites of mevalonic acid.

Kaneko *et al.* (1978) showed that compactin inhibits the growth of human fibroblasts grown in the presence of a source of LDL and that the inhibitory effect can be reversed by low doses of mevalonate ($\cong$0.3 m*M*). Since LDL-receptor activity is expressed by fibroblasts in the presence of compactin, this observation indicates that normal growth of fibroblasts requires some product of the metabolism of mevalonic acid that cannot be replaced by exogenous cholesterol. Kaneko *et al.* suggested that the essential growth factor was either ubiquinone, a dolichol derivative, or an isopentenyl nucleotide. However, these experiments did not exclude the remote possibility that growing cells require a supply of endogenously synthesized sterol which cannot be replaced by cholesterol delivered to the cell via the LDL-receptor pathway. In view of the later observations of Siperstein and co-workers, and of others (see Siperstein, 1984), this seems most unlikely.

Quesney-Huneeus *et al.* (1979) made a detailed study of the temporal relation between reductase activity and DNA synthesis in BHK-21 cells (a transformed fibroblast line) during the first two synchronous cycles after release from a double thymidine block. At each cycle, DNA synthesis increased to a peak during S phase[4] and then declined to the resting level. Reductase activity showed a similar rise and fall, the peak of enzyme activity preceding the peak of DNA synthesis by about 2 hours. When reductase was inhibited by compactin, the S-phase increase in DNA synthesis was abolished, although the rate of DNA synthesis throughout the remainder of the cycle was unchanged. The addition of LDL-rich lipoproteins to the incubation medium failed to reverse the inhibition of S-phase DNA synthesis, despite a 64% increase in the cholesterol content of the cells. On the other hand, the addition of 0.4 m*M* mevalonic acid completely reversed the inhibition of DNA synthesis.

Quesney-Huneeus *et al.* (1979) concluded that a product of the metabolism of mevalonate is required specifically for the burst of DNA synthesis that occurs during the S phase of the cell cycle. They argued that this was unlikely to be cholesterol because the amount of cholesterol formed from 0.4 m*M* mevalonate added to the compactin-treated cells could not have increased their cholesterol content by more the 0.01%. It is consistent with this interpretation that isopentenyladenine is 100 times as effective as mevalonic acid in reversing the inhibition of S-phase DNA synthesis by compactin (Quesney-Huneeus *et al.*, 1980). Isopentenyladenine has not been demonstrated in animal tissues, but it may be a breakdown product of isopentenyl tRNA (Panini *et al.*, 1985), shown by Faust *et al.* (1980) to derive its isopentenyl residue from mevalonic acid in human fibroblasts. Thus, it seems likely that isopentenyladenine, derived indirectly from mevalonic acid, is required in trace amounts for the S-phase increase in DNA synthesis. Observations reported from other laboratories (Habenicht *et al.*, 1980; Brown and Goldstein, 1980; Fairbanks *et al.*, 1984) support the conclusion that normal growth and division of cells in culture requires small amounts of one or more nonsterol metabolites of mevalonic acid.

The dual requirement for cholesterol in bulk and for trace quantities of other substances derived from mevalonate explains what might otherwise seem anomalous. In the experiments of Quesney-Huneeus *et al.* (1979), cells in which reductase was fully inhibited by compactin, and which were incubated in the absence of LDL, were able to undergo two cycles of S-phase DNA synthesis when they were supplied with mevalonate in amounts too small to make a significant contribution to their supply of cholesterol. Presumably, enough cholesterol was available from intracellular stores to sustain membrane formation for two divisions. For longer-term growth, cells in which reductase is completely inhibited require both LDL and small amounts of mevalonic acid (either alone

[4]The S (synthesis) phase of the cycle of a eukaryotic cell is the period (usually less than 10% of the total cycle) during which DNA is synthesized at a high rate in preparation for cell division. During the remainder of the cycle DNA is synthesized at a low rate.

will not suffice), or mevalonate at concentrations high enough to supply both cholesterol and essential nonsterols. As shown by Brown and Goldstein (1980), CHO cells incubated in a lipoprotein-deficient medium containing compactin at low concentration (2 μ*M*) are unable to grow, but they are able to grow normally if LDL, without mevalonate, is added to the medium. Under these conditions, inhibition of reductase is incomplete so that the cells can form enough mevalonate to support growth and division, provided that the requirement for cholesterol in bulk is met. An analogous situation arises with cells grown in the presence of oxysterols. In the absence of a source of cholesterol in the medium, and in the presence of an oxysterol, the cells eventually cease growing and die. However, they can be restored to normal growth if LDL is added to the medium, without the addition of mevalonic acid. The growth requirements of oxysterol-treated cells are readily explained in terms of the dual requirement for cholesterol and nonsterol metabolites of mevalonic acid. As mentioned below, oxysterols, like cholesterol, suppress reductase only partially. Hence, cells grown in the presence of an inhibitory oxysterol such as 25-hydroxycholesterol synthesize enough mevalonate to sustain growth and division if their requirement for cholesterol is satisfied. If there is no source of cholesterol in the medium, the sterol biosynthetic pathway competes for the small supply of endogenous mevalonic acid, resulting in deficiency of essential nonsterol metabolites.

2. Independent Regulation of the Synthesis of Sterol, Ubiquinone, and Dolichol

As we have seen, several end-products that fulfill different biological functions originate from a common precursor—mevalonic acid—and share a common rate-limiting enzyme—HMG-CoA reductase. Since the requirements for these products must vary independently, one would expect the cell to be capable of independent regulation of their rates of synthesis. Work carried out in the laboratories of Olson and of Kandutsch (summarized by James *et al.*, 1983, and Panini *et al.*, 1985) has shown that, under certain conditions, regulation of the synthesis of ubiquinone and dolichol is independent of that of sterol synthesis. The more recent studies of Brown and Goldstein and their co-workers on compactin-treated cells *in vitro* have gone some way toward explaining the mechanisms by which cells achieve this independent regulation.

Gold and Olson (1966) noted that the apparent K_m for ubiquinone synthesis in rat-liver slices was lower than that for sterol synthesis. Rao and Olson (1967) then showed that cholesterol feeding decreased incorporation of [^{14}C]mevalonate into hepatic sterols but increased incorporation into ubiquinone. In agreement with Gould and Swyryd (1966) they also found that prolonged cholesterol feeding suppressed enzymes in the sterol pathway beyond mevalonate. On the basis of these findings, Olson and his co-workers suggested that ubiquinone synthesis is conserved when reductase activity is diminished by cholesterol feeding because the ubiquinone pathway has a higher affinity than the sterol pathway for a common intermediate derived from mevalonic acid, and because an enzyme in the sterol pathway distal to the branch point is partially suppressed.

James and Kandutsch (1979, 1980) showed that dolichol synthesis can also undergo regulation independent of the regulation of cholesterol synthesis. Rates of incorporation of [^{14}C]acetate into dolichol and cholesterol in mouse tissues were compared under conditions in which reductase activity was varied experimentally over a wide range. Except at very low levels of reductase activity, dolichol synthesis remained more or less constant, while cholesterol synthesis varied in parallel with reductase activity. James and Kandutsch concluded that above a critical level of reductase activity the concentration of an intermediate at the branch point of the two pathways (probably FPP) was always high enough to saturate the first enzyme in the dolichol pathway, but was not high enough to saturate that in the sterol pathway.

Brown and Goldstein and their co-workers have extended these ideas and have given them more precision, especially in relation to the regulation of ubiquinone synthesis.

The earlier studies of Brown *et al.* (1978), already mentioned (p. 29), had shown that compactin-treated fibroblasts compensate for the inhibition of reductase by synthesizing increased amounts of enzyme. In these experiments, reductase in compactin-treated cells was not completely suppressed by LDL at a concentration high enough to satisfy the requirements of the cells for cholesterol. However, complete suppression was achieved when small amounts of mevalonic acid, in addition to LDL, were included in the medium. Brown *et al.* (1978) suggested that cells in culture require a residual level of reductase activity to provide trace amounts of mevalonate as precursor for an essential nonsterol metabolite. They also suggested that this residual activity was not suppressed until the cell's requirement for the nonsterol metabolite was met. In keeping with these proposals, reductase activity in cultured fibroblasts grown in the absence of an inhibitor of reductase cannot be completely suppressed by LDL (Brown *et al.*, 1974). In parenthesis, it is worth noting that oxysterols, like cholesterol, cannot suppress reductase completely in compactin-treated cells; full suppression occurs only when mevalonic acid, in addition to the inhibitory oxysterol, is present in the medium (Cohen *et al.*, 1982).

3. The Basis of Independent Regulation

Faust and co-workers have examined the way in which cells regulate the sharing of mevalonate carbon between the biosynthetic pathways for ubiquinone and cholesterol when the supply of mevalonate is varied. Incorporation of exogenous [^{3}H]mevalonate into ubiquinone and cholesterol was measured in fibroblasts in which the production of endogenous mevalonate was blocked by compactin. This facilitated interpretation of the results by eliminating variable entry of unlabeled endogenous mevalonate into the mevalonate pool.

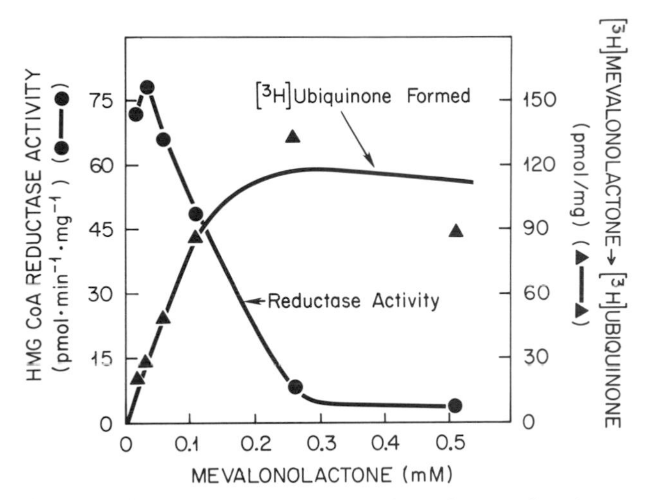

Figure 3.18. The effect of increasing concentrations of mevalonic acid on the incorporation of [^{3}H]mevalonic acid into ubiquinone (▲) and on the activity of HMG-CoA reductase (●) in fibroblast monolayers incubated in the presence of compactin and LDL. On the fifth day of cell growth the cells were incubated in a medium containing 10% human lipoprotein-deficient serum, 1.1 μ*M* compactin, and LDL (25 μg of protein/ml). After incubation at 37°C for 24 hours, the dishes were divided into two groups. Each dish in the first group received the indicated concentrations of unlabeled mevalonic acid. Each dish in the second group received the indicated concentration of [^{3}H]mevalonic acid. After incubation for 24 hours, reductase activity was assayed in the cells in the first group and [^{3}H] ubiquinone was assayed in the cells in the second group. (From Faust *et al.*, 1979a, with the permission of the authors.)

Faust *et al.* (1979a) found that when compactin-treated cells were incubated in the presence of tracer amounts of [^{3}H]mevalonate, and without LDL in the medium, most of the ^{3}H incorporated appeared in cholesterol. When LDL was added to the medium there was a marked decrease in the incorporation of tracer amounts of [^{3}H]mevalonate into cholesterol and an increase in incorporation into ubiquinone. When the mevalonate concentration was increased progressively, incorporation into ubiquinone rose to a plateau, indicating saturation of the ubiquinone pathway at a mevalonate concentration of about 0.2 m*M* (Fig. 3.18). At concentrations above this level, an increasing proportion of the incorporated ^{3}H appeared in cholesterol. As shown in Fig. 3.18, in the presence of LDL plus compactin residual reductase activity fell as the concentration of [^{3}H]mevalonate was raised, until complete suppression was achieved at the concentration of mevalonate at which the ubiquinone pathway was saturated.

The reciprocal effect of LDL on incorporation of [^{3}H]mevalonic acid into cholesterol and ubiquinone suggested that cholesterol delivered to cells by LDL suppresses an enzyme catalyzing a step in the synthesis of cholesterol immediately after the branch point for the two pathways. Faust *et al.* (1979b) confirmed this prediction by showing that LDL suppresses squalene synthetase (see Fig. 3.1) in

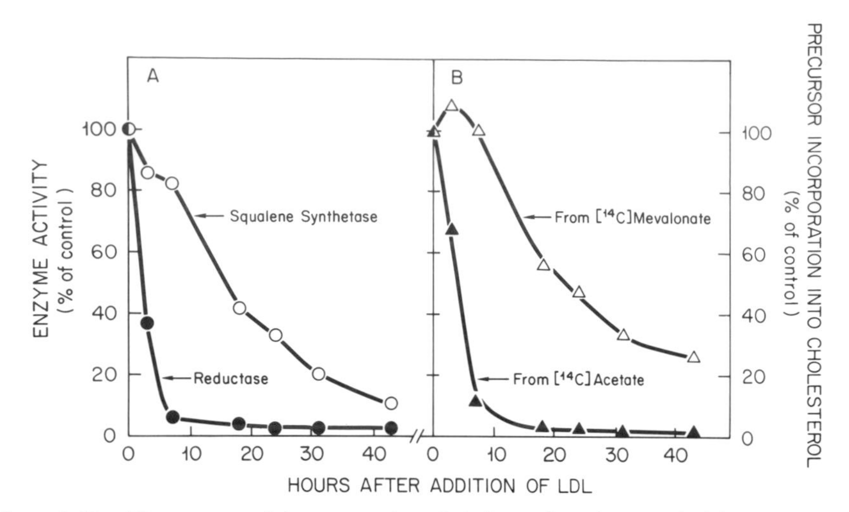

Figure 3.19. Time course of the suppression of cholesterol synthesis and of the activities of squalene synthetase and HMG-CoA reductase by LDL in fibroblast monolayers. On the fifth and sixth days of cell growth the medium was replaced by a medium containing 10% human lipoprotein-deficient serum. On day 6, 7, or 8, LDL (100 μg of protein) was added to the appropriate dish and the cells were incubated at 37°C. The time of addition of LDL was arranged so that all dishes were harvested at the same time on day 8. After incubation with LDL for the indicated time, duplicate dishes were harvested for measurement of reductase activity (A, ●) and squalene synthetase activity (A, ○), and duplicate dishes were pulse-labeled for 2 hours with either 0.5 m*M* [^{14}C]acetate (B, ▲) or 0.5 m*M* [^{14}C]mevalonate (B, △), after which the cell content of [^{14}C]cholesterol was determined. Note that all values are expressed as percentages of control values at zero time. (From Faust *et al.*, 1979b, with the permission of the authors.)

normal fibroblasts in culture, an effect analogous to the earlier observation of Gould and Swyryd (1966) that cholesterol feeding suppresses squalene synthetase in the liver. As with the hepatic enzyme in cholesterol-fed rats, suppression of squalene synthetase in fibroblasts by LDL takes longer to develop and is less marked than suppression of reductase (Fig. 3.19). Moreover, the capacity of squalene synthetase (pmol of mevalonate incorporated into squalene per minute) is several times greater than the capacity of reductase (pmol of mevalonate produced per minute), both in the presence and in the absence of LDL. Hence, the capacity of reductase is always rate-limiting for the synthesis of cholesterol from acetyl-CoA, even when squalene synthetase is maximally suppressed by LDL.

Using a different approach to the study of ubiquinone synthesis, Faust *et al.* (1979a) measured incorporation of the methyl group of [methyl ^{3}H]methionine into ubiquinone in fibroblasts incubated without compactin. The rate of synthesis of ubiquinone in the presence of LDL, when reductase was suppressed by more than 95%, was only about 20% less than that observed in the absence of LDL.

Faust *et al.* (1980) have also studied the incorporation of [^{3}H]mevalonate into isopentenyl tRNA in compactin-treated fibroblasts. In the absence of LDL, ^{3}H appeared mainly in cholesterol and very little was recovered in tRNA. In the presence of LDL, incorporation into tRNA was markedly increased at low concentrations of mevalonate. When the mevalonate concentration was increased, incorporation reached saturation at about 0.1 m*M*.

Brown and Goldstein (1980) have proposed the following hypothesis to explain the observations described in this section, including those relating to the role of reductase in dividing cells.

Mevalonic acid is required as precursor for the formation of small amounts of essential nonsterols as well as for the synthesis of cholesterol when this is not available in the external medium. When the cholesterol requirements of the cell are satisfied by the uptake of LDL, reductase activity falls to less than 5% of the level observed in the absence of an external source of cholesterol. The small amounts of mevalonic acid formed as a result of residual reductase activity are diverted preferentially to the production of essential nonsterol metabolites. Residual reductase activity is not fully suppressed until these metabolites are present in quantities sufficient to satisfy the needs of the cell.

Two factors combine to maintain a more or less constant supply of nonsterols in the face of wide variations in reductase activity. First, enzymes at the branch points to nonsterols have a higher affinity for mevalonate-derived intermediates than has the enzyme at the branch point leading to cholesterol. [In the ubiquinone and dolichol pathways the intermediate at the branch point is presumably FPP; for isopentenyl tRNA it may be IPP (see Fig. 3.1)]. Second, cholesterol delivered to the cell, in addition to suppressing reductase, also suppresses enzymes on the sterol pathway distal to mevalonic acid, including squalene synthetase. The relatively high affinity of ''nonsterol'' enzymes for their substrates helps to maintain the flow of mevalonate carbon into nonsterol pathways when the supply of mevalonate is drastically reduced by suppression of reductase. The partial suppression of squalene synthetase by LDL, by narrowing the path from FPP to squalene, minimizes the fall in intracellular FPP concentration that must occur when reductase activity falls.

In the scheme proposed by Brown and Goldstein (1980), reductase is regulated independently by two or more of its own products: cholesterol and mevalonic acid[5] or some nonsterol product(s) of its further metabolism. By analogy with the multivalent feedback repression that occurs in some branched biosynthetic pathways in bacteria, they suggest the term *multivalent feedback regulation* for the mechanism by which reductase activity is regulated in animal cells.

[5]Popják *et al.* (1985) have shown that suppression of HMG-CoA reductase in rat-hepatoma cells by mevalonate is abolished by enucleating the cells. This makes it unlikely that mevalonic acid itself suppresses reductase activity in compactin-treated fibroblasts or CHO cells.

4. The Nature of the Nonsterol Regulatory Molecule

The above scheme raises the question of the nature of the nonsterol feedback regulator of reductase and of its relation to the product of mevalonate that is required for S-phase DNA synthesis. Both ubiquinone and dolichol fail to suppress residual reductase activity (Brown and Goldstein, 1980) and to support S-phase DNA synthesis (Quesney-Huneeus *et al.*, 1980) in compactin-treated cells grown in the presence of LDL. As we have seen, isopentenyladenine is more effective than mevalonate in supporting DNA replication in the presence of compactin. Huneeus *et al.* (1982) have also reported an inhibitory effect of exogenous isopentenyladenine on residual reductase activity in compactin-treated BHK-21 cells supplied with LDL. If this effect were shown to be specific for reductase, it would suggest that endogenous isopentenyladenine mediates suppression of residual reductase activity by mevalonic acid, as well as promoting cell division. This would raise the further question as to whether or not the synthesis of ubiquinone and dolichol can be regulated independently of that of isopentenyladenine.

Popják *et al.* (1987) and Popják and Meenan (1987) have shown that suppression of reductase in rat-hepatoma (H4) cells by mevalonate does not occur in the presence of iminosqualene, an inhibitor of squalene-oxide cyclase. It would be of interest to test the effect of this inhibitor on the suppression of residual reductase activity by mevalonate in cells incubated in the presence of 25-hydroxycholesterol. If iminosqualene abolished suppression under these conditions, this would suggest that suppression of reductase by mevalonate in H4 cells is mediated entirely by metabolites distal to squalene-oxide on the pathway to cholesterol. Popják and Meenan suggest that the mevalonate-derived repressor that cannot be replaced by exogenous cholesterol is a subfraction of the sterol synthesized endogenously within the cell.

Brown *et al.* (1983) have developed a line of insect cells (Kc cells) that do not synthesize cholesterol and do not require cholesterol for growth. HMG-CoA reductase in these cells is not suppressed by LDL, cholesterol, or oxysterols, but is suppressed when mevalonic acid is added to the culture medium. In Kc cells, the metabolite of mevalonic acid that suppresses reductase activity is distal to IPP but is not squalene or a prenylated precursor of ubiquinone or ubiquinone (Watson *et al.*, 1985). Brown *et al.* (1983) suggest that suppression of reductase by mevalonate in Kc cells reflects the presence of a primitive regulatory mechanism common to all cells that require nonsterol derivatives of mevalonate, including those that also require cholesterol. If this suggestion is valid, the study of Kc cells should advance our understanding of the regulation of residual reductase activity by mevalonate in the cells of vertebrates.

5. The Mechanism by Which Nonsterol Metabolites of Mevalonate Suppress Reductase

Because mevalonic acid is the precursor of both sterol and nonsterol end products, investigation of the effect of nonsterol metabolites on HMG-CoA reductase in intact cells requires special experimental conditions. In the experiments of Luskey *et al.* (1983) on CHO cells adapted to growth in the presence of compactin (UT-1 cells), 10 m*M* mevalonate caused marked repression of reductase synthesis, accompanied by a fall in the rate of transcription of the reductase gene (see p. 42). However, the repression of transcription observed in these experiments could have been mediated by cholesterol, rather than by nonsterol products, derived from the added mevalonate. Peffley and Sinensky (1985) devised experimental conditions in which mutant CHO cells of the Mev-1 line were used. These cells lack HMG-CoA synthase and are therefore unable to synthesize mevalonic acid from endogenous substrates (see p. 27). In the absence of an external source of mevalonate the rate of synthesis of HMG-CoA reductase in Mev-1 cells is increased. Peffley and Sinensky showed that the addition of maximal doses of 25-hydroxycholesterol to the culture medium caused a moderate fall in reductase synthesis attributable to repression of transcription of the reductase gene. However, the further addition of mevalonate caused an additional 50-fold decrease in the rate of synthesis of reductase without any change in the rate of transcription of the gene. Since this effect occurred under conditions in which repression by sterols was already maximal, Peffley and Sinensky concluded that nonsterol metabolites of mevalonic acid repress reductase synthesis by inhibiting translation of the reductase message.

Nakanishi *et al.* (1988) have extended these observations, using different experimental conditions in which normal, wild-type CHO cells were used. The cells were incubated for 24 hours in medium containing LDL at a concentration high enough to satisfy their requirement for cholesterol, together with compactin at high concentrations. Under these conditions, HMG-CoA reductase was synthesized at a high rate, although the cells were saturated with cholesterol. When mevalonate was added to the culture medium, reductase activity fell by 99% without any change in the cell content of reductase mRNA. The fall in reductase activity was due to a combination of decreased translation of the reductase message and increased degradation of reductase protein. The results of parallel experiments with cells of the TR-36 and TR-70 lines (see Section V,E above) showed that the effect of mevalonate on the degradation of reductase required the presence of a normal membrane-spanning domain in the enzyme molecule. Taking these results in conjunction with the earlier findings discussed in this chapter, Nakanishi *et al.* (1988) suggest the following scheme for the multivalent regulation of HMG-CoA reductase. Sterols suppress reductase synthesis incompletely by partial repression of transcription of the gene. Nonsterol metabolites of mevalonic acid bring about a further reduction in enzyme synthesis by inhibiting translation of the reductase message. In addition, sterol and nonsterol metabolites accelerate degradation of the enzyme. The net effect of these mechanisms is to regulate HMG-CoA reductase activity over a several-hundred-fold range in animal cells.

As we saw in Section III of this chapter, the hamster reductase gene gives rise to multiple mRNAs with 5′-untranslated regions ranging from ≈70 to 670 nucleotides in length. Reynolds *et al.* (1985) have shown that cells grown in the absence of sterol produce an increased proportion of the shorter reductase messages. Nakanishi *et al.* (1988) suggest that in the presence of sterol, when transcription of the reductase gene is partially repressed, cells produce a preponderance of longer reductase mRNAs and that these are translated only when the cells are depleted of isopentenyl-tRNA derived from mevalonate.

References

Anderson, R. G. W., Orci, L., Brown, M. S., Garcia-Segura, L. M., and Goldstein, J. L. (1983). Ultrastructural analysis of crystalloid endoplasmic reticulum in UT-1 cells and its disappearance in response to cholesterol. *J. Cell Sci.* **63,** 1–20.

Arebalo, R. E., Hardgrave, J. E., and Scallen, T. J. (1981). The *in vivo* regulation of rat liver 3-hydroxy-3-methylglutaryl coenzyme A reductase. Phosphorylation of the enzyme as an early regulatory response following cholesterol feeding. *J. Biol. Chem.* **256,** 571–574.

Balasubramaniam, S., Goldstein, J. L., and Brown, M. S. (1977). Regulation of cholesterol synthesis in rat adrenal gland through coordinate control of 3-hydroxy-methylglutaryl coenzyme A synthase and reductase activities. *Proc. Natl. Acad. Sci. U.S.A.* **74,** 1421–1425.

Beg, Z. H., and Brewer, H. B. (1981). Regulation of liver 3-hydroxy-3-methylglutaryl coenzyme A reductase. *Curr. Top. Cell. Regul.* **20,** 139–184.

Beg, Z. H., Allmann, D. W., and Gibson, D. M. (1973). Modulation of 3-hydroxy-3-methylglutaryl coenzyme A reductase activity with cAMP and with protein fractions of rat liver cytosol. *Biochem. Biophys. Res. Commun.* **54,** 1362–1369.

Beg, Z. H., Stonik, J. A., and Brewer, H. G., Jr. (1984). *In vivo* modulation of rat liver 3-hydroxy-3-methylglutaryl coenzyme A reductase, reductase kinase, and reductase kinase kinase by mevalonolactone. *Proc. Natl. Acad. Sci. U.S.A.* **81,** 7293–7297.

Beg, Z. H., Reznikon, D. C., and Avigan, J. (1986). Regulation of 3-hydroxy-3-methylglutaryl coenzyme A reductase activity in human fibroblasts by reversible phosphorylation: modulation of enzymatic activity by low density lipoprotein, sterols, and mevalonate. *Arch. Biochem. Biophys.* **244,** 310–322.

Bell, J. J., Sargeant, T. E., and Watson, J. A. (1976). Inhibition of 3-hydroxy-3-methylglutaryl coenzyme A reductase activity in hepatoma tissue culture cells by pure cholesterol and several cholesterol derivatives. *J. Biol. Chem.* **251,** 1745–1758

Benoist, C., and Chambon, P. (1981). *In vivo* sequence requirements of the SV40 early promoter region. *Nature* **290,** 304–310.

Beytia, E. D., and Porter, J. W. (1976). Biochemistry of polyisoprenoid biosynthesis. *Ann. Rev. Biochem.* **45,** 113–142.

Blobel, G. (1980). Intracellular protein topogenesis. *Proc. Natl. Acad. Sci. U.S.A.* **77,** 1496–1500.

Breslow, J. L., Lothrop, D. A., Spaulding, D. R., and Kandutsch, A. A. (1975). Cholesterol, 7-ketocholesterol and 25-hydroxycholesterol uptake studies and effect on 3-hydroxy-3-methylglutaryl-coenzyme A reductase activity in human fibroblasts. *Biochim. Biophys. Acta* **398,** 10–17.

Brown, M. S., and Goldstein, J. L. (1974). Suppression of 3-hydroxy-3-methylglutaryl coenzyme A reductase activity and inhibition of growth of human fibroblasts by 7-ketocholesterol. *J. Biol. Chem.* **249,** 7306–7314.

Brown, M. S., and Goldstein, J. L. (1980). Multivalent feedback regulation of HMG CoA reductase, a control mechanism coordinating isoprenoid synthesis and cell growth. *J. Lipid Res.* **21,** 505–517.

Brown, D. A., and Simoni, R. D. (1984). Biogenesis of 3-hydroxy-3-methylglutaryl coenzyme A reductase, an integral glycoprotein of the endoplasmic reticulum. *Proc. Natl. Acad. Sci. U.S.A.* **81,** 1674–1678.

Brown, M. S., Dana, S.E., and Goldstein, J. L. (1974). Regulation of 3-hydroxy-3-methylglutaryl Coenzyme A reductase activity in cultured human fibroblasts. Comparison of cells from a normal subject and from a patient with homozygous familial hypercholeserolemia. *J. Biol. Chem.* **249,** 789–796.

Brown, A. G., Smale, T. C., King, T. J., Hasenkamp, R., and Thompson, R. H. (1976). Crystal and molecular structure of Compactin, a new antifungal metabolite from *Penicillium brevicompactum. J. Chem. Soc. Perkin Trans.* **I,** 1165–1170.

Brown, M. S., Faust, J. R., Goldstein, J. L., Kaneko, I., and Endo, A. (1978). Induction of 3-hydroxy-3-methylglutaryl coenzyme A reductase activity in human fibroblasts incubated with compactin (ML-236B), a competitive inhibitor of the reductase. *J. Biol. Chem.* **253,** 1121–1128.

Brown, M. S., Goldstein, J. L., and Dietschy, J. M. (1979). Active and inactive forms of 3-hydroxy-3-methylglutaryl coenzyme A reductase in the liver of the rat. *J. Biol. Chem.* **254,** 5144–5149.

Brown, K., Havel, C. M., and Watson, J. A. (1983). Isoprene synthesis in isolated embryonic *Drosophila* cells. II. Regulation of 3-hydroxy-3-methylglutaryl coenzyme A reductase activity. *J. Biol. Chem.* **258,** 8512–8518.

Calvin, M. (1969). "Chemical Evolution. Molecular Evolution Towards the Origin of Living Systems of the Earth and Elsewhere." Oxford University Press, Oxford.

Cavenee, W. K., Chen, H. W., and Kandutsch, A. A. (1981). Regulation of cholesterol biosynthesis in enucleated cells. *J. Biol. Chem.* **256,** 2675–2681.

Chang, T-Y. (1983). Mammalian HMG-CoA reductase and its regulation. *The Enzymes* **XVI,** 491–521.

Chang, T-Y., and Limanek, J. S. (1980). Regulation of cytosolic acetoacetyl coenzyme A thiolase, 3-hydroxy-3-methylglutaryl coenzyme A synthase, 3-hydroxy-3-methylglutaryl coenzyme A reductase, and mevalonate kinase by low density lipoprotein and by 25-hydroxycholesterol in Chinese hamster ovary cells. *J. Biol. Chem.* **255,** 7787–7795.

Chang, T-Y., Limanek, J. S., and Chang, C. C. Y. (1981). Evidence indicating that inactivation of 3-hydroxy-3-methylglutaryl coenzyme A reductase by low density lipoproteins or by 25-hydroxycholesterol requires mediator protein(s) with rapid turnover rate. *J. Biol. Chem.* **256,** 6174–6180.

Chen, H. W., Heiniger, H-J., and Kandutsch, A. A. (1975). Relationship between sterol synthesis and DNA synthesis in phytohemagglutinin-stimulated mouse lymphocytes. *Proc. Natl. Acad. Sci. U.S.A.* **72,** 1950–1954.

Chin, D. J., Luskey, K. L., Anderson, R. G. W., Faust, J. R., Goldstein, J. L., and Brown, M. S. (1982a). Appearance of crystalloid endoplasmic reticulum in compactin-resistant Chinese hamster cells with a 500-fold increase in 3-hydroxy-3-methylglutaryl-coenzyme A reductase. *Proc. Natl. Acad. Sci. U.S.A.* **79,** 1185–1189.

Chin, D. J., Luskey, K. L., Faust, J. R., MacDonald, R. J., Brown, M. S., and Goldstein, J. L. (1982b). Molecular cloning of 3-hydroxy-3-methylglutaryl coenzyme A reductase and evidence for regulation of its mRNA. *Proc. Natl. Acad. Sci. U.S.A.* **79,** 7704–7708.

Chin, D. J., Gil, G., Russell, D. W., Liscum, L., Luskey, K. L., Basu, S. K., Okayama, H., Berg, P., Goldstein, J. L., and Brown, M. S. (1984). Nucleotide sequence of 3-hydroxy-3-methylglutaryl coenzyme A reductase, a glycoprotein of endoplasmic reticulum. *Nature* **308,** 613–617.

Chin, D. J., Gil, G., Faust, J. R., Goldstein, J. L., Brown, M. S., and Luskey, K. L. (1985). Sterols accelerate degradation of HMG CoA reductase encoded by a constitutively expressed cDNA. *Mol. Cell. Biol.* **5,** 634–641.

Clarke, C. F., Edwards, P. A., Lan, S-F., Tanaka, R. D., and Fogelman, A. M. (1983). Regulation of 3-hydroxy-3-methylglutaryl-coenzyme A reductase mRNA levels in rat liver. *Proc. Natl. Acad. Sci. U.S.A.* **80,** 3305–3308.

Clarke, C. F., Fogelman, A. M., and Edwards, P. A. (1984). Diurnal rhythm of rat liver mRNAs encoding 3-hydroxy-3-methylglutaryl coenzyme A reductase. Correlation of functional and total mRNA levels with enzyme activity and protein. *J. Biol. Chem.* **259,** 10439–10447.

Clinkenbeard, K. D., Sugiyama, T., Reed, W. C., and Lane, M. D. (1975). Cytoplasmic 3-hydroxy-3-methylglutaryl Coenzyme A synthase from liver: purification, properties, and role in cholesterol synthesis. *J. Biol. Chem.* **250,** 3124–3135.

Cohen, D. C., Massoglia, S. L., and Gospodarowicz, D. (1982). Feedback regulation of 3-hydroxy-3-methylglutaryl coenzyme A reductase in vascular endothelial cells. *J. Biol. Chem.* **257,** 11106–11112.

Dawson, P. A., Hofmann, S. L., Van Der Westhuyzen, D. R., Südhof, T. C., Brown, M. S., and Goldstein, J. L. (1988). Sterol-dependent repression of low density lipoprotein receptor promoter mediated by 16-base-pair sequence adjacent to binding site for transcription factor Sp1. *J. Biol. Chem.* **263,** 3372–3379.

Dawson, P. A., Ridgway, N. D., Slaughter, C. A., Brown, M. S., and Goldstein, J. L. (1989). cDNA cloning and expression of oxysterol-binding protein, an oligomer with a potential leucine zipper. *J. Biol. Chem.* **264,** 16798–16803.

DeKruyff, B., Van Dijck, P. W. M., Goldbach, R. W., Demel, R. A., and Van Deenen, L. L. M. (1973). Influence of fatty acid and sterol composition on the lipid phase transition and activity of membrane-bound enzymes in *Acholeplasma laidlawii*. *Biochim. Biophys. Acta* **330,** 269–282.

DeKruyff, B., Van Dijck, P. W. M., Demel, R. A., Schuijff, A., Brants, F., and Van Deenen, L. L. M. (1974). Non-random distribution of cholesterol in phosphatidylcholine bilayers. *Biochim. Biophys. Acta* **356,** 1–7.

Dynan, W. S., and Tjian, R. (1983a). Isolation of transcription factors that discriminate between different promoters recognized by RNA polymerase II. *Cell* **32,** 669–680.

Dynan, W. S., and Tjian, R. (1983b). The promoter-specific transcription factor Sp1 binds to upstream sequences in the SV40 early promoter. *Cell* **35,** 79–87.

Easom, R. A., and Zammit, V. A., (1984). 3-hydroxy-3-methylglutaryl-CoA reductase in the active form in rat liver microsomal fractions. *Biochem. J.* **220,** 739–745.

Edwards, P. A., Lemongello, D., Kane, J., Shechter, I., and Fogelman, A. M. (1980). Properties of purified rat hepatic 3-hydroxy-3-methylglutaryl coenzyme A reductase and regulation of enzyme activity. *J. Biol. Chem.* **255,** 3715–3725.

Edwards, P. A., Lan, S-F., Tanaka, R. D., and Fogelman, A. M. (1983a). Mevalonolactone inhibits the rate of synthesis and enhances the rate of degradation of 3-hydroxy-3-methylglutaryl coenzyme reductase. *J. Biol. Chem.* **258,** 7272–7275.

Edwards, P. A., Lan, S-F., Fogelman, A. M. (1983b). Alterations in the rates of synthesis and degradation of rat liver 3-hydroxy-3-methylglutaryl coenzyme A reductase produced by cholestyramine and mevinolin. *J. Biol. Chem.* **258,** 10219–10222.

Edwards, P. A., Lan, S-F., and Fogelman, A. M. (1984). High density lipoprotein and lecithin dispersions increase the activity of 3-hydroxy-3-methylglutaryl coenzyme A reductase by increasing the rate of synthesis and decreasing the rate of degradation of the enzyme. *J. Biol. Chem.* **259,** 8190–8194.

Endo, A., Kuroda, M., and Tanzawa, K. (1976). Competitive inhibition of 3-hydroxy-3-methylglutaryl coenzyme A reductase by ML-236A and ML-236B fungal metabolites, having hypocholesterolemic activity. *FEBS Lett.* **72,** 323–326.

Erickson, S. K., Shrewsbury, M. A., Gould, R. G., and Cooper, A. D. (1980). Studies on the mechanisms of the rapid modulation of 3-hydroxy-3-methylglutaryl coenzyme A reductase in intact liver by mevalonolactone and 25-hydroxycholesterol. *Biochim. Biophys. Acta* **620,** 70–79.

Esfahani, M., Scerbo, L., and Devlin, T. M. (1984). A requirement for cholesterol and its structural features for a human macrophage-like cell line. *J. Cell Biochem.* **25,** 87–97.

Fairbanks, K. P., Witte, L. D., and Goodman, D. S. (1984). Relationship between mevalonate and mitogenesis in human fibroblasts stimulated with platelet-derived growth factor. *J. Biol. Chem.* **259,** 1546–1551.

Faust, J. R., Brown, M. S., and Goldstein, J. L. (1980). Synthesis of isopentenyl tRNA from mevalonate in cultured human fibroblasts. *J. Biol. Chem.* **255,** 6546–6548.

Faust, J. R., Goldstein, J. L., and Brown, M. S. (1979a). Synthesis of ubiquinone and cholesterol in human fibroblasts: regulation of a branched pathway. *Arch. Biochem. Biophys.* **192,** 86–99.

Faust, J. R., Goldstein, J. L., and Brown, M. S. (1979b). Squalene synthetase activity in human fibroblasts; regulation via the low density lipoprotein receptor. *Proc. Natl. Acad. Sci. U.S.A.* **76,** 5018–5022.

Faust, J. R., Luskey, K. L., Chin, D. J., Goldstein, J. L., and Brown, M. S. (1982). Regulation of synthesis and degradation of 3-hydroxy-3-methylglutaryl-coenzyme A reductase by low density lipoprotein and 25-hydroxycholesterol in UT-1 cells. *Proc. Natl. Acad. Sci. U.S.A.* **79,** 5205–5209.

Fimognari, G. M., and Rodwell, V. W. (1965). Substrate-competitive inhibition of bacterial mevalonate: nicotinamide-adenine dinucleotide oxidoreductase (acylating CoA). *Biochemistry* **4,** 2086–2090.

Fromm, M., and Berg, P. (1982). Deletion mapping of DNA regions required for SV40 early region promoter function *in vivo*. *J. Mol. Appl. Genet.* **1,** 457–481.

Frostegård, J., Hamsten, A., Gidlund, M., and Nilsson, J. (1990). Low density lipoprotein-induced growth of U937 cells—a novel method to determine the receptor binding of low density lipoprotein. *J. Lipid Res.* **31,** 37–44.

Gibbons, G. F. (1983). Molecular control of HMG-CoA reductase: the role of oxygenated sterols. *In* "3-Hydroxy-3-Methylglutaryl Coenzyme A Reductase" (J. R. Sabine, ed.), Vol. 1, pp. 153–168. CRC Press, Boca Raton, Florida.

Gibbons, G. F., Mitropoulos, K. A., and Myant, N. B. (1982). "Biochemistry of Cholesterol." Elsevier, Amsterdam.

Gibson, D. M. (1985). Reversible phosphorylation of hepatic HMG-CoA reductase in endocrine and feedback control of cholesterol biosynthesis. *In* "Regulation of HMG-CoA Reductase" (B. Preiss, ed.), pp. 79–132. Academic Press, New York.

Gibson, D. M., and Parker, R. A. (1987). Hydroxymethylglutaryl-coenzyme A reductase. *The Enzymes* **XVIII,** 179–215.

Gidoni, D., Dynan, W. S., and Tjian, R. (1984). Multiple specific contacts between a mammalian transcription factor and its cognate promoters. *Nature* **312,** 409–413.

Gil, G., Faust, J. R., Chin, D. J., Goldstein, J. L., and Brown, M. S. (1985). Membrane-bound domain of HMG-CoA reductase is required for sterol-enhanced degradation of the enzyme. *Cell* **41,** 249–258.

Gil, G., Osborne, T. F., Goldstein, J. L., and Brown, M. S. (1988). Purification of a protein doublet that binds to six TGG-containing sequences in the promoter for hamster 3-hydroxy-3-methylglutaryl-coenzyme A reductase. *J. Biol. Chem.* **263,** 19009–19019.

Gold, P. H., and Olsen, R. E. (1966). Studies on coenzyme Q. The biosynthesis of coenzyme Aq in rat tissue slices. *J. Biol. Chem.* **241,** 3507–3516.

Gould, R. G. (1951). Lipid metabolism and atherosclerosis. *Am. J. Med.* **11,** 209-227.

Gould, R. G. (1977). Some aspects of the control of hepatic cholesterol biosynthesis. *In* ''Cholesterol Metabolism and Lipolytic Enzymes'' (J. Polonovski, ed.), pp. 13–38. Masson, New York.

Gould, R. G., and Swyryd, E. A. (1966). Sites of control of hepatic cholesterol biosynthesis. *J. Lipid Res.* **7,** 698–707.

Gupta, A., Sexton, R. C., and Rudney, H. (1986). Modulation of regulatory oxysterol formation and low density lipoprotein suppression of 3-hydroxy-3-methylglutaryl coenzyme A (HMG-CoA) reductase activity by ketoconazole. A role for cytochrome P-450 in the regulation of HMG-CoA reductase in rat intestinal epithelial cells. *J. Biol. Chem.* **261,** 8348–8356.

Habenicht, A. J., Glomset, J. A., and Ross, R. (1980) Relation of cholesterol and mevalonic acid to the cell cycle in smooth muscle and Swiss 3T3 cells stimulated to divide by platelet-derived growth factor. *J. Biol. Chem.* **255,** 5134–5140.

Hall, R. H. (1970). N^6-(Δ^2-isopentenyl) adenosine: chemical reactions, biosynthesis, metabolism and significance to the structure and function of tRNA. *Prog. Nucleic Acid Res. Mol. Biol.* **10,** 57–86.

Hansen, U., and Sharp, P. A. (1983). Sequences controlling *in vitro* transcription of SV40 promoters. *EMBO J.* **2,** 2293–2303.

Hardeman, E. C., Janke, H-S., and Simoni, R. D. (1983). Overproduction of a M_r 92,000 promoter of 3-hydroxy-3-methylglutaryl coenzyme A reductase in compactin-resistant C100 cells. *Proc. Natl. Acad. Sci. U.S.A.* **80,** 1516–1520.

Hardeman, E. C., Endo, A., and Simoni, R. D. (1984). Effects of compactin on the levels of 3-hydroxy-3-methylglutaryl coenzyme A reductase in compactin-resistant C100 and wild-type cells. *Arch. Biochem. Biophys.* **232,** 549–561.

Hardie, D. G., Carling, D., and Sim, A. T. R. (1989). The AMP-activated protein kinase: a multisubstrate regulator of lipid metabolism. *Trends Biochem. Sci.* **14,** 20–23.

Heller, R. A., and Gould, R. G. (1973). Solubilization and partial purification of hepatic-3-hydroxy-3-methylglutaryl coenzyme A reductase from rat liver. *Biochem. Biophys. Res. Commun.* **50,** 859–865.

Heller, R. A., and Gould, R. G. (1974). Reversible cold inactivation of microsomal 3-hydroxy-3-methylglutaryl coenzyme A reductase from rat liver. *J. Biol. Chem.* **249,** 5254–5260.

Higgins, M., and Rudney, H. (1973). Regulation of rat liver β-hydroxy-β-methylglutaryl-CoA reductase activity by cholesterol. *Nature (New Biol.)* **246,** 60–61.

Higgins, M., Kawachi, T., and Rudney, H. (1971). The mechanisms of the diurnal variation of hepatic HMG-CoA reductase activity in the rat. *Biochem. Biophys. Res. Commun.* **45,** 138–144.

Huneeus, V. Q., Galick, H. A., and Siperstein, M. D. (1982). Fine tuning of HMG-CoA reductase by isopentenyl adenine. *Fed. Proc.* **41,** 881.

James, M. J., and Kandutsch, A. A. (1979). Inter-relationships between dolichol and sterol synthesis in mammalian cell cultures. *J. Biol. Chem.* **254,** 8442–8446.

James, M. J., and Kandutsch, A. A. (1980). Regulation of hepatic dolichol synthesis by β-hydroxy-β-methylglutaryl coenzyme A reductase. *J. Biol. Chem.* **255,** 8618–8622.

James, M. J., Potter, J. E. R., and Kandutsch, A. A. (1983). HMG-CoA reductase and the synthesis of ubiquinone and dolichol. *In* "3-Hydroxy-3-Methylglutaryl Coenzyme A Reductase" (J. R. Sabine, ed.), Vol. 1, pp. 19–28. CRC Press, Boca Raton, Florida.

Jingami, H., Brown, M .S., Goldstein, J. L., Anderson, R. G. W., and Luskey, K. L. (1987). Partial deletion of membrane-bound domain of 3-hydroxy-3-methylglutaryl coenzyme A reductase eliminates sterol-enhanced degradation and prevents formation of cyrstalloid endoplasmic reticulum. *J. Cell Biol.* **104,** 1693–1704.

Jones, K. A., Kodonaga, J. T., Rosenfeld, P. J., Kelly, T. J., and Tjian, R. (1987). A cellular DNA-binding protein that activates eukaryotic transcription and DNA replication. *Cell* **48,** 79–89.

Kadonaga, J. T., Jones, K. A., and Tjian, R. (1986). Promoter-specific activation of RNA polymerase II transcription by Sp1. *Trends Biochem. Sci.,* **11,** 20-23.

Kandutsch, A. A. (1982). A model for the regulation of HMG-CoA reductase by oxygenated sterols. *Fed. Proc.* **41,** 1387.

Kandutsch, A. A., and Chen, H. W. (1973). Inhibition of sterol synthesis in cultured mouse cells by 7α-hydroxycholesterol, 7β-hydroxycholesterol, and 7-ketocholesterol. *J. Biol. Chem.* **248,** 8408–8417.

Kandutsch, A. A., and Chen, H. W. (1974). Inhibition of sterol synthesis in cultured mouse cells by cholesterol derivatives oxygenated in the side chain. *J. Biol. Chem.* **249,** 6057–6061.

Kandutsch, A. A., and Chen, H. W. (1977). Consequences of blocked sterol synthesis in cultured cells. DNA synthesis and membrane competition. *J. Biol. Chem.* **252,** 409–415.

Kandutsch, A. A., and Thompson, E. B. (1980). Cytosolic proteins that bind oxygenated sterols. Cellular distribution, specificity, and some properties. *J. Biol. Chem.* **255,** 10813–10821.

Kandutsch, A. A., Chen, H. W., and Heiniger, H-J (1978). Biological activity of some oxygenated sterols. *Science* **201,** 498–501.

Kaneko, I., Hazama-Shimada, Y., and Endo, A. (1978). Inhibitory effects on lipid metabolism in cultured cells of ML-236B, a potent inhibitor of 3-hydroxy-3-methylglutaryl Coenzyme A reductase. *Eur. J. Biochem.* **87,** 313–321.

Kawaguchi, A. (1970). Control of ergosterol biosynthesis in yeast. *J. Biochem. (Toyko)* **67,** 219–227.

Kennelly, P. J., and Rodwell, V. W. (1985). Regulation of 3-hydroxy-3-methylglutaryl coenzyme A reductase by reversible phosphorylation-dephosphorylation. *J. Lipid Res.* **26,** 910–914.

Kirsten, E. S., and Watson, J. A. (1974). Regulation of 3-hydroxy-3-methylglutaryl coenzyme A reductase in hepatoma tissue culture cells by serum lipoproteins. *J. Biol. Chem.* **249,** 6104–6109.

Krieger, M., Goldstein, J. L., and Brown, M. S. (1978). Receptor-mediated uptake of low-density-lipoprotein reconstituted with 25-hydroxycholesteryl oleate suppresses 3-hydroxy-3-methlglutaryl-coenzyme A reductase and inhibits growth of normal fibroblasts. *Proc. Natl. Acad. Sci. U.S.A.* **75,** 5052–5056.

Krieger, M., Brown, M. S., and Goldstein, J. L. (1981). Isolation of Chinese hamster cell mutants defective in the receptor-mediated endocytosis of low density lipoprotein. *J. Mol. Biol.* **150,** 167–184.

Kyte, J., and Doolittle, R. F. (1982). A simple method for displaying the hydropathic character of a protein. *J. Mol. Biol.* **157,** 105–132.

Lerner, R. A. (1982). Tapping the immunological repertoire to produce antibodies of predetermined specificity. *Nature* **299,** 592–596.

Lewin, B. (1987). "Genes III." John Wiley, New York.

Linn, T. C. (1967). The effect of cholesterol feeding and fasting upon β-hydroxy-β-methylglutaryl coenzyme A reductase. *J. Biol. Chem.* **242,** 990–993.

Liscum, L., Cummings, R. D., Anderson, R. G. W., De Martino, G. N., Goldstein, J. L., and Brown, M. S. (1983a). 3-hydroxy-3-methylglutaryl-CoA reductase: a transmembrane glycoprotein of the endoplasmic reticulum with N-linked high-mannose. *Proc. Natl. Acad. Sci. U.S.A.* **80,** 7165–7169,

Liscum, L., Luskey, K., Chin, D. J., Ho, Y. K., Goldstein, J. L., and Brown, M. S. (1983b). Regulation of 3-hydroxy-3-methylglutaryl coenzyme A reductase and its mRNA in rat liver as studied with a monoclonal antibody and a cDNA probe. *J. Biol. Chem.* **258,** 8450–8455.

Liscum, L., Finer-Moore, J., Stroud, R. M., Luskey, K. L., Brown, M. S., and Goldstein, J. L. (1985). Domain structure of 3-hydroxy-3-methylglutaryl coenzyme A reductase, a glycoprotein of the endoplasmic reticulum. *J. Biol. Chem.* **260,** 522–530.

Luskey, K. L., Chin, D. J., MacDonald, R. J., Liscum, L., Goldstein, J. L., and Brown, M. S. (1982). Identification of a cholesterol-regulated 53,000 dalton cytosolic protein in UT-1 cells and cloning of its cDNA. *Proc. Natl. Acad. Sci. U.S.A.* **79,** 6210–6214.

Luskey, K. L., Faust, J. R., Chin, D. J., Brown, M. S., and Goldstein, J. L. (1983). Amplification of the gene for 3-hydroxy-3-methylglutaryl coenzyme A reductase, but not for the 53-kDa protein, in UT-1 cells. *J. Biol. Chem.* **258,** 8462–8469.

Maltese, W. A., and Volpe, J. J. (1979). Activation of 3-hydroxy-3-methylglutaryl-coenzyme A reductase in homogenates of developing rat brain. *Biochem. J.* **182,** 367–370.

Mitropoulos, K. A., and Venkatesan, S. (1977). The influence of cholesterol on the activity, on the isothermic kinetics and on the temperature-induced kinetics of 3-hydroxy-3-methylglutaryl coenzyme A reductase. *Biochim. Biophys. Acta* **489,** 126–142.

Mitropoulos, K. A., and Venkatesan, S. (1985). Membrane-mediated control of reductase activity. *In* ''Regulation of HMG-CoA Reductase'' (B. Preiss, ed.), pp. 1–48. Academic Press, New York.

Mitropoulos, K. A., Balasubramaniam, S., Venkatesan, S., and Reeves, B. E. A. (1978a). On the mechanism for the regulation of 3-hydroxy-3-methylglutaryl coenzyme A reductase, of cholesterol 7-alpha-hydroxylase and of acyl coenzyme A:cholesterol acyltransferase by free cholesterol. *Biochim. Biophys. Acta* **530,** 99–111.

Mitropoulos, K. A., Venkatesan, S., Balasubramaniam, S., and Peters, T. J. (1978b). The submicrosomal localization of 3-hydroxy-3-methylglutaryl-coenzyme-A reductase, cholesterol 7α-hydroxylase and cholesterol in rat liver. *Eur. J. Biochem.* **82,** 419–429.

Mosley, S. T., Brown, M. S., Anderson, R. G. W., and Goldstein, J. L. (1983). Mutant clone of Chinese hamster ovary cells lacking 3-hydroxy-3-methylglutaryl coenzyme A reductase. *J. Biol. Chem.* **258,** 13875–13881.

Myant, N. B. (1981). ''The Biology of Cholesterol and Related Steroids.'' Heinemann, London.

Nakanishi, M., Goldstein, J. L., and Brown, M. S. (1988). Multivalent control of 3-hydroxy-3-methylglutaryl coenzyme A reductase. Mevalonate-derived product inhibits translation of mRNA and accelerates degradation of enzyme. *J. Biol. Chem.* **263,** 8929–8937.

Ness, G. C., Way, S. C., and Wickham, P. S. (1981). Proteinase involvement in the solubilization of 3-hydroxy-3-methylglutaryl coenzyme A reductase. *Biochem. Biophys. Res. Commun.* **102,** 81–85.

Nordstrom, J. L., Rodwell, V. W., and Mitschelen, J. J. (1977). Interconversion of active and inactive forms of rat liver hydroxymethylglutaryl-CoA reductase. *J. Biol. Chem.* **252,** 8924–8934.

Orci, L., Brown, M. S., Goldstein, J. L., Garcia-Segura, L. M., and Anderson, R. G. W. (1984). Increase in membrane cholesterol: a possible trigger for degradation of HMG-CoA reductase and crystalloid endoplasmic reticulum in UT-1 cells. *Cell* **36,** 835–845.

Osborne, T. F., Goldstein, J. L., and Brown, M. S. (1985). 5′ End of HMG-CoA reductase gene contains sequences responsible for cholesterol-mediated inhibition of transcription. *Cell* **42,** 203–212.

Osborne, T. F., Gil, G., Brown, M. S., Kowal, R. C., and Goldstein, J. L. (1987). Identification of promoter elements required for *in vitro* transcription of hamster 3-hydroxy-3-methylglutaryl coenzyme A reductase gene. *Proc. Natl. Acad. Sci. U.S.A.* **84,** 3614–3618.

Osborne, T. F., Gil, G., Goldstein, J. L., and Brown, M. S. (1988). Operator constitutive mutation of 3-hydroxy-3-methylglutaryl coenzyme A reductase promoter abolishes protein binding to sterol regulatory element. *J. Biol. Chem.* **263,** 3380–3387.

Panini, S. R., Sexton, R. C., and Rudney, H. (1983). Role for endogenous oxysterols in the regulation of hydroxymethylglutaryl CoA reductase. *J. Lipid Res.* **24,** 1410.

Panini, S. R., Rogers, D. H., and Rudney, H. (1985). Regulation of HMG-CoA reductase and the biosynthesis of nonsteroid prenyl derivatives. *In* "Regulation of HMG-CoA Reductase" (B. Preiss, ed.), pp. 149–181. Academic Press, New York.

Parker, R. A., Miller, S. J., and Gibson, D. M. (1989). Phosphorylation of native, 97 kDa 3-hydroxy-3-methylglutaryl coenzyme A reductase from rat liver: impact on activity and degradation of the enzyme. *J. Biol. Chem.* **264,** 4877–4887.

Peffley, D., and Sinensky, M. (1985). Regulation of 3-hydroxy-3-methylglutaryl coenzyme A reductase synthesis by a non-sterol mevalonate-derived product of Mev-1 cells. *J. Biol. Chem.* **260,** 9949–9952.

Popják, G., and Meenan, A. (1987). Regulation of 3-hydroxy-3-methylglutaryl coenzyme A reductase: search for the enzyme's repressor derived from mevalonate. *Proc. Roy. Soc. B.* **231,** 391–414.

Popják, G., Clarke, C. F., Hadley, C., and Meenan, A. (1985). Role of mevalonate in regulation of cholesterol synthesis and 3-hydroxy-3-methylglutaryl coenzyme A reductase in cultured cells and their cytoplasts. *J. Lipid Res.* **26,** 831–841.

Popják, G., Meenan, A., and Ness, W. D. (1987). Effects of 2,3-iminosqualene on cultured cells. *Proc. Roy. Soc. B.* **232,** 273–287.

Quesney-Huneeus, V., Wiley, M. H., and Siperstein, M. D. (1979). Essential role for mevalonate synthesis in DNA replication. *Proc. Natl. Acad. Sci. U.S.A.* **76,** 5056–5060.

Quesney-Huneeus, V., Wiley, M. H., and Siperstein, M. D. (1980). Isopentenyladenine as a mediator of mevalonate-regulated DNA replication. *Proc. Natl. Acad. Sci. U.S.A.* **77,** 5842-5846.

Rajavashisth, T. B., Taylor, A. K., Andalibi, A., Svenson, K. L. and Lusis, A. J. (1989). Identification of a zinc finger protein that binds to the sterol regulatory element. *Science,* **245,** 640–643.

Rao, K. S., and Olson, R. E. (1967). The effect of exogenous cholesterol on the synthesis *in vivo* of cholesterol, ubiquinone, and squalene in rat liver. *Biochem. Biophys. Res. Commun.* **26,** 668–673.

Reynolds, G. A., Basu, S. K., Osborne, T. F., Chin, D. J., Gil, G., Brown, M. S., Goldstein, J. L., and Luskey, K. L. (1984). HMG-CoA reductase: a negatively regulated gene with unusual promoter and 5′ untranslated regions. *Cell* **38,** 375–285.

Reynolds, G. A., Goldstein, J. L., and Brown, M. S. (1985). Multiple mRNAs for 3-hydroxy-3-methylglutaryl coenzyme A reductase determined by multiple transcription initiation sites and intron slicing sites in the 5′-untranslated region. *J. Biol. Chem.* **260,** 10369–10377.

Rodwell, V. W., Nordstrom, J. L., and Mitschelen, J. J. (1976). Regulation of HMG-CoA reductase. *Adv. Lipid Res.* **14,** 1–74.

Rogers, D. H., Panini, S. R., and Rudney, H. (1983). Properties of HMG-CoA reductase and its mechanism of action. *In* "3-Hydroxy-3-Methylglutaryl Coenzyme A Reductase" (J. R. Sabine, ed.), pp. 58–75. CRC Press, Boca Raton, Florida.

Rothblat, G. H. (1969). Lipid metabolism in tissue culture cells. *Adv. Lipid Res.* **7,** 135–163.

Ryan, J., Hardeman, E. C., Endo, A., and Simoni, R. D. (1981). Isolation and characterization of cells resistant to ML 236B (compactin) with increased levels of 3-hydroxy-3-methylglutaryl coenzyme-A reductase. *J. Biol. Chem.* **256,** 6762–6768.

Sabine, J. R., and James, M. J. (1976). The intracellular mechanism responsible for dietary feedback control of cholesterol synthesis. *Life Sci.* **18,** 1185–1192.

Sabitini, D. D., Kreibich, G., Morimoto, T., and Adesnik, M. (1982). Mechanisms for the incorporation of proteins in membranes and organelles. *J. Cell Biol.* **92,** 1–22.

Saucier, S. E., and Kandutsch, A. A. (1979). Inactive 3-hydroxy-3-methylglutaryl coenzyme A reductase in broken cell preparations of various mammalian tissues and cell cultures. *Biochim. Biophys. Acta* **572,** 541–556.

Schimke, R. T., Kaufman, R. J., Alt, F. W., and Kelleman, R. F. (1978). Gene amplification and drug resistance in cultured murine cells. *Science* **202,** 1051–1056.

Schnitzer-Polokoff, R., von Gunten, C., Logel, J., Torget, T. and Sinensky, M. (1982). Isolation and characterization of a mammalian cell mutant defective in 3-hydroxy-3-methylglutaryl coenzyme A synthase. *J. Biol. Chem.* **257,** 472–476.

Shapiro, D. J., and Rodwell, V. W. (1969). Diurnal variation and cholesterol regulation of hepatic HMG-CoA reductase activity. *Biochem. Biophys. Res. Commun.* **37,** 867–872.

Sinensky, M. (1985). Somatic cell genetic analysis of cholesterol biosynthesis. *In* ''Regulation of HMG-CoA Reductase'' (B. Preiss, ed.), pp. 201–220. Academic Press, New York.

Sinensky, M., and Logel, J. (1983). Inhibition of degradation of 3-hydroxy-3-methylglutaryl-coenzyme A reductase by mevinolin. *J. Biol. Chem.* **258,** 8547–8549.

Sinensky, M., and Mueller, G. (1981). Cytosolic 25-hydroxycholesterol binding activity of Chinese hamster ovary cells. *Arch. Biochem. Biophys.* **209,** 314–320.

Sinensky, M., Torget, R., and Edwards, P. A. (1981). Radioimmune precipitation of 3-hydroxy-3-methylglutaryl coenzyme A reductase from Chinese hamster fibroblasts. *J. Biol. Chem.* **256,** 11774–11779.

Siperstein, M. D. (1984). Role of cholesterogenesis and isoprenoid synthesis in DNA replication and cell growth. *J. Lipid Res.* **25,** 1462–1468.

Stoffel, W., Tunggal, B. D., Zierenberg, O., Schreiber, E., and Binczek, E. (1974). 13C nuclear magnetic resonance studies of lipid interactions in single- and multi-component lipid vesicles. *Hoppe-Seyler's Z. Physiol. Chem.* **355,** 1367–1380.

Sundström, C., and Nilsson, K. (1976). Establishment and characterization of human histiocytic lymphoma cell-line (U-937). *Int. J. Cancer* **17,** 565–577.

Tanaka, R. D., Li, A. C., Fogelman, A. M., and Edwards, P. A. (1986). Inhibition of lysosomal protein degradation inhibits the basal degradation of 3-hydroxy-3-methylglutaryl coenzyme A reductase. *J. Lipid Res.* **27,** 261–273.

Thomas, D. D., and Hidalgo, C. (1978). Rotational motion of the sarcoplasmic reticulum Ca^{2+}-ATPase. *Proc. Natl. Acad. Sci. U.S.A.* **75,** 5488–5492.

Venkatesan, S., and Mitropoulos, K. A. (1982). 3-Hydroxy-3-methylglutaryl coenzyme A reductase. The difference in the mechanism of the *in vitro* modulation by phosphorylation and dephosphorylation to modulation of enzyme activity by non-esterfied cholesterol. *Biochim. Biophys. Acta* **710,** 446–455.

Watson, J. A., Havel, C. M., Lobos, D. V., Baker, F. C., and Morrow, C. J. (1985). Isoprenoid synthesis in isolated embryonic *Drosophila* cells. Sterol-independent regulatory signal molecule is distal to isopentenyl 1-pyrophosphates. *J. Biol. Chem.* **260,** 14083–14091.

Chapter 4

Acyl-CoA:Cholesterol Acyltransferase

I. Biological Functions

Acyl-CoA:cholesterol acyltransferase (ACAT) is an intracellular enzyme that catalyzes the transfer of a long-chain fatty acyl residue from acyl-CoA to the β-hydroxyl group of cholesterol to form a cholesteryl ester. ACAT activity has been demonstrated in many animal tissues and cells, including liver, small intestine, steroid-hormone-forming organs, artery wall, fibroblasts, macrophages, CHO cells, and ascites tumor cells. Indeed, in view of the important role that the enzyme appears to play in the homeostasis of intracellular cholesterol, it may well be as widely distributed as cholesterol itself. However, since ACAT has not yet been characterized, it remains possible that ACAT activity in different tissues is due not to one and the same enzyme, but to different enzymes with closely similar properties.

The most important general function of ACAT is to act in conjunction with HMG-CoA reductase and the LDL-receptor system to minimize fluctuations in the concentration of free cholesterol in intracellular membranes. When the input of cholesterol into the cell increases, reductase activity decreases, LDL-receptor activity decreases, and the activity of ACAT increases. The net effect of these coordinated changes is to reduce cholesterol input and to increase the flow of membrane cholesterol into the cytoplasmic pool of cholesteryl ester. These homeostatic mechanisms have been revealed most clearly in fibroblasts in culture, but a direct relationship between cholesterol input and cholesterol esterification has been demonstrated in a variety of other tissues (see Section III). In its homeostatic role, ACAT may be thought of as monitoring the amount of membrane cholesterol in the cell as a whole, or perhaps in a specific regulatory pool, which could be the regulatory pool of sterol shown in Fig. 3.11. The possible modes of regulation of the enzyme are considered in Section IV.

In addition to its general role in buffering changes in membrane cholesterol content, ACAT performs other specific functions in certain specialized cells. In liver and intestinal cells it provides esterified cholesterol for the nonpolar core of nascent very-low-density lipoproteins and (in intestinal cells) of chylomicrons. In adrenal cortex and other steroid-hormone-forming tissues it maintains the reservoir of cytoplasmic esterified cholesterol required as a source of free cholesterol that can be mobilized rapidly for hormone synthesis. In the small intestine it is responsible for the esterification of cholesterol absorbed from the intestinal lumen; when cholesterol esterification in the intestine of the intact rat is prevented by a specific inhibitor of ACAT, absorption of cholesterol is markedly impaired, though the intestinal mucosa continues to secrete lipoproteins into the mesenteric lymphatics (Bennett Clark and Tercyak, 1984).

II. Properties and Methods of Assay

Compared with what is known about HMG-CoA reductase our knowledge of the nature and properties of ACAT is meager in the extreme. ACAT has not been purified to the point where its molecular weight can be deduced, a cDNA for the ACAT message has not been isolated, and satisfactory monospecific or monoclonal antibodies against the enzyme have not been reported. Hence, investigation of its properties has had to be confined largely to the measurement of catalytic activity determined from the incorporation of radioactive substrates into esterified cholesterol under various experimental conditions.

ACAT is an integral membrane enzyme associated with the ribosome-rich (rough) endoplasmic reticulum (Balasubramaniam *et al.*, (1978a). It is active in the microsomal fraction of broken-cell preparations. It has also been solubilized by treating microsomes with detergent, but catalytic activity is not expressed unless the solubilized enzyme is incorporated into phospholipid-cholesterol vesicles (Doolittle and Chan, 1982a).

Thus, ACAT is active only when it is present, together with its cholesterol substrate, in a lipid environment. This complicates the problem of devising an assay system for measuring intrinsic enzyme activity in the presence of saturating concentrations of both of its substrates. Since this problem underlies much of the controversy surrounding the regulation of ACAT, something must be said here about its assay.

ACAT uses as its second substrate the CoA esters of fatty acids with a wide range of chain length and degree of saturation, but the preferred substrate for the enzyme in liver, and probably for that in most other tissues, is oleyl-CoA. Hence, many methods for assaying enzyme activity are based on incorporation of radioactive oleyl-CoA into esterified cholesterol in microsomal suspensions or in liposomes containing the solubilized enzyme. Incorporation of radioactive oleate into cholesteryl esters has also been used to provide an index of ACAT activity in intact cells. Provided that the cells do not contain an active cytoplasmic cholesteryl-ester hydrolase, changes in ACAT activity have been found to correspond closely to changes in the net rate of incorporation of radioactive oleate into cholesteryl oleate measured over short incubation periods (Brown *et al.*, 1975a; Drevon *et al.*, 1980a; Gavigan and Knight, 1983).

When oleate or oleyl-CoA is used as radioactive substrate, the substrate cholesterol is usually endogenous cholesterol present in the microsomal membrane in which the enzyme is embedded or, when the enzyme has been solubilized and reconstituted in liposomes, the exogenous cholesterol present in the liposomes. Exogenous radioactive cholesterol can also act as substrate for ACAT in microsomal suspensions. In this case the exogenous cholesterol is incorporated into the pool of endogenous substrate cholesterol by exchange or net influx. Hence, ACAT in microsomal suspensions can be assayed by measuring incorporation of added radioactive cholesterol into esterified cholesterol in the presence of saturating concentrations of oleyl-CoA.

When the cholesterol content of isolated microsomes from liver and other tissues is raised by preincubation with increasing concentrations of serum or cholesterol dissolved in organic solvent, ACAT activity rises progressively, eventually reaching a plateau several times higher than the baseline value. This indicates that in most tissues the concentration of endogenous cholesterol in the mocrosomal pool of substrate for ACAT is not high enough to saturate the enzyme (for references, see Suckling and Stange, 1985). In adrenal-cortex microsomes ACAT appears to be nearly saturated with endogenous cholesterol under normal conditions (Balasubramaniam *et al.*, 1977; Suckling *et al.*, 1983b).

The fact that the cholesterol content of untreated microsomes is usually too low to saturate ACAT with its first substrate raises difficulties in the investigation of the mode of regulation of this enzyme. In particular, in experiments in which a change in enzyme activity occurs in response to a change in the cholesterol content of microsomes it may be difficult to decide whether the change in activity is due merely to a change in the supply of substrate or whether it reflects a change in the amount or state of activation of the enzyme. Attempts have been made to avoid this problem by raising the concentration of substrate cholesterol in the microsomes to saturating levels by preincubation with a source of cholesterol before assaying the enzyme (Balasubramaniam *et al.*, 1978a; Gavigan and Knight, 1983). Another approach has been to assay the enzyme after it has been solubilized and reconstituted in liposomes with a standard cholesterol : phospholipid ratio (Doolittle and Chan, 1982b).

III. ACAT and the Metabolism of Intracellular Cholesteryl Esters

Changes in the rate of esterification of intracellular cholesterol, attributable to changes in ACAT activity, occur in conditions in which the cell content of free cholesterol is altered. Such conditions include the receptor-mediated entry of cholesterol-rich lipoproteins into cells *in vitro,* the uptake of cholesterol of dietary origin by liver and small intestine *in vivo,* and the uptake of plasma lipoproteins by adrenal-cortex cells *in vivo.* These examples will be considered in this section.

A. Uptake of LDL by Cells *in Vitro*

The effect of LDL uptake on cholesteryl-ester metabolism in cultured fibroblasts has been investigated by Brown and Goldstein and their co-workers (summarized by Goldstein and Brown, 1977).

When skin fibroblasts in culture are incubated in a lipoprotein-deficient medium, their cholesteryl-ester content falls to a low level and the rate of incorporation of [^{14}C]oleate into intracellular esterfied cholesterol is markedly reduced. If the medium is changed to one containing LDL, the cells take up LDL particles by the receptor-mediated process described in Chapter 8 and the cholesteryl esters in LDL are hydrolyzed by lysosomal enzymes, resulting in the release of free cholesterol into the cytoplasm. This leads to suppression of HMG-CoA reductase (see Chapter 3) and a marked increase in the cell content of esterified cholesterol with oleate as the major fatty acyl residue. The increase in esterified cholesterol is accompanied by increased incorporation of [^{14}C]oleate into cholesteryl esters in intact cells (Fig. 4.1B) and a rise in ACAT activity assayed in isolated microsomes. The consequence of these reciprocal changes in the activities of reductase and ACAT is that the influx of cholesterol into the cell causes a much smaller proportional increase in the content of free cholesterol than in that of esterified cholesterol (Fig. 4.1C,D). Inhibitory oxysterols that suppress reductase, such as 25-hydroxycholesterol, have the same effect as LDL on cholesteryl-ester metabolism in cultured fibroblasts—ACAT activity rises and there is an increase in the cholesteryl-ester content of the cells.

When the fibroblasts are derived from patients with homozygous FH, a condition in which the cells have a genetic deficiency of LDL receptors, addition of LDL to the medium has no effect on reductase activity or on cholesteryl-ester synthesis. The failure of LDL to stimulate synthesis of cholesteryl esters in FH cells could be due simply to inability of these cells to take up LDL. On the other hand, Attie *et al.* (1980) have shown that when LDL is taken up and catabolized by rat hepatocytes via a non-receptor-mediated pathway, reductase is not suppressed and cholesteryl-ester synthesis is not stimulated. They suggest that cholesterol released from LDL that enters cells other than by the LDL-receptor pathway does not have immediate access to the regulatory sites that control reductase activity and cholesterol esterification.

Free cholesterol added to the incubation medium in an organic solvent stimulates the synthesis of cholesteryl esters in normal and FH fibroblasts in culture. Mevalonic acid also stimulates cholesteryl-ester synthesis and ACAT activity in

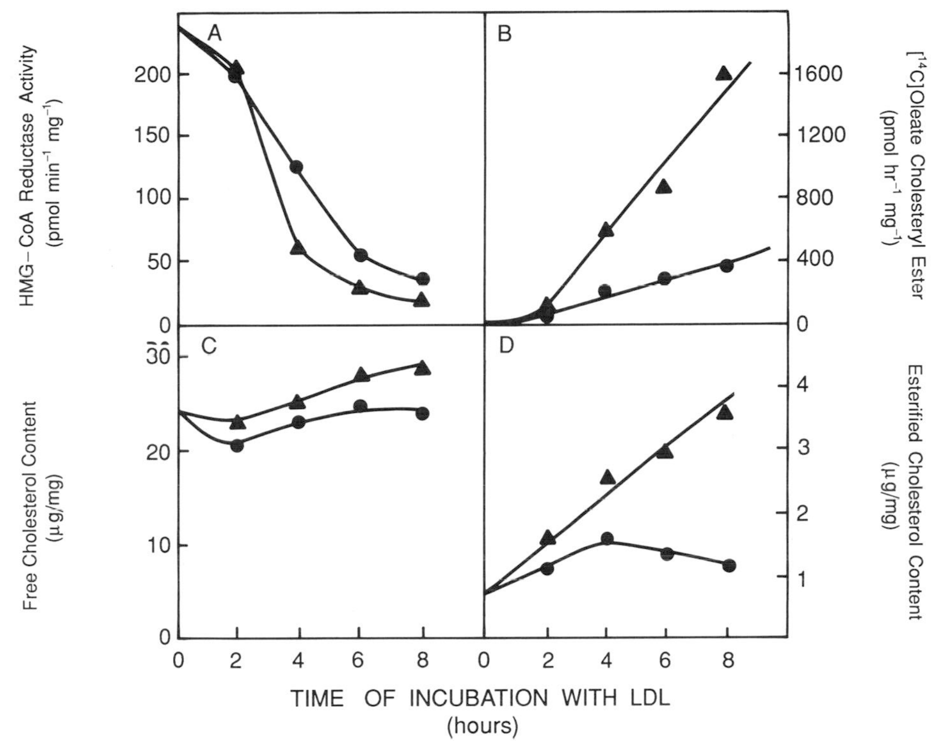

Figure 4.1. The effect of LDL-receptor-mediated uptake of LDL on HMG-CoA reductase activity (A), cholesteryl-ester synthesis (B), free cholesterol content (C), and esterfied cholesterol content (D) in normal human fibroblasts.

Fibroblasts in culture were preincubated at 37°C for 24 hours in a medium containing lipoprotein-deficient serum. LDL at 5 (●) and 25 (▲) μg protein/ml was then added to the incubation medium and the incubations continued for varying time intervals. HMG-CoA reductase activity was assayed in cell free extracts. Synthesis of cholesteryl esters was estimated from the amount of [^{14}C]oleate incorporated into esterfied cholesterol in the intact cells 30 minutes after adding albumin-bound [^{14}C]oleate to the incubation medium. Cellular sterols were assayed by quantitative gas-liquid chromatography. (From Brown *et al.*, 1975b, reproduced from the *Journal of Clinical Investigation,* 1975, Volume 55, pp. 783–793 by copyright permission of the American Society for Clinical Investigation, and with the permission of the authors.)

normal fibroblasts (Gavigan and Knight, 1983), presumably after its conversion into cholesterol.

B. Uptake of Lipoproteins by Macrophages

Macrophages usually express little or no LDL-receptor activity. However, macrophages and other cells of the reticuloendothelial (RE) system express receptors, known as acetyl-LDL receptors, that bind and internalize LDL that has been modified by acetylation or other treatments that increase the net negative charge on the apoB in LDL particles. The properties of these receptors and their possible role in lipoprotein metabolism *in vivo* are discussed in Chapter 8. Here we shall consider acetyl-LDL receptors only in relation to the intracellular metabolism of cholesteryl esters. (See Brown and Goldstein, 1983, for review.)

When macrophages are incubated in the presence of acetyl-LDL, the lipoprotein particles are bound by acetyl-LDL receptors and delivered to lysosomes by a process similar to that by which LDL is internalized by LDL receptors. As with internalized LDL, lysosomal hydrolysis of esterified cholesterol in the internalized acetyl-LDL results in the release of free cholesterol into the cytoplasm, reesterification of the free cholesterol, and a rise in the cholesteryl-ester content of the macrophages. If the incubation with acetyl-LDL is continued for several days, acetyl-LDL continues to be taken up by the cells, resulting in the appearance of cholesterol-rich droplets in the cytoplasm similar to those seen in the foam cells of atherosclerotic lesions. Despite the massive increase in the cell content of esterified cholesterol, the concentration of cellular free cholesterol increases only to a small extent. If the cholesterol-loaded macrophages are incubated in the absence of acetyl-LDL but in the presence of HDL or other acceptor for cellular free cholesterol, the cholesteryl esters in the cytoplasm undergo net hydrolysis and the free cholesterol so released is taken up from the cells by the extracellular acceptor.

The biochemical basis of these events has been elucidated by Brown *et al.* (1980), using a double-label technique that enabled them to measure simultaneously the rates of hydrolysis and synthesis of esterified cholesterol in macrophages. Cholesteryl esters in macrophages take part in a continual cycle of hydrolysis by a cytoplasmic cholesteryl-ester hydrolase and reesterification by ACAT. The rate of hydrolysis remains more or less constant under most conditions, but the rate of esterification of free cholesterol is variable. Under "resting" conditions the rates of esterification and hydrolysis are equal and the cell content of esterified cholesterol is constant. When the supply of intracellular cholesterol is increased, ACAT activity rises and the rate of esterification of cholesterol increases, resulting in an increase in the cholesteryl-ester content of the cells. When the supply of cholesterol is reduced, ACAT activity falls and the rate of esterification of cytoplasmic cholesterol decreases. Under these conditions an increasing proportion of the free cholesterol released by continual hydrolysis of cholesteryl esters is not reesterified and is removed from the cell by cholesterol acceptors in the medium. If reesterification falls to a low enough level, net hydrolysis of cytoplasmic cholesteryl esters eventually leads to the disappearance of lipid droplets in acetyl-LDL-treated macrophages.

When macrophages are loaded with esterified cholesterol they synthesize and secrete apoE in the form of apoE/phospholipid discs. Brown and Goldstein (1983) suggest that, when cholesterol-loaded macrophages excrete free cholesterol, the excreted cholesterol is taken up by HDL and that the apoE/phospholipid secreted concurrently by the cells then becomes associated with cholesterol-enriched HDL particles. Addition of apoE to HDL may serve to direct these particles to the liver for receptor-mediated uptake by hepatic receptors that recognize apoE-containing lipoproteins. Thus, secretion of apoE may facilitate the transport of surplus cholesterol from macrophages to the liver.

In addition to the uptake of acetyl-LDL and of other electronegative forms of LDL, macrophages also bind and internalize β-VLDL, a β-migrating VLDL present in the plasma of cholesterol-fed animals and human subjects with type III hyperlipoproteinemia. (The possibility that uptake of β-VLDL by macrophages is mediated by a separate receptor is discussed in Chapter 8.) Receptor-mediated uptake of β-VLDL by macrophages *in vitro* is followed by intracellular release of free cholesterol, activation of ACAT, and the formation of cholesteryl esters in the cytoplasm. Since β-VLDL is a naturally occurring lipoprotein, this pathway may contribute to the formation of foam cells in the arterial wall in man and in experimental animals *in vivo*. Other receptors on macrophages, probably distinct from acetyl-LDL and β-VLDL receptors, are responsible for the uptake of specific lipoproteins (Brown and Goldstein, 1983; and see Chapter 8, Section III,D). Uptake of lipoproteins by these other receptors leads to increased ACAT activity, increased cholesteryl-ester synthesis, and the accumulation of lipid droplets in the cytoplasm of macrophages.

As discussed in Chapter 8, foam cells in atherosclerotic lesions are thought to be derived predominantly from monocyte-derived macrophages. ACAT activity is markedly increased in microsomes isolated from atherosclerotic areas of the aorta in human subjects and in animals fed atherogenic diets. The increase in ACAT activity in the arterial wall occurs mainly in foam cells and is accompanied by increased esterification of free cholesterol derived from the intracellular hydrolysis of cholesteryl esters that have entered the cells by uptake of lipoproteins (see St. Clair, 1976; Kritchevsky and Kothari, 1978; Myant, 1981, Chapter 13). Thus, the intracellular accumulation of esterified cholesterol, resulting in the formation of foam cells, is probably due to a combination of increased entry of plasma lipoprotein into the arterial wall and increased ACAT activity in those cells in which lipid droplets eventually appear.

C. Esterification of Cholesterol in the Liver and Intestine

Cholesteryl-ester synthesis in the liver and intestine has been investigated in considerable detail, both *in vivo* and *in vitro*. (See Suckling and Stange, 1985), for review and references.)

The addition of cholesterol to the diet, an intravenous infusion of chylomicrons, or the administration of a single dose of mevalonic acid by stomach tube all lead to an increase in the cholesteryl ester content of the liver, with a relatively small increase in hepatic free cholesterol content. The increase in hepatic cholesteryl-ester content is accompanied by a rise in microsomal ACAT activity.

The effect of dietary cholesterol and of infusion of chylomicrons on hepatic cholesteryl-ester content is thought to be due to the uptake of chylomicron remnant particles by the liver. Liver cells bind and internalize these particles by a receptor-mediated process analogous to that described above for the uptake of lipoproteins by fibroblasts and macrophages. As in these other cases, the particles are delivered to lysosomes, resulting in the hydrolysis of their cholesteryl esters. The free cholesterol is then reesterified by ACAT.

The effect of intragastric mevalonate on hepatic ACAT is probably mediated in part by hepatic uptake of chylomicron remnant particles carrying cholesterol synthesized in the intestinal mucosa from the exogenous mevalonate. It is also possible that some of the dose is absorbed into the bloodstream without conversion into cholesterol in the intestine. Mevalonate would then be taken up by the liver from the plasma and converted into cholesterol in hepatocytes. As mentioned below, mevalonic acid taken up by isolated liver cells *in vitro* is capable of increasing the intracellular content of esterified cholesterol.

The metabolism of esterified cholesterol in relation to ACAT has been investigated in liver cells *in vitro* in several laboratories. Mevalonic acid added to the incubation medium increases cholesteryl-ester synthesis (Nilsson, 1975) and ACAT activity (Drevon *et al.*, 1980a) in isolated hepatocytes, resulting in an increase in the cell content of esterified cholesterol but essentially no change in free cholesterol content. These effects are not prevented by treating the cells with cycloheximide. Rothblat and co-workers (Rothblat *et al.*, 1976, 1977) have shown that uptake of cholesterol by rat hepatoma cells incubated in the presence of cholesterol-enriched serum or cholesterol-phospholipid vesicles stimulates the synthesis of cholesteryl esters. Incubation of hepatic microsomes with hyperlipidemic serum (Rothblat *et al.*, 1977) or with normal serum (Mitropoulos *et al.*, 1981) also stimulates cholesteryl-ester formation and increases ACAT activity in the microsomes. Addition of 25-hydroxycholesterol to the incubation medium has also been shown to increase cholesteryl-ester synthesis and ACAT activity in rat hepatocytes in culture (Drevon *et al.*, 1980a) and in rat-liver microsomes (Erickson *et al.*, 1980). The effects of hyperlipidemic serum and of

25-hydroxycholesterol on ACAT activity in intact liver cells are not prevented by cycloheximide.

The behavior of ACAT in intestinal cells is in many respects similar to that of hepatic ACAT. The activity of ACAT in microsomes isolated from intestinal mucosa is increased by incubation of the microsomes with cholesterol-enriched lipoproteins from patients with familial LCAT deficiency (Norum *et al.*, 1981) or with cholesterol-phospholipid liposomes or 25-hydroxycholesterol (Field and Mathur, 1983). ACAT activity in the intestinal mucosa of intact animals is also increased by cholesterol feeding (Norum *et al.*, 1983).

It was suggested in Section I above that ACAT serves to minimize fluctuations in the cholesterol content of cell membranes and that, in addition to this general function, it also performs other roles in certain types of cell. The buffering effect of ACAT in hepatocytes and intestinal cells is shown by the fact that changes in the input of cholesterol into these cells result in much smaller changes in the intracellular concentration of free cholesterol than in that of esterified cholesterol.

However, in liver cells the cholesteryl esters synthesized by ACAT are not a metabolic cul-de-sac, as they seem to be in fibroblasts and macrophages. The esterified cholesterol in the core of VLDL particles newly secreted by the liver is derived from the pool of cytoplasmic cholesteryl esters formed by ACAT rather than from unhydrolyzed cholesteryl esters entering the liver by uptake of lipoproteins from the plasma (hence the predominance of oleate in the cholesteryl esters of nascent hepatic VLDL). Thus, one of the functions of hepatic ACAT may be to provide VLDL, synthesized in the liver, with a regulated supply of esterified cholesterol.

In keeping with this, when ACAT activity and cholesteryl-ester synthesis in cultured hepatocytes a by stimulated by mevalonic acid or 25-hydroxycholesterol, the cholesteryl-ester content of the VLDL particles secreted by the cells increases (Drevon *et al.*, 1980b). However, it is by no means certain that esterified cholesterol is essential for the formation and secretion of stable VLDL particles. Thus, it remains possible that, as with ACAT in most cells, the primary function of hepatic ACAT is to keep the cell content of free cholesterol constant within narrow limits. If this is so, the amount of esterified cholesterol incorporated into VLDL may be determined by the amount of free cholesterol that must be esterified in order to maintain homeostasis of membrane cholesterol.

In addition to its role in membrane-cholesterol homeostasis, ACAT in the intestine facilitates cholesterol absorption by esterifying free cholesterol taken up by the villous cells from the lumen. The esterified cholesterol formed by ACAT is incorporated into the nonpolar core of chylomicrons before secretion into the intestinal lymphatics. As mentioned above, when ACAT is inhibited, absorption of cholesterol from the intestine is impaired. In keeping with its additional role in cholesterol absorption, the activity of intestinal ACAT in rats is higher in cells of the villi than in crypt cells (Norum *et al.*, 1983) and is higher in the segment of the jejunum from which cholesterol is absorbed than in any other segment of the intestine (Haugen and Norum, 1976).

D. Esterification of Cholesterol in the Adrenal Gland

The role of ACAT in adrenal-cortex cells is complicated by the need to maintain a reservoir of cholesteryl esters in the cytoplasm. When steroid-hormone synthesis is stimulated by ACTH, cytoplasmic cholesteryl esters are hydrolyzed by a hormone-sensitive ester hydrolase. This releases free cholesterol, which is then transferred to the inner mitochondrial membrane where the initial step in the conversion of cholesterol into corticosteroids takes place (see Myant, 1981, Chapter 8).

In rats and in some other species (including man) plasma lipoproteins, rather than cholesterol synthesized *in situ,* are the ultimate source of cholesterol used for hormone synthesis in the adrenal gland. The predominant lipoprotein taken up by the adrenals varies from species to species; the rat adrenal uses both LDL and HDL, while the human adrenal uses LDL. Uptake of lipoproteins occurs by receptor-mediated endocytosis, with hydrolysis of lipoprotein cholesteryl esters in lysosomes and the release of free cholesterol which is reesterified in the cytoplasm by ACAT.

If the supply of lipoprotein to the adrenals is adequate and hormone production is at its basal level, ACAT activity is high and the cell content of cholesteryl esters is maximal. When the rate of hormone synthesis increases in response to ACTH, free cholesterol arising from hydrolysis of the reservoir of esterified cholesterol in the cytoplasm is probably the major substrate for hormone production, but the mitochondria may also use cholesterol released from lysosomes without prior esterification by ACAT. The diversion of free cholesterol into the corticosteroid pathway in response to ACTH may lead to a fall in the rate of esterification of cholesterol (see Gwynne and Strauss, 1982), presumably by reducing the supply of substrate for ACAT.

Balasubramaniam *et al.* (1977) have examined the consequences of a profound fall in the supply of lipoprotein cholesterol to the adrenal glands of rats *in vivo.* When the plasma cholesterol concentration is lowered drastically by 4-APP, the cholesteryl-ester content of the adrenals falls sharply and ACAT activity declines to 10% of the baseline value. When the reservoir of intracellular esterified cholesterol reaches a very low level, HMG-CoA reductase activity increases by more than 100-fold and cholesterol synthesized *in situ* becomes the major source of cholesterol for the production of corticosteroids. When 4-APP-treated rats are given intravenous infusions of LDL or HDL, ACAT activity in the adrenals rises rapidly, the reservoir of esterified cholesterol in the cytoplasm is refilled, and reductase activity returns to the control level. In keeping with the general role of

ACAT in minimizing changes in cell-membrane cholesterol concentration, Balasubramaniam *et al.* (1977) noted that the free-cholesterol content of the adrenals did not change when the cholesteryl-ester content varied 20-fold.

IV. Regulation of ACAT

ACAT is clearly a highly regulated enzyme. However, the failure of cycloheximide to prevent the changes in enzyme activity observed in intact cells under a variety of conditions shows that it is not readily inducible.

As discussed in Section II, ACAT in several types of cells is not saturated with its cholesterol substrate. Hence, an increase in its activity in isolated microsomes, assayed with endogenous cholesterol as substrate, could be mediated by an increase in the size of the substrate pool of free cholesterol in the microsomal membranes. This is probably the mechanism responsible for increased ACAT activity in atherosclerotic lesions in the aorta (Hashimoto and Dayton, 1977) and in CHO cells (Doolittle and Chan, 1982b) and fibroblasts (Gavigan and Knight, 1983) incubated in the presence of LDL. In all three instances the increase in ACAT activity is no longer apparent when the enzyme is assayed in the presence of saturating concentrations of free cholesterol. The finding that a very small increase in microsomal free cholesterol content leads to a very large increase in ACAT activity (Hashimoto *et al.*, 1974) is easily explained on the basis of activation by substrate supply. The pool of cholesterol that acts as substrate for ACAT must be only a small fraction of the total cholesterol in microsomal preparations, since the specific activity of the cholesteryl esters formed from radioactive microsomal cholesterol is much higher than that of the total microsomal free cholesterol (Balasubramaniam *et al.*, 1978a; Synouri-Vrettakou and Mitropoulos, 1983). Hence, a large increase in the substrate pool of cholesterol could occur without significant change in the concentration of free cholesterol in the microsomal fraction as a whole.

A passive response to substrate supply may underlie many of the changes in ACAT activity that have been described in intact cells and isolated microsomes. However, the stimulatory effect of 25-hydroxycholesterol on ACAT in hepatocytes and fibroblasts is difficult to explain on this basis because the increase in ACAT activity is accompanied by decreased synthesis of endogenous cholesterol due to suppression of HMG-CoA reductase. It thus seems likely that the catalytic activity of the enzyme can be modulated under some conditions. One possibility is that the intrinsic activity of ACAT changes in response to alterations in the fluidity of the microsomal membrane in the vicinity of the enzyme (Suckling *et al.*, 1982; Doolittle and Chan, 1982a). Such changes in fluidity could be brought about by changes in the amount of sterol or oxysterol in a domain of the membrane adjacent to an enzyme molecule. It has also been suggested that the catalytic activity of ACAT in liver and intestine can be modulated reversibly by phosphorylation (Basheeruddin *et al.*, 1982; Suckling *et al.*, 1983a; Skrzypczak and Higgins, 1985). Much of the evidence for this suggestion is based on the observation that ACAT activity in isolated microsomes increases when the microsomes are incubated under conditions in which phosphorylation would be expected to occur. However, covalent addition of phosphate to enzyme protein has not been demonstrated. A conclusive test of the hypothesis that ACAT activity can be modulated by phosphorylation–dephosphorylation will only be possible when the enzyme has been purified.

References

Aqel, N. M., Ball, R. Y., Waldman, H., and Mitchinson, M. J. (1984). Monocytic origin of foam cells in human atherosclerotic plaques. *Atherosclerosis* **53,** 265–271.

Attie, A. D., Pittman, R. C., and Steinberg, D. (1980). Metabolism of native and lactosylated human low density lipoprotein: evidence for two pathways for catabolism of exogenous proteins in rat hepatocytes. *Proc. Natl. Acad. Sci. U.S.A.* **77,** 5923–5927.

Balasubramaniam, S., Goldstein, J. L., Faust, J. R., Brunschede, G. Y., and Brown, M. S. (1977). Lipoprotein-mediated regulation of 3-hydroxy-3-methylglutaryl coenzyme A reductase activity and cholesteryl ester metabolism in the adrenal gland of the rat. *J. Biol. Chem.* **252,** 1771–1779.

Balasubramaniam, S., Mitropoulos, K. A., and Venkatesan, S. (1978a). Rat liver acyl-CoA:cholesterol acyltransferase. *Eur. J. Biol.* **90,** 377–383.

Balasubramaniam, S., Venkatesan, S., Mitropoulos, K. A., and Peters, T. J. (1978b). The submicrosomal localisation of acyl-coenzyme A:cholesterol acyltransferase and its substrate, and of cholesteryl esters in rat liver. *Biochem. J.* **174,** 863–872.

Basheeruddin, K., Rawsthorne, S., and Higgins, M. J. P. (1982). Reversible activation of rat liver acyl-coA:cholesterol acyltransferase *in vitro*. *Biochem. Soc. Trans.* **10,** 390–391.

Bennett Clark, S., and Tercyak, A. M. (1984). Reduced cholesterol transmucosal transport in rats with inhibited mucosal acyl CoA:cholesterol acyltransferase and normal pancreatic function. *J. Lipid Res.* **25,** 148–159.

Brown, M. S., and Goldstein, J. L. (1983). Lipoprotein metabolism in the macrophage: implications for cholesterol deposition in atherosclerosis. *Ann. Rev. Biochem.* **52,** 223–261.

Brown, M. S., Dana, S. E., and Goldstein, J. L. (1975a). Cholesterol ester formation in cultured human fibroblasts. Stimulation by oxygenated sterols. *J. Biol. Chem.* **250,** 4025–4027.

Brown, M. S., Faust, J. R., and Goldstein, J. L. (1975b). Role of the low density lipoprotein receptor in regulating the content of free and esterified cholesterol in human fibroblasts. *J. Clin. Invest.* **55,** 783–793.

Brown, M. S., Ho, Y. K, and Goldstein, J. L. (1980). The cholesteryl ester cycle in macrophage foam cells. Continual hydrolysis and re-esterification of cytoplasmic cholesteryl esters. *J. Biol. Chem.* **255,** 9344–9352.

Doolittle, G. M., and Chan, T-Y. (1982a). Solubilization, partial purification, and reconstitution in phosphatidyl-cholesterol liposomes of acyl-CoA:cholesterol acyltransferase. *Biochemistry* **21,** 674–679.

Doolittle, G. M., and Chan, T-Y. (1982b). Acyl-CoA:cholesterol acyltransferase in Chinese hamster ovary cells. Enzyme activity determined after reconstitution in phospholipid/cholesterol liposomes. *Biochim. Biophys. Acta* **713,** 529–537.

Drevon, C. A., Weinstein, D. B., and Steinberg, D. (1980a). Regulation of cholesterol esterification and biosynthesis in monolayer cultures of normal adult rat hepatocytes. *J. Biol. Chem.* **255,** 9128–9137.

Drevon, C. A., Engelhorn, S. C., and Steinberg, D. (1980b). Secretion of very low density lipoproteins enriched in cholesteryl esters by cultured rat hepatocyes during stimulation of intracellular cholesterol esterification. *J. Lipid Res.* **21,** 1065–1071.

Erickson, S. K., Shrewsbury, M. A., Brooks, C., and Meyer, D. J. (1980). Rat liver acyl-coenzyme A:cholesterol acyltransferase: its regulation *in vivo* and some of its properties *in vitro*. *J. Lipid Res.* **21,** 930–941.

Field, F. J., and Mathur, S. N. (1983). Regulation of acyl CoA:cholesterol acyltransferase by 25-hydroxycholesterol in rabbit intestinal microsomes and absorptive cells. *J. Lipid Res.* **24,** 1042–1059.

Gavigan, S. J. P., and Knight, B. L. (1983). The effects of low-density lipoprotein and cholesterol on acyl-coenzyme A:cholesterol acyltransferase activity in membranes from cultured human fibroblasts. *Biochem. J.* **216,** 93–100.

Goldstein, J. L., and Brown, M. S. (1977). The low-density lipoprotein pathway and its relation to atherosclerosis. *Ann. Rev. Biochem.* **46,** 897–930.

Gwynne, J. T., and Strauss, J. F. (1982). The role of lipoproteins in steroidogenesis and cholesterol metabolism in steroidogenic glands. *Endocr. Rev.* **3,** 299–329.

Hashimoto, S., and Dayton, S. (1977). Studies of the mechanism of augmented synthesis of cholesteryl ester in atherosclerotic rabbit aortic microsomes. *Atherosclerosis* **28,** 447–452.

Hashimoto, S., Dayton, S., Alfin-Slater, R. B., Bui, P. T., Baker, N., and Wilson, L. (1974). Characteristics of the cholesterol-esterifying activity in normal and atherosclerotic rabbit aortas. *Circ. Res.* **34,** 176–183.

Haugen, R., and Norum, K. R. (1976). Coenzyme-A-dependent esterification of cholesterol in rat intestinal mucosa. *Scand. J. Gastroenterol.* **11,** 615–621.

Kritchevsky, D., and Kothari, H. V. (1978). Arterial enzymes of cholesteryl ester metabolism. *Adv. Lipid Res.* **16,** 221-266.

Mitropoulos, K. A., Venkatesan, S., Reeves, B. E. A., and Balasubramaniam, S. (1981). Modulation of 3-hydroxy-3-methylglutaryl-CoA reductase and of acyl-CoA-cholesterol acyltransferase by the transfer of non-esterified cholesterol to rat liver microsomal vesicles. *Biochem. J.* **194,** 265–271.

Myant, N. B. (1981). "The Biology of Cholesterol and Related Steroids." Heinemann, London.

Nilsson, A. (1975). Increased cholesterol-ester formation during forced cholesterol synthesis in rat hepatocytes. *Eur. J. Biochem.* **51,** 337–342.

Norum, K. R., Helgerud, P., and Lilljeqvist, A-C. (1981). Enzymic esterification of cholesterol in rat intestinal mucosa catalyzed by acyl-CoA:cholesterol acyltransferase. *Scand. J. Gastroenterol.* **16,** 401–410.

Norum, K. R., Helgerud, P., Petersen, L. B., Groot, P. H. E., and DeJonge, H. R. (1983). Influence of diets on acyl-CoA:cholesterol acyltransferase and on acyl-CoA:retinol acyltransferase in villous and crypt cells from rat small intestinal mucosa and in the liver. *Biochim. Biophys. Acta* **751,** 153–161.

Rothblat, G. H., Arbogast, L., Kritchevsky, D., and Naftulin, M. (1976). Cholesteryl ester metabolism in tissue culture cells: II. Source of accumulated esterfied cholesterol in Fu5A rat hepatoma cells. *Lipids* **11,** 97–108.

Rothblat, G. H., Naftulin, J., and Arbogast, L. Y. (1977). Stimulation of acyl-CoA:cholesterol acyltransferase activity by hyperlipemic serum lipoproteins. *Proc. Soc. Exp. Biol. Med.* **155,** 501–506.

St. Clair, R. W. (1976). Metabolism of the arterial wall and atherosclerosis. *Atherosclerosis Rev.* **1,** 61–117.

Skrzypczak, K. H., and Higgins, M. J. A. (1985). ATP-dependent activation of acyl-coenzyme A: cholesterol acyltransferase; catalysis by a cytoplasmic enzyme. *Biochem. Soc. Trans.* **13,** 150–151.

Suckling, K. E., Boyd, G. S., and Smellie, C. G. (1982). Properties of a solubilized and reconstituted preparation of acyl-CoA:cholesterol acyltransferase from rat liver. *Biochim. Biophys. Acta.* **710,** 154–163.

Suckling, K. E., and Stange, E. F. (1985). Role of acyl-CoA:cholesterol acyltransferase in cellular cholesterol metabolism. *J. Lipid Res.* **26,** 647–671.

Suckling, K. E., Stange, E. F., and Dietschy, J. M. (1983a). Dual modulation of hepatic and intestinal acyl-CoA:cholesterol acyltransferase activity by (de-)phosphorylation and substrate supply *in vitro. FEBS Lett.* **151,** 111–116.

Suckling, K. E., Tocher, d. R., Smellie, C. G., and Boyd, G. S. (1983b). *In vitro* regulation of bovine adrenal cortex acyl-CoA:cholesterol acyltransferase and comparison with the rat liver enzyme. *Biochim. Biophys. Acta* **753,** 422–429.

Synouri-Vrettakou, S., and Mitropoulos, K. A. (1983). Acyl-coenzyme A:cholesterol acyltransferase. Transfer of cholesterol to its substrate pool and modulation of activity. *Eur. J. Biochem.* **133,** 299–307.

Chapter 5

LDL: Physical and Chemical Characteristics

I. Composition and Physical Characteristics

Plasma lipoproteins are particles formed by the noncovalent association of lipids with specific proteins called apolipoproteins or apoproteins (apoA, apoB, etc.). Five classes of lipoproteins in normal plasma are defined operationally in terms of their hydrated densities or electrophoretic mobilities, as shown in Table 5.1. Within each lipoprotein class the size and composition of the particles vary continuously across the whole density range, though there is at least one major peak in the distribution of densities, indicating that some combinations of lipid and protein within each class are more stable than others. Because particle characteristics are distributed over a range of values, the density, particle diameter, molecular weight, and percentage composition of each class of lipoprotein are expressed as ranges or as mean or modal values. As discussed below, two or more subpopulations of particle, with different mean densities, particle sizes, and chemical composition, are present in human LDL.

Human plasma LDL is here defined as the total population of lipoproteins in normal human plasma within the density range 1.019–1.063 g/ml. Normal lipoproteins within this range have β mobility on zonal electrophoresis on agarose gel or other solid supports and have flotation rates in the range S_f 0–12.[1] The above definition of LDL, adopted here for convenience, excludes abnormal lipoproteins, such as Lp-X, whose density extends into the LDL range. Lipoproteins analogous to human LDL with respect to their composition, origin, and

[1]When a lipoprotein is centrifuged in a solvent with density greater than that of the lipoprotein, the particles float toward the meniscus at a rate proportional to their buoyant density, the least dense particles having the highest flotation rate. The flotation rate of lipoproteins of density <1.063 g/ml is determined by centrifugation in the analytical ultracentrifuge in a solution of NaCl of density 1.063 g/ml and is expressed as the S_f value in Svedberg units (10^{-13}cm sec^{-1}dyne^{-1}g^{-1}).

Table 5.1
Density, Flotation Rate, and Electrophoretic Mobility of Lipoproteins of Normal Human Plasma

Lipoprotein class	Density (g/ml)	Flotation rate[a]		Electrophoretic mobility in agarose gel
		$F_{1.2}$	S_f	
High density (HDL)	1.063–1.21	0.9	—	α
Low density (LDL)	1.019–1.063	—	0–12	β
Intermediate density (IDL)	1.006–1.019	—	12–20	β or Pre-β_1
Very low density (VLDL)	0.95–1.006	—	20–400	Pre-β
Chylomicrons	<0.95	—	>400	Origin

[a]The flotation rate of HDL is expressed as an $F_{1.2}$ value determined by ultracentrifugation in a solution of NaCl and NaBr of density 1.20 g/ml. For a definition of S_f see footnote on p. 99.

metabolism are present in the plasma of mammals other than man, although the density range and apoprotein composition of these ''LDLs'' may differ slightly from those of human LDL (see Soutar and Myant, 1979). In human LDL the predominant species of particle has apoB as its sole apoprotein. However, LDL isolated from the plasma of normal subjects by sequential ultracentrifugation at density 1.019–1.063 g/ml usually includes traces of apoE- and apoC-containing particles. Lipoprotein-a [Lp(a)] is also present in small amounts in LDL at the high end of the density range, though the bulk of this lipoprotein has a density >1.063 g/ml. Lp(a) is metabolically distinct from the predominant LDL species; it contains, in addition to apoB, a sialic-acid-rich apoprotein [apo(a)] that imparts pre-β mobility to the particles (see Chapter 6 for details).

Table 5.2 shows the molecular weights, particle diameters, and chemical composition of human plasma lipoproteins. Values for molecular weight are based on measurement of sedimentation velocity or sedimentation equilibrium; values for particle diameter are derived from electron microscopy, usually after negative staining. Note that molecular weight and size increase with decreasing density of the lipoproteins. This is mainly due to the fact that the density of the particles decreases as the ratio of lipid to protein increases, the larger particles having a greater relative and absolute amount of lipid than the smaller ones. Note also that chylomicrons and VLDL, in keeping with their role as carriers of neutral fat, have triglyceride as their major constituent. Hence, they are known collectively as triglyceride-rich lipoproteins.

It will be seen from Table 5.2 that esterified cholesterol is the major lipid in LDL. The composition and plasma concentration of LDL in normal human subjects are such that more than 60% of the total cholesterol in the plasma is carried in LDL. If the mean MW of LDL is taken to be 2.2×10^6, each particle would contain up to 0.88×10^6 Da of esterified cholesterol. This is equivalent to about 1300 molecules of esterified cholesterol (mean MW ≅660) per particle. A similar calculation shows that each LDL particle contains between 5.0×10^5 and 5.5×10^5 Da of protein.

Table 5.2
Molecular Weight, Particle Diameter, and Composition of Human Plasma Lipoproteins[a]

Lipoprotein class	Molecular weight	Diameter (Å)	Composition (% dry weight)					
			Protein	Triglyceride	Phospholipid	Free cholesterol	Esterified cholesterol	Total lipid
HDL	$1.9–3.9 \times 10^5$	50–120	40–55	2–5	20–35	3–5	10–12	45–60
LDL	$2.0–2.5 \times 10^6$	190–250	20–25	7–10	15–20	7–10	35–40	75–80
IDL	$3.5–4.5 \times 10^6$	250–350	15–20	20–30	20	7–10	22	80–85
VLDL	$5 \times 10^6–1 \times 10^7$	250–750	5–10	50–65	15–20	5–10	10–15	90–95
Chylomicrons	$10^9–10^{10}$	$10^3–10^4$	1.5–2.5	85–90	7–9	1–3	3–5	97–99
β-VLDL			18	19	26	10	27	82

[a]Values are taken from various sources. Values for β-VLDL, shown for comparison with IDL, are from Patsch *et al.* (1975). Molecular weights are based on sedimentation analysis in the analytical ultracentrifuge; diameters are determined by electron microscopy.

Table 5.3
Relation between Density, Molecular Weight, and Composition of Human LDL Particles[a]

Subfraction	Density (g/ml)	Molecular weight ($\times 10^{-6}$)	Number of molecules per LDL particle				
			Phospholipid	Free cholesterol	Esterified cholesterol	Triglyceride	Amino acids ($\times 10^{-3}$)
1	1.027	2.89	811	616	1790	178	7.0
2	1.031	2.61	707	537	1715	108	6.5
3	1.036	2.38	677	474	1488	82	6.3
4	1.041	2.13	604	403	1281	74	6.1
5	1.049	1.93	496	362	1139	110	5.8
6	1.060	1.88	383	351	1077	123	6.5

[a] Values assembled from Shen *et al.* (1981).

The relationship shown in Table 5.2 between density, molecular weight, and size of the lipoproteins of different classes also holds for subfractions of LDL. Shen *et al.* (1981) have analyzed six subfractions of normal human LDL isolated at densities between 1.027 and 1.060 g/ml after equilibrium density-gradient ultracentrifugation (see Section III). Their results are shown in Table 5.3. As the density of the subfractions increases (1→6) molecular weight decreases, the total number of lipid molecules per particle decreases, and there is essentially no change in the amount of protein per particle. Shen *et al.* (1981) also noted that particle diameter, determined by electron microscopy, decreased in each successive fraction as density increased. Thus, the progressive decrease in particle size, accompanied by a rise in density, is due to a decrease in the amount of lipid in particles that maintain a more or less constant amount of protein.

II. Structure

LDL shares a common basic structure with all other plasma lipoproteins (see Shen *et al.*, 1977). The generally accepted model for this structure is based on a variety of unrelated techniques including electron microscopy, small-angle X-ray scattering, chemical analysis, and studies of the interaction of intact lipoprotein particles with hydrolytic enzymes and antibodies to the apoproteins. The mature lipoprotein particle present in the circulation is a sphere comprising a core of nonpolar lipids (esterified cholesterol and triglyceride) surrounded by a polar shell of phospholipid, free cholesterol, and protein. The phospholipids and free cholesterol in the outer shell are arranged in a monolayer with which the protein is associated. In the model for LDL structure proposed by Deckelbaum *et al.* (1977), up to 15% of the total free cholesterol in the particle is dissolved in the nonpolar core.

The precise orientation of apoB in the outer shell of LDL has not been established. However, the ability of anti-apoB antibodies to react immunologically with native LDL shows that at least some of the antigenic determinants[2] of apoB are exposed to the aqueous medium. In agreement with this, apoB in native LDL interacts specifically with concanavalin A (a lectin that binds to the oligosaccharide residues of a glycoprotein) and is susceptible to partial hydrolysis by proteolytic enzymes. Several apoproteins other than apoB have α-helical regions, known as amphipathic helices, in which polar and nonpolar amino acid residues are clustered on opposite sides of the helix. It is thought that the nonpolar face of an α-helix penetrates the lipid shell of the lipoprotein particle, while the polar face interacts with the aqueous medium (Segrest *et al.*, 1974). A

[2]An *antigenic determinant,* or *epitope,* is a site on a macromolecule, usually a protein, that combines with a specific antibody. A protein has multiple epitopes, each reacting with a single monoclonal antibody.

protein orientated in this way would contribute to the stabilization of a lipoprotein particle and would facilitate specific interactions of the particle with proteins and other elements in its environment. The orientation of apoB in LDL is discussed in Chapter 6.

With regard to the core lipids in LDL, Deckelbaum *et al.* (1977) have shown that the cholesteryl esters in intact LDL undergo a reversible change from a partially ordered liquid-crystalline state to a liquid state at temperatures between 20 and 40°C, the transition temperature rising as the concentration of triglyceride in the particle decreases. Deckelbaum *et al.* conclude that at 37°C the cholesteryl esters in some of the LDL particles in some individuals are in a liquid-crystalline state. This is likely to be the case, for example, in patients with FH, in whom the ratio of triglyceride to esterified cholesterol in LDL is decreased (see Chapter 10).

III. Heterogeneity and Genetic Polymorphism

A. Heterogeneity

As we have seen, LDL can be subdivided at arbitrary cut-off points into fractions of increasing density. This heterogeneity could merely reflect the presence of a single "monodisperse" population of particles whose densities are distributed on either side of a single mean value. Indeed, Schlieren curves given by normal LDL in the analytical ultracentrifuge usually have single symmetrical peaks (Hammond and Fisher, 1971; Krauss and Burke, 1982). However, methods for separating lipoproteins with greater resolving power than is possible by centrifugation in solutions of uniform density have revealed the presence of two or more overlapping, or possibly discrete, subpopulations of particles in LDL from some normal individuals and from many hyperlipidemic patients.

The study of the heterogeneity of LDL, especially in man, is a field of growing importance in view of the possibility that different subpopulations within the LDL class of lipoproteins have different origins and metabolic fates and may differ with respect to their atherogenic potential and their ability to interact with cells and extracellular elements. Heterogeneity of LDL arising from differences in the physical, chemical, and immunological properties of different LDL particles is considered in the following summary. Heterogeneity of LDL with respect to metabolic behavior is considered in Chapter 7. As we shall see, the presence of heterogeneity in LDL may to some extent be a consequence of the heterogeneity of IDL and VLDL, the precursors of LDL.

1. Methodology

Methods that can be used to demonstrate LDL polydispersity include the following.

1. Equilibrium banding, in which LDL is separated in the analytical ultracentrifuge into bands of different density by centrifugation to equilibrium in a

continuous density gradient, the particles forming bands at positions corresponding to their hydrated densities (Adams and Schumaker, 1970). Banding of LDL subpopulations may also be achieved on a preparative scale by centrifugation in swingout buckets or in zonal rotors.

2. Separation by centrifugation in discontinuous density gradients (for theory, see Kelley and Kruski, 1986).

3. Electrophoresis in a polyacrylamide gradient gel. Here, separation is based on particle size, the electrophoretic current providing the driving force that moves the charged particles to their final positions in the gel.

4. Gel filtration, in which the particles are separated on the basis of their size. This method has been used successfully in the study of size heterogeneity of human VLDL (Sata *et al.*, 1970, 1972) and LDL (Rudel *et al.*, 1986), but has not been used systematically to isolate subpopulations of LDL.

5. Immunoaffinity columns, in which particles are separated by differential adsorption to an antibody coupled to the solid phase of the column. Particles expressing an antigen that reacts with the antibody are selectively retained on the column. Milne *et al.* (1984) have adapted this method to the separation of particles containing apoB-48 from those containing apoB-100 in human VLDL, using monoclonal antibodies that react with apoB-100 but not with apoB-48. Human LDL does not normally contain apoB-48, but the method of Milne *et al.* (1984) should be applicable to the study of LDL in rats and other species in which LDL contains apoB-48 and apoB-100 on separate particles.

2. Subfractionation by Density

Adams and Schumaker (1970), using the equilibrium banding technique, showed that LDL from some normal subjects contains two or more subpopulations with different mean densities and molecular weights. The LDL of Adams himself was monodisperse while he was eating a normal diet but developed an additional subpopulation when fat was added to his diet. This suggests that the LDL pattern in a given individual may depend partly upon habitual diet. The presence of two or more LDL subpopulations, differing in density and molecular weight, has also been demonstrated in normal subjects by Krauss and Burke (1982), who analyzed LDL by density gradient ultracentrifugation and by gradient gel electrophoresis. Polydispersity of LDL appears to be more marked in patients with hypertriglyceridemia than in normal subjects. Thus, Hammond and Fisher (1971) found two, and sometimes three, peaks in the Schlieren curves given by the LDL of diabetics with raised plasma triglyceride levels. In each individual the Schlieren pattern was constant on repeated analysis over many months. When LDL subfractions corresponding to each of the peaks were isolated and recombined, the recombined subfractions gave the same Schlieren pattern in the analytical ultracentrifuge as that given by the original, unfractionated LDL. Hence, the multiple peaks were not artifacts produced by ultracentrifugation in strong salt solutions. Analysis of the three LDL subpopulations showed that as the mean density increased, the molecular weight and lipid content of the particles decreased but the protein content remained essentially unchanged. This is consistent with conversion of the larger particles into the smaller ones by loss of lipid without loss of protein.

Teng *et al.* (1983) have shown that LDL, when submitted to prolonged ultracentrifugation in a discontinuous density gradient, separates into two visible subfractions—a less dense ("light") fraction at the meniscus and a more dense ("heavy") fraction about 1 cm below the meniscus. The particles in the heavy fraction are smaller and have a higher protein : lipid ratio than those in the light fraction.

It is worth noting that LDL of the rhesus monkey, a species with a plasma lipoprotein pattern very similar to that of man, has three subpopulations separable on the basis of hydrated density (Fless and Scanu, 1979). The fraction with the highest density has an unusually high sialic acid content and migrates with pre-β mobility in agarose gel.

The formation of lipoprotein bands when LDL is centrifuged in a continuous density gradient is often assumed to indicate the presence of completely separate subpopulations of particles. However, densitometric scans of the LDL bands formed by equilibrium banding show a considerable degree of overlap between adjacent peaks (see, for example, Fisher *et al.* (1980). This suggests that particle densities are distributed continuously over the whole LDL density range and that subpopulations, characterized by their mean densities, overlap one another. A distinction between discrete and nondiscrete subpopulations of LDL may or may not be worth pursuing. However, it should be noted that nondiscrete subpopulations would be expected if, as seems to be the case, one aspect of LDL metabolism is the stepwise conversion of the larger, less dense particles into the smaller ones. Hammond and Fisher (1971) have drawn attention to the fact that two subfractions of LDL can be completely separated by centrifugation at a single density between the densities of the subfractions. But if particles with all possible intermediate densities were present in the LDL sample, all those less dense than the solvent would move to the meniscus and all those more dense would eventually appear at the bottom of the tube. The complete separation of LDL into two separate bands after centrifugation in a discontinuous density gradient may be explained in a similar way.

3. Protein Heterogeneity

LDL is also heterogeneous with respect to its protein component. This heterogeneity may be due either to differences in the species of apoprotein present in the particles or to differences in the extent to which immunoreactive sites (epitopes) in the particle are expressed.

Particles in the upper range of the density spectrum of human LDL contain small amounts of apoC as well as apoB (Lee, 1976). ApoC-III and apoE have also been detected in human LDL of density 1.025–1.029 g/ml (Lee and Alaupovic, 1986). It is not known whether apoB, apoC, and apoE are present in the same or in separate particles of human LDL. The presence of Lp(a) at the upper end of the LDL density spectrum has already been referred to. ApoE is also present in the LDL of rats and rabbits. Yamada *et al.* (1986) have shown that the LDL of rabbits contains some particles with apoB as their only apoprotein and others that contain apoE in addition to apoB. The apoE-containing subpopulation of particles can be isolated by passing the LDL through an immunoaffinity column of anti-apoE antibody coupled to Sepharose; as discussed in Chapter 7, the two types of LDL particle are metabolized at different rates. The presence of ApoB-48 and apoB-100 on separate LDL particles in rats and other species has already been mentioned. Human LDL contains little or no apoB-48. However, particles containing two complementary fragments of apoB-100 are present in the LDL isolated from some normal human subjects. The possible significance of this heterogeneity is discussed in Chapter 6.

Heterogeneity in the immunoreactivity of apoB in LDL and other apoB-containing lipoproteins has been studied with monospecific and monoclonal antibodies raised against intact LDL. All these antibodies react specifically with apoB. Schonfeld *et al.* (1979) showed that the immunoreactivity of VLDL to a monospecific anti-LDL antibody increased as the particles underwent increasing lipolysis by lipoprotein lipase. Since there was no change in the apoB content of the particles, this observation suggests that a decrease in the size of VLDL particles, associated with loss of lipid, leads to exposure of more apoB epitopes on the particle surface or to a change in the conformation of apoB that enhances their reactivity with antibodies. Other workers (Tsao *et al.*, 1982; Marcel *et al.*, 1984) have shown that VLDL, IDL, and LDL can be distinguished by their patterns of immunoreactivity with groups of monoclonal antibodies to apoB. Two of the monoclonal antibodies used by Marcel *et al.* (1984) reacted with epitopes close to the LDL-receptor-binding domain of apoB. The reactivities of both antibodies with VLDL increased when the particles were partially delipidated, and both reacted more strongly with LDL than with VLDL. These differences in immunoreactivity are in parallel with differences in the ability of apoB-containing lipoproteins to react with the LDL receptor. In particular, LDL binds to the fibroblast receptor much more strongly than does VLDL. Thus, it seems likely that an important determinant of the binding affinity of a lipoprotein particle for the LDL receptor is the extent to which the binding domain of apoB is unmasked or is oriented in a favorable conformation.

Antigenic heterogeneity within the LDL class of lipoproteins has also been demonstrated by Teng *et al.* (1985), who compared the reactivities of a panel of six monoclonal antibodies with three subfractions of LDL isolated by density gradient ultracentrifugation. With three of the antibodies, reactivity was highest with the least dense LDL fraction, lowest with the fraction of greatest density, and intermediate with the fraction of intermediate density. Thus, three subpopulations of LDL could be identified by the pattern of apoB epitopes they expressed.

B. Genetic Polymorphism

The presence, in a given individual, of LDL subfractions that differ from each other in physical characteristics or immunoreactivity raises the question of a possible relationship between heterogeneity and genetic polymorphism of LDL (for a definition of polymorphism, see Chapter 6, Section VII). Can one discern differences in the distribution of densities or in the expression of apoB epitopes in the LDL of different individuals within a population? If such differences exist, to what extent are they inherited? The answers to these questions could provide clues to the causes of intrapopulation variability of the plasma LDL concentration—a major risk factor for coronary heart disease.

Much of the variability of LDL-cholesterol level within populations is due to environmental causes, but there is also a substantial genetic component (see Myant, 1981, Chapter 12). We already know a good deal about the mechanism underlying the abnormal plasma lipid levels in a number of inborn errors of lipoprotein metabolism. However, these disorders make a comparatively small contribution to the total burden of coronary heart disease in the population as a whole. What we now need is information about the way in which genes contribute to the relatively small differences in plasma LDL concentration between apparently healthy members of the general population. One possibility is that there are genetically determined differences in the characteristics of LDL particles that affect their susceptibility to cellular uptake and degradation. As we shall see in Chapter 6, the human apoB gene exhibits a considerable degree of polymorphism, due mainly to single-base substitutions. Some of these substitutions result in an amino acid change in apoB-100 that might be expected to affect the binding properties, or the immunoreactivity, of LDL particles carrying the variant apoB.

Evidence for genetic polymorphism of human LDL was first obtained by Allison and Blumberg (1961), who identified alloantibodies[3] to LDL in the serum of a patient who had been given multiple blood transfusions. The antigens reacting to the alloantibodies were called the Ag antigens. Subsequent studies showed that there are at least 10 different Ag antigens in the human population. The most likely arrangement of the Ag polymorphic system is one with five closely linked loci, each with two codominant alleles (Bütler *et al.*, 1974, and

[3]*Alloantibodies* (also called isoantibodies) are antibodies that distinguish between genetic variants of a protein or glycoprotein (*alloantigens* or isoantigens) within a single species, e.g., the human ABO blood group system. Each set of alloantigens constitutes a genetic polymorphic system.

see Kostner, 1976, for review). It has now been shown that the Ag system is the phenotypic expression of polymorphism in the *apoB* gene (see Chapter 6, Section VII, C).

Fisher *et al.* (1975) have shown that the average molecular weight of LDL varies within populations and that there is a genetic contribution to this variance. Family studies led Fisher *et al.* to propose that LDL molecular weight is determined partly by a single gene with two codominant alleles. Austin and Krauss (1986) have also reported observations on families that suggest the presence of two different patterns of the size and density of LDL particles in the normal population. In pattern A the predominant species of LDL particle is relatively large and of low density. In pattern B there is a predominance of smaller, denser particles. Austin and Krauss postulate that the two patterns are determined by a single gene with two alleles, the allele specifying pattern B having a frequency of about 0.15 in their study population. It would clearly be of interest to know whether or not the polymorphic genes postulated by Fisher *et al.* (1975) and by Austin and Krauss (1986) correspond to any of the *Ag* loci or to any of the loci identified by monoclonal antibodies to LDL apoB. Young *et al.* (1986) could find no association between the molecular weight, lipid composition, or mean density of LDL and either of the epitopes identified by antibody MB19. The possible association between plasma lipid concentration and apoB polymorphisms, identified by means of restriction endonucleases, is dealt with in Chapter 6.

In conclusion, it should be noted that genetic polymorphism of LDL is not confined to man. A similar polymorphism, also identified by means of alloantibodies, is present in many nonhuman species, including rabbits, pigs, sheep, monkeys, cows, and mink (see Rapacz *et al.*, 1977, for references). In pigs, the genetic system underlying this polymorphism is highly complex, involving eight allelic loci giving 36 phenotypically different groups of pigs. Since some of these loci are now known to be in the *apoB* gene, this genetic system will be discussed in more detail in the next chapter.

References

Adams, G. H., and Schumaker, V. N. (1970). Equilibrium banding of low-density lipoproteins. II. Analysis of banding patterns. *Biochim. Biophys. Acta* **202,** 315–324.

Allison, A. C., and Blumberg, B. S. (1961). An isoprecipitation reaction distinguishing human serum-protein type. *Lancet* **1,** 634–647.

Austin, M. A., and Krauss, R. M. (1986). Genetic control of low-density lipoprotein subclasses. *Lancet* **2,** 592–595.

Bütler, R., Brunner, E., and Morganti, P. E. (1974). Contribution to the inheritance of the Ag groups. *Vox Sang.* **26,** 485–496.

Deckelbaum, R. J., Shipley, G. G., and Small, D. M. (1977). Structure and interactions of lipids in human plasma low density lipoproteins. *J. Biol. Chem.* **252,** 744–754.

Fisher, W. R., Hammond, M. G., Mengel, M. C., and Warmke, G. L. (1975). A genetic determinant of the phenotypic variance of the molecular weight of low density lipoprotein. *Proc. Natl. Acad. Sci. U.S.A.* **72,** 2347–2351.

Fisher, W. R., Zech, L. A., Bardalaye, P., Warmke, G., and Berman, M. (1980). The metabolism of apolipoprotein B in subjects with hypertriglyceridemia and polydisperse LDL. *J. Lipid Res.* **21,** 760–774.

Fless, G. M., and Scanu, A. M. (1979). Isolation and characterization of the three major low density lipoproteins from normolipidemic rhesus monkeys (*Macaca mulatta*). *J. Biol. Chem.* **254,** 8653–8661.

Hammond, M. G., and Fisher, W. R. (1971). The characterization of a discrete series of low density lipoproteins in the disease, hyper-preβ-lipoproteinemia. Implications relating to the structure of plasma lipoproteins. *J. Biol. Chem.* **246,** 5454–5465.

Kelley, J. L., and Kruski, A. W. (1986). Density gradient ultracentrifugation of serum lipoproteins in a swinging bucket rotor. *Methods Enzymol.* **128,** 170–181.

Kostner, G. M. (1976). Lp(a) lipoproteins and the genetic polymorphisms of lipoprotein B. *In* "Low Density Lipoproteins" (C. E. Day and R. S. Levy, eds.), pp. 229–269. Plenum, New York.

Krauss, R. M., and Burke, D. J. (1982). Identification of multiple subclasses of plasma low density lipoproteins in normal humans. *J. Lipid Res.* **23,** 97–104.

Lee, D. M. (1976). Isolation and characterization of low density lipoproteins. *In* "Low Density Lipoproteins" (C. E. Day and R. S. Levy, eds.), pp. 3–47. Plenum, New York.

Lee, D. M., and Alaupovic, P. (1986). Apolipoproteins B, CIII and E in two major subpopulations of low-density lipoproteins. *Biochim. Biophys. Acta* **879,** 126–133.

Marcel, Y., Hogue, M., Weech, P. K., and Milne, R. W. (1984). Characterization of antigenic determinants on human solubilized apolipoprotein B. Conformational requirements for lipids. *J. Biol. Chem.* **259,** 6952–6957.

Milne, R. W., Weech, P. K., Blanchette, L., Davignon, J., Alaupovic, P., and Marcel, Y. L. (1984). Isolation and characterization of apolipoprotein B-48 and B-100 very low density lipoproteins from type III hyperlipiproteinemic subjects. *J. Clin. Invest.* **73,** 816–823.

Myant, N. B. (1981). "The Biology of Cholesterol and Related Steroids." Heinemann, London.

Patsch, J. R., Sailer, S., and Braunsteiner, H. (1975). Lipoprotein of the density 1.006–1.020 in the plasma of patients with type III hyperlipoproteinaemia in the postabsorptive state. *Eur. J. Clin. Invest.* **5,** 45–55.

Rapacz, J., Elson, C. E., and Lalich, J. J. (1977). Correlation of an immunologically defined lipoprotein type with aortic intimal lipidosis in swine. *Exp. Mol. Pathol.* **27,** 249–261.

Rudel, L. L., Marzetta, C. A., and Johnson, F. L. (1986). Separation and analysis of lipoprotein by gel filtration. *Methods Enzymol.* **129,** 45–57.

Sata, R., Estrich, D. L., Wood, P. D. S., and Kinsell, L. W. (1970). Evaluation of gel chromatography for plasma lipoprotein fractionation. *J. Lipid Res.* **11,** 331–340.

Sata, T., Havel, R. J., and Jones, A. L. (1972). Characterization of subfractions of triglyceride-rich lipoproteins separated by gel chromatography from blood plasma of normolipemic and hyperlipemic humans. *J. Lipid Res.* **13,** 757–768.

Schonfeld, G., Patsch, W., Pfleger, B., Witztum, J. L., and Weidman, S. W. (1979). Lipolysis produces changes in the immunoreactivity and cell reactivity of very low density lipoproteins. *J. Clin. Invest.* **64,** 1288–1297.

Segrest, J. P., Jackson, R. L., Morrisett, J. D., and Gotto, A. M. (1974). A molecular theory of lipid–protein interactions in the plasma lipoproteins. *FEBS Lett.* **38,** 247–253.

Shen, B. W., Scanu, A. M., and Kézdy, F. J. (1977). Structure of human serum lipoproteins inferred from compositional analysis. *Proc. Natl. Acad. Sci. U.S.A.* **74,** 837–841.

Shen, M. M. S., Krauss, R. M., Lindgren, R. T., and Forte, T. M. (1981). Heterogeneity of serum low density lipoproteins in normal human subjects. *J. Lipid Res.* **22,** 236–244.

Soutar, A. K., and Myant, N. B. (1979). Plasma lipoproteins. *In* "Chemistry of Macromolecules IIB. Macromolecular Complexes." (R. E. Offord, ed.), pp. 55–119. University Park Press, Baltimore.

Teng, B., Thompson, G. R., Sniderman, A. D., Forte, T. M., Krauss, R. M., and Kwiterovich, P. O. (1983). Composition and distribution of low density lipoprotein fractions in hyperapobetalipoproteinemia, normolipidemia, and familial hypercholesterolemia. *Proc. Natl. Acad. Sci. U.S.A.* **80,** 6662–6666.

Teng, B., Sniderman, A., Krauss, R. M., Kwiterovich, P. O., Milne, R. W., and Marcel, Y. L. (1985). Modulation of apolipoprotein B antigenic determinants in human low density lipoprotein subclasses. *J. Biol. Chem.* **260,** 5067–5072.

Tsao, B. P., Curtiss, L. K., and Edgington, T. S. (1982). Immunochemical heterogeneity of human plasma apolipoprotein B. II. Expression of apolipoprotein B epitopes on native lipoproteins. *J. Biol. Chem.* **257,** 15222–15228.

Yamada, N., Shames, D. M., Stoudemire, J. B., and Havel, R. J. (1986). Metabolism of lipoproteins containing apoprotein B-100 in blood plasma of rabbits: heterogeneity related to the presence of apolipoprotein E. *Proc. Natl. Acad. Sci. U.S.A.* **83,** 3479–3483.

Young, S. G., Bertics, S. J., Curtiss, L. K., Casal, D. C., and Witztum, J. L. (1986). Monoclonal antibody MB 19 detects genetic polymorphism in human apolipoprotein B. *Proc. Natl. Acad. Sci. U.S.A.* **83,** 1101–1105.

Chapter 6

Apolipoprotein B (ApoB)

I. LDL ApoB: Definition and Some Properties

The protein component of human LDL, when first isolated and characterized, was called apolipoprotein B (apoB) because LDL was then commonly referred to as β-lipoprotein. The apoB isolated from LDL is insoluble in 4.2 *M* tetramethylurea and in aqueous buffers in the absence of strong detergents or denaturing agents. Attempts to determine its molecular weight have long been frustrated by its extreme insolubility and also by its susceptibility to cleavage by mechanical shear and by proteolytic enzymes of endogenous or bacterial origin. Estimates of the molecular weight of LDL apoB have ranged from less than 10,000 to more than 500,000. In the light of the most recent estimates deduced from cDNAs corresponding to the complete apoB message (see Section V,C), it is likely that all but the highest values obtained by standard methods for determining the molecular weight of a macromolecule were too low because the native protein was degraded during its isolation.

II. ApoB Species in Human Plasma

A. Distribution in Lipoprotein Fractions

When the protein components of plasma lipoproteins began to be studied, VLDL and chylomicrons were found to contain a protein with properties similar to those of LDL apoB and with immunoreactivity to antibodies to LDL. It was therefore concluded that there is a single molecular species of apoB common to LDL, VLDL, and chylomicrons. However, Kane *et al.* (1980) have isolated four different proteins from human LDL, VLDL, and chylomicrons by polyacrylamide gel electrophoresis of the tetramethylurea-insoluble protein obtained from these lipoproteins. These four proteins were assumed by Kane *et al.* to be different species of apoB and they were accordingly designated B-100, B-74, B-48, and B-26 on a centile system based on their apparent molecular weights determined by polyacrylamide gel electrophoresis.

Thus, in this system of nomenclature ''apoB'' is a generic term, to be used only when one is not referring to any particular apoB species. The molecular weight of the largest apoB (apoB-100) was estimated by Kane *et al.* to be about 550,000 and that of the smallest (apoB-26) to be about 144,500. LDL from most normal people contains apoB-100 as essentially its only protein. However, LDL isolated from some normal individuals also contains variable, but equimolar, amounts of apoB-74 and apoB-26, as well as apoB-100. VLDL from human plasma obtained after fasting contains apoB-100 as its only apoB species, while chylomicrons contain apoB-48 as their predominant apoB, with small amounts of apoB-100. Each chylomicron particle contains two molecules of apoB-48 [estimated from the mass of apoB per particle (Bhattacharya and Redgrave, 1981) and the molecular weight of apoB-48].

B. ApoB-74 and ApoB-26 Are Cleavage Products of ApoB-100

Since the summed amino acid composition of apoB-74 and apoB-26 was the same as that of apoB-100, Kane *et al.* (1980) suggested that the two lower-molecular-weight proteins were complementary fragments produced by proteolytic cleavage of the parent apoB-100 at a single site. This suggestion was confirmed by Marcel *et al.* (1982), who have described a panel of monoclonal anti-LDL antibodies that react with apoB-100. Some of these also react with either apoB-74 or apoB-26, but none reacts with both. This, of course, would be expected if apoB-74 and apoB-26 represent different segments of the apoB-100 molecule. In further confirmation of the suggestion of Kane *et al.* (1980), Cardin *et al.* (1984) have shown that LDL prepared from blood taken in the presence of EDTA and inhibitors of proteolytic enzymes never contains apoB-74 or apoB-26. They have also shown that when LDL containing only apoB-100 is submitted to limited digestion with kallikrein (a proteolytic enzyme of the intrinsic coagulation pathway present in plasma and tissue) the apoB-100 is cleaved to two fragments that comigrate with apoB-74 and apoB-26 on polyacrylamide gel electrophoresis.

The susceptibility of apoB-100 to cleavage into two fragments must depend upon the conformation adopted by the apoprotein molecule in the native LDL particle, since digestion of delipidated apoB-100 with kallikrein gives rise to multiple fragments of low molecular weight (Yamamoto *et al.*, 1985). Thus, although cleavage of apoB-100 at the B-26/B-74 site may have some bearing on the apoB domains that are exposed at the surface of a native LDL particle, the presence of apoB-74 and apoB-26 in LDL removed from the body is probably of little physiological significance. In any case, LDL particles in which apoB-100 has been cleaved to give the two complementary fragments bind normally to the LDL receptors on human fibroblasts (Yamamoto *et al.*, 1985).

C. Sites of Synthesis of ApoB-48 and ApoB-100

An apoB with MW similar to that of human apoB-48 has been isolated from rats, mice, and other nonhuman species (see Kane, 1983, and Hardman and Kane, 1986, for references). In most mammalian species, including man (Glickman *et al.*, 1986), apoB-48 is the only apoB made in the intestine and apoB-100 is the only apoB made in the liver. (The traces of apoB-100 usually found in chylomicrons isolated from plasma or intestinal lymph are probably due to contamination with other lipoproteins.) In rats and mice the liver secretes particles containing apoB-48 as well as those that contain apoB-100. As already noted (p. 105), apoB-48 and apoB-100 must be present in different particles, since the two kinds of apoB, when present together in a sample of plasma lipoprotein, can be completely separated by immunoaffinity columns that selectively bind particles containing apoB-100. Further evidence for segregation of apoB-48 and apoB-100 on different particles is provided by the observation of Elovson *et al.* (1981) that the fractional rate of turnover *in vivo* of apoB-48 in the VLDL of rat plasma is nearly twice that of VLDL apoB-100. Similar observations have been made by other workers (Sparks and Marsh, 1981). The two apoBs would be very unlikely to have different fractional rates of turnover if they were integral components of the same particles.

Although, as noted above, the intestine of adult human subjects makes apoB-48 as the only apoB species, Glickman *et al.* (1986) have shown that the predominant species made by the 11-week human fetal intestine is apoB-100. In the later stages of fetal life there is a progressive increase in the relative rate of synthesis of apoB-48, intestinal synthesis of apoB-100 ceasing altogether before birth. At no stage in prenatal development is apoB-48 synthesis detectable in the human liver.

A detailed investigation of the distribution and time course of expression of the apoB message during development has been carried out in rats by Demmer *et al.* (1986). In adult animals, apoB mRNA was found in significant amounts only in liver and intestine. The apoB mRNA level in fetal liver was relatively high at the eighteenth day of gestation and remained high until 4 days after birth, when there was a fall to lower levels. Very high levels of apoB mRNA were noted by Demmer *et al.* in the placenta and in the yolk-sac membrane of the 19-day fetus. Measurements of the incorporation of [^{35}S]methionine into apoB of fetal and adult tissues showed that fetal and adult rat intestine synthesize only apoB-48, fetal and adult liver make both species, while placenta and yolk sac make only apoB-100.

III. Composition and Structure of ApoB-100

A. Composition

For a review of composition, see Scott (1989). Human apoB-100 is a glycoprotein, estimates of its carbohydrate content ranging from 5 to 10%. The carbohydrate of LDL apoB is present as N-linked high-mannose oligosaccharide units and N-linked complex oligosaccharides (see p. 12), the structure of the high-mannose oligosaccharides resembling that of the high-mannose units present in integral membrane proteins (Vauhkonen *et al.*, 1985). The complete amino acid sequence of the mature apoB in LDL has been deduced from apoB cDNAs spanning the complete apoB-100 message (see Section VI). ApoB-100 is an exceptionally large protein. The precursor has 4563 amino acids, from which a signal sequence of 27 residues is removed co-translationally to give a mature protein with 4536 amino acids (calculated $M_r \simeq 513K$). The discrepancy between the calculated molecular weight and the value determined by gel electrophoresis ($\simeq$550K) may be explained by glycosylation of the translation product. The N-terminal (Protter *et al.*, 1986b) and C-terminal (Knott *et al.*, 1985) amino acid sequences of plasma apoB-100 are identical with those predicted from the complete cDNA sequence. This excludes the possibility that amino acids are removed posttranslationally from either end of the 4536-residue protein. The value of 513K, deduced from the amino acid sequence, is in reasonable agreement with the earlier estimate of 5.0×10^5 to 5.5×10^5 daltons of protein per LDL particle and thus confirms the conclusion that each LDL particle contains one molecule of apoB-100 monomer. This was implicit in the earlier observation of Milne and Marcel (1982) that one molecule of monoclonal anti-LDL antibody binds to one particle of LDL.

The complete amino acid sequence of apoB-100 shows several features of interest (see Knott *et al.*, 1986 and Yang *et al.*, 1986).

There are 20 potential sites for N-glycosylation, clustered mainly between amino acid residues 3000 and 3500. Direct sequence analysis indicates that at least four potential sites are not glycosylated and that there is a cluster of six glycosylated sites in a segment of apoB-100 (residues 3050–3450) close to the probable position of the LDL-receptor recognition site. Since glycosylated regions of a protein are, in general, hydrophilic, it is possible that the glycosylated sites in apoB-100 are distributed within domains of the protein that are exposed at the surface of an LDL particle.

There are 25 cysteine residues in apoB-100, 12 of which occur in the first 500 amino acids. Since several of these 12 residues are crosslinked by S-S bonds, Knott *et al.* (1986) suggest that the N-terminal portion of apoB-100 is more globular than the rest of the molecule. One of the 25 cysteine residues must be involved in the formation of an S-S bridge with apo(a) (see below). Two intramolecular thiolesters are formed in apoB-100 by a reaction between Cys 51 and Glu 54 and between Cys 3734 and Asp 3737 (Lee and Singh, 1988). Other cysteine residues form thiolesters with stearate and palmitate (Huang *et al.*, 1988a). Davis *et al.* (1984) have shown that rat hepatocytes, when incubated *in vitro,* secrete a phosphorylated form of apoB-48 in which the phosphate groups are linked to serine residues. In the presence of serum, the phosphate groups are removed. ApoB-100 secreted by rat hepatocytes is not phosphorylated.

Analysis of the amino acid sequence by the "dot matrix" method (Dayhoff *et al.*, 1983) shows the presence of at least 37 pairs of internally repeated sequences, each more than 70 residues long and with homologies ranging from 13 to 26%. DeLoof *et al.* (1987) have identified several 22-residue repeats predicted to form amphipathic helices. These are clustered mainly in two regions near the middle and C-terminal end of apoB-100 (see Fig. 6.1). Scattered throughout apoB-100 there are also proline-rich regions containing repeated short homologous sequences, each terminating in a proline residue. These have a high probability of forming amphipathic β-sheets[1], in which a proline residue occurs at each turn. Both amphipathic α-helices and amphipathic β-sheets interact with lipids.

Knott *et al.* (1986) and Boguski *et al.* (1986) found no homology between apoB-100 sequences and those of other apoproteins, apart from homology between the LDL-receptor-binding domain in apoE and an 11-residue sequence in apoB-100 (see below). However, DeLoof *et al.* (1987) have reported significant homology between the 22-residue repeats they found in apoB-100 and the 22-residue tandem repeats previously described in the lipid-binding domain of several other apoproteins (see Breslow, 1985 and Li *et al.*, 1988). The hydrophobic proline-rich repeats, on the other hand, appear to be unique to apoB.

B. Secondary Structure and Orientation in LDL

Prediction of the overall secondary structure from the amino acid sequence (Chou and Fasman, 1978) shows that apoB-100 has about 40% α-helical structure and about 20% β-sheet structure (Knott *et al.*, 1985, 1986; Yang *et al.*, 1986). The presence of a high proportion of β-sheet structure distinguishes apoB-100 from all other apoproteins, suggesting that the orientation of apoB-100 in its lipoprotein particle differs from that of other apoproteins. The "hydrophobicity profile" of apoB-100, obtained by plotting hydrophobicity as a function of amino acid residue number (Kyte and Doolittle, 1982), shows many regions in which hydrophobic sequences alternate with hydrophilic sequences

[1]A β-sheet is formed when a polypeptide chain folds back and forth upon itself, adjacent segments of the sheet being held together by hydrogen bonds.

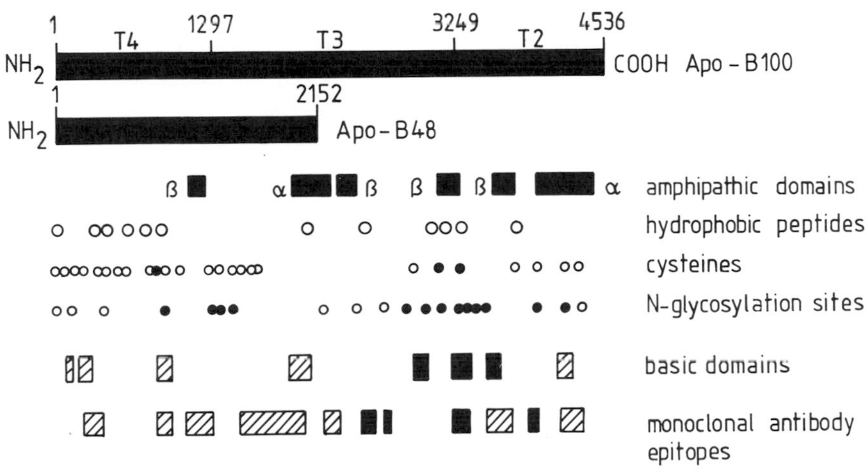

Figure 6.1. Structural aspects of apoB, showing apoB-100 and apoB-48 aligned. Numbers above apoB-100 show the amino acid residues at the T4/T3 and T3/T2 cleavage points. Amphipathic α-helices (α) and β-sheets (β) are shown as solid rectangles. Hydrophobic sequences are shown as open circles below the amphipathic domains. The two adjacent cysteine residues shown as solid circles form the disulfide bridge shown in Fig. 6.6. N-glycosylation sites shown as solid circles are known to be glycosylated. Three basic sequences in the region of apoB-100 thought to comprise the LDL-receptor-binding domain are shown as solid rectangles. Epitopes reacting with monoclonal antibodies that block binding of LDL to the LDL receptor are shown as solid rectangles. Epitopes that react with monoclonal antibodies with little or no effect on LDL binding are shown as hatched boxes. The positions of the 25 cysteine residues are taken from the complete sequence of apoB-100 published by Cladaras *et al.* (1986). (From Scott, 1989, with slight modification, with the permission of the author.)

(Knott *et al.*, 1985; Olofsson *et al.*, 1987). On the basis of these findings, Knott *et al.* (1985) suggested that apoB-100 is woven in and out of the LDL particle at irregular intervals. This arrangement is quite different from the manner in which other apoproteins associate with the lipid in their lipoprotein particles via a single lipid-binding domain (see Chapter 5). Interaction of apoB with LDL lipid at multiple sites could explain why apoB-100 does not exchange between lipoproteins and why it is the only protein component of VLDL that is retained in LDL.

Olofsson *et al.* (1987) have put forward a more detailed model for the orientation of apoB-100 in a lipoprotein particle. This takes into account both the predicted secondary structure of apoB-100 and the probable sequence of events by which it is synthesized in the rough ER of a hepatocyte and is then incorporated into a VLDL particle. In formulating their proposal, Olofsson *et al.* point out that most of the hydrophobic sequences in apoB-100 are only about half the 20-residue length required to span the phospholipid bilayer of a biological membrane such as the ER membrane. As shown in Fig. 6.2, Olofsson *et al.* propose

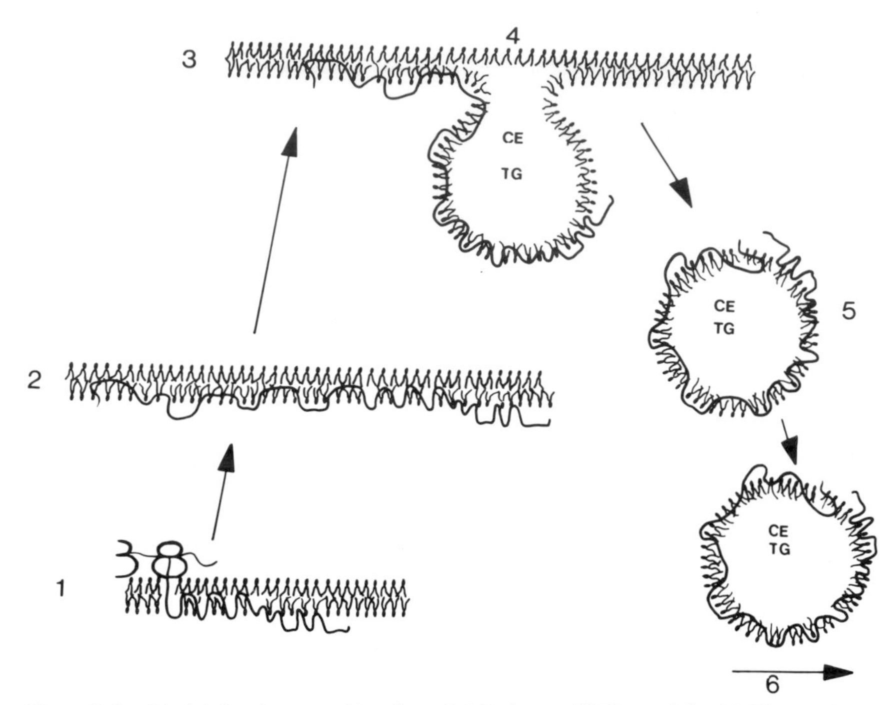

Figure 6.2. Model for the assembly of apoB-100 into a VLDL particle. (1) The apoB-100 molecule is synthesized on a polyribosome attached to the cytoplasmic surface of the phospholipid bilayer of the rough ER. The N-terminal signal sequence passes through the ER membrane and is cleaved co-translationally in the ER lumen. (2) The growing peptide chain weaves in and out of the phospholipid monolayer forming the luminal face of the ER membrane, the short hydrophobic sequences of apoB interacting with the nonpolar fatty acyl chains of the phospholipids. (3) The interwoven apoB molecule diffuses laterally to the junction of the rough and smooth ER, where it forms a VLDL particle by enclosing a droplet of triglyceride (TG) and cholesteryl ester (CE) synthesized by enzymes attached to the cytoplasmic face of the ER (4). (5) The VLDL particle is extruded into the ER lumen and is then transported to the Golgi apparatus for processing of apoB (6). (From Olofsson *et al.*, 1987, with the permission of the authors.)

that apoB-100 is synthesized on polyribosomes attached to the cytoplasmic surface of the rough ER, and that as the growing polypeptide chain enters the lumen of the ER the presence of short hydrophobic sequences causes the chain to weave in and out of the phospholipid monolayer forming the luminal face of the ER membrane. The completed apoB-100 molecule, while retaining its orientation in the ER membrane, then diffuses laterally to the site of synthesis of triglyceride and cholesteryl ester (probably the junction between the rough and smooth ER). Olofsson *et al.* suggest that these lipids form a droplet between the two leaflets of the bilaminar ER membrane and that the droplet, surrounded by the luminal leaflet of the ER into which apoB-100 is woven, is extruded into the ER lumen as a VLDL particle. The orientation of apoB in the phosopholipid shell of the VLDL particle would presumably be maintained throughout the conversion of VLDL into LDL. It should be noted that, in this model, no structural role is assigned to the potential amphipathic helices in apoB-100.

C. Proteolytic Fragments of ApoB and the Mapping of Epitopes

Defined fragments of apoB have played an essential role in the mapping of apoB epitopes. Cardin *et al.* (1984) have shown that digestion of LDL with thrombin cleaves apoB-100 into three fragments, which they call T4, T3, and T2. Knott *et al.* (1986) showed that these fragments are derived from the parent molecule in the order shown in Fig. 6.3 and that the N-terminal peptide (T4) is identical with apoB-26 of Kane *et al.* (1980). When their positions within the apoB molecule had been established, the three fragments could be used to determine the approximate positions of epitopes detected with monoclonal antibodies to LDL or delipidated apoB (see Discussion in Marcel *et al.*, 1987). More detailed linear maps of apoB epitopes have been constructed by observing the ability of monoclonal antibodies to react with smaller fragments of apoB. These are generated by proteolytic cleavage, by expressing short sequences of the *apoB* gene as fusion proteins, or by chemical synthesis (Krul *et al.*, 1988; Marcel *et al.*, 1987).

The positions of some monoclonal-antibody-binding sites in apoB-100, determined by these methods, are shown in Fig. 6.3. As discussed below, epitope maps have been used in the localization of functional domains in apoB and in studies of the structural relationship between apoB-48 and apoB-100. It should also be noted that the linear order of epitopes in apoB may be compared with

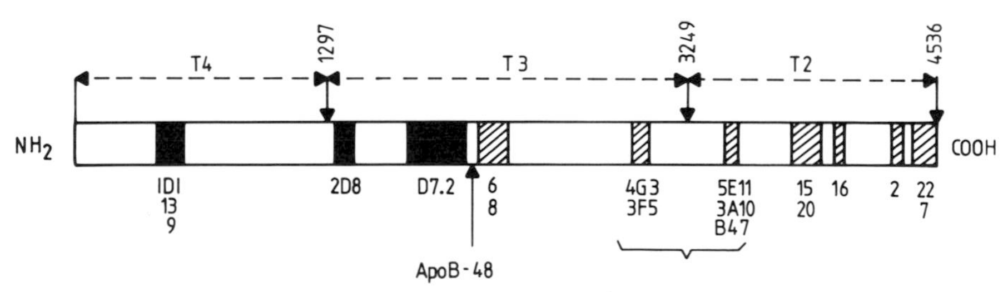

Figure 6.3. The three thrombolytic fragments of apoB-100 (T4, T3, and T2), showing the positions of epitopes detected with 18 monoclonal anti-apoB antibodies. Numbers above the broken lines are amino acid residue numbers. Solid boxes denote epitopes for antibodies that bind to apoB-100 and apoB-48. Hatched boxes show epitopes for antibodies that bind to apoB-100 but not to apoB-48. Antibodies that inhibit the binding of LDL to the LDL receptor on fibroblasts are included within the bracket. The arrow between D7.2 and 6/8 shows the position of the C-terminus of apoB-48. The positions of other apoB epitopes are shown in Fig. 10 of Marcel *et al.* (1987). (From Powell *et al.*, 1987, with the permission of the authors.)

their relative positions in native LDL or relipidated apoB-100, as determined by competition between monoclonal antibodies for binding to apoB. This provides information about the conformation of the apoB-100 molecule under physiological conditions.

IV. Lp(a): A Lipoprotein Containing ApoB-100

A. Background, Clinical Significance, and Genetics

Lipoprotein(a) [Lp(a)] was first identified by Berg (1963) as a genetic variant of LDL, detected with a rabbit antiserum, in about 30% of a Norwegian population. Lp(a) is now thought to be present in the plasma of virtually all human subjects (except those with abetalipoproteinemia), though the concentration varies between different individuals from <1 to >100 mg/100 ml of plasma. The subsequent demonstration that a high plasma concentration of Lp(a) is a positive, independent risk factor for coronary atherosclerosis in Western populations (see Kostner, 1976, and Armstrong *et al.*, 1986) focused attention on the chemical composition and structure of Lp(a), especially with regard to the nature of the Lp(a) antigenic determinant, now called apolipoprotein(a) or apo(a). As noted below, a Lp(a) particle consists essentially of a normal LDL particle in which a molecule of apo(a) is linked covalently to the apoB-100 molecule. It was the apo(a) in Lp(a) that enabled Berg to distinguish Lp(a) immunochemically from LDL in plasma.

Berg (1963) concluded that the presence of Lp(a) in plasma is inherited as an autosomal dominant trait. However, the inheritance of Lp(a) is now known to be more complex than this. Family studies of Utermann *et al.* (1987) have shown that apo(a) is highly polymorphic with respect to molecular weight and that the different apo(a) isoforms are encoded in at least seven alleles, including a null allele, at a single locus. Utermann *et al.* noted a significant inverse correlation between apo(a) molecular weight and Lp(a) concentration, suggesting that each allele determines both the molecular weight of the apo(a) phenotype and its rate of synthesis or secretion, or that the different isoforms are catabolized at different rates. Whatever the mode of inheritance of Lp(a), there can be little doubt that variation in plasma Lp(a) concentration contributes significantly to the genetic component of risk for coronary artery disease, at least in Western populations.

B. Composition and Physical Properties of Lp(a)

Table 6.1 shows the composition and some physical properties of Lp(a) compared with those of LDL. The two lipoproteins resemble each other in lipid composition and in having apoB-100 as their major apoprotein. However, owing

Table 6.1
Composition and Properties of Lp(a) and LDL[a]

Component	Lp(a)	LDL	Property	Lp(a)	LDL
Lipoprotein			Hydrated density (g/ml)	1.050–1.120	1.019–1.063
(% by weight)					
Protein	26	22			
Carbohydrates	8	2			
Lipids	66	76	Diameter (mm)	25	21
Protein[b]					
(% by weight)					
ApoB-100	50	100	MW $\times 10^{-6}$	5.5	2.4
Apo(a)	50	0			
Carbohydrates			Electrophoretic mobility in agarose	pre-β_1	β
(μg/mg protein)					
Hexose	150	56			
Hexose-NH_2	85	30	Isoelectric point	4.5	5.5
Sialic acid	66	11			
Total	301	97			
Lipids					
(% by weight)					
Phospholipids	30	31			
Free cholesterol	17	15			
Cholesteryl ester	48	46			
Triglycerides	5	8			

[a]Modified from Kostner (1983).
[b]Values based on the assumption that each Lp(a) particle contains one molecule of apoB-100 and one of apo(a) of approximately equal MW. Note that although Lp(a) was first detected in the LDL fraction of plasma, its density overlaps with that of HDL (1.063–1.210 g/ml).

to the presence of apo(a), Lp(a) has a higher protein content, a larger particle diameter, a larger molecular weight, and a higher density than LDL. The carbohydrate content of Lp(a) is unusually high (Enholm *et al.*, 1972). In particular, Lp(a) has six times as much sialic acid per milligram of protein as LDL. Because of its larger particle size, Lp(a) migrates more slowly than LDL on polyacrylamide-gel electrophoresis. However, on agarose gel, Lp(a) has pre-β mobility owing to the strong negative charge contributed by sialic acid residues.

The protein component of Lp(a) has been characterized by Utermann and Weber (1983) and by Gaubatz *et al.* (1983). When purified Lp(a) is submitted to SDS–polyacrylamide-gel electrophoresis in the absence of reducing agent, a single protein band is observed. In the presence of mercaptoethanol, the protein is split into two components, one identical with apoB-100 and the other a glycoprotein that reacts with anti-Lp(a) antiserum but not with antibody to apoB. Thus, the protein of Lp(a) consists of apo(a) (a glycoprotein) attached covalently in disulfide linkage to apoB-100. As noted above, apo(a) exhibits a high degree of polymorphism, the apparent molecular weights of the apo(a)s obtained from different subjects ranging from ~400,000 to ≃700,000 (Fless *et al.*, 1984; Utermann *et al.*, 1987). In confirmation of the conclusion that Lp(a) differs from LDL only by the addition of apo(a), Armstrong *et al.* (1985) have shown that chemical reduction of Lp(a) yields a lipoprotein indistinguishable from LDL, together with lipid-free apo(a).

C. Metabolism of Lp(a)

Studies of the metabolism of radioiodinated lipoproteins in normal human subjects have shown that Lp(a) is not derived from plasma VLDL or LDL (Krempler *et al.*, (1980). This raises the possibility that Lp(a) is secreted, as such, directly into the circulation. In a single FH homozygote investigated by Krempler *et al.* (1982), the fractional catabolic rate (FCR) of Lp(a) was within the normal range, suggesting that Lp(a) is not catabolized efficiently by the LDL-receptor pathway. In agreement with this, in normal subjects the FCR of Lp(a) is about 30% lower than that of LDL (Krempler *et al.*, 1980).

D. Amino Acid Sequence of Apo(a)

McLean *et al.* (1987) have deduced the amino acid sequence of apo(a) from a series of overlapping cDNAs complementary to the complete apo(a) mRNA. The sequence deduced from this composite cDNA has 4548 amino acids including a signal sequence of 19 residues, giving a mature protein with 4529 residues. The amino acid sequence of apo(a) has a high degree of homology with plasminogen, the precursor of fibrinolysin (plasmin), a trypsinlike enzyme that digests fibrin. Plasminogen consists of a short N-terminal tail followed by five homologous repeats in tandem, each with about 100 residues, and a C-terminal protease domain (Fig. 6.4). Each repeat is folded by three S-S bonds into a structure resembling a Danish cake called a kringle (see Lerch *et al.*, 1980). The protease in plasminogen is inactive until it has been cleaved by plasminogen activators at a single arginine to give plasmin, the proteolytically active enzyme. Kringles 1 and 4 have binding sites for fibrin.

Figure 6.4 shows the cDNAs and predicted protein domains of plasminogen and apo(a). The mature apo(a) studied by McLean *et al.* (1987) has 37 copies of kringle 4, followed by a single copy of kringle 5 and a protease domain homologous to that in plasminogen. Kringle 36 has a free cysteine residue that could be involved in S-S linkage to apoB-100. In apo(a) the arginine residue at the N-terminus of the protease domain is replaced by serine, so that apo(a) is not a substrate for plasminogen activators and cannot, therefore, generate a fibrinolytic enzyme *in vivo*.

The cysteine residue in apoB-100 that is involved in disulfide linkage to apo(a) has not yet been identified. Armstrong *et al.* (1986) have shown that maximal

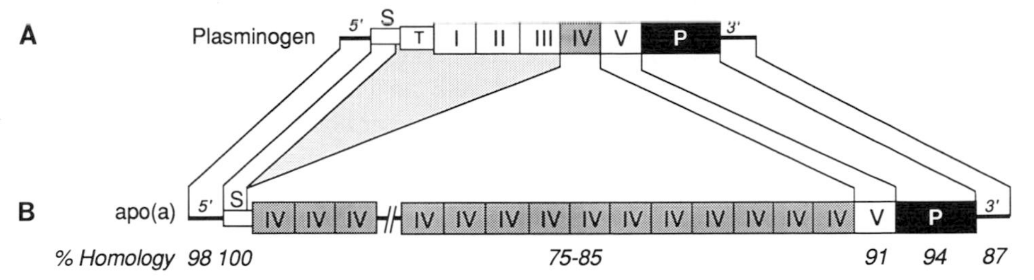

Figure 6.4. Comparison between plasminogen and apo(a) cDNAs, showing protein domains in both proteins. (A) Plasminogen cDNA; (B) apo(a) cDNA. The 5′ and 3′ noncoding sequences are shown as thick lines at each end of the cDNA. Coding sequences are shown for the signal sequence (S), the N-terminal tail of plasminogen (T), the 5 kringles (I–V), and the C-terminal protease (P). The broken line of the apo(a) sequence represents 13 kringle 4s. The shaded triangle shows the plasminogen domains not represented in apo(A) (T and kringles I–III). Numbers below apo(a) show the percentage homology between plasminogen and apo(a) nucleotide sequences. Note the 98% homology of the 5′-noncoding sequences and the 100% homology of the first 16 codons of the signal sequences. This suggests that in this region the two genes exchanged sequences by gene conversion (see Lewin, 1987, Chapter 19) about 7 million years ago (McLean *et al.*, 1987). (From McLean *et al.*, 1987. Reprinted by permission from the authors and from *Nature*, Vol. 330, pp. 132–137. Copyright © 1987 Macmillan Magazines Ltd.)

degradation of Lp(a) via the LDL-receptor pathway in skin fibroblasts *in vitro* is lower than that of LDL. This suggests that the site at which apo(a) is linked to apoB-100 is sufficiently close to the receptor-recognition domain to affect its binding to the LDL receptor. The observations of Krempler *et al.* (1982) on the metabolism of Lp(a) *in vivo* are in agreement with this (see above).

The molecular basis of the heterogeneity of apo(a) has not been determined. One possibility is that the different isoforms have different numbers of carbohydrate residues per molecule of protein. It is also possible that the alleles at the apo(a) locus have different numbers of exons encoding kringle 4. (See, for example, Hixson *et al.*, 1989.)

E. Evolution and Functional Significance of Apo(a)

A comparison of the nucleotide sequences in apo(a) and plasminogen mRNAs indicates that the two genes diverged from a common ancestral gene about 40 million years ago (McLean *et al.*, 1987). The apo(a) gene probably evolved from the common ancestor by repeated duplication of the exon encoding kringle 4 and loss of the exons encoding the N-terminal tail and kringles 1–3 of plasminogen.

In view of the strong association between plasma Lp(a) concentration and coronary heart disease, one may well ask what biologically useful function led to the evolution of this remarkable protein. Kringle 4 in plasminogen has weak binding affinity for fibrin (Lerch *et al.*, 1980). Possibly Lp(a), by virtue of the kringle 4s in apo(a), binds *in vivo* to newly formed fibrin at the site of an injury to the arterial wall and thus provides the regenerating artery with an immediate supply of cholesterol. Harpel *et al.* (1989) have shown that Lp(a) competes with plasminogen for binding to lysine residues in fibrin or fibrinogen. This raises the possibility that Lp(a) inhibits the degradation of arterial thrombi. The atherogenicity of Lp(a) may have been too weak to have exerted negative selection pressure at a time when the plasma LDL concentration was far below that prevalent in modern Western populations. There is ample evidence that many potentially atherogenic agents do not give rise to atherosclerosis unless the plasma LDL-cholesterol concentration exceeds a threshold of less than about 200 mg/100 ml (see Myant, 1981, Chapter 13).

V. The LDL-Receptor-Binding Site on ApoB

A. Relevance of the ApoE Receptor-Binding Domain

ApoE, an apoprotein present in several lipoproteins other than LDL, shares with apoB-100 the ability to bind specifically to the LDL receptor when each apoprotein is suitably oriented in a lipoprotein particle. Indeed, Mahley and co-workers (see Mahley, 1979) have isolated a lipoprotein (HDL_c) from the plasma of cholesterol-fed animals that contains apoE as its sole apoprotein and that binds to the LDL receptor with an affinity several times that of LDL. In view of the similarity between the receptor-binding properties of apoB-100 and apoE, it would be reasonable to expect the amino acid sequence of the LDL-receptor-binding site in apoB-100 to be similar, or closely homologous, to that in apoE. The complete amino acid sequence of apoE, a protein with only 299 amino acids, has been determined directly, and the receptor recognition site has been localized to a small segment near the midpoint of the peptide chain. Since the amino acid sequence of this segment has provided important clues to the nature of the binding domain in apoB-100, a brief discussion of this domain in apoE will be helpful at this point.

Information on the position of the LDL-receptor-binding site in apoE was derived initially from the study of mutant forms of apoE in man. ApoE in the human population exists in three major isoforms (E2, E3, and E4) defined in terms of their isoelectric points. These three isoforms differ from each other in series by unit charge, apoE2 being the most electronegative and apoE4 the most electropositive. Family studies have shown that the genes coding for the three major isoforms are allelic. Hence, the three alleles ($\epsilon 2$, $\epsilon 3$, and $\epsilon 4$) give rise to six phenotypes, three of which are homozygous (E2/E2, E3/E3, and E4/E4). Determination of the amino acid sequences of apoE from individuals homozygous for each of the three alleles has shown that the differences in isoelectric point are due to interchanges between charged and uncharged amino acid residues at single positions in the peptide (see Mahley *et al.*, 1984; and Table 6.2).

Identification of the three major isoforms of apoE was originally based on their separation by isoelectric focusing in polyacrylamide gel. However, this does not distinguish between isoforms that have the same net charge but different amino acid sequences. It is now known that apoE2 is heterogeneous and that apoE3 can exist in at least two forms, one common and the other probably uncommon (McLean *et al.*, 1984). Since the more common of the two ϵ3 alleles appears to be by far the commonest ϵ allele in the human population, McLean *et al.* (1984) designate the common ϵ3 allele ϵ3(0) and the corresponding E3 phenotype E3(0).

Table 6.2 shows the known isoforms of apoE that are of interest in relation to the receptor-binding site, with the positions at which amino acids in apoE3(0) (the wild-type isoform) are substituted. All isoforms other than apoE3(0) are designated by their amino acid difference from the wild type. Thus, in apoE4 arginine replaces cysteine at position 112, while in the less common of the two apoE3 phenotypes arginine replaces cysteine at position 112 and cysteine replaces arginine at position 142. In the latter example the double substitution results in no change in net charge. Note that in all three forms of apoE2 a basic amino acid is replaced by a neutral one, the substitutions occurring at positions 145, 146, and 158.

As shown in Table 6.2, some apoE isoforms, when incorporated into phospholipid vesicles, have a markedly reduced ability to bind to LDL receptors on fibroblasts in culture. Binding is diminished by substitution of a neutral amino

Table 6.2
Genetic Isoforms of ApoE[a,b,c]

ApoE isoform	Amino acid residue number					Charge relative to E3(0)	Relative binding ability
	112	142	145	146	158		
E4 (Cys$_{112}$ → Arg)	Arg	Arg	Arg	Lys	Arg	+1	100
E3(0)	Cys	Arg	Arg	Lys	Arg	0	100
E3 (Cys$_{112}$ → Arg, Arg$_{142}$ → Cys)	Arg	Cys	Arg	Lys	Arg	0	<20
E2 (Arg$_{158}$ → Cys)	Cys	Arg	Arg	Lys	Cys	−1	<2
E2 (Arg$_{145}$ → Cys)	Cys	Arg	Cys	Lys	Arg	−1	45
E2 (Lys$_{146}$ → Gln)	Cys	Arg	Arg	Gln	Arg	−1	40

[a]Shown for each isoform are the positions of amino acid substitutions, the net charge relative to that on apoE3, and their relative abilities to bind to the LDL receptor when incorporated into phospholipid vesicles.

[b]ApoE3(0) is taken to be the normal, wild-type, isoform. Relative binding abilities were determined by comparing the abilities of apoE isoforms incorporated into phospholipid vesicles to compete with LDL for LDL receptors. Substituted residues are underlined.

[c]Modified from Mahley *et al.*, 1984.

acid for arginine at positions 142, 146, or 158 and for lysine at position 146. On the other hand, interchange between arginine and cysteine at position 112 has no effect on binding. These findings suggest that the ability of apoE to bind specifically to the LDL receptor depends upon the integrity of a relatively short segment, rich in positively charged amino acids, that includes residues 142 to 158. The requirement for a basic amino acid at position 158 has been confirmed by the observation that the diminished binding ability of apoE2 ($Arg_{158} \rightarrow Cys$) can be partially restored by converting the cysteine at 158 into a positively charged analog of lysine (Weisgraber *et al.*, 1982). A requirement for a cluster of positively charged amino acids in the receptor-binding site is consistent with the observation that binding of apoE by the receptor is abolished by chemical modification of a limited number of lysyl and arginyl residues in the apoprotein (Weisgraber *et al.*, 1982). It is also worth noting that the amino acid sequence from 142 through 158 has a high proportion of basic residues (see Fig. 6.5). A positively charged domain in apoE could provide the basis for specific interaction with the clusters of negatively charged amino acids in the ligand-binding domains of the LDL receptor (Brown and Goldstein, 1986).

The above evidence for the position of the apoE binding site, based on the study of receptor-defective mutants, has been strengthened by experimental observations on isolated fragments of the protein. Innerarity *et al.* (1983), using proteolytic fragments of apoE with known amino acid sequences, have shown that the ability of apoE to bind to the LDL receptor is restricted to a segment of the molecule extending from residue 126 to residue 191. Weisgraber *et al.* (1983) have also investigated the binding of proteolytic and synthetic fragments of apoE to monoclonal anti-apoE antibodies. One of these antibodies that blocked the binding of intact apoE to the receptor reacted specifically with a segment of apoE comprising residues 139–144. The ability of a monoclonal anti-apoE antibody to inhibit binding of apoE to the receptor presumably depends

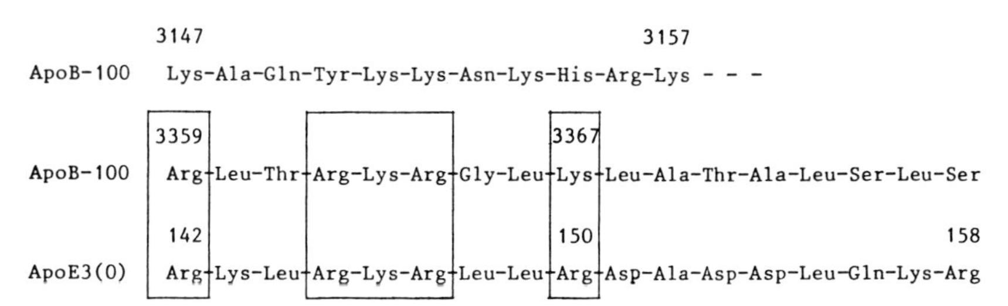

Figure 6.5. Two apoB-100 sequences containing a high proportion of basic amino acids (residues 3147–3157 and residues 3359–3367). ApoB-100 residues 3359–3375 are aligned with the sequence in apoE3(0) that is thought to contain the recognition site for the LDL receptor. Positions at which there is a basic amino acid in apoB-100 (3359–3367) and apoE3(0) (142–150) are enclosed in boxes.

upon interaction of the antibody with an apoE epitope that is close to or within the receptor-binding site. Hence, the observation of Weisgraber *et al.* (1983) is consistent with all the other evidence indicating that the binding domain on apoE includes residue 144. Residue 158 apparently does not interact directly with the LDL receptor, but is necessary for maintaining the binding domain in the correct conformation. Innerarity *et al.* (1987b) concluded that the binding domain extends from residue 142 to 150; this has now been confirmed by the studies of Lalazar *et al.* (1988) on the receptor-binding behavior of synthetic apoE mutants with single amino acid substitutions.

B. The ApoB Receptor-Binding Domain

1. Evidence for a Binding Domain Near the T3/T2 Junction

Inspection of the complete amino acid sequence of apoB-100 shows that there are two regions containing high proportions of basic residues (3147–3157 and 3359–3367) on either side of the T3/T2 junction (Knott *et al.*, 1986; Yang *et al.*, 1986). The sequence nearer the C-terminus of apoB-100 has a significant degree of homology with the receptor-binding domains of apoE. Figure 6.5 shows the two sequences of apoB aligned with the apoE sequence thought to include the receptor-binding domain (residues 142–150). Six of the nine residues in the apoE sequence are basic and at five of these six positions there is a basic residue in the corresponding apoB sequence. The consensus for the apoE sequence and the 3359–3367 sequence is Arg-X-X-Arg-Lys-Arg-X-X-Arg/Lys. In support of the suggestion that the 3359–3367 sequence is a component of the receptor-binding domain, Yang *et al.* (1986) have synthesized a short peptide that includes this sequence and have shown that, when incorporated into a modified lipoprotein, it binds with high affinity to the LDL receptor.

Figure 6.6 shows the main features of the binding domain postulated by Milne *et al.* (1989). Between the two basic amino acid sequences there is a disulfide bridge linking the two cysteines at positions 3167 and 3297. The two cysteines are separated by a proline-rich cluster of amino acids with high probability of forming amphipathic β-sheets. Several monoclonal antibodies, whose epitopes have been mapped to regions close to the T3/T2 junction, inhibit the binding of LDL to the LDL receptor. In the model shown in Fig. 6.6 these regions are brought into apposition by folding of the apoB amino acid chain in native LDL. Since LDL containing the apoB variant Arg_{3500}→Gln does not bind to the LDL receptor (see Section VIII), Milne *et al.* postulate that residue 3500 is involved, either directly in the interaction of apoB-100 with the receptor, or in maintaining the receptor-binding domain in the correct conformation.

In the model put forward by Milne *et al.* (1989) there is only one binding domain in an apoB-100 molecule present in a native LDL particle. This view was based mainly on extensive studies of the behavior of monoclonal anti-apoB

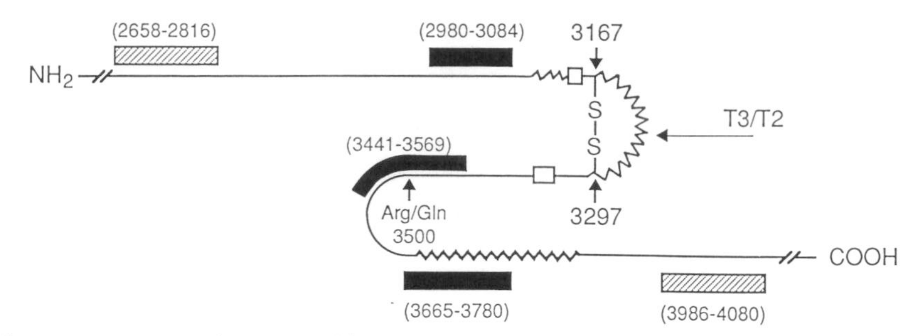

Figure 6.6. Suggested structure of the LDL-receptor-binding domain of human apoB-100. The model shows the two regions on either side of the T3/T2 junction (horizontal arrow). Open boxes are the basic amino acid sequences 3147–3157 and 3359–3367 brought together by a proline-rich segment with potential for forming amphipathic β-sheets (zigzag line) and a disulfide bridge (present only in man) between cysteines 3167 and 3297. Black rectangles show regions containing epitopes of monoclonal antibodies that totally block the binding of LDL to the LDL receptor. Hatched boxes show regions reacting with monoclonal antibodies that block LDL binding only partially. The position of residue 3500 is shown by a vertical arrow. Note that the second loop is introduced into the model to take into account the functional proximity of epitopes in the 3665–3780 segment to the other elements of the receptor-binding domain. (From Scott, 1989, with the permission of the author.)

antibodies (Marcel *et al.*, 1982; Marcel *et al.*, 1987; Krul *et al.*, 1988; Milne *et al.*, 1989). These studies showed that antibodies that react with epitopes close to the T3/T2 junction block the binding of LDL to its receptor, whereas those reacting with epitopes in other regions of apoB-100 have no effect, or only a partial effect, on LDL binding. Furthermore, antibodies reacting with epitopes close to the postulated receptor-binding domain do not bind to LDL particles that are already bound to the receptor. On the other hand, antibodies reacting with regions on either side of this domain bind to receptor-bound LDL with a 1 : 1 molar ratio.

Hospattankar *et al.* (1986) have used computer analysis of the apoB-100 amino acid sequence to identify a total of 12 potential LDL-receptor domains enriched with basic amino acids. These domains are distributed throughout the length of the apoB-100 molecule and have the following consensus sequence.

[Lys/Arg/His]-X-X-X[Lys/His]-X-X[Lys/Arg/His]----[Lys/Arg]-[Lys/Arg/His]

The broken line represents a variable loop of 8–11 residues. The basic amino acids in this sequence are complementary to acidic amino acids in the consensus sequence of the LDL-receptor-binding domains identified by Südhof *et al.* (1985). However, the suggestion that there are receptor-binding domains throughout apoB-100 is contrary to the above evidence derived from studies with monoclonal antibodies. It is also inconsistent with the failure of apoB-48 (Hui *et al.*, 1984) and apoB-37 (Young *et al.*, 1987c) to mediate the binding of lipoproteins to the LDL receptor. In any case, the functional significance of potential binding domains, deduced from their amino acid sequences, needs to be tested experimentally. A possible approach is to synthesize specific peptides containing the derived sequences, either chemically or by expression of DNA sequences cloned in expression vectors, and then to incorporate the peptides into LDL-like particles by recombination with lipids (see Chapman *et al.*, 1986). The reconsituted particles may then be tested for their ability to compete with native LDL for binding to cultured fibroblasts or adrenal-cortex membranes. However, it should be noted that apoB fragments in a reconsituted particle may expose amino acid sequences that are not exposed at the surface of a particle of native LDL (see Discussion in Corsini *et al.*, 1987).

2. Evidence from Comparisons between Species

Amino acid sequences in functionally important regions of a protein are usually conserved by natural selection and they therefore tend to be closely similar in the protein from different species. In agreement with this, the amino acid sequences around the T3/T2 junction in apoB-100 from human subjects and from pigs (Maeda *et al.*, 1988; Ebert *et al.*, 1988), rabbits, rats, mice, Syrian hamsters, and chicks (Milne *et al.*, 1989) show a high degree of homology, especially with respect to the proline-rich domains. However, the disulfide bridge between residues 3167 and 3297 is not present in species other than man. Ebert *et al.* (1988) suggest that, in the absence of the S-S bridge, the two basic amino acid sequences in the postulated receptor-binding domain of apoB-100 are brought together by the conformation of the proline-rich segment between these two sequences.

Ebert *et al.* (1988) have identified another significant difference between the binding domains in pig and human apoB-100. In pig apoB-100 the arginine present at position 3359 in human apoB (see Fig. 6.5) is replaced by serine. Yet, despite the loss of a positive charge in this region of the binding domain, the affinity of pig LDL for the human LDL receptor is more than five times that of human LDL. It is worth noting that replacement of arginine by cysteine at the equivalent position in the apoE receptor-binding domain (residue 142) markedly diminishes the affinity of apoE for LDL receptors (see Table 6.2). This raises the possibility that the effect of the $Arg_{142} \rightarrow Cys$ substitution on the binding affinity of apoE for the receptor is due not to the loss of a basic residue, but to the introduction of a cysteine.

In conclusion, it seems very probable that there is a single receptor-binding domain in a region between residues 3000 and 3800 of apoB-100. However, we are still some way from understanding, in terms of three-dimensional structure, how the binding domains of apoB-100 and the LDL receptor fit into each other.

VI. The ApoB Gene and Its mRNA

A. The Background

The apoB gene was the last of the major human apoprotein genes to be cloned. Complementary DNA clones for apoproteins AI, AII, CI, CII, CIII, and E were all identified before the first report of an apoB cDNA clone. A cDNA clone for each of the above six apoproteins was selected from a liver library by hybridization with a radioactive synthetic oligonucleotide whose sequence was chosen to give the smallest number of coding ambiguities. (These ambiguities arise because for most amino acids there is more than one codon.) This choice was possible because the complete amino acid sequences of all six apoproteins had already been determined directly. In the case of apoB, however, chemical studies had been hampered by the unusual properties already referred to. As a consequence of this, elucidation of the complete amino acid sequence of apoB-100 had to await the isolation of cDNA clones from which it could be deduced. In this respect, apoB-100 resembles HMG-CoA reductase and the LDL receptor. In both cases, the amino acid sequence was deduced from the nucleotide sequences of their mRNAs, rather than by direct sequencing of their peptide chains.

B. Isolation of Partial cDNA Clones

Lusis *et al.* (1985) isolated an apoB cDNA clone from a rat-liver cDNA library and showed that it encoded a sequence of 74 amino acids. Within a year of the report by Lusis *et al.* the isolation of partial cDNA clones for human apoB-100 had been achieved in several laboratories. Three of these clones were identified by screening bacterial colonies with radioactive synthetic oligonucleotides based on the amino acid sequences of proteolytic fragments of apoB-100 (Knott *et al.*, 1985; Deeb *et al.*, 1985; Protter *et al.*, 1986b). The others were identified by immunoselection from cDNA expression libraries (Shoulders *et al.*, 1985; Carlsson *et al.*, 1985; Wei *et al.*, 1985; Law *et al.*, 1985). In this procedure, colonies expressing amino acid sequences present in the protein of interest are identified by their ability to react *in situ* with a specific antibody (in the present case, an antibody to apoB or to LDL).

Radioactive apoB cDNAs were shown to hybridize with a liver mRNA at least 12 kb long. A message of this length should code for a protein not less than 400K. It was clear, therefore, that previous estimates of 400–550-K for the M_r of apoB-100 (Kane *et al.*, 1980) were not, as some had claimed, artifacts due to self-association of smaller monomeric units. The cDNA clone isolated by Knott *et al.* (1985) spanned a total of 5 kb at the 3′ end of the apoB-100 message. This included the two sequences rich in basic amino acids and thus provided the first clues to the site and structure of the LDL-receptor-binding domain in apoB (see the previous section).

The clone isolated by Protter *et al.* (1986b) spanned part of the 5′-untranslated region of the apoB-100 message and the nucleotide sequence encoding a 27-residue signal peptide plus the first 264 amino acids of the mature protein. An antibody raised against a synthetic 17-residue peptide encoded by a segment of this clone reacted with apoB-100, apoB-48, and apoB-26, but not with apoB-74. This showed that apoB-48 and apoB-26 have amino acid sequences in common with those of the N-terminal, rather than the C-terminal, end of apoB-100. In agreement with this, Wei *et al.* (1985) showed that their cDNA clone, which spanned 2.8 kb at the 3′ end of the apoB-100 message, encoded a protein that failed to react with a monoclonal antibody previously shown to react with apoB-48. These two observations were consistent with other evidence for the localization of apoB-26 and apoB-48 to the N-terminal portion of apoB-100 (see the previous section).

By hybridizing a radioactive apoB-100 cDNA clone to the chromosomes of human cells in metaphase, Knott *et al.* (1985) were able to show that the human apoB gene is near the tip of the short arm of chromosome 2. In agreement with this, Law *et al.* (1985), using an apoB cDNA probe, localized the apoB gene to chromosome 2 in human–mouse somatic cell hybrids.

Shoulders *et al.* (1985) determined the nucleotide sequences of three overlapping apoB-100 cDNA clones isolated from the same liver cDNA library. One clone contained an *Eco*RI restriction site (GAA TTC) which was replaced by the sequence AAA TTC in the other two. Since the library was constructed from a single human liver, Shoulders *et al.* concluded that the apoB gene is polymorphic at this site and that the individual from whom the liver was obtained was heterozygous for the polymorphism. The *Eco*RI polymorphism, which was subsequently reported by Priestley *et al.* (1985), is of potential clinical relevance because the substitution of A for G in the sequence GAA TTC change the amino acid encoded by the first triplet from glutamic acid (GAA) to lysine (AAA). The probable relationship between this polymorphism and the LDL Ag system has already been mentioned in Chapter 5. Other aspects of the *Eco*RI polymorphism are considered below.

C. The Complete ApoB-100 Message

Overlapping cDNA clones spanning the whole of the human apoB-100 message have now been isolated by Knott *et al.* (1986), Yang *et al.* (1986), Law *et al.*, (1986a) and Cladaras *et al.* (1986). (For the complete cDNA and amino acid sequences, see Chen *et al.*, 1986.) The human apoB gene has also been isolated as a series of overlapping genomic clones by Blackhart *et al.* (1986). The apoB-100 message is 14.1 kb long, the longest mRNA reported so far. It contains a 5′-untranslated region of 128 nucleotides, a 3′-untranslated region of 301 nucleotides, and a region coding for 4563 amino acids. As noted in Section III above, the N-terminal 27 of these amino acids are removed before the mature protein is secreted in VLDL.

D. The ApoB gene

The apoB gene is about 43 kb long and is divided into 29 exons separated by 28 introns (Fig. 6.7). The distribution of introns is asymmetrical, 25 of them occurring in the 5′ half of the gene. The first intron interrupts the codon for the first amino acid of the mature protein and thus separates the coding region for the signal peptide from the remainder of the gene. The lengths of most of the exons (150–250 nucleotides) are similar to those of exons in a typical mammalian gene. However, the human apoB gene contains one exon (exon 26, 7.6 kb) that is far longer than any other so far reported for a mammalian gene. The apoB gene has at least six Alu repeat sequences (see Chapter 9, Section III), all within introns, a TATA box 29 nucleotides 5′ to the startpoint for transcription, and a CAAT box 31 nucleotides 5′ to the TATA box. It also has two repeats of the sequence CCGCCC in the segment of the gene coding for the 5′-untranslated region of the apoB message, i.e., 3′ to the startpoint for transcription (see Fig. 6.9). As we saw in Chapter 3, this sequence and the reverse of its complement are present in the 5′-flanking region of the HMG-CoA reductase gene and are thought to be involved in the regulation of transcription of the gene.

There is no obvious correspondence between the exon–intron organization of the human apoB-100 gene and any of the probable structural or functional domains of the gene product. In particular, segments of the gene coding for the structural domains postulated by Yang *et al.* (1986) do not coincide with the positions of exons. Moreover, the nucleotide sequence encoding the amino acids thought to constitute the receptor recognition site lies within exon 26, at about 1.6 kb from its 3′ end.

E. The Message for ApoB-48

1. Evidence That ApoB-48 Is Colinear with the N-Terminal Half of ApoB-100

Before the complete amino acid sequence of apoB-48 had been deduced (see Section VI,E,4 below), there were already strong indications that it is colinear with the N-terminal half of apoB-100. Much of the evidence for this was derived from studies with monoclonal antibodies to LDL or solubilized apoB-100 (Marcel *et al.*, 1982, 1987; Theolis *et al.*, 1984). Some monoclonal antibodies were

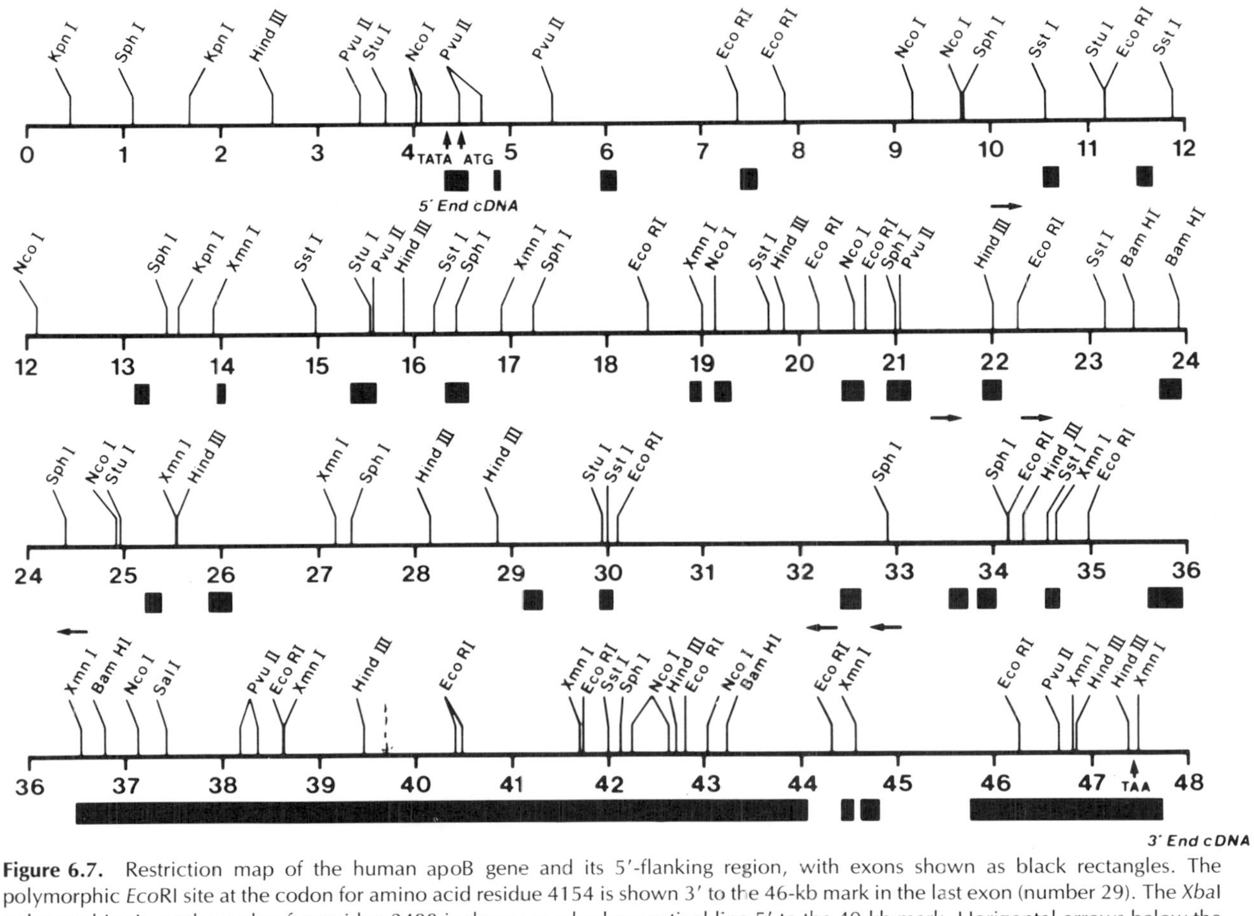

Figure 6.7. Restriction map of the human apoB gene and its 5′-flanking region, with exons shown as black rectangles. The polymorphic *Eco*RI site at the codon for amino acid residue 4154 is shown 3′ to the 46-kb mark in the last exon (number 29). The *Xba*I polymorphic site at the codon for residue 2488 is shown as a broken vertical line 5′ to the 40-kb mark. Horizontal arrows below the exons show six Alu repeats, indicating the length and orientation (→=5′→3′) of each repeat. TATA, TATA box 29 bp upstream from the startpoint for transcription; ATG, the first translated codon. (From Blackhart *et al.*, 1986, with the permission of the authors.)

found to react with both apoB-48 and apoB-100, whereas others reacted only with apoB-100. As shown in Fig. 6.3, those that react with both apoB species are directed against epitopes in the N-terminal half of apoB-100, while the epitopes for those reacting only with apoB-100 are in the C-terminal half. Assignment of a single receptor-binding domain to a C-terminal segment of apoB-100 was consistent with these observations, since it had been shown that apoB-48 does not bind to LDL receptors on fibroblasts (see the previous section). Hardman *et al.* (1987) had also shown that all the peptides released by proteolytic digestion of apoB-48 are homologous to sequences within a region of the apoB-100 molecule extending from the N-terminal amino acid to a position close to residue 2144 (corresponding to about 47% of the apoB-100 amino acid sequence).

2. ApoB-48 and ApoB-100 Are Encoded in the Same Gene

Taken together, these observations pointed to the conclusion that apoB-48 and apoB-100 are encoded either in the same gene or in two different genes with virtually identical coding sequences. Analysis of the DNA restriction fragments that hybridized with apoB cDNA probes gave no evidence for the presence of more than one apoB gene in the human haploid genome (Blackhart *et al.*, 1986; Higuchi *et al.*, 1987). Moreover, Young *et al.* (1986a) showed that the MB19 polymorphism (mentioned in Chapter 5) is expressed in both apoB-48 and apoB-100, a given individual expressing the same MB19 allotypes in the two species of apoB. It was recognized that the chance of this occurring if apoB-48 and apoB-100 are encoded in different genes is vanishingly small, unless the two genes arose by duplication very recently.

3. Evidence for a Separate ApoB-48 Message

From a consideration of the known mechanisms by which one gene produces more than one protein, there seemed to be two possible explanations for the production of apoB-48 and apoB-100 from a single gene. ApoB-100 could be produced by the same gene in liver and intestine, followed by tissue-specific cleavage of apoB-100 in intestinal cells to give apoB-48. Alternatively, differential splicing of the primary transcript of the apoB gene in the intestine might give rise to a shortened mRNA encoding apoB-48.

Pulse-chase experiments with radioactive amino acids appeared to rule out conversion of apoB-100 into apoB-48 in human (Glickman *et al.*, 1986) or rat (Demmer *et al.*, 1986) intestinal cells. If apoB-48 were produced by differential splicing of mRNA, the apoB mRNA in intestine should be about half the length of that in liver. A half-length message for the N-terminal half of apoB-100 should be detectable by Northern blotting of intestinal RNA with probes that include sequences from the 5′ half of apoB-100 cDNA. Experiments to test this point were suggestive but not conclusive. In several studies with a variety of cDNA probes an abundant full-length apoB message (12–18 kb) was detected in human and rabbit intestinal cells (Deeb *et al.*, 1985; Knott *et al.*, 1985; Law *et al.*, 1985; Mehrabian *et al.*, 1985; Shoulders *et al.*, 1985; Cladaras *et al.*, 1986; Glickman *et al.*, 1986). In three of these studies, in which a 5′ cDNA probe was used, small amounts of an additional 7- to 8-kb message were also detected in intestinal cells but not in liver (Deeb *et al.*, 1985; Law *et al.*, 1985; and Cladaras *et al.*, 1986). It was difficult to exclude the possibility that the short messages were produced by the action of intestinal ribonuclease upon a full-length mRNA. Nevertheless, these observations seemed to support the view that apoB-48 is translated from a separate message.

However, this raised two puzzling questions. First, the presence of what appeared to be an apoB-100 message in human intestine was difficult to reconcile with the fact that the adult human intestine synthesizes apoB-48 but not apoB-100. What, then, is the functional significance of the full-length message? Second, the coding sequence of a message encoding the N-terminal half of apoB-100 should end at a position corresponding to a codon within exon 26 of the apoB gene, at least 2 kb from the nearest intron (intron 25) (see Fig. 6.7). Hence, the normal mechanism of "splicing out" at an exon–intron junction would not generate a message encoding apoB-48. So how is the apoB-48 message produced? The answers to both questions, obtained independently in three laboratories, turned out to involve a mechanism for which there was no known precedent.

4. The Stop Codon in Intestinal ApoB mRNA

Chen *et al.* (1987), Powell *et al.* (1987), and Hospattankar *et al.* (1987) prepared cDNA libraries from the mRNAs of human and rabbit small intestine, using the standard procedure in which mRNAs are transcribed into complementary DNA (cDNA) by the enzyme reverse transcriptase and the cDNAs are then cloned. A total of more than 20 clones, selected by hybridization with apoB cDNA probes, were sequenced over the region assumed to include the codon for the C-terminal amino acid of apoB-48. In all the clones examined, a stop codon (TAA) at position 2153 replaced the codon for glutamine (CAA) present at this position in the apoB gene and in hepatic apoB-100 cDNA. Thus, as shown in Fig. 6.8, the apoB message in small intestine is read by reverse transcriptase as though cytosine at nucleotide position 6666 has been replaced by uracil (the base in RNA that is equivalent to thymine in DNA).

The base at nucleotide position 6666 in the apoB gene in intestinal cells is cytosine (Powell *et al.*, 1987), as in the hepatic apoB gene. Hence, the change from C to U in intestinal apoB mRNA must occur during or after transcription of the gene, rather than as a result of somatic mutation in the intestinal apoB gene. It is not known how this change takes place. However, since C and U differ only by the presence of an amino group in cytosine, it is possible that the cytosine

	Liver					Intestine				
Codon		2152	2153	2154			2152	2153	2154	
mRNA	5'	AUA	CAA	UUU	3'	5'	AUA	UAA	UUU	3'
cDNA (1st strand)	3'	TAT	GTT	AAA	5'	3'	TAT	ATT	AAA	5'
cDNA (2nd strand)	5'	ATA	CAA	TTT	3'	5'	ATA	TAA	TTT	3'
Amino Acid		Ile	Gln	Phe			Ile	STOP	—	

Figure 6.8. Nucleotide sequence at codons 2152–2154 in apoB mRNA from liver and small intestine, deduced from cDNA sequences. Codons are numbered relative to the codon for the N-terminal amino acid of mature apoB-100 (glutamic acid). The base at nucleotide position 6666 (relative to the startpoint for transcription) is enclosed in a box. The amino acids encoded by the cDNA coding strand are shown in the bottom row. Owing to the presence of the stop codon in intestinal mRNA, translation comes to an end after amino acid 2152. Amino acid 2151 is methionine (not shown). Note that in the construction of a cDNA library, reverse transcriptase produces a complementary DNA copy of mRNA (cDNA first strand) and that the first strand is used as template for synthesis of a second strand containing the coding sequence in the original mRNA template (T, replacing U).

residue is deaminated enzymically. If this is so, the enzyme must be specific for a particular cytosine in apoB mRNA, and in adult human subjects its activity must be restricted to the small intestine.

An apoB mRNA with a stop at codon 2153 should generate a mature apoB-48 with an amino acid sequence identical to that of the N-terminal 2152 residues of apoB-100, with MW ≃241K before glycosylation and with isoleucine as the C-terminal amino acid (see Fig. 6.8). Chen *et al.* (1987) have, in the main, confirmed these predictions by sequencing the peptides produced by tryptic digestion of apoB-48 obtained from a patient with chylous ascites. The C-terminal residue in the sample of apoB-48 that they examined was methionine, not isoleucine. This may be explained by cleavage of the C-terminal isoleucine by a proteolytic enzyme in the ascitic fluid, leaving methionine 2151 as the C-terminus.

The presence of the ≃14-kb apoB mRNA in intestinal cells is readily explained by the formation of a full-length transcript of the apoB gene, from which apoB-48 is produced after introduction of a stop codon at position 2153. The shorter apoB mRNAs in rabbit and human intestine, ranging from ≃7 to 8 kb, are not so easy to explain. The apoB message in both liver and intestine contains several potential polyadenylation signals[2] at various positions about 7 kb from its

[2]Most eukaryotic mRNAs have poly(A) tails at their 3′ ends. These are added to the primary transcript and are not encoded in the gene. AU-rich hexamers, whose consensus is AAUAAA, are found 10–30 bases upstream from the polyadenylation site. These sequences, called polyadenylation signals, are encoded in the gene.

5′ end and within about 500 bases downstream from the site of the stop codon in intestinal apoB mRNA (codon 2153). These signals do not, of course, initiate polyadenylation in hepatic apoB-100 mRNA, since the apoB-100 message extends for ≃14 kb. However, Powell *et al.* (1987) and Chen *et al.* (1987) have shown that some 7- to 8-kb apoB mRNAs in rabbit and human intestine have poly(A) tails. Chen *et al.* (1987) suggest that the introduction of the stop codon into the message in some way allows nearby potential signals for polyadenylation to initiate the addition of a poly(A) tail a few bases downstream, and thus to generate 7- to 8-kb mRNAs. The presence of 14-kb apoB mRNAs in the intestine, all of which appear to have the stop codon (Powell *et al.*, 1987), shows that such an effect can only be a partial one.

5. Biological Significance of the Stop Codon

In most mammalian species, apoB-48 is essentially the only form of apoB synthesized in the small intestine postnatally. Furthermore, in most mammals little or no apoB-48 is synthesized in any other tissue. On the other hand, developmental studies have shown that apoB-100 is the predominant apoB species synthesized in the intestine throughout most of prenatal life. All this suggests that the tissue-specific mechanisms responsible for the switch from apoB-100 to apoB-48 in the intestine have evolved in conjunction with the need to handle chylomicrons produced during the absorption of dietary fat.

In Chapter 8 (Section III) I discuss the evidence for a separate, nonregulated pathway for the hepatic uptake of chylomicron remnants, in which apoE (not apoB-100) is the recognition element. I also consider the possible biological advantages of such a pathway. If a separate pathway for chylomicron remnants does exist, it could have arisen by the parallel evolution of two components: one a hepatic receptor that recognizes apoE but not apoB-100; the other a modified intestinal apoB lacking an LDL-receptor-binding domain but retaining the lipid-binding domains that enable it to form a stable lipoprotein particle. (As discussed in Section VII of this chapter, unless an apoB at least as long as apoB-39 is synthesized in the intestine, the assembly and secretion of chylomicrons are defective.) Absence of apoB-100 from the remnant particle may serve to reduce the ability of hepatic LDL receptors to compete with remnant receptors for the uptake of remnants. It should be noted that this hypothesis does not explain why at least half the VLDL particles secreted by the rat's liver have apoB-48 as their only apoB protein.

To judge from the strategies that are used by cells to make more than one protein from a single gene, a modified apoB could have evolved by any one of a number of possible routes. The route actually followed seems to have been a highly unusual one. It will be possible to test some of the above speculations when a mutation has been identified that abolishes the mechanism for introducing the stop codon into the apoB message.

F. Regulation of the ApoB Gene

1. Three Genes Compared

The genes for HMG-CoA reductase (Chapter 3) and the LDL receptor (Chapter 9) have two features in common. Both are highly regulated, the level of transcription varying from complete repression to maximal induction, and both are expressed in all tissues in the body. In contrast, activity of the apoB gene varies only over a comparatively narrow range and, in adult animals, its expression is limited to the liver and intestine (Mehrabian *et al.*, 1985; Demmer *et al.*, 1986). As discussed in Chapters 3 and 4, the major problem in relation to regulation of the reductase and LDL-receptor genes is to identify the *cis*-acting sequences[3] and protein transcription factors that interact to bring about sterol-dependent repression of gene activity. Much less attention has been paid to metabolic regulation of the apoB gene. A more challenging problem is to explain why this gene, though present in every cell in the body, is expressed only in hepatocytes and enterocytes. Metabolic regulation and tissue-specific expression of the apoB gene are considered in the next two Sections.

2. The ApoB Gene Promoter and the Regulation of Transcription

As noted in Section VI,D above, the organization of the apoB gene promoter region resembles that of most eukaryotic genes. Figure. 6.9 shows the sequence of the 220 bases 5′ to the startpoint for translation of the message (A of the start codon, ATG). The startpoint for transcription is at nucleotide -128. A TATA box and a CAAT box are present 29 and 60 bases, respectively, upstream from position -128. The sequence CCGCCC occurs twice in the 5′-untranslated region at positions -20 and -81, i.e., downstream from the startpoint for transcription. As we saw in Chapter 3, multiple copies of this sequence (the GC box) or of its reverse complement are present in the 5′-flanking region of many genes, including the HMG-CoA reductase gene. GC boxes are binding sites for the protein transcription factor Sp1.

Promoter and enhancer sequences whose activities are inducible by extracellular factors are present in or near many viral and eukaryotic genes (reviewed in Maniatis *et al.*, 1987). The activities of these "inducible" sequences are modified by specific interaction with cellular transcription factors, several of which (including Sp1) have been purified to homogeneity (see Jones *et al.*, 1988 for review). Extracellular inducers, such as hormones and growth factors, are thought to act in many instances by influencing the binding of transcription factors to promoter or enhancer sequences.

Little progress has been made in the study of inducible *cis*-acting sequences in the apoB gene and of the transcription factors that influence them. This is due

[3]A nucleotide sequence that influences the transcription of a gene in the same molecule of DNA is called *cis*-acting. Nuclear transcription factors encoded in a chromosome other than the one bearing the gene upon which they act are called *trans*-acting factors.

```
                                -200
TCAGGCCCGGGAGGCGCCCTTTGGACCTTTTGCAATCCTGGCGCTCTTGCAGCCTGGGCTTCCTATAAAT
                                                       -100
GGGGTGCGGGCGCCGGCCGCGCATTCCCACCGGGACCTGCGGGGCTGAGTGCCCTTCTCGGTTGCTGCCG

CTGAGGAGCCCGCCCAGCCAGCCAGGGCCGCGAGGCCGAGGCCAGGCCGCAGCCCAGGAGCCGCCCCACC
           1
GCAGCTGGCGATG
```

Figure 6.9. The 5′-untranslated and 5′-flanking regions of the human apoB gene. The vertical arrow shows the startpoint for transcription at position -128 from the startpoint for translation (A of ATG). The two GC boxes in the 5′-untranslated region are underlined with thin lines. The TATA box and the CAAT box, beginning at positions 29 and 60 bp upstream from the startpoint for transcription, are underlined with thick lines. (From Blackhart *et al.*, 1986.)

mainly to the difficulty in finding experimental conditions in which the rate of transcription of the gene is altered. As noted in Chapter 7, when the production of VLDL by hepatocytes or of chylomicrons by enterocytes *in vitro* is altered by changing the nutrient composition of the medium, apoB synthesis does not change. Moreover, observations on transformed intestinal (CaCo-2) and hepatoma (Hep G2) cell lines in culture have shown that experimentally induced changes in apoB production are not accompanied by parallel changes in cell content of apoB mRNA (Moberly *et al.*, 1988; Pullinger *et al.*, 1988; Dashti *et al.*, 1988). Figure 6.10 shows the effects of albumin, oleate, and insulin on apoB production and cell mRNA content in Hep G2 cells in culture. ApoB production

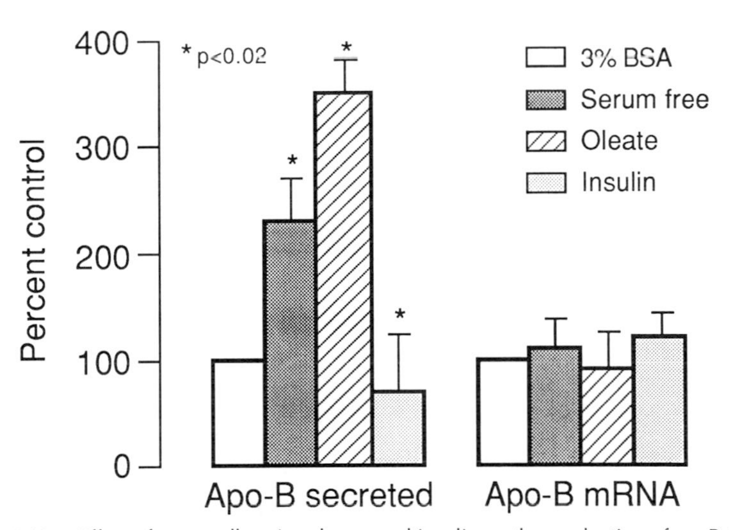

Figure 6.10. Effect of serum albumin, oleate, and insulin on the production of apoB-100 and the cell content of apoB mRNA in human hepatoma (Hep-G2) cells in culture. Results are shown as percentages of the values obtained with cells incubated in the presence of bovine serum albumin (BSA). The concentrations were: oleate, 0.8 m*M;* insulin, 10^{-8} *M;* albumin, 3% (w/v). (C. R. Pullinger and J. Scott, personal communication.)

was stimulated by the addition of oleate to the medium and was inhibited by adding albumin or insulin. Despite these changes, the cell content of apoB mRNA did not change significantly.

From the evidence available at present, it seems likely that hepatocytes and intestinal cells *in vivo* do not respond to rapid fluctuations in their requirement for apoB by regulating the rate of transcription of the apoB gene. An increased demand for apoB may be met partly by posttranscriptional effects on apoB synthesis and partly by drawing upon an intracellular store of apoB, which is replenished when the need for apoB declines. In keeping with these suggestions, Pullinger *et al.* (1988) have found that the half-life of apoB mRNA is at least 20 hours. With a half-life as long as this, the cell content of the apoB message would not change rapidly in response to a change in the rate of transcription of the apoB gene.

3. Tissue-Specific Expression of the ApoB Gene

The mechanisms by which the expression of certain genes is restricted to one or a few cell types in the mature organism are beginning to be understood. Tissue-specific expression of several genes has been shown to involve multiple enhancer sequences, usually about 10 bp long, that are recognized by *trans*-acting transcription factors. Maniatis *et al.* (1987) have discussed how the binding of transcription factors to the sequences they recognize in a particular gene (resulting in expression of the gene) could be limited to a single cell type. One possibility is that an essential transcription factor is synthesized only in those cells in which the gene is expressed. Another is that a particular transcription factor is made in cells of many types, but its target sequences in the gene in question are accessible only in cells expressing the gene.

Some progress has already been made toward identifying transcription factors and their binding sites in the apoB gene that are necessary for its specific expression in liver and intestine. Levy-Wilson *et al.* (1988) isolated a 1-kb fragment from the 5′-untranslated and 5′-flanking regions of the apoB gene and ligated it 5′ to the *CAT* gene (see Chapter 3). The *CAT* gene in this construct was active in Hep G2 and CaCo-2 cells, both of which normally express the apoB gene, but was inactive in HeLa cells, which do not express the apoB gene.

A possible explanation for these findings is that enhancer sequences in the 5′-flanking region of the apoB gene are accessible to binding proteins in Hep G2 and CaCo-2 cells but not in HeLa cells. A fruitful approach to the study of gene sequences accessible to proteins *in vivo* is to search for regions in nuclear DNA that are hypersensitive to cleavage by DNases (see Eissenberg *et al.*, 1985 for review). DNase-hypersensitive (DH) sites in the DNA of intact nuclei are found in the 5′-flanking regions of many active genes, often corresponding closely to binding sites for transcription factors. Moreover, the presence of a DH site near a gene may be correlated with its tissue-specific expression. This is exemplified by the rat insulin II gene, which has a DH region near its 5′ end in chromatin from pancreatic β cells but not in chromatin from other rat cells (Wu and Gilbert, 1981). DH sites are thought to correspond either to nucleosome-free[4] regions in a chromatin strand or to protein-free linker regions between nucleosomes.

Levy-Wilson *et al.* (1988) have examined nuclei from Hep G2, CaCo-2, and HeLa cells for the presence of DH and micrococcal nuclease-hypersensitive (MH) sites near the 5′ end of the apoB gene. They identified at least seven sites in the apoB gene in nuclei from Hep G2 and CaCo-2 cells, none of which were present in the HeLa-cell gene. Since nucleosomes were shown to be present along the whole 5′-flanking region of the apoB gene in all three cell lines, Levy-Wilson *et al.* suggest that the spacing of nucleosomes in the three types of cells differs in a manner such as to expose MH and DH sites in linker regions only in Hep G2 and CaCo-2 cells.

Using a different approach, Das *et al.* (1988) have compared the abilities of different segments of the 5′-flanking region of the apoB gene to activate the *CAT* gene in a tissue-specific manner. They identified two sequences between nucleotides -70 and -128 (see Fig. 6.9) that initiate expression of the *CAT* gene at a high level in Hep G2 cells but not in HeLa cells. A short sequence within this segment exhibited 70% or more homology with sequences in the 5′-flanking regions of four other genes expressed only in liver, suggesting that all five genes share liver-specific regulatory elements. The DNase I footprinting assay (see Chapter 9, Section III for explanation) was used to demonstrate the presence of proteins in mouse-liver nuclei that bind to sequences between nucleotides -70 and -86 in the apoB gene. These proteins were not detected in HeLa-cell extracts.

Taken together, the experiments of Levy-Wilson *et al.* and of Das *et al.* suggest that more than one mechanism contributes to tissue-specific expression of the apoB gene. First, certain binding sites for transcription factors may be accessible only when the gene is in a liver or intestinal cell. Second, there may be specific sequences in the 5′-flanking region of the apoB gene that are recognized by transcription factors synthesized only in the liver (and ? intestine).

The tissue-specific expression of apoB-48 has not been explained. Two possibilities are worth considering. The mechanism for inserting a stop codon into the apoB message may be confined to those tissues in which apoB-48 is produced (the intestine in most species). Alternatively, the stop-insertion mechanism may be ubiquitous but capable of generating apoB-48 only in tissues in which the apoB gene is transcribed. In the latter case it would be necessary to explain why apoB-48 is not produced in the liver in most species. The stop-insertion mechanism appears to be inducible, since Davidson *et al.* (1988) have shown that

[4]The DNA in cell nuclei is bound tightly to basic proteins (mainly histones) arranged in discoidal units, called nucleosomes, separated from each other by short protein-free strands of DNA called linkers. Each nucleosome plus one adjacent linker contains about 200 bp of DNA.

administration of triiodothyronine to rats increases the proportion of hepatic apoB mRNAs that contain the stop codon.

G. Evolution of the ApoB Gene

Comparison between the nucleotide and amino acid sequences of seven of the apoproteins (A-I, A-II, A-IV, C-I, C-II, C-III, and E) shows that their genes are derived from a common ancestral gene, i.e., they constitute a *multigene family*. In each of the seven apoproteins it is possible to discern a repeat unit of 22 amino acids (which may be the structural unit of the amphipathic α-helix), with significant sequence homology between the repeat units of different apoproteins (see Li *et al.*, 1988). Any doubts as to the common evolutionary origin of these genes are dispelled by a consideration of their exon–intron organizations, seven of which are shown in Fig. 6.11. In all except apoA-IV (which appears to have lost the first intron), the gene is divided into four exons, one in the 5′-untranslated region and the other three encoding essentially the same functional domains of the protein in all the genes. These similarities reflect the common origin of this group of genes and could not have arisen by chance. Li *et al.* (1988) have proposed an evolutionary tree which explains how the present family of genes might have arisen from a single ancestor by a succession of complete gene duplications, each one followed by internal duplication of repeat units and other mutational events. It is worth noting that recruitment of exons from other genes does not seem to have occurred at any stage in the evolution of this multigene family.

The apoB gene differs in so many respects from the other apoprotein genes that it clearly does not belong to the apoprotein multigene family. The molecular weight of apoB-100 is more than 12 times that of the largest member of the family (apoA-IV) and its amino acid sequence has little or no homology with that of any other apoprotein (see Section III,A above). There are also striking differences between the structural organization of the apoB gene and that of the typical apoprotein gene. Whereas the genes encoding the smaller apoproteins are divided into three or four exons, each encoding a more or less distinct domain, the apoB is divided into 28 exons, none of which encodes a separate domain in the protein product. The exceptionally long twenty-sixth exon, presumably the result of loss of introns during evolution of the gene, is quite unlike any of the exons in other apoprotein genes. It should be noted that the length of exon 26 in the pig apoB gene is similar to that of the human exon (Maeda *et al.*, 1988).

Comparative studies indicating the presence of an apoB-like LDL protein in lower vertebrates (Chapman, 1980) suggest that the progenitor of the apoB gene may have arisen by duplication of an older gene at the beginning of vertebrate evolution (400–500 million years ago). The presence of long repeats with a low degree of homology (DeLoof *et al.*, 1987) suggests that elongation of the ancestral gene by internal duplication occurred at an early stage in its evolution. A

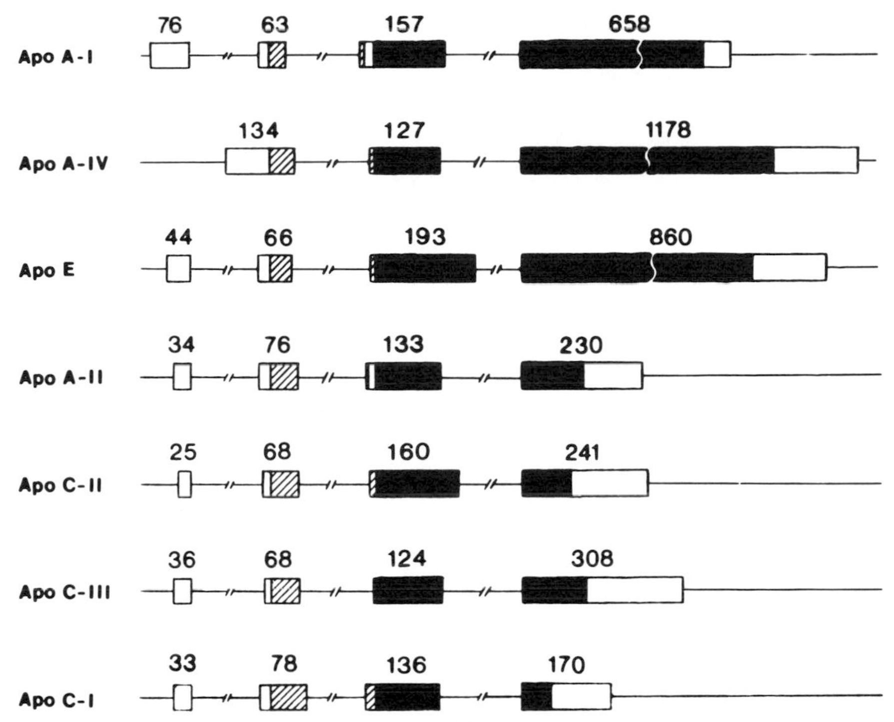

Figure 6.11. The exon–intron organization of the human apoA-I, apoA-IV, apoE, apoA-II, apoC-I, apoC-II, and apoC-III genes. Exons are shown as rectangles, open for 5′- and 3′-untranslated regions, closed for regions encoding the mature protein plus proprotein, and hatched for signal peptides. Introns and flanking regions are shown as thin lines. Numbers above each gene are numbers of nucleotides in the corresponding exon. Note that, except in the apoA-IV gene, the first intron interrupts the 5′-untranslated region and the second intron separates virtually the whole of the nucleotide sequence encoding the signal peptide from the remainder of the gene. Note also that differences in total coding length are due almost entirely to differences in the length of the last exon. (From Chan, 1989, with the permission of the author.)

successful search for amino acid homologies between apoB and other vertebrate proteins would help to throw light on the ancestry of the apoB gene and might show whether or not it has acquired exons from other genes.

VII. Polymorphism in the Human ApoB Gene

A. Genetic Polymorphism: Definition and Detection by RFLP

Genetic polymorphism (Gr. *polloi,* many; *morphe,* shape) refers to a situation in which two or more alleles at a given gene locus are present in the population at appreciable frequencies, as in the well-known ABO blood-group polymorphism. For practical purposes, population geneticists restrict the term arbitrarily to loci at which there are at least two alleles with frequencies greater than 1% (Bodmer and Cavalli-Sforza, 1976; Harris, 1980). For such loci, the frequency of heterozygotes in a population in Hardy-Weinberg equilibrium would be at least 2%. Until the development of recombinant DNA technology, genetic polymorphism in human populations could only be detected at the level of the phenotype. Examples already referred to are the phenotyping of variants of the Ag system by means of alloantibodies and of apoE isoforms by isoelectric focusing. However, the methods available for detecting phenotypic variants are very limited. For example, isoelectric focusing, a method capable of separating proteins of the same molecular weight that differ by only a single charge, cannot be used to separate polymorphic forms of a protein that have the same charge.

It is now possible to examine the genome directly and to detect polymorphism in the DNA in coding regions of a gene (whether or not the difference at the polymorphic site leads to an amino acid change), in introns within genes, and in the long stretches of DNA between genes. Occasionally a DNA polymorphism is picked up by direct sequencing of cDNA or genomic clones from a single donor, as in the detection of the G→A substitution in the apoB gene by Shoulders *et al.* (1985), or by comparing the sequences of cDNAs from different libraries (Ludwig *et al.*, 1987). However, DNA polymorphism is more usually discovered by examining the fragments produced by digesting genomic DNA from random samples of the population with *restriction endonucleases*. These are enzymes that cut double-stranded DNA at specific sequences of nucleotides (restriction sites). The *Eco*RI site, for example, has the sequence G/AATTC in the sense strand (the vertical line showing the cleavage site). Hence, when a segment of DNA is digested with *Eco*RI the DNA is cut wherever this sequence occurs in the genome, giving rise to a number of fragments whose lengths depend on the distances between adjacent *Eco*RI sites. A single-base change that abolishes an existing site or creates a new one, or the insertion or deletion of a sufficient stretch of DNA between two sites, will alter the pattern of fragments produced by digestion with the restriction endonuclease, producing what is called a *restriction fragment length polymorphism* (RFLP). The two alternative sequences at a restriction site on homologous chromosomes are inherited as codominant alleles.

The pattern of restriction fragments generated from a specified region of genomic DNA is determined by "Southern blotting," a procedure devised by Southern (1979). After digesting the DNA with restriction enzymes, the fragments in the total digest are separated according to length by electrophoresis in agarose gel and are then blotted from the gel onto a filter. The filter is incubated with a radioactive DNA probe designed to hybridize with the fragments of interest and the positions of the hybridized fragments are detected by autoradiography.

Figure 6.12 shows a map of the *Eco*RI and *Hin*dIII restriction sites in the apoB gene near the polymorphic *Eco*RI site, the fragments produced from this region by digestion of total genomic DNA with *Eco*RI and *Hin*dIII, and the cDNA probe used to detect the restriction fragments. The *Eco*RI site includes the base triplet coding for amino acid residue 4154 (lysine or glutamic acid) in the 5′ half of exon 29. The probe includes sequences complementary to the 3′ end of exon 26, the whole of exons 27 and 28, and the 5′ end of exon 29. When genomic DNA is digested with the two endonucleases, a vast number of fragments is generated, but only a few of these are detected by the probe. There are three possible genotypes, $E^{+,+}$, $E^{-,-}$, and $E^{+,-}$ (in which the site is present on one homologous chromosome but absent from the other). With all three, the probe hybridizes with an invariant 1.8-kb fragment (E_2-E_3). The probe also "sees" two fragments 2.0 and 0.5 kb long with $E^{+,+}$, a single 2.5-kb fragment with $E^{-,-}$, and all three fragments (0.5, 2.0, and 2.5 kb) with $E^{+,-}$ (see Fig. 6.13). If *Hind*III were omitted from the digest, the 2.5-kb fragment would not be produced and the fragment patterns detected by the probe would be different. E_1-E_2 is not detected by the probe shown in Figure 6.12, but could be detected by a probe extending further upstream. Thus, the fragment pattern observed is determined by the restriction enzymes used and the nature of the probe, as well as the presence or absence of the restriction site in the subject's genomic DNA.

Mendelian inheritance of the E^+ and E^- alleles may be demonstrated by tracking the fragment patterns through two or more generations in families in which both alleles are present, as in the kinship shown in Fig. 6.14. This kinship also illustrates the way in which some haplotypes (see next section) may be deduced from the diploid genotype at two or more polymorphic sites.

B. Genetic Linkage, Linkage Disequilibrium, and Association: Some General Remarks

When two genes are close to each other on the same chromosome they do not segregate independently in the progeny of a mating but are inherited together as a haplotype.[5] Such genes are said to be linked. Linkage is detected by analyzing the inheritance of pairs of alleles in the progeny of parents heterozygous at the two loci. (Several examples of the use of linkage analysis in the study of monogenic disorders will be found at the end of this chapter and in Chapter 9).

If two linked genes are separated by an appreciable stretch of DNA, a proportion will cross from one chromosome to the other when segments of DNA are

[5]In Mendelian genetics a haplotype (Gr. *haplous*, single) is defined as a combination of two or more alleles on a single chromosome. The term is now used to include combinations of genetic markers at variable sites in the DNA (such as those detected on RFLPs) that are not, strictly speaking, alleles. Examples of haplotypes based on RFLPs are shown in Fig. 6.14. Other examples will be found elsewhere in this chapter and in Chapter 9.

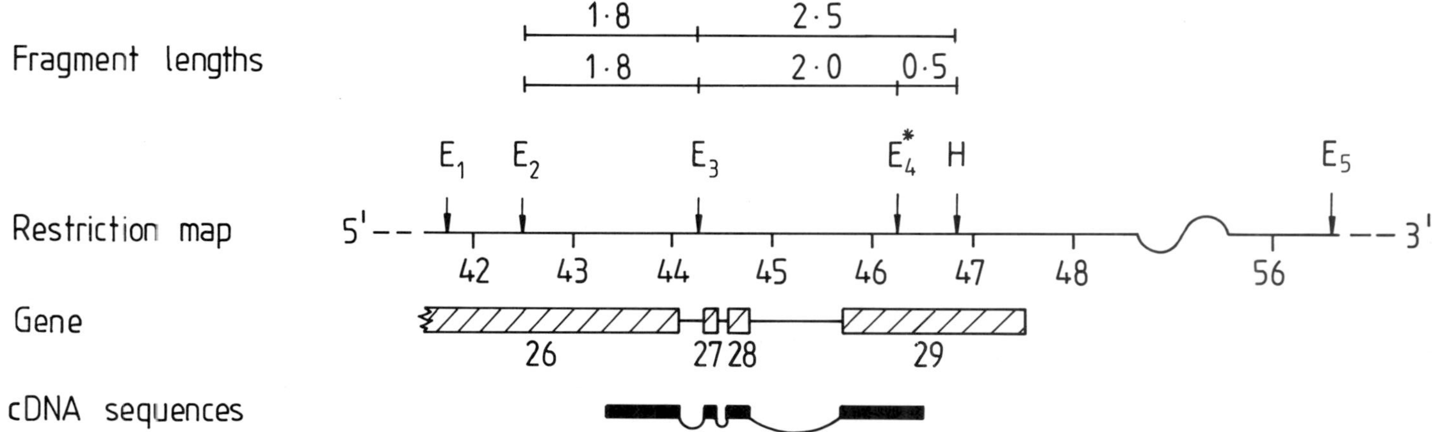

Figure 6.12. Restriction fragments generated at the *Eco*RI polymorphic site near the 3′ end of the apoB gene. Fragment lengths: The lengths of the fragments (kb) detected with a radioactive cDNA probe when $E^{-,-}$ DNA (upper line) and $E^{+,+}$ DNA are digested with *Eco*RI and *Hin*dIII. Restriction map: The *Eco*RI (E) and *Hin*dIII (H) restriction sites in the region of the *Eco*RI polymorphic site (E_4^*). Distances are in kilobases from a point about 4.5 kb 5′ from the startpoint for transcription of the gene. Gene: The positions of exons 26–29 (hatched boxes) and of introns 26–28 in relation to the restriction map. cDNA sequences: The sequences in the cDNA probe shown in relation to the exons in the gene. Note that in the mRNA template from which cDNA was synthesized by reverse transcription, the introns have been excised (shown by loops joining the four cDNA segments). The cDNA probe, approximately 1.9 kb long, was obtained by excising a segment of apoB clone pB4 (Shoulders *et al.*, 1985) with *Bam*HI.

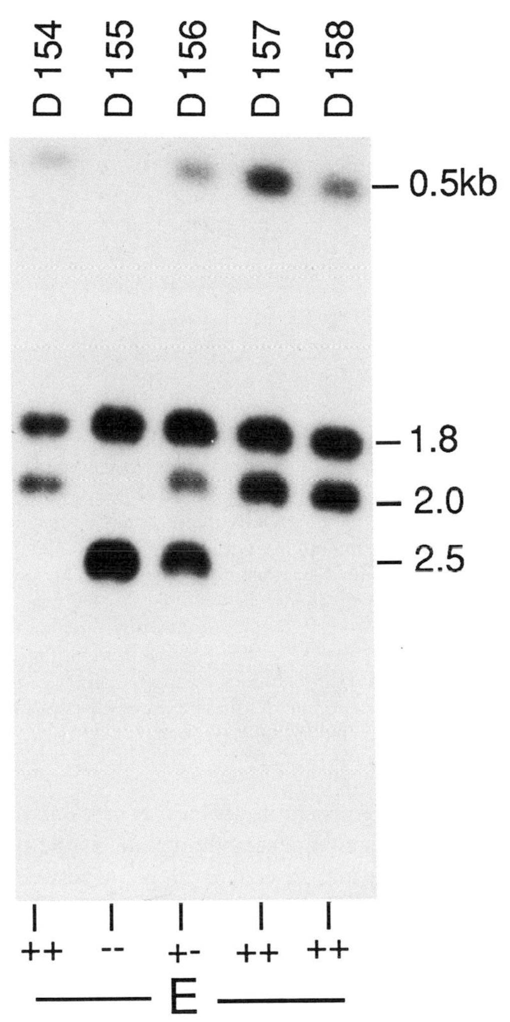

Figure 6.13. *Eco*RI polymorphism in the apoB gene detected by Southern blot analysis. Samples of genomic DNA, obtained from three individuals whose genotypes at the *E* locus were $E^{+,+}$, $E^{-,-}$, and $E^{+,-}$, were digested with *Eco*RI and *Hin*dIII (see text). The DNA fragments were separated by electrophoresis in agarose gel and were then blotted from the gel onto a nylon filter (Hybond N). The filter was incubated with the radioactive probe shown in Fig. 6.11, washed, and submitted to autoradiography to reveal the positions of the fragments that hybridized to the probe. The lengths of the fragments were determined from the positions of standard markers run in parallel with the samples. The probe was labeled with [^{32}P]dCTP, using a random oligonucleotide primer (Feinberg and Vogelstein, 1984). Note that DNA fragments derived from the total genomic DNA were present as smears extending throughout the length of each lane; only those fragments with sequences complementary to sequences in the probe are revealed on the autoradiograph. (I am indebted to John Gallagher for this analysis.)

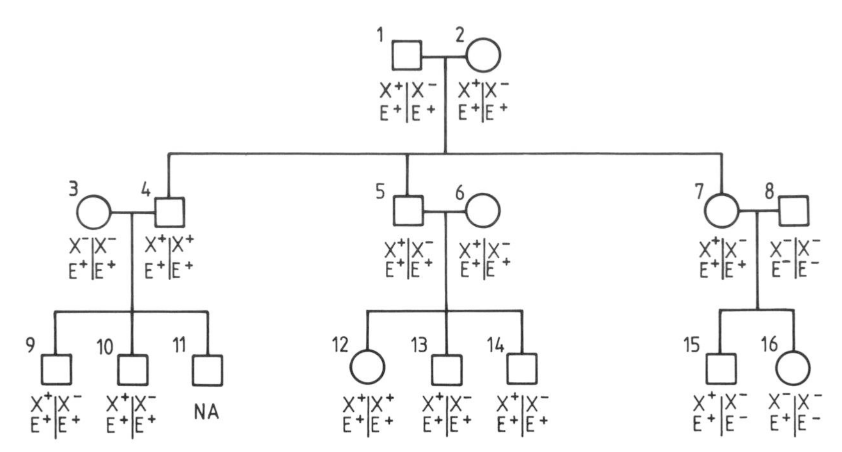

Figure 6.14. Inheritance of RFLPs detected with the restriction enzymes *Xba*I (*X*) and *Eco*RI (*E*) in a three-generation kindred. The *X* and *E* genotypes were determined by Southern blotting of DNA digests. Of the 15 examined subjects, 14 were heterozygous at one locus and homozygous at the other. In each of these the haplotypes could be deduced from the subject's combined genotype without information from other members of the family, e.g., in subject 1 (genotype $X^{+,-}/E^{+,+}$) one chromosome must have been $X^{+}E^{+}$ and the other $X^{-}E^{+}$. Subject 15 is a double heterozygote with genotype $X^{+,-}/E^{+,-}$. On this information alone his haplotypes could have been $X^{+}E^{+}$ and $X^{-}E^{-}$ (as shown in the figure) or $X^{+}E^{-}$ and $X^{-}E^{+}$. Since he can only have inherited an $X^{-}E^{-}$ chromosome from his father, his haplotypes must have been as shown. In genetic terms, the mating between 7 and 8 is "informative" for the *X*/*E* haplotypes. NA, not available. (I am grateful to members of my family for allowing me to take blood samples and to John Gallagher for the analyses.)

exchanged between homologous chromosomes by crossing over at meiosis. The greater the distance between two genes, the greater is the probability that they will be separated by a chiasma (a crossover point between two homologous chromosomes at meiosis) and, hence, that they will come to lie on opposite chromosomes. Thus, the extent of recombination between two linked genes can be used as an index of the chromosomal distance by which they are separated.

The unit of chromosomal length determined from the frequency of recombination is the centiMorgan (cM), defined as the minimum length over which recombination between two genes occurs in 1% of meiotic events. One cM is equivalent to approximately 1000 kb of DNA, and since the human genome contains about 3×10^9 nucleotide pairs, there would be about 3000 cM in the whole genome if the probability of crossing over were the same at all points in the genome.

Crossing over between linked genes is a rare enough event (on average, 1% per 1000 kb per generation) for it to be possible to demonstrate linkage in man by examining a comparatively small number of families in which the parental genotypes are informative. However, given a sufficient number of generations, the alleles at any pair of linked polymorphic loci should become randomized between the two homologous chromosomes by repeated recombination. This state is called *linkage equilibrium*. For example, consider two linked polymorphic loci with alleles *Aa* and *Bb*. The population frequencies of the four possible haplotypes *AB, Ab, aB,* and *ab* at linkage equilibrium would be those expected if the two pairs of alleles were randomly associated. (The frequency of each haplotype would be equal to the product of the frequencies of its two alleles.)

Many pairs of linked genes at polymorphic loci in the human population are known to be in linkage equilibrium. However, some alleles at loci on the same chromosome occur together in the population more frequently, or less frequently, than would be expected by chance. This state, known as *linkage disequilibrium*, may arise for several reasons. Among the more important are selection pressure favoring the survival of a particular haplotype, and insufficient time for the establishment of equilibrium since the occurrence of the mutation giving rise to the more recent polymorphism. Other factors that may affect the rate at which two linked loci equilibrate are nonrandom mating and the mixing or separation of populations. It is for the last two reasons that the extent of linkage disequilibrium between two loci may differ widely from one population to another. The following rough calculation gives an idea of the rate at which linkage equilibrium between two loci might occur in a randomly mating human population if there were no advantage for carriers of any one haplotype. If the two loci were separated by 1 cm (100 kb) the extent of linkage disequilibrium would fall to 1/10 of its initial value in about 230 generations (about 7000 years). (For the basis of this estimate, see Vogel and Motulsky, 1986.)

In general, the closer are two linked genes, the greater the number of generations required to bring about linkage equilibrium. Hence, the occurrence of linkage disequilibrium between two alleles at linked loci suggests that the two loci are near to each other or that the mutation at one of the loci occurred relatively recently. However, a haplotype may persist in the population for a surprisingly long time. Murray *et al.* (1984) have described a haplotype based on two RFLPs in the human albumin gene that appears to have been maintained since its probable origin before the evolutionary divergence of hominids from their prehominid ancestors, a time sufficient for more than 3×10^5 generations. Murray *et al.* suggest that one reason for the remarkable persistence of this haplotype is a low rate of recombination in the region of the albumin gene.

The term *population association,* as used in a genetic sense, should not be confused with linkage. Two alleles at different loci, or an allele and a phenotypic trait such as a disease, are said to be associated when they occur together *in unrelated individuals* more often than would be expected by chance. A striking example of association is the unexpectedly high frequency of the HLA allele *B27* in people who have ankylosing spondylitis. An allele could be associated with a disease because the product of the allele itself increases susceptibility to the disease. Alternatively, due to linkage disequilibrium, the allele in question could be associated with a nearby allele whose product is causally related to the disease, as in the hypothetical case shown in Fig. 6.15. It should be clear from this figure that the strength of a population association between an allele at a polymorphic site and a disease will vary with time and between different populations. Initially, the association will be strong in the population in which the second of the two polymorphic sites appears. At complete linkage equilibrium it will disappear altogether. Moreover, the strength of the association at a given time will vary from one population to another, depending upon the extent to which linkage equilibrium has been achieved. This, in turn, will depend upon the distance between the two variable sites and the genetic history of the population. There is also the possibility that the mutation creating the second polymorphic site could occur independently in two separate populations at different times.

C. RFLPs in the ApoB Gene

1. Sites and Population Frequencies

In addition to the *Eco*RI polymorphic site described in Section VII,A above, several other RFLPs in the apoB gene have been reported. These are listed in

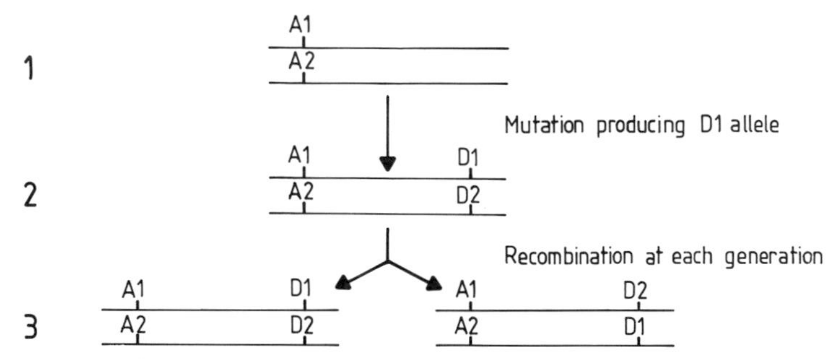

Figure 6.15. Possible evolution of an association between coronary artery disease (CAD) and an allele at a polymorphic site in a human population. (1) A pair of homologous chromosomes with clinically silent alleles *A1* and *A2* at a polymorphic site (A) detected as an RFLP. (2) A mutation D1, whose product increases susceptibility to CAD, occurs in a single individual on the chromosome bearing the *A1* allele (*D2* is the wild-type allele). This creates a new polymorphic site (D). When *D1* begins to spread through the population, there will be linkage disequilibrium between the two sites, with a significant association between *A1* and CAD in unrelated individuals (*A1* allele frequency higher in CAD patients than in normal people). (3) The two sites approach linkage equilibrium due to repeated recombination at crossover points between the *A* and *D* loci. At complete equilibrium there is no population association between *A1* and CAD (*A1* allele frequency is the same in people with and without CAD). Note that at stage (2) most people in the population who have the *A1* allele will not have *D1* and, hence, will not have increased susceptibility to CAD due to the presence of this allele. Note also that because CAD is multifactorial, not all people with *D1* will develop CAD. Conversely many who do not carry *D1* will develop CAD. The combined effect of these factors is to reduce the strength of the population association between *A1* and clinical CAD.

Table 6.3. In view of the large size of the gene and the expected frequency of RFLPs in human DNA (see Cooper and Schmidtke, 1984), a further search is likely to reveal additional restriction-site polymorphisms. Blackhart *et al.* (1986) found 13 single-base differences between the nucleotide sequences of apoB cDNAs from a human cDNA library and the sequences in the coding portions of the apoB gene (determined in human genomic clones). Eight of these changes result in an amino acid substitution and several involve loss of a restriction site, including the *Eco*RI site already referred to. However, the existence of polymorphism at these sites, other than that detected with *Eco*RI, has not been demonstrated in the general population. Extensive variation in the human apoB gene is also revealed by comparing the cDNA sequences reported from different sources. Ludwig *et al.* (1987) noted differences at 60 nucleotide positions and at 39 amino acid positions in the cDNAs sequenced in seven different laboratories. These sequence variations include some of the RFLPs shown in Table 6.3.

The first nine RFLPs in this table involve single-base substitutions, three of which lead to an amino acid change. The polyallelic system detected with three separate enzymes is due to the presence of a variable number of tandem repeats of a 30-bp AT-rich sequence in the 3′-flanking region of the gene (Berg *et al.*,

Table 6.3
Some Restriction Fragment Length Polymorphisms in the ApoB Gene

Restriction enzyme used	Position of site[a]	Amino acid change	Frequency of + allele	Reference[b]
*Msp*I	ApoB promoter	—	0.79	11
*Pvu*II	Intron 3	—	0.07	1, 2, 3
*Hinc*II	5′ end of gene	—	0.12	1
*Eco*RV	5′ end of gene	—	0.02	3
*Alu*I	591 (GTT-GCT)	Val-Ala	?	4
*Eco*RV	3′ end of gene	—	0.80	5
*Xba*I	2488 (ACC-ACT)	—	0.50	5, 6
*Msp*I	3611 (CGG-CAG)	Arg-Gln	0.88	7
*Eco*RI	4154 (GAA-AAA)	Glu-Lys	0.85	5, 8
Hypervariable (*Msp*I, *Bam*HI, *Eco*RV)	3′-flanking	Polyallelic system due to variable number of repeats		9, 10

[a]Numbers refer to amino acid residues in mature apoB-100; — represents no change in amino acid or not known. See text for polyallelic system detected with *Msp*I or *Bam*HI or *Eco*RV. The restriction site is present in the + allele. The first codon in each pair in parentheses is present in the + allele. Frequencies are based on samples taken from normal white European or North American populations. Note that the allele frequency of an RFLP may vary between geographical regions and between different subgroups within a populaton, e.g., the frequency of the *Xba*I+ allele in Japan is 0.04 (Aburatani *et al.*, 1987).

[b]References (1) Darnfors *et al.*, 1986; (2) Frossard *et al.*, 1986; (3) Protter *et al.*, 1986a; (4) Wang *et al.*, 1988; (5) Priestley *et al.*, 1985; (6) Berg *et al.*, 1986; (7) Huang *et al.*, 1988b; (8) Shoulders *et al.*, 1985; (9) Berg *et al.*, 1986; (10) Jenner *et al.*, 1988; (11) Jones *et al.*, 1989. See also Boerwinkle and Chan, 1989.

1986) about 200 bases downstream from the polyadenylation signal (Jenner *et al.*, 1988). At least seven alleles at this locus are present in the population, producing restriction fragments with seven different numbers of repeats ranging from 1 to about 10. Since there are *Msp*I, *Bam*HI, and *Eco*RI sites on either side of the 30-bp repeat region, this polymorphism can be detected with any one of the three enzymes.

The population frequencies of the alleles in the polymorphic systems shown in Table 6.3 are such that there should be a reasonable chance of finding informative genotypes for each locus in any family chosen at random from the population; this is necessary if a polymorphic locus is to be of practical value in linkage studies. Analysis of haplotype frequencies in small samples from European populations suggests that there is linkage equilibrium between the *Pvu*II and *Xba*I RFLPs (Darnfors *et al.*, 1986). The *Eco*RI and *Xba*I polymorphisms, on the other hand, are in linkage disequilibrium (Berg *et al.*, 1986; Myant *et al.*, 1989). Failure of the *Eco*RI and *Xba*I variable sites to have reached linkage equilibrium in many populations would not be surprising, since these sites are separated by only about 6 kb in the apoB gene. Linkage equilibrium between two such close loci would be expected to require at least 46,000 generations, or not less than a million years.

2. ApoB RFLPs and the Ag System

As we saw in Chapter 5, the antigen group (Ag) of Allison and Blumberg (1961) is a polymorphic system detected with alloantibodies to human LDL. Family studies, combined with analysis of the Ag genotypes present in the population, have shown that the molecular basis of the Ag system is a set of five closely linked loci, each with two codominant alleles (Bütler and Brunner, 1974; Kostner, 1976). The pairs of epitopes encoded by these alleles are designated Ag(a_1/d), Ag(c/g), Ag(h/i), Ag(t/z), and Ag(x/y).

Two independent lines of evidence have shown that the Ag system is the phenotypic expression of polymorphism in the apoB gene. Berg *et al.* (1986) showed that the Ag(x) epitope cosegregates within families with the X^- allele (restriction site absent) at the polymorphic *Xba*I site in the apoB gene, and that there is a strong association between Ag(x) and the X^- allele in unrelated individuals. They concluded that the *Ag(x/y)* locus is within the apoB gene and is in linkage disequilibrium with the *Xba*I RFLP locus. In agreement with these findings, Tikkanen *et al.* (1986) showed that a monoclonal anti-apoB antibody ($MB19_1$) reacts specifically with the Ag(c) epitope in LDL. Since the epitope detected by MB19 is shared by apoB-26 and apoB-100 (Curtiss and Edgington, 1982), this showed that the *Ag(c/g)* locus is in the 5′ region of the apoB gene. In view of the close linkage between the five Ag loci, these results indicated that the Ag system as a whole reflects polymorphism in the amino acid sequence of apoB, suggesting that each pair of Ag epitopes represents a pair of alternative amino acids.

The strong population association between an Ag(x/y) epitope and an allele at the *Xba*I RFLP locus shows that the two loci are close to each other. However, the nucleotide substitution at the RFLP locus cannot be the site of the *Ag(x/y)* locus because the base change giving rise to the RFLP (C→T in the third postion of codon 2488) does not change the amino acid. Nor is it possible to deduce the amino acid position at the *Ag(c/g)* locus from the observation that the Ag(c) epitope reacts with an antibody to an epitope known to be in the N-terminal portion of apoB-100. However, there is now little doubt as to the amino acid positions corresponding to the other three *Ag* loci.

Ma *et al.* (1987) observed perfect correspondence between the Ag(t/z) phenotype and the *Eco*RI genotype in an ethnically mixed group of 17 unrelated North Americans, Ag(+) corresponding to E^+ (restriction site present) at the RFLP locus. This result could have been obtained if the *Ag(t/z)* and *Eco*RI RFLP loci are separate but so close together on a chromosome carrying the apoB gene that they have undergone zero recombination. This is very unlikely, the more so since Dunning *et al.* (1988) have noted a similar perfect correspondence between *Ag(t)* and the E^+ allele in a group of unrelated individuals from the Finnish population, a population known to have a long history of genetic isolation (Norio *et al.*, 1973). It seems more likely that the *Ag(t/z)* and *Eco*RI RFLP loci are one and the same, in which case the structural determinants of the Ag(t) and Ag(z) epitopes are, respectively, glutamic acid and lysine at amino acid position 4154 in mature apoB-100.

Wang *et al.* (1988) have also demonstrated perfect correspondence in unrelated individuals between the Ag(a_1/d) phenotype and the genotype at an *Alu*I RFLP site in the apoB gene. The presence of this restriction site results in a change from valine to alanine at residue 591 in apoB-100. Wang *et al.* sequenced the segment of DNA containing the *Alu*I RFLP locus in two $Ag(a_1/a_1)$ individuals and in two who were *Ag(d/d)*. The codon at position 591 corresponded to valine in both $Ag(a_1/a_1)$ homozygotes and to alanine in both the *Ag(d/d)* individuals. Wang *et al.* concluded that the determinants of the Ag(a_1) and Ag(d) epitopes are, respectively, valine and alanine at residue 591. Xu *et al.* (1989) have investigated the *Ag(h/i)* locus in a Finnish population. In 106 unrelated individuals they noted perfect correspondence between Ag(h/i) phenotype and the genotype at the *Msp*I RFLP locus at the 3′ end of the apoB gene. *Ag(i)* corresponded to the M^+ allele (*Msp*I site present). As shown in Table 6.3, loss of this restriction site, due to a change from CGG to CAG in codon 3611, results in an amino acid change from arginine to glutamine. This finding provides strong evidence that the change from arginine to glutamine creates the Ag(h/i) polymorphism.

Figure 6.16 summarizes the probable positions in apoB-100 corresponding to four of the five Ag polymorphic loci in the apoB gene. The Ag(c/g) epitope pair is placed at an arbitrary position within the region of apoB corresponding to

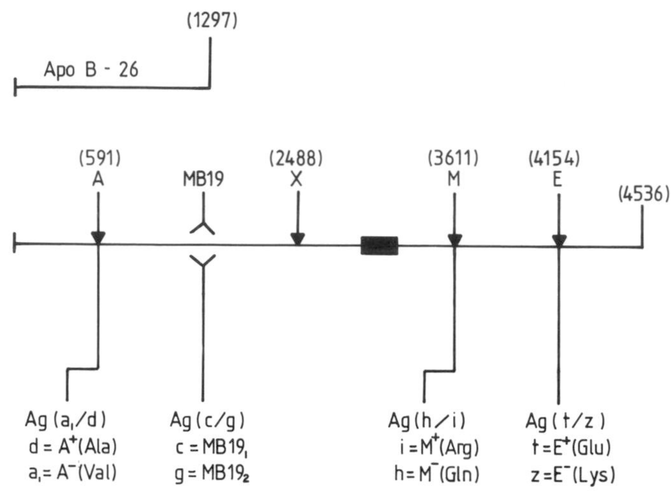

Figure 6.16. Probable positions of four of the five Ag epitope pairs on apoB-100. Vertical arrows show the positions of RFLPs detected with the enzymes *Alu*I (A), *Xba*I (X), *Msp*I (M), and *Eco*RI (E). The numbers above each RFLP site are amino acid residue numbers in mature apoB-100. A plus sign indicates that the restriction site is present. The site of the epitope detected with monoclonal antibody MB19 is placed at an arbitrary position within the apoB-26 segment of apoB-100. The closed rectangle between X and M indicates the probable site of the LDL-receptor-recognition domain.

apoB-26 because monoclonal antibody MB19 detects an epitope in this fragment of the apoB molecule. The *Xba*I polymorphism is more strongly associated with *Ag*(*c*/*g*) than with *Ag*(*x*/*y*) (Ma *et al.*, 1987; Dunning *et al.*, 1988), suggesting that the Ag(c/g) epitope pair is closer to amino acid 2488 than is Ag(x/y). However, as pointed out above, distance of separation is not the only factor that determines the strength of a population association between two adjacent polymorphic loci.

3. RFLPs in the ApoB Gene and the Genetic Contribution to Coronary Artery Disease

There is a substantial genetic component in the etiology of coronary artery disease (CAD) (see Berg, 1983, for review). A small proportion of the genetic contribution to CAD in the population as a whole is due to rare disorders, such as FH, in which a mutation at a single locus has a clinically overwhelming effect on the phenotype, irrespective of other environmental or genetic factors that may be present. In the great majority of cases, however, CAD develops as a consequence of the presence of two or more genes acting in conjunction with environmental factors, i.e., it is multifactorial. This is why CAD tends to cluster in families but rarely segregates as a Mendelian trait.

It is likely that at least some of the variance in liability to CAD in the general population is due to polymorphism at several gene loci whose products are related directly or indirectly to known CAD risk factors or to risk factors that have yet to be identified. The genes at such loci may be called *candidate genes*. On this hypothesis it is assumed that the variant allele at a candidate locus is relatively common in the normal population and that the function or rate of synthesis of its product differs only to a small extent from that of the wild-type allele. It might be supposed, for example, that some alleles at polymorphic sites in the apoB gene are associated with slight changes in the rate of synthesis of apoB or with changes in its primary structure such as to affect the rate of metabolism of LDL or the affinity of LDL for structural elements in the arterial wall. Likewise, it is conceivable that polymorphism at the LDL-receptor locus is associated with minor variations in the functional efficiency of the LDL receptor and that these variations are responsible for some of the population variance in plasma LDL concentration other than that due to FH.

Current strategy for detecting genes that contribute to heart disease in the general population is to search for population associations between CAD (or a risk factor for CAD) and RFLPs at candidate gene loci (see Humphries, 1988; Lusis, 1988; and Cooper and Clayton, 1988, for discussion of the value and limitations of this approach). An association between an allele at an RFLP locus and CAD in a group of unrelated people may mean that the product of the allele itself is causally related to CAD. For example, the RFLP might cause an amino acid change in a functional domain of a protein involved in the metabolism of plasma lipoproteins or in their interaction with the arterial wall. Alternatively, as discussed in Section VII,B above, the association may reflect linkage disequilibrium between the RFLP locus and another polymorphic locus that is directly related to CAD. In either case, the association is unlikely to be a strong one. Moreover, the strength of an association is unlikely to be the same in two different populations. In the first place, since CAD is multifactorial, the combination of factors that lead to clinical expression of the disease will not be the same in all individuals with CAD. Hence, the contribution of a particular gene to CAD may vary between different groups of patients. Secondly, since the extent of linkage disequilibrium between two polymorphic loci in a given population is determined partly by its genetic history, the strength of an association due to linkage disequilibrium may vary from one population to another. This is so even if, in each study, the CAD patients and control subjects belong to the same genetically homogeneous population. It is partly for these reasons that there has been so much disagreement between different studies of the association between CAD and RFLPs in candidate genes, including those in the apoB gene.

Groups of unrelated normal individuals from different parts of the world have been tested for the presence of an association between RFLPs in the apoB gene and plasma lipid concentrations. A significant positive association between plasma cholesterol concentration and the X^+ allele at the *Xba*I polymorphic locus (see Table 6.3) has been observed in some populations (Law *et al.*, 1986b; Berg *et al.*, 1986; Talmud *et al.*, 1987), but in other populations no such association was detected (Deeb *et al.*, 1986; Hegele *et al.*, 1986; Aburatani *et al.*, 1987; Rajput-Williams *et al.*, 1988; Darnfors *et al.*, 1989; Myant *et al.*, 1989). The most likely explanation for these differences is that a polymorphic site in the apoB gene that influences plasma cholesterol concentration is in linkage disequilibrium with the *Xba*I locus in some populations but not in others.

As noted above, the *Eco*RI RFLP involves a change from glutamic acid to lysine at residue 4154, while the *Msp*I RFLP changes residue 3611 from arginine to glutamine (see Table 6.3). Since these changes alter the charge on apoB, both RFLPs might be expected to affect the metabolism of LDL and, hence, to influence the plasma cholesterol concentration. Several groups of workers have sought an association between the *Eco*RI RFLP and plasma cholesterol concentration in unrelated individuals, but the results reported are contradictory. No association was found in the studies of Hegele *et al.* (1986), Talmud *et al.* (1987), and Myant *et al.* (1989). However, Rajput-Williams *et al.* (1988) found significantly higher frequencies of the E^+ allele (*Eco*RI site present) and of the M^- allele (*Msp*I site absent) in men with high plasma cholesterol concentrations than in those with low concentrations. When men with different RFLP haplotypes were compared, a more marked effect was observed. The mean plasma cholesterol concentration in men homozygous for the haplotype E^+/M^- was 56 mg/100 ml higher than that in men homozygous for the haplotype E^-/M^+. Both RFLPs are in the 3′ segment of the apoB gene containing sequences thought to encode the LDL-receptor-binding domain. Hence, these findings raise the possibility that LDL particles containing the apoB variant with Glu 4514 and Gln 3611 are metabolized more slowly by the LDL-receptor pathway than are particles containing the Lys 4514/Arg 3611 variant. This could be tested by comparing the rates of degradation of the two LDL species by fibroblasts *in vitro*.

There are several reports of an association between the E^- and X^- alleles and CAD (Hegele *et al.*, 1986; Paulweber *et al.*, 1989; Myant *et al.*, 1989). However, in two of these studies there was no association between either allele and the plasma LDL concentration. In any case, the X^+ allele has usually been found to be associated with high plasma cholesterol levels (see above). Thus, it seems unlikely that insofar as there is a positive effect of the E^- and X^- alleles on CAD risk in human populations, this is mediated by an effect on the plasma cholesterol level.

4. The Polymerase Chain Reaction (PCR) in the Detection of Genomic Polymorphism

As noted above, a mutation will affect the restriction-fragment pattern of DNA only if it removes an existing restriction site or creates a new one, or if it involves an insertion or deletion of appreciable length. Hence, there must be many polymorphic sites in the human genome that are not detectable as RFLPs. The recent development of the polymerase chain reaction (see footnote, p. 369) has now made it possible to search for polymorphic sites in any region of the genome whose nucleotide sequence is known. In principle, a short segment of DNA from a small number of unrelated people is amplified and the amplified DNA samples are analyzed for differences in nucleotide sequence. Using this approach, Boerwinkle and Chan (1989) have detected a 9-bp insertion/deletion polymorphism in the region of the human apoB gene encoding the signal sequence.

Amplification of selected segments of a gene by PCR may also be combined with *mismatch analysis* (Myers *et al.*, 1985; White *et al.*, 1989) in the detection of point mutations or larger sequence variations. The segment of DNA to be tested (the target DNA) is amplified by PCR and the amplified segment is hybridized with a radiolabeled RNA probe complementary to the corresponding region of wild-type DNA. The RNA/DNA duplex is then digested with RNase A, an enzyme that cleaves RNA sequences only if they are single stranded or mismatched. If the target DNA is identical to wild-type DNA, the RNA probe will not be cleaved. If the target has a single-base substitution, the probe will be cleaved at the site of the mutation, generating two RNA fragments of total length equal to that of the probe. The mutation is characterized by sequencing short segments of the target DNA on either side of the cleavage point, determined roughly from the lengths of the RNA fragments.

RNase A cleavage of mismatched duplexes has been used successfully in the study of some human monogenic disorders (see Stout and Caskey, 1988), but the method detects only about 60% of all point mutations (Myers *et al.*, 1985). Montandon *et al.* (1989) have described a modification that detects all sequence variations in genomic DNA. The amplified target DNA is hybridized with a DNA probe obtained by amplifying the corresponding region of wild-type DNA. The DNA/DNA duplexes are treated with hydroxylamine or osmium tetroxide, followed by piperidine. This results in cleavage of the DNA at all mutant sites. Cleavage sites are determined from the lengths of the DNA fragments, and the mutations are fully characterized by direct sequencing. This procedure should be applicable to the detection of polymorphic sites in specific regions of the genome by identifying mismatches between a standard target DNA and DNA from groups of unrelated members of the general population. In combination with segregation analysis, the method may also provide preliminary, and in some cases definitive, information on the causal mutation in families with a monogenic disorder. It has been used successfully in the study of hemophilia (Montandon *et al.*, 1989) and should be applicable to the study of families with hypobetalipoproteinemia in which the causal mutation is known to be in the apoB gene (see Section X). The possible use of PCR combined with mismatch analysis in the diagnosis of heterozygous FH is mentioned in Chapter 10.

VIII. Polymorphism in the ApoB Gene in Animals

As we saw in Chapter 5, LDL polymorphism detected with alloantibodies has been demonstrated in animals of several species. Polymorphism in the apoB gene detected with restriction enzymes has also been found in mice (Lusis *et al.*, 1987) and pigs (Maeda *et al.*, 1988). Of particular interest is the genetic polymorphism of LDL in pigs involving multiple allelic loci in the apoB gene. Rapacz *et al.* (1986) have shown that pigs with one copy of the Lpb^5 allele, when fed a normal diet, develop raised plasma cholesterol levels due to increased concentrations of LDL and IDL. Moreover, animals with a double dose of the Lpb^5 allele plus another allele (Lpu^1) have plasma cholesterol levels more than twice the normal level. These animals develop severe premature atherosclerosis of the coronary arteries resembling human coronary atherosclerosis. The raised plasma cholesterol level in these pigs appears to be due to a structural abnormality in apoB that affects the catabolism of LDL, since the FCR of "mutant" LDL in normal pigs is about 30% lower than that of normal pig LDL (Checovich *et al.*, 1988). Rapacz *et al.* (1986) have also shown that pigs carrying the Lpb^7 allele have abnormally low plasma cholesterol levels and that their LDL is catabolized by macrophages *in vitro* more efficiently than is Lpb^5 LDL.

The molecular basis of the apoB polymorphism in pigs has been investigated by Maeda *et al.* (1988). A total of six RFLPs was identified in the apoB genes from pigs with eight different *Lpb* alleles. Each *Lpb* allele could be characterized by a specific RFLP haplotype, but none of the polymorphic sites in the apoB gene corresponded to an epitope in apoB-100 detected with an alloantibody. In several *Lpb* alleles Maeda *et al.* (1988) determined the nucleotide sequence across the region thought to encode the LDL-receptor-binding domain. There were no sequence differences that could explain the differences in plasma cholesterol concentration in pigs with alleles Lpb^2 (the wild-type), Lpb^3, Lpb^5, Lpb^7. Maeda *et al.* suggest that polymorphic epitopes in apoB at a distance from the receptor-recognition site can influence the binding of LDL to the LDL receptor.

IX. ApoB (Arg_{3500}→Gln): A Rare Variant

A. The Binding Defect in LDL

In a search for possible causes of hypercholesterolemia other than FH, Vega and Grundy (1986) found five hypercholesterolemic individuals in whom autologous LDL was catabolized more slowly than LDL from a normal donor. Innerarity *et al.* (1987a) went on to show that the LDL from one of the five patients (G.R.)

was bound and degraded by human fibroblasts *in vitro* only about 30% as efficiently as normal LDL. Observations on the LDL from G.R. and his first-degree relatives suggested that the LDL-receptor-binding defect was due to a codominantly inherited abnormality of apoB-100 and that G.R. and his affected relatives were heterozygous for the defect. In agreement with this, Innerarity *et al.* showed that G.R.'s LDL comprised two subpopulations of particles, one with and the other without the binding defect. Since each particle of LDL contains only one molecule of apoB-100, heterozygotes for a mutation in the apoB gene should produce LDL particles containing either one molecule of normal or one molecule of mutant apoB-100. Innerarity *et al.* (1987a) designated this clinical abnormality *familial defective apoB-100* (FDB).

Innerarity *et al.* (1988) identified two FDB patients who were heterozygous for the polymorphic apoB-100 site detected with the monoclonal antibody MB19 (see p. 152). This enabled them to separate the patients' LDL into normal and defective particles by affinity chromatography with MB19. (Normal LDL bound strongly to the MB19 in the column, while defective LDL bound weakly.) Examination of the purified defective LDL showed that it had less than 5% of normal LDL-receptor-binding activity. Since the defective LDL particles are catabolized abnormally slowly *in vivo* they reach a higher concentration in the plasma than normal LDL. In consequence, the LDL of FDB heterozygotes consists of about 70% abnormal particles and about 30% normal particles.

B. Identification of the Mutation

Weisgraber *et al.* (1988) noted that defective LDL from FDB patients bound monoclonal antibody MB47 with abnormally high affinity. The epitope recognized by MB47 was known to be in the C-terminal portion of apoB-100, close to the receptor-recognition domain (Young *et al.*, 1986b), suggesting that the mutation responsible for defective binding is in a segment of the apoB gene encoding this region of apoB-100. Soria *et al.* (1989) examined this region of the gene in eight affected members of two FDB families and in four other FDB individuals unrelated to either family. All the affected individuals were shown to be heterozygous for a single-base substitution (CGG→CAG) in the codon for residue 3500 in mature apoB-100, resulting in an amino acid change from arginine to glutamine. All unaffected individuals were homozygous for the normal allele specifying arginine at position 3500. Since the G→A substitution was the only difference, in this region, between the *FDB* gene and the normal gene in G.R., it is almost certain that the apoB ($Arg_{3500} \rightarrow Gln$) mutation is the cause of the defective receptor binding of LDL, and the consequent hypercholesterolemia, in patients with FDB.

C. Clinical Aspects

The existence of the FDB mutation in the human population raises several questions of considerable interest. What is the frequency of the allele and what is its geographical distribution? If it is a very rare allele, are affected families in different regions of the world descended from a single founder family, or has the mutational event occurred more than once? What is the range of clinical expression of the mutation; in particular, what is the clinical picture in homozygotes? All these questions are being studied by several groups of workers, but it is too early to give any definitive answers.

The mutation was first detected in the North American white population, but it has since been found in Germany, Austria, and England. Four individuals with the mutation were detected by screening about 100 patients with a diagnosis of heterozygous FH at the Hammersmith Hospital Lipid Clinic (Tybjaerg-Hansen *et al.*, 1989). This suggests that a significant proportion of patients who have the clinical signs and family history characteristic of FH are, in fact, carriers of the FDB mutation at codon 3500. The extent to which different affected families are descended from a common founder will not be known until haplotypes at multiple loci in the apoB gene have been compared in FDB patients from different families. Soria *et al.* (1989) showed that in the G.R. family and in two other unrelated index patients the FDB mutation was on a chromosome with the X^- allele at the *Xba*I polymorphic site. As noted in Section VII, in some populations there is a positive association between the X^+ allele and plasma cholesterol concentration. This makes it unlikely that the high plasma cholesterol level associated with the FDB mutation is due to linkage disequilibrium between the locus encoding amino acid residue 3500 and the polymorphic site detected with *Xba*I.

As regards the clinical expression of the mutation, in all FDB families reported so far, hypercholesterolemia was present in every heterozygote. In the family shown in Fig. 6.17 the plasma cholesterol concentration was raised in a 2-year-old child who had inherited the mutation from her mother. The 64-year-old heterozygous father of the index subject had hypercholesterolemia but no tendon xanthomas and no evidence of heart disease on routine clinical examination.[6] As noted in Chapter 10, most, though not all, FH heterozygotes have tendon xanthomas by age 50. These limited observations suggest that FDB is expressed at an early age, perhaps at birth, but that the disease is less pronounced than FH. If the clinical consequences of the FDB mutation are, in fact, less severe than are those of the FH mutation, it would be reasonable to attribute this to the difference between a receptor defect and a ligand defect. Defective LDL-receptor function will affect the catabolism of apoB-containing and apoE-containing lipoproteins, whereas a defect in the receptor-binding domain of apoB will only affect the catabolism of LDL.

[6]Note added in proof. This man had a myocardial infarct at age 65.

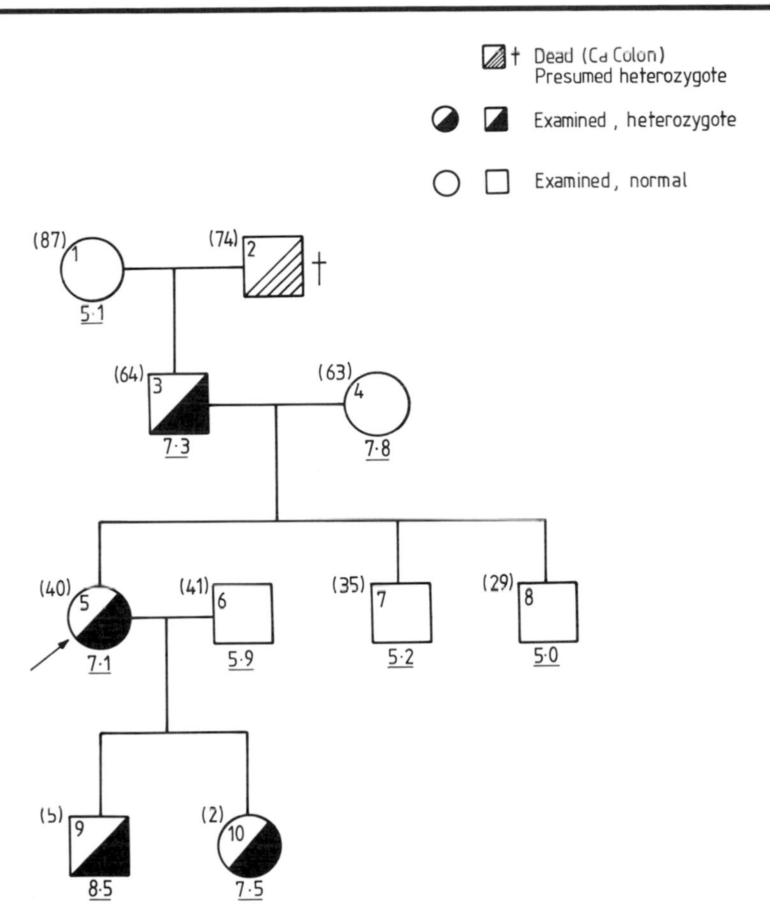

Figure 6.17. The pedigree of a four-generation family with heterozygous familial defective apoB-100 (FDB). Numbers in parentheses are ages. Numbers below each subject are plasma total cholesterol concentrations (mmol/l); arrow indicates index patient. For determination of genotypes, DNA was prepared from the white cells of 10 ml of whole blood. The polymerase chain reaction (see Chapter 9, p. 369) was used to amplify a 345-bp segment of the apoB gene that includes the codon for residue 3500 of apoB-100. The mixture containing the amplified DNA was applied to a nylon membrane by slot blotting and the membranes were incubated with radioactive 13-base synthetic oligonucleotides complementary to either the wild-type or the mutant sequence encoding residue 3500 and adjacent amino acids. The oligonucleotides used were (1) GCACAC [G] GTCTTC (complementary to the noncoding strand of the wild-type sequence) and (2) GCACAC [A] GTCTTC (complementary to the mutant noncoding strand of the sequence containing the G→A substitution). After incubation, the membranes were washed and the radioactive bands detected by autoradiography. Amplified segments from subjects homozygous for the normal allele hybridized only with oligonucleotide (1); those from subjects heterozygous for the mutant allele hybridized with oligonucleotides (1) and (2). ▨† represents dead (Ca colon), presumed heterozygote; ◕, ◪, examined, heterozygote; ○, □, examined, normal. (J. G. Gallagher, N. B. Myant, S. E. Humphries, and A. M. Dunning, unpublished observations.)

X. Mutations Causing Absence or Deficiency of Plasma ApoB

Several primary disorders of lipoprotein metabolism are characterized by a selective absence or deficiency of plasma lipoproteins containing apoB. In many cases, the abnormality is inherited as a Mendelian trait, showing that the underlying lesion is a mutation at a single gene locus. In others, such as normotriglyceridemic abetalipoproteinemia (see below), the condition is assumed to be monogenic, although the number of known affected individuals is too small for pedigree analysis. In some families carrying a monogenically inherited deficiency of apoB, the molecular lesion has been shown to be a mutation in the apoB gene; in others, an apoB gene mutation has been excluded by segregation analysis.

A. Abetalipoproteinemia and Familial Hypobetalipoproteinemia

1. Modes of Inheritance

The first familial apoB-deficiency state to be recognized was an inherited disorder, occurring in both sexes and characterized by (1) the complete absence of LDL (β-lipoprotein) from the plasma, (2) malabsorption of fat due to failure to form chylomicrons, (3) retinitis pigmentosa, (4) acanthocytosis (abnormal spiny red cells in the circulation), and (5) a progressive neurological disorder. Since the absence of β-lipoprotein from the plasma was thought to underlie all the other manifestations of the disease, it was called *abetalipoproteinemia* (ABL) (see Salt *et al.*, 1960; Lamy *et al.*, 1960). Subsequently, it was shown that the plasma of ABL patients contains no immunodetectable apoB and that all lipoproteins containing apoB-100 or apoB-48 are absent from plasma taken both in the fasting state and after a fatty meal. Analysis of most of the families of patients with ABL showed no detectable biochemical or clinical abnormality in obligate heterozygotes, a high frequency of consanguinity in the parents of affected individuals, and no vertical transmission. Hence, it was concluded that the mode of inheritance of ABL in most families is autosomal recessive.

Within a few years of the recognition of ABL as an inborn error of lipoprotein metabolism, a familial abnormality characterized by *partial* deficiency of plasma β-lipoprotein was reported from several laboratories. This disorder was called *familial hypobetalipoproteinemia* (HBL) (see Herbert *et al.*, 1983, for historical background and for clinical and biochemical details). In contrast to ABL, apoB-containing lipoproteins are present in the plasma in HBL, though the LDL concentration is usually about half the age-adjusted normal. Most affected individuals have no clinical or hematological abnormalities; in particular, fat absorption is usually normal and chylomicrons appear in the plasma after a fatty meal.

The life expectancy of affected members of HBL families is several years longer than that of the general population, the difference being due to a reduced risk of death from coronary artery disease in HBL (Glueck *et al.*, 1976). In view of the heterozygous advantage of HBL, the low population frequency of mutations giving rise to this condition calls for explanation.

Pedigree analysis of HBL kinships has shown that the disorder is inherited as an autosomal dominant trait and that affected individuals are heterozygous for the mutation. In agreement with this conclusion, vertical transmission through two (Richet *et al.*, 1969) and three generations (Mars *et al.*, 1969; Cottrill *et al.*, 1974; Young *et al.*, 1987b) has been reported. Figure 6.18 shows the inheritance of HBL in a kinship investigated by Mars *et al.* (1969). Of the 31 examined members, 13 expressed the abnormality (plasma cholesterol concentration <150 mg/100 ml). Note that the trait was expressed in three successive generations and that in several families within the kinship a parent with HBL had at least one normal offspring. These features are consistent with the conclusion that every individual with HBL in this kinship was heterozygous for a dominant mutation.

The existence of a dominantly inherited disorder expressed as HBL in the heterozygous state raised the question as to how a double dose of the mutant gene would be expressed in homozygotes. Cottrill *et al.* (1974) were the first to give an unequivocal description of a kinship containing one or more individuals homozygous for the HBL mutation. (As noted below, the ABL patient of Salt *et al.*, 1960, was probably homozygous for this mutation, but a full pedigree analysis of the patient's family is not available.)

Figure 6.19 shows the pedigree of the kinship reported by Cottrill *et al.* (1974). HBL (plasma LDL-cholesterol concentration <70 mg/100 ml) was present in three successive generations and in the two matings between HBL (II,1 and III,1) and normal parents, half the examined offspring were normal and half had HBL. This ratio would be expected if both HBL parents were heterozygotes. The index patient (aged 6 months) and her brother (aged 6 years) had the plasma lipoprotein abnormality and its associated clinical syndrome previously described in patients with ABL. There was no immunoreactive LDL in the plasma, fat absorption was grossly impaired, and acanthocytes were present in the circulation. Neither had retinitis pigmentosa or neurological signs, but both were too young for these to have developed as a consequence of ABL. The parents of the two patients were related, insofar as the father's great-grandmother was the sister of the mother's grandfather. These two ancestors were presumably heterozygous carriers of the *HBL* gene. Thus, we may conclude that the mutation was transmitted from a heterozygous brother and sister along two branches of the kinship, eventually giving rise to two homozygous offspring of a consanguineous marriage.

Since the report of Cottrill *et al.* (1974), at least three other HBL families have been described in which one or more members have ABL, together with a

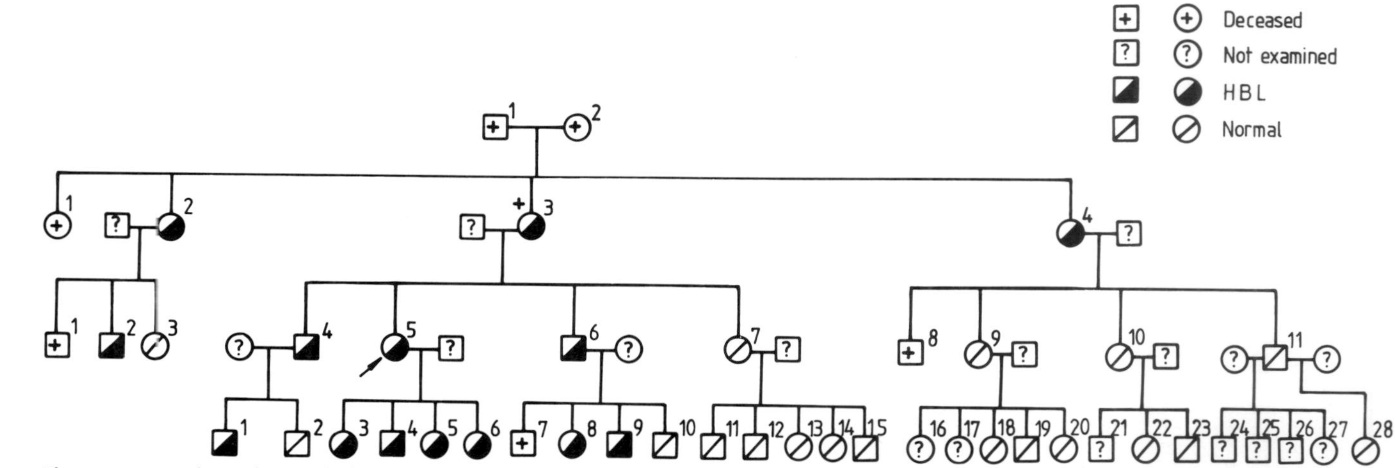

Figure 6.18. The pedigree of a kinship with familial hypobetalipoproteinemia (HBL). The lower limit of normal serum total cholesterol concentration was taken to be 150 mg/100 ml. Arrow shows the index subject. Note that subject 3 in generation II was not examined. She was assumed to be a heterozygous carrier of the HBL mutation because of the presence of HBL in two of her sibs and in three of her four offspring. (Redrawn from Mars *et al.*, 1969.)

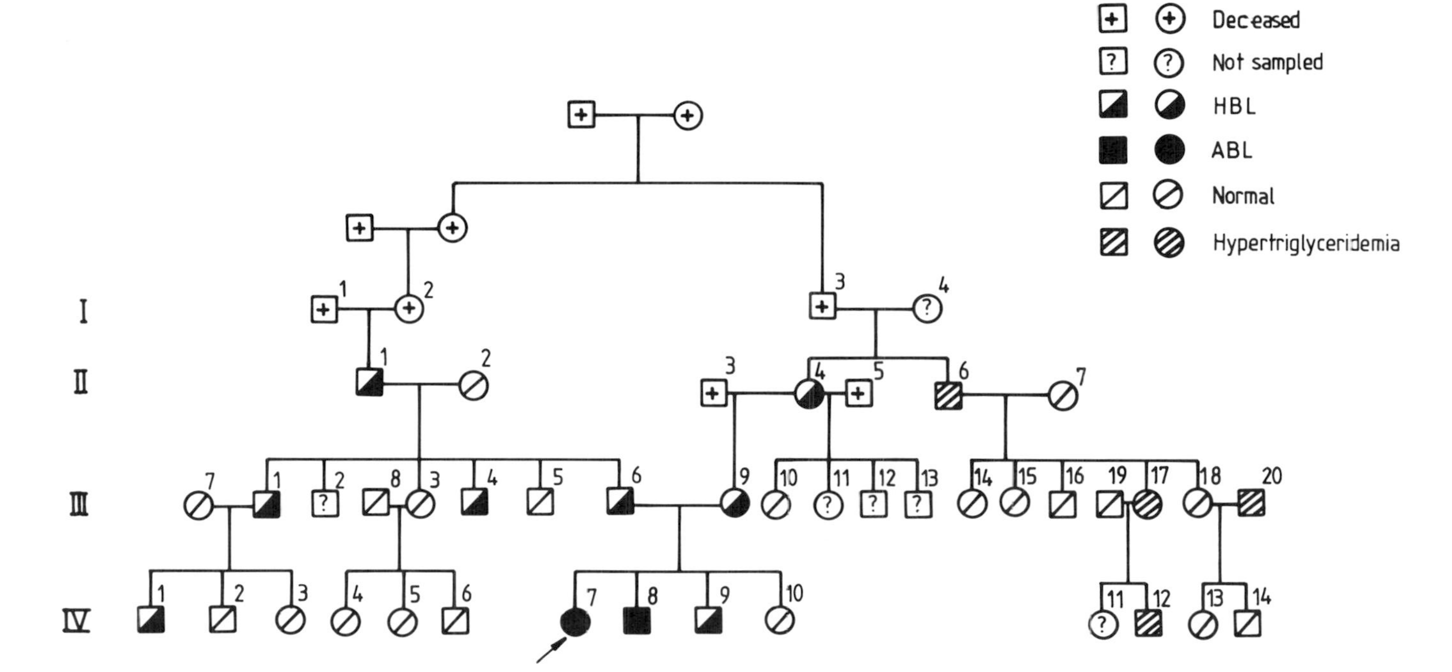

Figure 6.19. The pedigree of a kinship with familial hypobetalipoproteinemia (HBL), showing two members with abetalipoproteinemia (ABL). In the subjects diagnosed as having HBL, the plasma LDL-cholesterol concentration ranged from 17 to 69 mg/100 ml. No immunoreactive LDL was detectable in the plasma of IV, 7 or IV, 8. (Redrawn from Cottrill *et al.*, 1974.)

clinical syndrome indistinguishable from that seen in patients with ABL in whom the abnormality is inherited as an autosomal recessive trait (see Herbert *et al.*, 1983). In every HBL kinship in which a patient with ABL has been identified, both parents have had HBL and in five of the seven reported cases the parents were related. Thus, the ABL patients in these kinships appear to be homozygous for a dominant mutation. In retrospect, the patient with ABL described by Salt *et al.* (1960) was probably homozygous for the dominantly inherited HBL mutation, since the plasma LDL concentration was low in both her parents and in her paternal grandfather.

The existence of two forms of ABL, distinguishable only by their different modes of inheritance, poses a problem in nomenclature. When ABL was first shown to occur in families carrying the HBL mutation, the condition was called *familial homozygous hypobetalipoproteinemia* to distinguish it from recessively inherited ABL (sometimes referred to as classical ABL because of its earlier recognition). This is not a satisfactory solution to the problem because patients who have abetalipoproteinemia (Gr. *a*-, prefix meaning *without* and implying *absence of*) do not, by definition, have HBL. It would be better to use the term ABL to refer to an inherited absence of apoB-containing lipoproteins from the plasma, irrespective of the mode of inheritance. If obligate heterozygotes can be examined, it should then be possible to determine whether inheritance is dominant or recessive in a particular patient. As we shall see below, the HBL mutation is heterogeneous and so, probably, is the mutation giving rise to the less uncommon, recessive form of ABL.

The fact that HBL is dominantly inherited would be understandable if, as seems to be the case, all HBL mutations are in the gene encoding the product (apoB) by which the phenotype is recognized. In the presence of one normal and one mutant allele it is likely that only half the normal amount of product would be formed and, hence, that the abnormality would be expressed in heterozygotes. As discussed below, the molecular basis of the recessive form of ABL, at least in some families, is a mutation in a gene, other than the apoB gene, whose product is required for the normal processing of apoB. In heterozygotes the product of the one normal allele might be sufficient for normal processing of apoB. In this case, inheritance would be recessive.

2. Molecular Basis

In some patients with ABL, immunoreactive apoB is not detectable in hepatocytes or intestinal cells (Schwartz *et al.*, 1978; Glickman *et al.*, 1979; Green *et al.*, 1982; Levy *et al.*, 1987; Infante, 1987) or is detectable only in greatly reduced amounts (Ross *et al.*, 1988); in the two unrelated abetalipoproteinemic patients of Levy *et al.* (1987) no synthesis of apoB-48 or apoB-100 could be detected in primary cultures of jejunum. In other ABL patients, intracellular apoB has been demonstrated in normal or increased amounts in the liver and/or small intestine (Dullaart *et al.*, 1986; Lackner *et al.*, 1986; Infante, 1987). In two of the four patients examined by Lackner *et al.*, the concentration of normal-length apoB mRNA in hepatocytes was six times the normal level, suggesting that transcription of the apoB gene was increased owing to lack of feedback repression by plasma apoB-containing lipoproteins.

These results show that ABL is genetically heterogeneous. It is difficult to say how far this heterogeneity is related to the manner in which the mutation is inherited, because in most of the above reports no details are given from which the mode of inheritance could be deduced. However, it is worth noting that among the seven ABL patients in whom the mutation was shown to be recessive (Dullaart *et al.*, 1986; Lackner *et al.*, 1986; Levy *et al.*, 1987), apoB was present in normal or increased amounts in the liver or intestine in five (those of Dullaart *et al.* and Lackner *et al.*). This is consistent with evidence discussed below that the mutation in recessively inherited ABL usually does not involve the apoB gene.

It should be possible to tell whether or not the mutation in a family with ABL or HBL is in the apoB gene by testing for linkage between the expression of the abnormality and alleles at polymorphic sites in or close to the apoB gene, provided that the matings are informative. Talmud *et al.* (1988) have used this approach in a study of two unrelated ABL families, each with two affected homozygous offspring. In both families inheritance of the mutation was recessive. The two affected offspring inherited different alleles of the apoB gene from both parents in one family (Fig. 6.20) and from one parent in the other. Thus, in both families the mutation failed to segregate with the apoB gene. Since the ABL in these families was recessive, affected sibs should have inherited the same gene at the mutant locus in each heterozygous parent. Hence, the mutation causing ABL in these families must have been in a gene other than the apoB

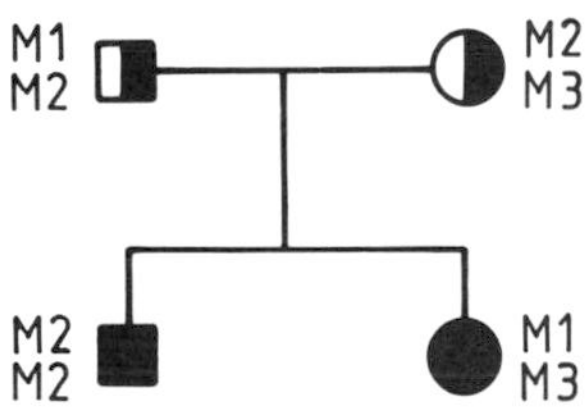

Figure 6.20 Segregation analysis of a family with abetalipoproteinemia due to a recessive mutation. Both parents were obligate heterozygotes for the recessive mutation and both had normal plasma LDL concentrations. The two offspring were homozygotes and had abetalipoproteinemia. The family carried three different alleles of the apoB gene, detected with the restriction enzyme *Msp*I. These are designated *M1*, *M2*, and *M3*. The male sib inherited an *M2* apoB allele from both parents. The female inherited *M1* from her father and *M3* from her mother. Since the two sibs inherited different apoB genes from both parents, the mutation cannot have been in the apoB gene. (Redrawn from Talmud *et al.*, 1988. Reproduced from the *Journal of Clinical Investigation*, 1988, Vol. 82, pp. 1803–1806 by copyright permission of the American Society for Clinical Investigation.)

gene. This is in line with the suggestion, discussed above, that in the majority of patients with recessively inherited ABL the ability to synthesize immunoreactive apoB is retained. Many steps, each requiring a separate gene product, must be involved in the posttranslational modification of apoB, in its incorporation into a lipoprotein particle, and in the secretion of the particle into the external medium. Mutations in any of the genes encoding these proteins might lead to accumulation of apoB in hepatocytes and intestinal cells and its consequent absence from the plasma.

In contrast to the observations of Talmud *et al.* (1988), Leppert *et al.* (1988) have described an HBL family with 17 affected members in which the lipoprotein abnormality cosegregated with the apoB gene. These findings suggest that in families with HBL the mutation is in the apoB gene itself or its promoter region. This is borne out by the recent identification, in several unrelated HBL families, of apoB-gene mutations giving rise to truncated apoBs (see Table 6.4). The first to be identified was apoB-37. The events leading to the discovery of this mutation and to the elucidation of the molecular lesion at the DNA level are worth recording in some detail.

In 1979, Steinberg *et al.* described a family with atypical HBL in which the index patient (H.J.B.) and two of his sibs had marked HBL (plasma LDL-

Table 6.4
Mutations in the ApoB Gene Giving Rise to Truncated ApoBs in Familial Hypobetalipoproteinemia

Predicted amino acid chain length	MW of abnormal apoB on centile scale[a]	Molecular basis	Reference
1728	37	Frameshift due to 4-bp deletion. Gives STOP at codon 1730.	Young *et al.* (1988)
1799	39	Frameshift due to single-bp deletion. Gives STOP at codon 1800.	Collins *et al.* (1988)
1305	(26) Not detected in plasma	C → T in codon 1306. Changes Arg to STOP.	Collins *et al.* (1988)
1830	40[b]	Deletion of TG gives STOP after codon 1830.	Krul *et al.* (1989); Talmud *et al.* (1989)
2057	46	C → T in codon 2058 gives STOP after codon 2057.	Young *et al.* (1989)
4040	89[b]	Deletion of G gives STOP after codon 4040.	Krul *et al.* (1989); Talmud *et al.* (1989)

[a]MW of B100 ≃513K.

[b]Two mutant apoBs inherited in the same family, giving a genetic compound with apoB-40 and apoB-89 in plasma.

cholesterol concentration <10 mg/100 ml), mild fat malabsorption, and normal plasma triglyceride levels. In several other members of the family the plasma LDL-cholesterol level was about half the normal value. Metabolic studies showed that the rate of production of LDL apoB in H.J.B. was less than 10% of normal. Steinberg *et al.* (1979) suggested that the inherited abnormalities in this family were due to a dominant mutation causing more severe deficiency of LDL in homozygotes (including the index patient) than in heterozygotes.

Young *et al.* (1987b), in a reexamination of this family, identified an abnormal 203K apoB (apoB-37) in all the plasma lipoproteins of H.J.B. (subject 1, Fig. 6.21), in his two severely hypobetalipoproteinemic sibs (subjects 2 and 3), and in several of their relatives. Although apoB-37 was present in all lipoproteins of these subjects, the majority was found in the HDL fraction, in which it was the only apoB species. The apoB-37 in HDL was restricted to a subfraction of large particles (Lp-B37) containing no apoA-I. ApoB-100 and apoB-48 were also present in the plasma of subjects 1, 2, and 3, but at concentrations less than 1% of normal. Characterization of apoB-37, using partial amino acid sequencing and mapping of antigenic sites detected with monoclonal antibodies to specific segments of apoB-100, indicated that it is colinear with the N-terminal 203K portion of apoB-100 and that it contains no sequences from the C-terminal end of apoB-100. The monoclonal antibodies that bound to apoB-37 included MB19, the antibody reacting with the apoB allotypes $MB19_1$ and $MB19_2$ (see p. 152).

Figure 6.21 shows the findings in 41 members of the H.J.B. family. In addition to the three subjects with severe HBL and apoB-37 (subjects 1, 2, and 3), 16 others had moderately reduced plasma LDL-cholesterol concentrations. Six of these had apoB-37 in their plasma; in the other 10, apoB-37 was not detectable. To explain this pedigree, Young *et al.* (1987a) postulated the presence of two mutant *apoB* alleles, one leading to the production of apoB-37 present at very low concentration in plasma (the *apoB-37* allele), and the other associated with hypobetalipoproteinemia due to a marked reduction in the plasma concentrations of normal apoB-100 and apoB-48 (the *H* allele). In this scheme, subjects 1, 2, and 3 have both mutant alleles and are therefore genetic compounds; subjects 13, 14, 18, 28, 32, and 34 have inherited one copy of the *apoB-37* allele and one normal *apoB* allele, while the remaining 10 subjects with HBL but not apoB-37 have inherited one copy of the *H* allele and one normal *apoB* allele. Direct analysis of the binding of antibody MB19 to apoB-37 showed that, in all subjects in whom apoB-37 was present, the abnormal protein was the product of the *apoB* allele encoding the allotype $MB19_1$. Hence, this allele was called $MB19_1$*-37*. In contrast, the small amounts of normal apoB-100 and apoB-48 in subjects 1, 2, and 3 were shown to be products of the *apoB* allele encoding allotype $MB19_2$. This allele was called $MB19_2$*-H*. In agreement with these findings, linkage analysis of informative matings within this pedigree showed that the inheritance of moderate HBL without apoB-37 always segregated with the allele encoding allotype $MB19_2$.

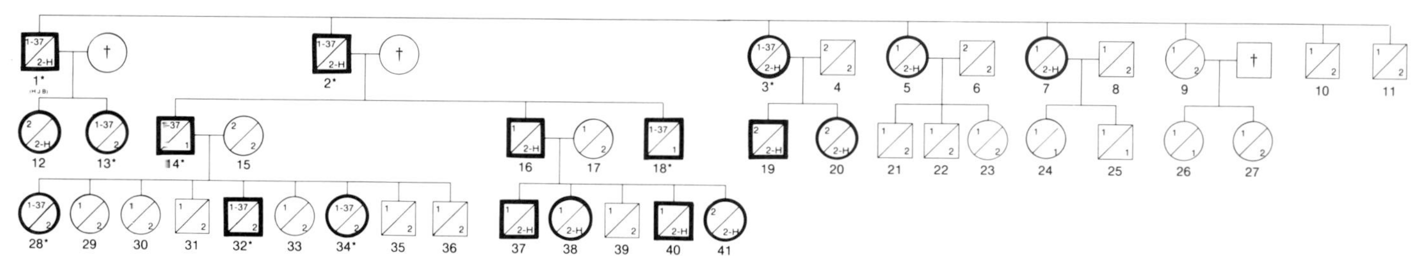

Figure 6.21. The H.J.B. Kindred. H.J.B.is subject 1. All other subjects are referred to by numbers. A square (male) or circle (female) with a thick line indicates that the plasma LDL-cholesterol concentration was below the fifth percentile of age- and sex-matched controls. A dagger indicates that the subject is dead. An asterisk shows that the plasma contained apoB-37. The *apoB* genotype is shown for each subject. 1, normal *MB19₁* allele; 2, normal *MB19₂* allele; 1-37, *MB19₁–37* allele; 2-H, *MB19₂-H* allele. (From Young *et al.*, 1987a, with the permission of the authors. Reproduced from the *Journal of Clinical Investigation*, 1987, Vol. 79, pp. 1842–1851 by copyright permission of the American Society for Clinical Investigation.)

The linkage between the MB19 polymorphism and the *apoB-37* and *H* alleles showed that both alleles are in, or closely adjacent to, the apoB gene itself. In keeping with this, the two alleles failed to complement each other when present together in the genetic compounds.

Young *et al.* (1987c) have isolated Lp-B37 from the HDL of members of the H.J.B. family, using immunoaffinity chromatography with an antibody to apoB-37. Measurement of high-affinity binding and degradation by human fibroblasts *in vitro* showed that Lp-B37 is not recognized by the LDL receptor. Since apoB-37 (comprising only the N-terminal portion of apoB-100) is the sole apoB species in Lp-B37, this observation supports the suggestion of Knott *et al.* (1986) that apoB-100 contains a single receptor-binding domain near its C-terminal end.

From a consideration of the molecular weight of apoB-37 and its relationship to apoB-100, Young *et al.* (1988) concluded that the C-terminal amino acid of apoB-37 lies between residues 1725 and 1750 of apoB-100, a region encoded by the 5′ end of exon 26 of the apoB gene. Sequencing of cloned segments from this region of genomic DNA, taken from a member of the H.J.B. family heterozygous for the apoB-37 mutation, revealed the presence of a 4-bp deletion beginning in codon 1728. This results in a frameshift and a premature stop codon after a single additional amino acid, giving rise to a truncated mature apoB protein with 1728 residues (MW $\simeq$193K).

The apoB-39 described by Collins *et al.* (1988) was identified in a patient with dominantly inherited ABL (both her parents had HBL). She had virtually no LDL in her plasma, but small amounts of an abnormal apoB with MW $\simeq$201K were present in her triglyceride-rich lipoproteins. Sequencing of amplified segments of the *apoB* gene, encoding the presumed carboxyl end of the mutant apoB, revealed a single-bp deletion in codon 1794. This resulted in a frameshift and a stop codon after five more codons. The mutant gene encodes a mature apoB protein with only the first 1799 residues of apoB-100. Young *et al.* (1989) have also identified an HBL family heterozygous for a mutant *apoB* allele encoding a truncated apoB equivalent to apoB-46. The mutant allele has a single-base substitution that changes codon 2058 from arginine (CGA) to STOP (TGA). ApoB-46 was present in all lipoprotein fractions of affected subjects, but the total plasma concentration of apoB-46 was much less than that of apoB-100.

In another HBL family investigated by Collins *et al.* (1988), an abnormal *Taq*I fragment of the apoB gene was detected with a radioactive cDNA probe hybridizing with exon 25. Sequencing of this region of the patient's apoB gene showed the presence of a single-bp substitution in codon 1306, converting it to a stop codon. The resulting apoB mutant, apoB ($Arg_{1306}\rightarrow$Term), has only the N-terminal 1305 residues of apoB-100 and is equivalent to apoB-26. This truncated protein was not detectable in the patient's plasma. Krul *et al.* (1989) have described an HBL family with two mutations in the apoB gene, one encoding apoB-40 and the other encoding apoB-89. Talmud *et al.* (1989) have shown that the *apoB-40* allele has a deletion of 2 bp, giving rise to a stop codon at position 1830, and that the *apoB-89* allele has a stop codon at position 4040 due to deletion of a single bp from codon 4034. Figure 6.22 shows the pedigree of this family. The index patient and two of her sisters are compound heterozygotes, each carrying both the mutant *apoB* alleles. One brother has one copy of the *apoB-89* allele and another brother has one copy of the *apoB-40* allele.

Huang *et al.* (1988a) have reported a patient with dominantly inherited ABL who was homozygous for an apoB gene from which exon 21 was deleted. Linkage analysis of the patient's family showed that the mutation segregated with HBL, indicating that it was the cause of HBL in this family. The apoB in the patient's plasma was not referred to by the authors.

The identification of at least six apoB-gene mutations giving rise to premature termination of translation suggests that a search among families with HBL would bring to light other mutations that result in the production of truncated apoBs of various lengths. Naturally occurring lipoproteins in which the apoB component is of abnormal length may provide information about the sites of functional domains of apoB. As already noted, the failure of Lp-B37 to bind to the LDL receptor supports the view that there are no receptor-recognition sites in the N-terminal portion of apoB-100. Mutant apoBs terminating at positions closer to the receptor-recognition domain postulated by Knott *et al.* (1986) would be of even greater interest. ApoB-39 was present in LDL and triglyceride-rich lipoproteins in the subjects studied by Collins *et al.* (1988), whereas most of the plasma apoB-37 in the H.J.B. family was present in lipoproteins with the density of

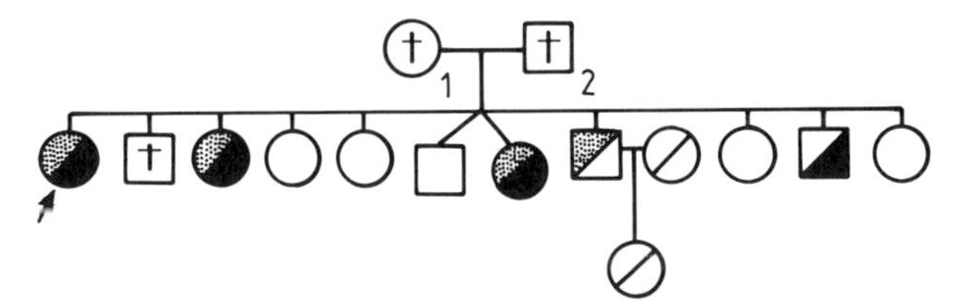

Figure 6.22. The pedigree of a family with familial hypobetalipoproteinemia due to the presence of two mutant *apoB* alleles, one encoding apoB-89 and the other encoding apoB-40. The index patient (shown by an arrow) and two of her sisters are compound heterozygotes. Two of her brothers are heterozygotes. (From Krul *et al.*, 1989, with the permission of the authors.)

HDL. This suggests that the amino acid sequence between residues corresponding to the C-termini of apoB-37 and apoB-39 includes a domain required for the formation of lipoprotein particles of density less than 1.063 g/ml.

It is not known why truncated apoB mutants are present in plasma only at very low concentration. Lipoprotein particles containing the abnormal apoB may be cleared very rapidly from the circulation, although the slow rate of catabolism of Lp-B37 by skin fibroblasts *in vitro* (Young *et al.*, 1987c) argues against this. Another possibility is that the rate of secretion of lipoproteins containing the abnormal apoB is greatly reduced, perhaps because much of the truncated protein synthesized is degraded intracellularly without incorporation into lipoprotein. Measurement of the fractional and absolute catabolic rates of lipoproteins containing the mutant apoB species should help to distinguish between these two possibilities. Levy *et al.* (1970) and Sigurdsson *et al.* (1977) observed normal FCRs and greatly reduced rates of production of radioactive isologous LDL in heterozygotes with HBL from two unrelated families. However, these measurements must have reflected predominantly the behavior of the subjects' normal LDL apoB-100 present in plasma at half the normal concentration.

B. Normotriglyceridemic Abetalipoproteinemia

Malloy *et al.* (1981) have described a variant form of ABL in an 8-year-old girl. In this patient, lipoproteins containing apoB-100 were absent from the plasma, but chylomicrons containing normal amounts of apoB-48 were secreted into the plasma after the ingestion of fat. Moreover, there was no evidence of fat malabsorption and acanthocytes were not present in significant numbers in the blood circulation. No first-degree relatives of the patient could be examined, but Malloy *et al.* suggested that she had an inherited disease leading to selective deletion of apoB-100, with normal production of apoB-48. They called the disease *normotriglyceridemic abetalipoproteinemia.*

Two other unrelated patients have been reported with lipoprotein abnormalities and clinical features similar to those in the patient of Malloy *et al.* In both patients, one described by Takashima *et al.* (1985) and the other by Herbert *et al.* (1985), apoB-100 was not detectable in the plasma, but apoB-48-containing chylomicrons appeared in the plasma after fat ingestion, though in subnormal amounts. Both patients had fat malabsorption and moderate acanthocytosis. In both families the parents were clinically normal.

The lipoprotein abnormality has not been proved to be genetic in any of these three patients. However, when taken together the findings are consistent with the existence of a group of recessive defects that abolish apoB-100 but spare apoB-48 production to varying degrees—completely in the patient of Malloy *et al.* and partially in the other two. One of several possibilities is that the three families have mutations in the 3′ half of the apoB gene at different distances from the site corresponding to the apoB-48 stop codon in intestinal apoB mRNA.

References

Aburatani, H. Murase, T., Takaku, F., Itoh, H., Matsumoto, A., and Itakura, H. (1987). Apolipoprotein B-gene polymorphism and myocardial infarction. *N. Engl. J. Med.* **317,** 52–53.

Allison, A. C., and Blumberg, B. S. (1961). An isoprecipitation reaction distinguishing human serum-protein types. *Lancet* **1,** 634–637.

Armstrong, V. W., Walli, A. K., and Seidel, D. (1985). Isolation, characterization, and uptake in human fibroblasts of an apo(a)-free lipoprotein obtained on reduction of lipoprotein(a). *J. Lipid Res.* **26,** 1314–1323.

Armstrong, V. W., Cremer, P., Eberle, E., Manke, A., Schulze, F., Wieland, H., Kreuzer, H., and Seidel, D. (1986). The association between serum Lp(a) concentrations and angiographically assessed coronary atherosclerosis. Dependence on serum LDL levels. *Atherosclerosis* **62,** 249–257.

Berg, K. (1963). A new serum type system in man—the Lp system. *Acta Pathol. Microbiol. Scand.* **59,** 369–382.

Berg, K. (1983). Genetics of coronary heart disease. *In* "Progress in Medical Genetics" (A. G. Steinberg, A. G. Bearn, A. Motulsky, and B. Childs, eds.), Vol. 5A, pp. 35–90. Saunders, Philadelphia.

Berg, K., Powell, L. M., Wallis, S. C., Pease, R., Knott, T. J., and Scott, J. (1986). Genetic linkage between the Ag antigenic variation and the apolipoprotein B gene: Assignment of the Ag locus. *Proc. Natl. Acad. Sci.* U.S.A. **83,** 7367–7370.

Bhattacharya, S., and Redgrave, T. G. (1981). The content of apolipoprotein B in chylomicron particles. *J. Lipid Res.* **22,** 820–828.

Blackhart, T. B. D., Ludwig, E. M., Pierotti, V. R., Caiati, L., Onasch, M. A., Wallis, S. C., Powell, L., Pease, R., Knott, T. J., Chu, M-L., Mahley, R. W., Scott, J., McCarthy, B. J., and Levy-Wilson, B. (1986). Structure of the human apolipoprotein B gene. *J. Biol. Chem.* **261,** 15364–15367.

Bodmer, W. F., and Cavalli-Sforza, L. L. (1976). "Genetics, Evolution and Man." Freeman, San Francisco.

Boerwinkle, E., and Chan, L. (1989). A three codon insertion/deletion polymorphism in the signal peptide region of the human apolipoprotein B (apoB) gene directly typed by the polymerase chain reaction. *Nucleic Acids Res.* **17,** 4003.

Boguski, M. S., Freeman, M., Elshourbagy, N. A., Taylor, J. M., and Gordon, J. I. (1986). On computer-assisted analysis of biological sequences: proline punctuation, consensus sequences, and apolipoprotein repeats. *J. Lipid Res.* **27,** 1011–1034.

Breslow, J. L. (1985). Human apolipoprotein molecular biology and genetic variation. *Ann. Rev. Biochem.* **54,** 699–727.

Brown, M. S., and Goldstein, J. L. (1986). A receptor-mediated pathway for cholesterol homeostasis. *Science* **232,** 34–47.

Bütler, R., and Brunner, E. (1974). Ag(i): detection of an antithetical factor to Ag(h). *Vox Sang.* **27,** 550–555.

Cardin, A. D., Witt, K. R., Chao, J., Margolius, H. S., Donaldson, V. H., and Jackson, R. L. (1984). Degradation of apolipoprotein B-100 of human plasma low density lipoproteins by tissue and plasma kallikreins. *J. Biol. Chem.* **259,** 8522–8528.

Carlsson, P., Olofsson, S. O., Bondjers, G., Darnfors, C., Wiklund, O., and Bjursell, G. (1985). Molecular cloning of human apolipoprotein B cDNA. *Nucleic Acids Res.* **13,** 8813–8826.

Chan, L. (1989). The apoprotein multigene family: structure, expression, evolution, and molecular genetics. *Klin. Wochenschr.* **67,** 225–237.

Chapman, M. J. (1980). Animal lipoproteins: chemistry, structure, and comparative aspects. *J. Lipid Res.* **21,** 789–853.

Chapman, M. J., Forgez, P., and Innerarity, T. L. (1986). Structure–function studies of apolipoprotein B100 and the search for mutants. *In* "Human Apolipoprotein Mutants" (C. R. Sirtori, A. V. Nichols, and G. Franceschini, eds.), pp. 169–178. Plenum, New York.

Checovich, W. J., Fitch, W. L., Krauss, R. M., Smith, M. P., Rapacz, J., Smith, C. L., and Attie, A. D. (1988). Defective catabolism and abnormal composition of low density lipoproteins from mutant pigs with hypercholesterolemia. *Biochemistry* **27,** 1934–1941.

Chen, S-H., Habib, G., Yang, C-Y., Gu, Z-W., Lee, B. R. Weng, S-A., Silberman, S. R., Cai, S-J., Deslypere, J. P., Rosseneu, M., Gotto, A. M., Li, W-H., and Chan, L. (1987). Apolipoprotein B-48 is the product of a messenger RNA with an organ-specific in-frame stop codon. *Science* **238,** 363–366.

Chen, S-H., Yang, C-Y., Chen, P-F., Setzer, D., Tanimura, M., Li, W-H., Gotto, A. M., and Chan, L. (1986). The complete cDNA and amino acid sequence of human apolipoprotein B-100. *J. Biol. Chem.* **261,** 12918–12921.

Chou, P. Y., and Fasman, G. D. (1978). Empirical predictions of protein conformation. *Ann. Rev. Biochem.* **47,** 251–276.

Cladaras, C., Hadzopoulou-Cladaras, M., Nolte, R. T., Atkinson, D., and Zannis, V. I. (1986). The complete sequence and structural analysis of human apolipoprotein B-100: relationship between apoB-100 and apoB-48 forms. *EMBO J.* **5,** 3495–3507.

Collins, D. R., Knott, T. J., Pease, R. J., Powell, L. M., Wallis, S. C., Robertson, S., Pullinger, C. R., Milne, R. W., Marcel, Y. L., Humphries, S. E., Talmud, P. J., Lloyd, J. K., Miller, N. E., Muller, D., and Scott, J. (1988). Truncated variants of apolipoprotein B cause hypobetalipoproteinaemia. *Nucleic Acids Res.* **16,** 8361–8375.

Cooper, D. N., and Clayton, J. F. (1988). DNA polymorphism and the study of disease association. *Hum. Genet.* **78,** 299–312.

Cooper, D. W., and Schmidtke, J. (1984). DNA restriction fragment length polymorphisms and heterozygosity in the human genome. *Hum. Genet.* **66,** 1–16.

Corsini, A., Spilman, C. H., Innerarity, T. L., Arnold, K. S., Rall, S. C. Boyles, J. K., and Mahley, R. W. (1987). Receptor binding activity of lipid recombinants of apolipoprotein B-100 thrombolytic fragments. *J. Lipid Res.* **28,** 1410–1423.

Cottrill, C., Glueck, C. J., Leuba, V., Millett, F., Puppione, D., and Brown, W. V. (1974). Familial homozygous hypobetalipoproteinemia. *Metabolism* **23,** 779–791.

Curtiss, L. K., and Edgington, T. S. (1982). Immunochemical heterogeneity of human plasma apolipoprotein B. II. Expression of apolipoprotein-B epitopes on native lipoproteins. *J. Biol. Chem.* **257,** 15213–15221.

Darnfors, C., Nilsson, J., Protter, A. A., Carlsson, P., Talmud, P. J., Humphries, S. E., Whalström, J., Wiklund, O., and Bjursell, G. (1986). RFLPs for the human apolipoprotein B gene: *Hinc*II and *Pvu*II. *Nucleic Acids Res.* **14,** 7135.

Darnfors, C., Wiklund, O., Nilsson, J., Gerard, B., Carlsson, P., Johansson, S., Bondjers, G., and Bjursell, G. (1989). Lack of correlation between the apolipoprotein *Xba*I polymorphism and blood lipid levels in a Swedish population. *Atherosclerosis* **75,** 183–188.

Das, H. K., Leff, T., and Breslow, J. L. (1988). Cell type-specific expression of human apoB gene is controlled by two *cis*-acting regulatory regions. *J. Biol. Chem.* **263,** 11452–11458.

Dashti, N., Alaupovic, P., and Williams, D. L. (1988). Direct effects of oleate and insulin on apolipoprotein B mRNA levels in HepG2 cells. *Circulation* **78,** II-96.

Davidson, N. O., Powell, L. M., and Scott, J. (1988). Thyroid hormone (T3) mediates insertion of a stop codon in rat liver apoB100 mRNA. *Circulation* **78,** II-95.

Davis, R. A., Clinton, G. M., Borchardt, R. A., Malone-McNeal, M., Tan, T., and Lattier, G. R. (1984). Intrahepatic assembly of very low density lipoproteins. Phosphorylation of small molecular weight apolipoprotein B. *J. Biol. Chem.* **259,** 3383–3386.

Dayhoff, M. O., Barker, W. C., and Hunt, L. T. (1983). Establishing homologies in protein sequences. *Methods Enzymol.* **91,** 524–545.

Deeb, S. S., Motulsky, A. G., and Albers, J. J. (1985). A partial cDNA clone for human apolipoprotein B. *Proc. Natl. Acad. Sci. U.S.A.* **82,** 4983–4986.

Deeb, S., Failor, A., Brown, B. G., Brunzell, J. D., Albers, J. J., and Motulsky, A. G. (1986). Molecular genetics of apolipoproteins and coronary heart disease. *Cold Spring Harbor Symp. Quant. Biol.* **L1,** 403–409.

DeLoof, H., Rosseneu, M., Yang, C-Y., Li, W-H., Gotto, A. M., and Chan, L. (1987). Human apolipoprotein B: analysis of internal repeats and homology with other apolipoproteins. *J. Lipid Res.* **28,** 1455–1465.

Demmer, L. A., Levin, M. S., Elovson, J., Reuben, M. A., Lusis, A. J., and Gordon, J. I. (1986). Tissue-specific expression and developmental regulation of the rat apolipoprotein B gene. *Proc. Natl. Acad. Sci. U.S.A.* **83,** 8102–8106.

Dullaart, R. P. F., Speelberg, B., Schuurman, H-J., Milne, R. W., Havekes, L. M., Marcel, Y. L., Geuze, H. J., Hulshof, M. M., and Erkelens, D. W. (1986). Epitopes of apolipoprotein B-100 and B-48 in both liver and intestine. Expression and evidence for local synthesis in recessive abetalipoproteinemia. *J. Clin. Invest.* **78,** 1397–1404.

Dunning, A. M., Tikkanen, M. J., Ehnholm, C., Bütler, R., and Humphries, S. E. (1988). Relationships between DNA and protein polymorphisms of apolipoprotein B. *Hum. Genet.* **78,** 325–329.

Ebert, D. L., Maeda, N., Lowe, S. W., Hasler-Rapacz, J., Rapacz, J., and Attie, A. D. (1988). Primary structure comparison of the proposed low density lipoprotein (LDL) receptor binding domain of human and pig apolipoprotein B: implications for LDL–receptor interactions. *J. Lipid Res.* **29,** 1501–1509.

Eissenberg, J. C., Cartwright, I. L., Thomas, G. H., and Elgin, S. C. R. (1985). Selected topics in chromatin structure. *Ann. Rev. Genet.* **19,** 485–536.

Elovson, J., Huang, Y. O., Baker, N., and Kannan, R. (1981). Apolipoprotein B is structurally and metabolically heterogeneous in the rat. *Proc. Natl. Acad. Sci. U.S.A.* **78,** 157–161.

Enholm, C., Garoff, H., Renkonen, O., and Simons, K. (1972). Protein and carbohydrate composition of Lp(a) lipoprotein from human plasma. *Biochemistry* **11,** 3229–3232.

Feinberg, A. P. and Vogelstein, B. (1984). A technique for radiolabeling DNA restriction endonuclease fragments to high specific activity. *Anal. Biochem.* **137,** 266–267.

Fless, G. M., Rolik, C. A., and Scanu, A. M. (1984). Heterogeneity of human plasma lipoprotein(a). Isolation and characterization of the lipoprotein subspecies and their apoproteins. *J. Biol. Chem.* **259,** 11470–11478.

Frossard, P. M., Gonzalez, P. A., Protter, A. A., Coleman, R. T., Funke, H., and Assmann, G. (1986). *Pvu*II RFLP in the 5′ of the human apolipoprotein B gene. *Nucleic Acids Res.* **14,** 4373.

Gaubatz, J. W., Heideman, C., Gotto, A. M., Morrisett, J. D., and Dahlen, G. H. (1983). Human plasma lipoprotein(a): structural properties. *J. Biol. Chem.* **258,** 4582–4589.

Glickman, R. M., Green, D. H. R., Lees, R. S., Lux, S. E., and Kilgrove, A. (1979). Immunofluorescence studies of apolipoprotein B in intestinal mucosa. Absence in abetalipoproteinemia. *Gastroenterology* **76,** 288–292.

Glickman, R. M., Rogers, M., and Glickman, J. N. (1986). Apoprotein B synthesis by human liver and intestine *in vitro*. *Proc. Natl. Acad. Sci. U.S.A.* **83,** 5296–5300.

Glueck, C. J., Gartside, P., Fallat, R. W., Sielski, J., and Steiner, P. M. (1976). Longevity syndromes: Familial hypobeta and familial hyperalpha lipoproteinemia. *J. Lab. Clin. Med.* **88,** 941–957.

Green, P. H., Lefkowitch, J. H., Glickman, R. M., Riley, J. W., Quinet, E., and Blum, C. B. (1982). Apolipoprotein localization and quantitation in the human intestine. *Gastroenterology* **83,** 1223–1230.

Hardman, D. A., and Kane, J. P. (1986). Isolation and characterization of apolipoprotein B-48. *Methods Enzymol.* **128,** 262–272.

Hardman, D. A., Protter, A. A., Chen, G. C., Schilling, J. W., Sato, K. Y., Lau, K., Yamanaka, M., Mikita, T., Miller, J., Crisp, T., McEnroe, G., Scarborough, R. N., and Kane, J. P. (1987). Structural comparison of human apolipoproteins B-48 and B-100. *Biochemistry* **26,** 5478–5486.

Harpel, P. C., Gordon, B. R., and Parker, T. S. (1989). Plasmin catalyzes binding of lipoprotein(a) to immobilized fibrinogen and fibrin. *Proc. Natl. Acad. Sci., U.S.A.* **86,** 3847–3851.

Harris, H. (1980). "The Principles of Human Biochemical Genetics," 3rd ed. Elsevier, Amsterdam.

Hegele, R. A., Huang, L-S., Herbert, P. N., Blum, C. B., Buring, J. E., Hennekens, C. H., and Breslow, J. L. (1986). Apolipoprotein B-gene DNA polymorphisms associated with myocardial infarction. *N. Engl. J. Med.* **315,** 1509–1515.

Herbert, P. N., Assmann, G., Gotto, A. M., and Fredrickson, D. S. (1983). Familial lipoprotein deficiency: abetalipoproteinemia, hypobetalipoproteinemia, and Tangier disease. *In* "The Metabolic Basis of Inherited Disease" (J. B. Stanbury, J. B. Wyngaarden, D. S. Fredrickson, J. L. Goldstein, and M. S. Brown, eds.), pp. 589–623. McGraw-Hill, New York.

Herbert, P. N., Hyams, J. S., Bernier, D. N., Berman, M. M., Saritelli, A. L., Lynch, A. M., Nichols, A. V., and Forte, T. M. (1985). Apolipoprotein B-100 deficiency. Intestinal steatosis despite apolipoprotein B-48 synthesis. *J. Clin. Invest.* **76,** 403–412.

Higuchi, K., Monge, J. G., Lee, N., Law, S. W., and Brewer, H. B. (1987). The human apoB-100 gene: ApoB-100 is encoded by a single copy gene in the human genome. *Biochem. Biophys. Res. Commun.* **144,** 1332–1339.

Hixson, J. E., Britten, M. L., Manis, G. S., and Rainwater, D. L. (1989). Apolipoprotein (a) (apo(a)) glycoprotein isoforms result from size differences in Apo(a) mRNA in baboons. *J. Biol. Chem.* **264,** 6013–6016.

Hospattankar, A. V., Law, S. W., Lackner, K., and Brewer, H. B. (1986). Identification of low density lipoprotein receptor binding domains of human apolipoprotein B-100: a proposed consensus LDL receptor binding sequence of apoB-100. *Biochem. Biophys. Res. Commun.* **139,** 1078–1085.

Hospattankar, A. V., Higuchi, K., Law, S. W., Meglin, N., and Brewer, H. B. (1987). Identification of a novel in-frame translational stop codon in human intestine apoB mRNA. *Biochem. Biophys. Res. Commun.* **148,** 279–285.

Huang, G., Lee, D. M., and Singh, S. (1988a). Identification of the thiol ester linked lipids in apolipoprotein B. *Biochemistry* **27,** 1395–1400.

Huang, L-S., de Graaf, J., and Breslow, J. L. (1988b). ApoB gene *Msp*1 RFLP in exon 26 changes amino acid 3611 from Arg to Gln. *J. Lipid Res.* **29,** 63–67.

Huang, L-S., Ripps, M. E., Korman, S. H., Deckelbaum, R., and Breslow, J. L. (1988c). ApoB gene exon 21 deletion in familial hypobetalipoproteinemia (HPLP). *Circulation* **78,** II-389.

Hui, D. Y., Innerarity, T. L., Milne, R. W., Marcel, Y. L., and Mahley, R. W. (1984). Binding of chylomicron remnants and beta-very-low-density lipoproteins to hepatic and extrahepatic lipoprotein receptors: a process independent of apolipoprotein B48. *J. Biol. Chem.* **259,** 15060–15068.

Humphries, S. E. (1988). DNA polymorphisms of the apolipoprotein genes—their use in the investigation of the genetic component of hyperlipidaemia and atherosclerosis. *Atherosclerosis* **72,** 89–108.

Infante, R. (1987). Apolipoprotein-B deficiency syndromes. *In* "European Lipoprotein Club. The First Ten Years," pp. 97–104. Department of Biochemistry, Royal Infirmary, Glasgow, United Kingdom.

Innerarity, T. L., Friedlander, E. J., Rall, S. C., Weisgraber, K. H., and Mahley, R. W. (1983). The receptor-binding domain of human apolipoprotein E. Binding of apolipoprotein E fragments. *J. Biol. Chem.* **258,** 12341–12347.

Innerarity, T. L., Weisgraber, K. H., Arnold, K. S., Mahley, R. W., Krauss, R. M., Vega, G. L., and Grundy, S. M. (1987a). Familial defective apolipoprotein B-100: low density lipoproteins with abnormal receptor binding. *Proc. Natl. Acad. Sci. U.S.A.* **84,** 6919–6923.

Innerarity, T. L., Weisgraber, K. H., Rall, S. C., and Mahley, R. W. (1987b). *Acta Med. Scand. Suppl.* **715,** 51–59.

Innerarity, T. L., Balestra, M. E., Arnold, K. S., Mahley, R. W., Vega, G. L., Grundy, S. M., and Young, S. G. (1988). Isolation of defective receptor-binding low density lipoproteins from subjects with familial defective apolipoprotein B100. *Circulation* **78,** II-390.

Jenner, K., Sidoli, A., Ball, M., Rodriguez, J. R., Pagani, F., Giudici, G., Vergani, C., Mann, J., Baralle, F. E., and Shoulders, C. C. (1988). Characterization of genetic markers in the 3′ end of the apoB gene and their use in family and population studies. *Atherosclerosis* **69,** 39–49.

Jones, N. C., Rigby, P. W. J., and Ziff, E. B. (1988). *Trans*-acting protein factors and the regulation of eukaryotic transcription: lessons from studies on DNA tumor viruses. *Genes Develop.* **2,** 267–281.

Jones, T., Rajput-Williams, J., Knott, T. J., and Scott, J. (1989). An *Msp*I RFLP in the ApoB promoter. *Nucleic Acids Res.* **17,** 472.

Kane, J. P. (1983). Apolipoprotein B: structural and metabolic heterogeneity. *Ann. Rev. Physiol.* **45,** 637–650.

Kane, J. P., Hardman, D. A., and Paulus, H. E. (1980). Heterogeneity of apolipoprotein B: isolation of a new species from human chylomicrons. *Proc. Natl. Acad. Sci. U.S.A.* **77,** 2465–2469.

Knott, T. J., Rall, S. C., Innerarity, T. L., Jacobson, S. F., Urdea, M. S., Levy-Wilson, B., Powell, L. M., Pease, R. J., Eddy, R., Nakai, H., Byers, M., Priestley, L. M., Robertson, E., Rall, L. B., Betsholtz, C., Shows, T. B., Mahley, R. W., and Scott, J. (1985). Human apolipoprotein B: structure of carboxy-terminal domains, sites of gene expression, and chromosomal localization. *Science* **230,** 37–43.

Knott, T. J., Pease, R. J., Powell, L. M., Wallis, S. C., Rall, S. C., Innerarity, T. L., Blackhart, B., Taylor, W. H., Marcel, Y., Milne, R., Johnson, D., Fuller, M., Lusis, A. J., McCarthy, B. J., Mahley, R. W., Levy-Wilson, B., and Scott, J. (1986). Complete protein sequence and identification of structural domains of human apolipoprotein B. *Nature (London)* **323,** 734–738.

Kostner, G. M. (1976). Lp(a) lipoproteins and the genetic polymorphisms of lipoprotein B. *In* "Low Density Lipoproteins" (C. E. Day and R. S. Levy, eds.), pp. 229–269. Plenum, New York.

Kostner, G. M. (1983). Apolipoproteins and lipoproteins of human plasma. *Adv. Lipid Res.* **20,** 1–43.

Krempler, F., Kostner, G. M., Bolzano, K., and Sandhofer, F. (1980). Turnover of lipoprotein(a) in man. *J. Clin. Invest.* **65,** 1483–1490.

Krempler, F., Kostner, G. M., Roscher, A., Haslauer, F., Bolzano, K., and Sandhofer, F. (1982). Studies on the role of specific cell surface receptors in the removal of lipoprotein(a) in man. *J. Clin. Invest.* **71,** 1431–1441.

Krul, E. S. Kleinman, Y., Kinoshita, M., Pfleger, B., Oida, K., Law, A., Scott, J., Pease, R., and Schonfeld, G. (1988). Regional specificities of monoclonal anti-human apolipoprotein B antibodies. *J. Lipid Res.* **29,** 937–947.

Krul, E. S., Kinoshita, M., Talmud, P., Humphries, S. E., Turner, S., Goldberg, A. C., Cook, K., Boerwinkle, E., and Schonfeld, G. (1989). Two distinct truncated apolipoprotein B species in a kindred with hypobetalipoproteinemia. *Arteriosclerosis,* **9,** in press.

Kyte, J., and Doolittle, R. F. (1982). A simple method for displaying the hydropathic character of a protein. *J. Mol. Biol.* **157,** 105–132.

Lackner, K. J., Monge, J. C., Gregg, R. E., Hoeg, J. M., Triche, T. J., Law, S. W., and Brewer, H. B. (1986). Analysis of the apolipoprotein B gene and messenger ribonucleic acid in abetalipoproteinemia. *J. Clin. Invest.* **78,** 1707–1712.

Lalazar, A., Weisgraber, K. H., Rall, S. C., Giladi, H., Innerarity, T. L., Levanon, A. Z., Boyles, J. K., Amit, B., Gorecki, M., Mahley, R. W., and Vogel, T. (1988). Site-specific mutagenesis of human apolipoprotein E. Receptor binding activity of variants with single amino acid substitutions. *J. Biol. Chem.* **263,** 3542–3545.

Lamy, M., Frézal, J., Polonovski, J., and Rey, J. (1960). L'absence congénitale de β-lipoprotéines. *C. R. Soc. Biol. (Paris)* **154,** 1974–1978.

Law, S. W., Lackner, K. J., Hospattankar, A. V., Anchors, J. M., Sakaguchi, A. Y., Naylor, S. L., and Brewer, H. B. (1985). Human apolipoprotein B-100: cloning, analysis of liver mRNA, and assignment of the gene to chromosome 2. *Proc. Natl. Acad. Sci. U.S.A.* **82,** 8340–8344.

Law, S. W. Grant, S. M., Higuchi, K., Hospattankar, A., Lackner, K., Lee, N., and Brewer, H. B. (1986a). Human liver apolipoprotein B-100 cDNA: complete nucleic acid and derived amino acid sequence. *Proc. Natl. Acad. Sci. U.S.A.* **83,** 8142–8146.

Law, A., Powell, L. M., Brunt, H., Knott, T. J., Altman, D. G., Rajput, J., Wallis, S. C., Pease, R. J., Priestley, L. M., Scott, J., Miller, G. J., and Miller, N. E. (1986b). Common DNA polymorphism within coding sequence of apolipoprotein B gene associated with altered lipid levels. *Lancet* **1,** 1301–1303.

Lee, D. M., and Singh, S. (1988). Presence and localization of two intramolecular thiolester linkages in apolipoprotein B. *Circulation* **78,** II-286.

Leppert, M., Breslow, J. L., Wu, L., Hasstedt, S., O'Connell, P., Lathrop, M., Williams, R. R., White, R., and Lalouel, J-M. (1988). Inference of a molecular defect of apolipoprotein B in hypobetalipoproteinemia by linkage analysis in a large kindred. *J. Clin. Invest.* **82,** 847–851.

Lerch, P. G., Rickli, E. E., Lergier, W., and Gillessen, D. (1980). Localization of individual lysine-binding regions in human plasminogen and investigations on their complex-forming properties. *Eur. J. Biochem.* **107,** 7–13.

Levy, R. I., Langer, T., Gotto, A. M., and Fredrickson, D. S. (1970). Familial hypobetalipoproteinemia, a defect in lipoprotein synthesis. *Clin. Res.* **18,** 539.

Levy, E., Marcel, Y. L., Milne, R. W. Grey, V. L., and Roy, C. C. (1987). Absence of intestinal synthesis of apolipoprotein B-48 in two cases of abetalipoproteinemia. *Gastroenterology* **93,** 1119–1126.

Levy-Wilson, B., Fortier, C., Blackhart, B. D., and McCarthy, B. J. (1988). DNase I- and micrococcal nuclease-hypersensitive sites in the human apolipoprotein B gene are tissue specific. *Mol. Cell. Biol.* **8,** 71–80.

Lewin, B. (1987). "Genes III." Wiley, New York.

Li, W-H., Tanimura, M., Luo, C-C., Datta, S., and Chan, L. (1988). The apolipoprotein multigene family: biosynthesis, structure, structure–function relationships, and evolution. *J. Lipid Res.* **29,** 245–271.

Ludwig, E. H., Blackhart, B. D., Pierotti, V. R., Caiati, L., Fortier, C., Knott, T., Scott, J., Mahley, R. W., Levy-Wilson, B., and McCarthy, B. J. (1987). DNA sequence of the human apolipoprotein B gene. *DNA* **6,** 363–372.

Lusis, A. J. (1988). Genetic factors affecting blood lipoproteins: the candidate gene approach. *J. Lipid Res.* **29,** 397–429.

Lusis, A. J., West, R., Mehrabian, M., Reuben, M. A., LeBoeuf, R. C., Kaptein, J. S., Johnson, D. F., Schumaker, V. N., Yuhasz, M. P., Schotz, M. C., and Elovson, J. (1985). Cloning and expression of apolipoprotein B, the major protein of low and very low density lipoproteins. *Proc. Natl. Acad. Sci. U.S.A.* **82,** 4597–4601.

Lusis, A. J., Taylor, B. A. Quon, D., Zollman, S., and LeBeouf, R. C. (1987). Genetic factors controlling structure and expression of apolipoproteins B and E in mice. *J. Biol. Chem.* **262,** 7594–7604.

Ma, Y., Schumaker, V. N., Bütler, R., and Sparkes, R. S. (1987). Two DNA restriction fragment length polymorphisms associated with Ag(t/z) and Ag (g/c) antigenic sites of human apolipoprotein B. *Arteriosclerosis* **7,** 301–305.

McLean, J. W., Elshourbagy, N. A., Chang, D. J., Mahley, R. W., and Taylor, J. M. (1984). Human apolipoprotein E mRNA cDNA cloning and nucleotide sequencing of a new variant. *J. Biol. Chem.* **259,** 6498–6504.

McLean, J. W. Tomlinson, J. E., Kuang, W-J., Eaton, D. L., Chen, E. Y., Fless, G. M., Scanu, A. N., and Lawn, R. M. (1987). cDNA sequence of human apolipoprotein (a) is homologous to plasminogen. *Nature (London)* **330,** 132–137.

Maeda, N., Ebert, D. L., Doers, T. M., Newman, M., Hasler-Rapacz, J., Attie, A. D., Rapacz, J., and Smithies, O. (1988). Molecular genetics of the apolipoprotein B gene in pigs in relation to atherosclerosis. *Gene* **70,** 213–229.

Mahley, R. W. (1979). Dietary fat, cholesterol and accelerated atherosclerosis. *Atherosclerosis Rev.* **5,** 1–34.

Mahley, R. W., Innerarity, T. L., Rall, S. C., and Weisgraber, K. H. (1984). Plasma lipoproteins: apolipoprotein structure and function. *J. Lipid Res.* **25,** 1277–1294.

Malloy, M. J., Kane, J. P., Hardman, D. A., Hamilton, R. L., and Dalal, K. B. (1981). Normotriglyceridemic abetalipoproteinemia. Absence of the B-100 apolipoprotein. *J. Clin. Invest.* **67,** 1441–1450.

Maniatis, T., Goodbourn, S., and Fischer, J. A. (1987). Regulation of inducible and tissue-specific gene expression. *Science* **236,** 1237–1244.

Marcel, Y. L., Hogue, M., Theolis, R., and Milne, R. W. (1982). Mapping of antigenic determinants of human apolipoprotein B using monoclonal antibodies against low density lipoproteins. *J. Biol. Chem.* **257,** 13165–13168.

Marcel, Y. L., Innerarity, T. L., Spilman, C., Mahley, R. W., Protter, A. A., and Milne, R. W. (1987). Mapping of human apolipoprotein B antigenic determinants. *Arteriosclerosis* **7,** 166–175.

Mars, H., Lewis, L. A., Robertson, A. L., Butkus, A., and Williams, G. H. (1969). Familial hypo-β-lipoproteinemia: a genetic disorder of lipid metabolism with nervous system involvement. *Am. J. Med.* **46,** 886–900.

Mehrabian, M., Schumaker, V. N., Fareed, G. C., West, R., Johnson, D. F., Kirchgessner, T., Lin, H-C., Wang, X., Ma, Y., Mendiaz, E., and Lusis, A. J. (1985). Human apolipoprotein B: identification of cDNA clones and characterization of mRNA. *Nucleic Acids Res.* **13,** 6937–6953.

Milne, R. W., and Marcel, Y. L. (1982). Monoclonal antibodies against human low density lipoprotein. Stoichiometric binding studies using Fab fragments. *FEBS Lett.* **146,** 97–100.

Milne, R., Théolis, R., Maurice, R., Pease, R. J. Weech, P. K., Rassart, E., Fruchart, J-C., Scott, J., and Marcel, Y. L. (1989). The use of monoclonal antibodies to localize the low density lipoprotein receptor-binding domain of apolipoprotein B. *J. Biol. Chem.* **264**.

Moberly, J. B., Cole, T. G., and Schonfeld, G. (1988). Oleic acid regulates apolipoprotein-B production by Caco-2 and Hep-G2 cells. *Circulation* **78,** II-95.

Montandon, A. J., Green, P. M., Gianelli, F., and Bentley, D. R. (1989). Direct detection of point mutations by mismatch analysis-application to haemophilia B. *Nucleic Acids Res.* **17,** 3347–3358.

Murray, J. C., Mills, K. A., Demopulos, C. M., Hornung, S., and Motulsky, A. G. (1984). Linkage disequilibrium and evolutionary relationships of DNA variants (restriction fragment length polymorphisms) at the serum albumin locus. *Proc. Natl. Acad. Sci. U.S.A.* **81,** 3486–3490.

Myant, N. B. (1981). "The Biology of Cholesterol and Related Steroids." Heinemann, London.

Myant, N. B., Gallagher, J. G., Barbir, M., Thompson, G. R., Wile, D., and Humphries, S. E. (1989). Restriction fragment length polymorphisms in the apoB gene in relation to coronary artery disease. *Atherosclerosis* **77,** 193–201.

Myers, R. M., Larin, Z., and Maniatis, T. (1985). Detection of single base substitutions by ribonuclease cleavage at mismatches in RNA:DNA duplexes. *Science* **230,** 1242–1246.

Norio, R., Nevanlinna, H. R., and Peerheentupa, J. (1973). Hereditary diseases in Finland; rare flora in rare soil. *Ann. Clin. Res.* **5,** 109–141.

Olofsson, S-O., Bjursell, G., Boström, K., Carlsson, P., Elovson, J., Protter, A. A., Reuben, M. A., and Bondjers, G. (1987). Apolipoprotein B: structure, biosynthesis and role in the lipoprotein assembly process. *Atherosclerosis* **68,** 1–17.

Paulweber, B., Friedl, W., Krempler, F., Humphries, S. E., and Sandhofer, F. (1990). Association of DNA polymorphism at the apoB gene locus with coronary heart disease and serum VLDL levels. *Arteriosclerosis,* **10,** in press.

Powell, L. M., Wallis, S. C., Pease, R. J., Edwards, Y. H., Knott, T. J., and Scott, J. (1987). A novel form of tissue-specific RNA processing produces apolipoprotein-B48 in intestine. *Cell* **50,** 831–840.

Priestley, L., Knott, T., Wallis, S., Powell, L., Pease, R., Simon, A., and Scott, J. (1985). RFLP for the human apolipoprotein B gene: I, *Bam*HI; II, *Eco*RI; III, *Eco*RV; IV, *Msp*I; V, *Xba*I. *Nucleic Acids Res.* **13,** 6789–6793.

Protter, A. A., Coleman, R. T., Sato, K. Y., Dillan, N. A., Lim, D., and Frossard, P. M. (1986a). Very low nucleotide diversity in the 5′ portion of the human apolipoprotein B gene. *Am. J. Hum. Genet.* **39,** A216.

Protter, A. A., Hardman, D. A., Schilling, J. W., Miller, J., Appleby, V., Chen, G. C., Kirsher, S. W., McEnroe, G., and Kane, J. P. (1986b). Isolation of a cDNA clone encoding the amino-terminal region of human apolipoprotein B. *Proc. Natl. Acad. Sci. U.S.A.* **83,** 1467–1471.

Pullinger, C. R., North, J. D., Medina, P., Teng, B. B., Rifici, V. A., deBrito, A. E. R., and Scott, J. (1988). The apolipoprotein (apo-B) gene is constitutively expressed in HepG2 cells. *Circulation* **78,** II-96.

Rajput-Williams, J., Knott, T. J., Wallis, S. C., Sweetnam, P., Yarnell, J., Cox, N., Bell, G. I., Miller, N. E., and Scott, J. (1988). Variation of apolipoprotein-B gene is associated with obesity, high blood cholesterol levels, and increased risk of coronary heart disease. *Lancet* **2,** 1442–1446.

Rapacz, J., Hasler-Rapacz, J., Talor, K. M., Checovich, W. J., and Attie, A. D. (1986). Lipoprotein mutations in pigs are associated with elevated plasma cholesterol and atherosclerosis. *Science* **234,** 1573–1577.

Richet, G., Durepaire, H., Hartmann, L., Ollier, M-P., Polonovski, J., and Maitrot, B. (1969). Hypolipoprotéinémie familiale asymptomatique prédominant sur les beta-lipoprotéines. *Presse Med.* **77,** 2045–2048.

Ross, R. S., Gregg, R. E., Law, S. W., Monge, J. C., Grant, S. M., Higuchi, K., Triche, T. J., Jefferson, J., and Brewer, H. B. (1988). Homozygous hypobetalipoproteinemia: a disease distinct from abetalipoproteinemia at the molecular level. *J. Clin. Invest.* **81,** 590–595.

Salt, H. B., Wolff, O. H., Lloyd, J. K., Fosbrooke, A. S., Cameron, A. H., and Hubble, D. V. (1960). On having no beta-lipoprotein. A syndrome comprising abetalipoproteinaemia, acanthocytosis and steatorrhoea. *Lancet* **2,** 325–329.

Schwartz, D. E., Liotta, L., Schaefer, E., and Brewer, H. B. (1978). Localization of apoliproteins A-I, A-II, and B in normal, Tangier and abetalipoproteinemia intestinal mucosa. *Circulation* **57/58,** II-90.

Scott, J. (1989). The molecular and cell biology of apolipoprotein-B. *Mol. Biol. Med.* **6,** 65–80.

Shoulders, C. C., Myant, N. B., Sidoli, A., Rodriguez, J. C., Cortese, C., Baralle, F. E., and Cortese, R. (1985). Molecular cloning of human LDL apolipoprotein B cDNA. Evidence for more than one gene per haploid genome. *Atherosclerosis* **58,** 277–289.

Sigurdsson, G., Nicoll, A., and Lewis, B. (1977). Turnover of apolipoprotein-B in two subjects with familial hypobetalipoproteinemia. *Metabolism* **26,** 25–31.

Soria, L. F., Ludwig, E. H., Clarke, H. R. G., Vega, G. L., Grundy, S. M., and McCarthy, B. J. (1989). Association between a specific apolipoprotein B mutation and familial defective apolipoprotein B-100. *Proc. Natl. Acad. Sci. U.S.A.* **86,** 587–591.

Southern, E. (1979). Gel electrophoresis of restriction fragments. *Methods Enzymol.* **68,** 152–176.

Sparks, C. E. and Marsh, J. B. (1981). Metabolic heterogeneity of apolipoprotein B in the rat. *J. Lipid Res.*, **22,** 519–527.

Steinberg, D., Grundy, S. M., Mok, H. Y. I., Turner, J. D., Weinstein, J. B., Brown, W. V., and Albers, J. J. (1979). Metabolic studies in an unusual case of asymptomatic familial hypobetalipoproteinemia with hypoalphalipoproteinemia and fasting chylomicronemia. *J. Clin. Invest.* **64,** 292–301.

Stout, J. T., and Caskey, C. T. (1988). The Lesch–Nyhan syndrome: clinical, molecular and genetic aspects. *Trends Genet.* **4,** 175–178.

Südhof, T. C., Goldstein, J. L., Brown, M. S., and Russell, D. W. (1985). The LDL receptor gene: a mosaic of exons shared with different proteins. *Science* **228,** 815–822.

Takashima, Y., Kodama, T., Lida, H., Kawamura, M., Aburatani, H., Itakura, H., Akanuma, Y., Takaku, F., and Kawade, M. (1985). Normotriglyceridemic abetalipoproteinemia in infancy: an isolated apolipoprotein B-100 deficiency. *Pediatrics* **75,** 541–546.

Talmud, P. J., Barni, N., Kessling, A. M., Carlsson, P., Darnfors, C., Bjurselll, G., Galton, D., Wynn, V., Kirk, H., Hayden, M. R., and Humphries, S. E. (1987). Apolipoprotein B gene variants are involved in the determination of serum cholesterol levels: a study in normo- and hyperlipidaemic individuals. *Atherosclerosis* **67,** 81–89.

Talmud, P., King-Underwood, L., Krul, E. S., Dunning, A., Schonfeld, G., and Humphries, S. E. (1989). The molecular basis of truncated forms of apolipoprotein B in a kindred with compound heterozygous hypobetalipoproteinemia. *J. Lipid Res.* **30,** 1773–1779.

Talmud, P. J., Lloyd, J. K., Muller, D. P. R., Collins, D. R., Scott, J., and Humphries, S. (1988). Genetic evidence from two families that the apolipoprotein B gene is not involved in abetalipoproteinemia. *J. Clin. Invest.* **82,** 1803–1806.

Theolis, R., Weech, P. K., Marcel, Y. L., and Milne, R. W. (1984). Characterization of antigenic determinants of human apolipoprotein B. Distribution on tryptic fragments of low density lipoprotein. *Arteriosclerosis* **4,** 498–509.

Tikkanen, M. J., Ehnholm, C., and Kovanen, P. T. (1986). Monoclonal antibody detects Ag polymorphism of apolipoprotein B. *FEBS Lett.* **202,** 54–58.

Tybjaerg-Hansen, A., Gallagher, J., Vincent, J. L., Houlston, R., Talmud, P., Seed, M., Humphries, S., and Myant, N. (1989). Screening for the apoB ($Arg_{3500} \rightarrow Gln$) mutation. *Atherosclerosis* **80,** 235–242.

Utermann, G., and Weber, W. (1983). Protein composition of Lp(a) lipoprotein from human plasma. *FEBS Lett.* **154,** 357–361.

Utermann, G., Menzel, H. J., Kraft, H. G., Kuba, H. C., Kemmler, H. G., and Seitz, C. (1987). Lp(a) glycoprotein phenotypes. Inheritance and relation to Lp(a)-lipoprotein concentrations in plasma. *J. Clin. Invest.* **80,** 458–465.

Vauhkonen, M., Viitala, J., Parkinnen, and Rauvala, H. (1985). High mannose structure of apolipoprotein-B from low-density lipoproteins of human plasma. *Eur. J. Biochem.* **152,** 43–50.

Vega, G. L., and Grundy, S. M. (1986). *In vivo* evidence for reduced binding of low density lipoprotein to receptors as a cause of primary moderate hypercholesterolemia. *J. Clin. Invest.* **78,** 1410–1414.

Vogel, F., and Motulsky, A. G. (1986). ''Human Genetics. Problems and Approaches,'' 2nd ed. Springer-Verlag, Berlin.

Wang, X., Schlapfer, P., Ma, Y., Dütler, R., Elovson, J., and Schumaker, V. N. (1988). Apolipoprotein B. The Ag(a_1/d) immunogenetic polymorphism coincides with a T to C substitution at nucleotide 1981, creating an *Alu*I restriction site. *Arteriosclerosis* **8,** 429–435.

Wei, C-F., Chen, S-H., Yang, C-Y., Marcel, Y. L., Milne, R. W., Li, W-H., Sparrow, J. T., Gotto, A. M., and Chan, L. (1985). Molecular cloning and expression of partial cDNAs and deduced amino acid sequence of a carboxy-terminal fragment of human apolipoprotein B-100. *Proc. Natl. Acad. Sci. U.S.A.* **82,** 7265–7269.

Weisgraber, K. H., Innerarity, T. L., and Mahley, R. W. (1982). Abnormal lipoprotein receptor-binding activity of the human E apoprotein due to cysteine–arginine interchange at a single site. *J. Biol. Chem.* **257,** 2518–2521.

Weisgraber, K. H., Innerarity, T. L., Harder, K. J., Mahley, R. W., Milne, R. W., Marcel, Y. L., and Sparrow, J. T. (1983). The receptor-binding domain of human apolipoprotein E. Monoclonal antibody inhibition of binding. *J. Biol. Chem.* **258,** 12348–12354.

Weisgraber, K. H., Innerarity, T. L., Newhouse, Y. M., Young, S. G., Arnold, K. S., Krauss, R. M., Vega, G. L., Grundy, S. M., and Mahley, R. W. (1988). Familial defective apolipoprotein B-100: enhanced binding of monoclonal antibody MB47 to abnormal low density lipoproteins. *Proc. Natl. Acad. Sci. U.S.A.* **85,** 9758–9762.

White, T. J., Arnheim, N., and Erlich, H. A. (1989). The polymerase chain reaction. *Trends Genet.* **5,** 185–189.

Wu, C., and Gilbert, W. (1981). Tissue-specific exposure of chromatin structure at the 5′ terminus of the rat preproinsulin II gene. *Proc. Natl. Acad. Sci. U.S.A.* **78,** 1577–1580.

Xu, C., Nanjee, N., Tikkanen, M. J., Huttunen, J. K., Pietinen, P., Bütler, R., Angelico, F., Del Ben, M., Mazzarella, B., Antonio, R., Miller, N. E., Humphries, S., and Talmud, P. (1989). Apolipoprotein B amino acid 3611 substitution from arginine to glutamine creates the Ag(h/i) epitope: the polymorphism is not associated with differences in serum cholesterol and apolipoprotein B levels. *Hum. Genet.* **82,** 322–326.

Yamamoto, M., Ranganathan, S., and Kottke, B. A. (1985). Structure and function of human low density lipoproteins. Studies using proteolytic cleavage by plasma kallikrein. *J. Biol. Chem.* **260,** 8509–8513.

Yang, C-Y., Cheng, S-H., Gianturco, S. H., Bradley, W. A., Sparrow, J. T., Tanimura, M., Li, W-H., Sparrow, D. A., DeLoof, H., Rosseneu, M., Lee, F. S., Gu, Z-W., Gotto, A. M., and Chan, L. (1986). Sequence, structure, receptor-binding domains and internal repeats of human apolipoprotein B-100 *Nature (London)* **323,** 738–742.

Young, S. G., Bertics, S. J., Scott, T. M., Dubois, B. W., Curtiss, L. K., and Witztum, J. L. (1986a). Parallel expression of the MB 19 genetic polymorphism in apoprotein B-100 and apoprotein B-48. Evidence that both apoproteins are products of the same gene. *J. Biol. Chem.* **261,** 2995–2998.

Young, S. G., Witztum, J. L., Casal, D. C., Curtiss, L. K., and Bernstein, S. (1986b). Conservation of the low density lipoprotein receptor-binding domain of apoprotein B. Demonstration by a new monoclonal antibody MB47. *Arteriosclerosis* **6,** 178–188.

Young, S. G., Bertics, S. J., Curtiss, L. K., Dubois, B. W., and Witztum, J. L. (1987a). Genetic analysis of a kindred with familial hypobetalipoproteinemia. Evidence for two separate gene defects; one associated with an abnormal apolipoprotein B species, apolipoprotein B-37, and a second associated with low plasma concentrations of apolipoprotein B-100. *J. Clin. Invest.* **79,** 1842–1851.

Young, S. G., Bertics, S. J., Curtiss, L. K., and Witztum, J. L. (1987b). Characterization of an abnormal species of apolipoprotein B, apolipoprotein B-37, associated with familial hypobetalipoproteinemia. *J. Clin. Invest.* **79,** 1831–1841.

Young, S. G., Peralta, F. P., Dubois, B. W., Curtiss, L. K., Boyles, J. K., and Witztum, J. L. (1987c). Lipoprotein B37, a naturally occurring lipoprotein containing the amino-terminal portion of apolipoprotein B100, does not bind to the apolipoprotein B,E (low density lipoprotein) receptor. *J. Biol. Chem.* **262,** 16604–16611.

Young, S. G., Northey, S. T., and McCarthy, B. J. (1988). Low plasma cholesterol levels caused by a short deletion in the apolipoprotein B gene. *Science* **241,** 591–592.

Young, S. G., Hubl, S. T., Chappell, D. A., Smith, R. S., Claiborne, F., Snyder, S. M., and Terdiman, J. F. (1989). Familial hypobetalipoproteinemia associated with a mutant species of apolipoprotein B (B-46). *N. Engl. J. Med.* **320,** 1604–1610.

Chapter 7

LDL: Origin and Metabolism

I. The Production of LDL

A. Qualitative Aspects

1. The Major Pathway

In most mammals the protein component of LDL is normally derived almost exclusively from hepatic VLDL, one particle of VLDL generating one of LDL. A substantial proportion of the lipid component of LDL is also derived from VLDL. Hence, the starting point for the production of LDL is the assembly of VLDL particles in the liver and their secretion into the circulation. As soon as it enters the circulation, VLDL undergoes a sequence of changes resulting in the formation of particles that are either removed from the plasma or converted into LDL.

2. The Secretion of Nascent VLDL

Observations on isolated hepatocytes by electron microscopy, combined with information derived from other approaches, suggest the following steps by which hepatic VLDL is assembled and secreted. The apoproteins are synthesized on polyribosomes bound to the rough ER. During translation of the message the growing protein chain enters the lumen of the ER, with co-translational cleavage of the signal sequence followed by N-linked glycosylation of apoB and other apoproteins. The newly synthesized proteins then move to the interior of cisternae at the junctions of the rough and smooth ER. The lipid components of VLDL are synthesized on the cytoplasmic surface of the smooth ER, from whence they move to the interior of the junctional cisternae by an unknown route. Within the cisternae the proteins and lipids associate to form lipoprotein particles that are the precursors of nascent VLDL. These newly formed particles are transported to the

Golgi apparatus, possibly within cytoplasmic vesicles formed by budding from the ER. Within the Golgi, O-linked oligosaccharides are added to the proteins, N-linked oligosaccharides are processed, and the nascent VLDL particles are concentrated within vesicles which move to the plasma membrane of the hepatocyte. The vesicles fuse with the plasma membrane and release their contents into the external medium.

The time course of synthesis and intracellular transport of apoB has been studied in cultured chick-liver cells (Siuta-Mangano *et al.*, 1982) and in Hep G2 cells (Boström *et al.*, 1986). Synthesis of an apoB-100 molecule takes 10–15 minutes. Assembly of the newly synthesized protein into a lipoprotein particle, processing in the Golgi apparatus, and secretion of the nascent particle into the external medium take a total of about 30 minutes.

The steps by which hepatic VLDL is thought to be synthesized and secreted in man are shown in Fig. 7.1. In man, as in most other mammals, apoB-100 is essentially the only apoB present in hepatic VLDL. In rats, on the other hand, VLDL synthesized by hepatocytes consists of a mixture of particles, some with one molecule of apoB-100 and some with one of apoB-48 (Elovson *et al.*, 1988).

3. The Formation of Mature VLDL

In addition to apoB, nascent hepatic VLDL contains apoE and a small amount of apoC. However, on entering the circulation the VLDL particles acquire some additional apoE and a considerable quantity of apoC from HDL (Fig. 7.2). VLDL secreted by the rat's liver contains substantial amounts of esterified cholesterol synthesized by hepatic ACAT, but in man most of the cholesteryl ester present in mature circulating VLDL is derived from HDL. In human plasma, cholesteryl esters are generated on HDL by the LCAT reaction and are then transferred to VLDL and LDL by a cholesteryl-ester transfer protein. Rat plasma contains little or no transfer protein. Hence, rat VLDL cannot acquire esterified cholesterol from HDL. The function of the apoC present in mature VLDL is twofold. It provides lipoprotein lipase with its necessary cofactor (apoC-II) and it inhibits the uptake of VLDL by hepatic receptors. The function of the apoE in VLDL is to facilitate the removal of VLDL remnants from the circulation by hepatic LDL receptors, as discussed below.

4. Regulation of Synthesis and Secretion of Hepatic VLDL

The effects of diet and hormones on the synthesis and secretion of VLDL in the liver have been investigated in isolated perfused livers and in hepatocytes and Hep G2 cells in culture (for references, see Sparks and Sparks, 1985). Hepatocytes *in vitro* have been studied either by treating the donor animals before preparing the cells for culture or by changing the composition of the incubation medium.

The addition of long-chain fatty acids to the perfusing fluid of perfused rat

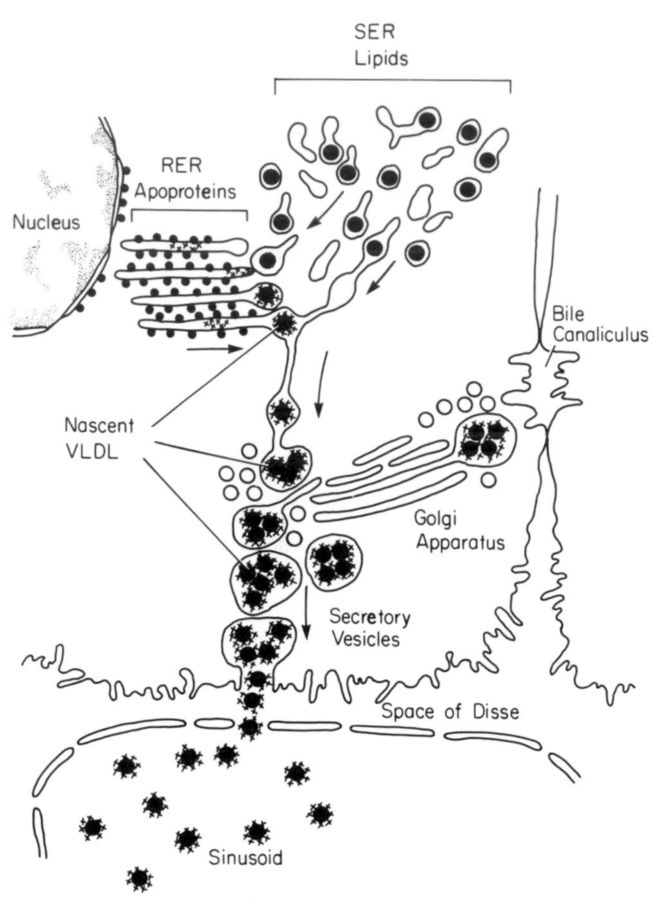

Figure 7.1. Diagram showing the probable steps in the synthesis, assembly, and secretion of VLDL particles in a hepatocyte. The apoproteins are synthesized on the cytoplasmic face of the RER. During synthesis they pass into the RER lumen. VLDL lipids are synthesized in the SER. Apoproteins and lipids move to the junctions between the RER and SER, where they are assembled into nascent VLDL particles (see also Fig. 6.1). The nascent particles move to the Golgi apparatus, where the apoproteins are modified and the nascent particles are packed into secretory vesicles. The secretory vesicles fuse with the hepatocyte plasma membrane and discharge their contents into the space of Disse. The VLDL particles cross the sinusoidal membrane to enter the blood circulation. RER, rough endoplasmic reticulum; SER, smooth endoplasmic reticulum. (From Alexander *et al.*, 1976. Reproduced from the *Journal of Cell Biology,* 1976, Vol. 69, pp. 241–263 by copyright permission of the Rockefeller University Press and with the permissions of the authors.)

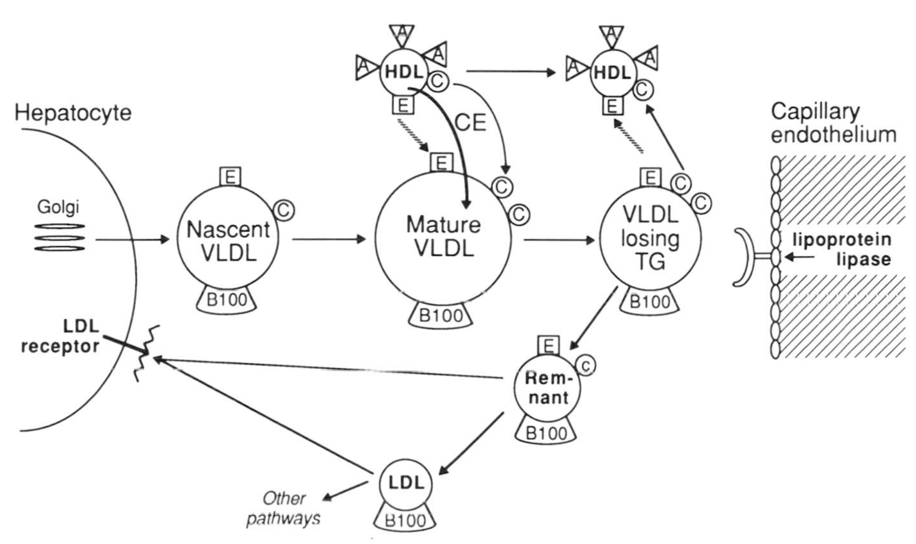

Figure 7.2. Diagram showing the main features of the intravascular metabolism of VLDL and LDL. VLDL particles containing apoB-100 (one molecule per particle), apoC, and apoE are secreted into the circulation by the liver. Esterified cholesterol, apoC, and apoE are transferred to VLDL from HDL. Triglycerides in the core of the VLDL particle are hydrolyzed by lipoprotein lipase attached to the luminal surface of the capillary endothelium. Shrinkage of the VLDL particle is accompanied by loss of some apoC and apoE, which are transferred back to HDL. The action of lipoprotein lipase results in the formation of a remnant particle containing the original apoB-100 molecule, apoE, and a reduced amount of apoC. Remnant particles are either taken up by the liver via the LDL-receptor pathway or are converted into LDL by a process that may involve hepatic lipase (see text). LDL particles, each one containing the original apoB-100 molecule as its only apoprotein, are catabolized via the LDL-receptor pathway in the liver and other tissues, or by other pathways. A, apoA; B100, apoB-100; C, apoC; E, apoE; CE, cholesteryl esters; TG, triglyceride.

livers stimulates triglyceride synthesis and VLDL secretion (Heimberg and Wilcox, 1972). Oleic acid stimulates production and secretion of VLDL by rat hepatocytes in culture without affecting the synthesis of apoB-48 or apoB-100 (Davis and Boogaerts, 1982; Patsch *et al.*, 1983b). Increased production of VLDL without a parallel increase in apoB synthesis may be mediated partly by the formation of larger particles with an increased lipid : protein ratio and partly by utilization of a preformed intracellular store of apoB. The uncoupling of triglyceride synthesis from apoB synthesis seen in short-term experiments cannot be complete because Davis and Boogaerts (1982) found that increased production of VLDL in response to oleate does not occur if apoB synthesis is selectively inhibited by cycloheximide. This suggests that an increase in the production of VLDL requires a continuous supply of newly synthesized apoB molecules.

Fasting decreases VLDL production and apoB synthesis in perfused livers (Marsh and Sparks, 1982) and in hepatocytes in culture (Boogaerts *et al.*, 1982) prepared from the treated animals. Conversely, lipogenesis, VLDL secretion, and apoB synthesis are all increased in hepatocytes prepared from sucrose-fed rats (Boogaerts *et al.*, 1984). However, when glucose is added to a medium containing cultured hepatocytes from normally fed rats, triglyceride synthesis and VLDL production are stimulated but apoB synthesis remains unchanged (Boogaerts *et al.*, 1982). Thus, although synthesis of VLDL lipids is not coordinated with that of apoB in hepatocytes in short-term experiments *in vitro,* the two processes are coupled in the long term *in vivo.* Coupling between VLDL and apoB synthesis is also revealed by the developmental changes that take place in hepatic VLDL secretion in rats. During the change from fetal to adult life, there is a coordinate increase in the synthesis and secretion of VLDL and the synthesis of apoB-48 in hepatocytes (Coleman *et al.*, 1988). The feeding of cholesterol to rats in amounts sufficient to stimulate cholesteryl-ester synthesis in hepatocytes *in vitro* has no effect on apoB synthesis (Davis and Malone-McNeal, 1985).

The effects of insulin on VLDL production and secretion in isolated hepatocytes are complex. Synthesis of triglyceride and other VLDL lipids is stimulated, synthesis of apoB is decreased, and secretion of VLDL is inhibited (Durrington *et al.*, 1982; Patsch *et al.*, 1983a; Sparks *et al.*, 1986). The net effect of these changes is that VLDL triglycerides and apoB accumulate in insulin-stimulated hepatocytes. Sparks *et al.* (1986) suggest that inhibition of VLDL secretion by insulin may create a store of VLDL within the liver during the absorption of a meal, when the plasma insulin concentration rises. This store may then be released when the plasma insulin level falls in the fasting state.

5. Intravascular Metabolism of VLDL

Soon after their entry into the circulation, VLDL particles are acted upon by lipoprotein lipase bound to the luminal surface of the endothelium of the blood capillaries. The action of this enzyme results in the hydrolysis of more than 90% of the triglycerides present in VLDL and the recycling, to HDL, of most of its apoC and some of its apoE (Fig. 7.2). The net result of these changes is the formation of particles smaller than the average VLDL particle and with a high cholesteryl ester : triglyceride ratio and a low apoC : apoE ratio. These particles have densities corresponding to IDL and the higher-density fraction of VLDL and are known as VLDL remnants (see Section I,C,4). In man, a considerable fraction of the remnants of VLDL is converted into LDL, estimates from different laboratories ranging from about 50% to more than 90% (see Section I,C,5). The remainder is taken up by the liver, and possibly by other tissues, via the LDL-receptor pathway. In rats and rabbits less than 10% of VLDL remnants are converted into LDL.

The final stage in the formation of LDL involves the loss of all apoproteins except apoB from the VLDL remnant and, in those species in which a cholesteryl-ester transfer protein is present in the plasma, the acquisition of esterified cholesterol from HDL. In the resulting particle about 60% of the total mass is esterified cholesterol.

6. Conversion of VLDL Remnants into LDL

Little is known about where or how LDL is formed from VLDL remnants. One possibility is that the continuing action of lipoprotein lipase on VLDL remnants leads ultimately to the formation of LDL. In favor of this, Deckelbaum *et al.* (1979) have shown that more than 97% of the triglyceride of VLDL can be hydrolyzed by lipoprotein lipase under appropriate conditions *in vitro*. However, the particles produced in this way differ from LDL in size and composition.

Another possibility is that VLDL remnants are converted into LDL by the action of heparin-releasable hepatic lipase, an enzyme attached to the luminal surface of the liver sinusoids. The observation of Goldberg *et al.* (1982) that antisera to hepatic lipase increase the plasma concentration of S_f 12–20 lipoproteins in monkeys is consistent with this. Familial absence of hepatic lipase is accompanied by a marked decrease in the fractional rate of transfer of apoB from IDL to LDL (Demant *et al.*, 1988). Furthermore, Turner *et al.* (1981) have demonstrated transfer of the apoB of radiolabeled IDL to LDL within the splanchnic bed of human subjects. As noted below, remnants of VLDL are present in S_f 12–20 lipoproteins (those with the density of IDL). However, additional processes not dependent on hepatic lipase must also contribute to LDL production because some LDL is present in the plasma of individuals with a familial deficiency of the enzyme (Breckenridge *et al.*, 1982). Moreover, in the experiments of Goldberg *et al.* (1982) on monkeys, radioactive apoB was transferred from VLDL to LDL when hepatic lipase was inhibited, though the rate of transfer was markedly reduced.

Packard *et al.* (1985) detected some delay in the formation *in vivo* of LDL from IDL (S_f 12–100) in which arginine residues of apoB and apoE had been modified by treatment with 1,2-cyclohexanedione (CHD). Since arginine residues are required for recognition of a lipoprotein particle by the LDL receptor, Packard *et al.* concluded that LDL receptors are involved in the conversion of IDL into LDL. However, for the reasons mentioned below it seems unlikely that LDL receptors are required for LDL production in man.

7. Evidence from Type III Hyperlipoproteinemia

In an attempt to explain the lipoprotein abnormalities in familial type III hyperlipoproteinemia in terms of a single genetic defect, it has been suggested that a reversible interaction between VLDL remnants and hepatic LDL receptors takes place at some stage in the production of LDL. This could involve binding of an LDL precursor without internalization or binding with internalization followed by retroendocytosis of an LDL particle. The basis of these suggestions is as follows. In the type III disorder, remnants of VLDL and chylomicrons accumulate in the plasma owing to defective clearance by their respective receptors in the liver, VLDL remnants being cleared via LDL receptors and chylomicron remnants via chylomicron-remnant receptors (Mahley, 1985). Failure to remove these lipoproteins from the plasma gives rise to the presence of two species of β-VLDL in the plasma—fraction I derived from chylomicrons and fraction II from VLDL (Fainaru *et al.*, 1982)—and to an increased plasma IDL concentration. The conversion of VLDL and of IDL into LDL is also defective (Berman *et al.*, 1978; Chait *et al.*, 1977), with a consequent decrease in plasma LDL concentration. The defective clearance of remnants in type III patients is due to the inheritance of the E2 variant of apoE (rather than the normal E3 or E4 isoforms), apoE2-containing remnants having a much reduced affinity for hepatic lipoprotein receptors (Schneider *et al.*, 1981; see also Table 6.2).

One way of explaining defective formation of LDL on the basis of the inheritance of the *E2* mutant is to suppose that the defect is due to the reduced binding affinity of VLDL remnants for the LDL receptor. However, an apoE-dependent interaction of a lipoprotein precursor of LDL with LDL receptors cannot be necessary for the production of LDL because LDL is formed in increased amounts in FH homozygotes, who have no LDL receptors. Indeed, Sigurdsson (1982) found that 99% of the apoB in an intravenous dose of ^{131}I-labeled VLDL appeared in the LDL of a group of subjects with heterozygous FH and normal plasma triglyceride levels. This indicates that the efficient formation of LDL from VLDL does not depend on the presence of normally functioning LDL receptors. Moreover, LDL is present in the plasma of patients with an inherited absence of apoE in their plasma (Ghiselli *et al.*, 1981).

Enholm *et al.* (1984) have investigated the conversion of fraction II β-VLDL, isolated from type III patients, into LDL-like particles during incubation with lipoprotein lipase *in vitro*. Lipoproteins that separate in the ultracentrifuge within the LDL density range are not formed unless apoE3 is added to the incubation mixture. Enholm *et al.* (1984) suggest that the presence of apoE2 in fraction II β-VLDL in some way impedes the lipolytic steps involved in its conversion into LDL. Although defective formation of LDL in type III patients has yet to be explained fully, there can be little doubt that this disorder holds important clues to the steps by which VLDL remnants are normally converted into LDL and to the manner in which these steps are regulated.

8. Chylomicrons Compared with VLDL

The components of chylomicrons are synthesized, assembled, and secreted by cells of the intestinal mucosa by processes analogous to those responsible for the formation and secretion of VLDL by hepatocytes. The nascent chylomicrons secreted into the intestinal lymphatics differ in size and composition from newly secreted VLDL. In particular, nascent chylomicrons are larger than VLDL, they are enriched with esterified cholesterol derived from dietary cholesterol by the action of intestinal ACAT, and they have apoB-48 as their only apoB species. As discussed in Chapter 6, the formation of a chylomicron particle requires apoB of a size not less than about 205K. The MW of apoB-48 is ≃241K.

Absorption of fat is accompanied by a marked increase in the rates of intestinal synthesis of triglyceride and other lipid components of chylomicrons. However, absorption of a single fatty meal has no effect on the synthesis of apoA-I (Davidson and Glickman, 1985) or apoB (Davidson *et al.*, 1986) in the small intestine of the rat. Presumably, the apoB and apoA-I required to sustain increased formation and secretion of chylomicrons are derived from preformed intracellular stores. There is disagreement as to the effects of long-term dietary changes on synthesis of apoproteins in the small intestine. Davidson *et al.* (1987) found no increase in intestinal synthesis of apoB or apoA-I in rats fed a high-fat diet or a cholesterol-rich atherogenic diet for 3 weeks. Go *et al.* (1988), on the other hand, observed increased intestinal synthesis of apoA-I in cholesterol-fed rats. In the experimental animals the intestinal content of apoA-I mRNA was decreased, indicating an increased rate of translation of the message.

After their entry into the circulation, chylomicrons are acted upon by lipoprotein lipase, with the formation of remnant particles. Like VLDL remnants, chylomicron remnants have a high apoE : apoC ratio and a high cholesteryl ester : triglyceride ratio. However, they are larger than VLDL remnants and have apoB-48 in place of apoB-100. Owing to their apoB composition and, possibly, to their larger particle size, chylomicron remnants follow a metabolic pathway different from that of VLDL remnants. As we have seen, VLDL remnants are either converted into LDL or are removed from the plasma by hepatic LDL receptors. Chylomicron remnants, on the other hand, are taken up rapidly and completely by the liver via a pathway mediated by receptors that are probably distinct from LDL receptors (see Chapter 8). These receptors, already mentioned in Section I,A,6, are called chylomicron-remnant receptors; they are discussed in more detail in Chapter 8. As a consequence of the complete removal of chylomicron remnants by the liver, none of their apoB-48 appears in LDL. Although the failure of chylomicron remnants to be converted into LDL must be due in part to their very rapid and irreversible removal from the circulation, this does not seem to be the whole explanation. Enholm *et al.* (1984), in the work mentioned above, noted that fraction I β-VLDL from type III patients could not be hydrolyzed by lipoprotein lipase to particles within the IDL or LDL density ranges, even in the presence of apoE3. This suggests that the chylomicron remnant, as well as having a high affinity for its hepatic receptor, is not a substrate for the lipolytic enzyme(s) responsible for the final step in the formation of LDL.

B. The Remnant Concept

The term ''VLDL remnant'' is worth discussing briefly at this point because it has been the source of some confusion. VLDL remnants are sometimes referred to as IDL (density 1.006–1.019 g/ml) or, less frequently, as β-VLDL[1] (density <1.006 g/ml). Chylomicron remnants were first recognized as the end products of chylomicron metabolism that accumulate in the plasma of animals in which the liver is excluded from the circulation. These remnant particles have a characteristic range of particle size and composition and, when injected intravenously into intact animals, they are taken up rapidly by the liver. Thus, chylomicron remnants correspond to a clearly definable stage in the normal metabolism of chylomicrons. They may be defined as the end products of the intravascular catabolism of chylomicrons and are characterized by physical and chemical properties that confer upon them a high affinity for specific chylomicron-remnant receptors in the liver.

The definition of VLDL remnants, on the other hand, is to some extent arbitrary. Particles whose electrophoretic mobility and composition are similar to those of chylomicron remnants are produced from VLDL in the circulation of functionally hepatectomized rats (Mjøs *et al.*, 1975). However, as discussed in Section I,C,4, VLDL in human plasma is catabolized through a chain of delipidations resulting in the formation of a series of particles of diminishing size and increasing density, all of which may spend some time in the circulation. Some of the particles generated at the end of this chain are removed from the plasma by the liver and are therefore the equivalent of chylomicron remnants, while others act as the immediate precursors of LDL (LDL is never referred to as a VLDL remnant). Particles at the end of the delipidation chain that are destined for hepatic uptake cannot be distinguished from those destined to become LDL. Until this uncertainty is resolved, the best course may be to define VLDL remnants as VLDL-derived particles within the density range that includes all the immediate precursors of LDL. On this definition, VLDL remnants in normal human subjects extend over the flotation range S_f 12–60 (Reardon *et al.*, 1978). This includes IDL (S_f 12–20, or density 1.006–1.019 g/ml) and the smaller, denser particles of VLDL (S_f 20–60).

It should be noted that the VLDL fraction with the highest density is thought to include particles that are secreted directly by the liver and are therefore in no sense remnants. Particles that are almost certainly VLDL remnants with density <1.006 g/ml may be separated from the bulk of VLDL by agarose-gel electrophoresis of total VLDL from normal human subjects (Pagnan *et al.*, 1977). The ''remnants'' appear as a faint band with slower mobility than that of the major

[1]Remnants of VLDL and chylomicrons with β or slow pre-β electrophoretic mobility are thought to be present in normal VLDL. To minimize confusion, the term β-VLDL is restricted in this book to the abnormal VLDL (density <1.006 g/ml) with β mobility present in the plasma of patients with type III hyperlipoproteinemia and of cholesterol-fed animals (see Mahley, 1982).

pre-β band. The composition of these particles is similar to that of IDL; they are smaller than VLDL with pre-β mobility and are relatively enriched with cholesteryl esters and apoE.

Despite the difficulty of defining VLDL remnants, what Havel (1984) calls "the remnant concept" has been of considerable help in deepening our understanding of human VLDL metabolism. VLDL remnants, as defined above, are situated at an important branch point in the metabolism of VLDL under normal conditions, and the extent to which these particles follow one or another of their alternative routes must help to determine the rate of production of LDL. Moreover, the idea that the formation of remnants is a normal step in the catabolism of VLDL has been helpful in the study of lipoprotein disorders. In particular, accumulation of VLDL remnants in the plasma of patients with type III hyperlipoproteinemia helps to explain the increased concentration of S_f 12–20 lipoproteins and the presence of β-VLDL in this disorder.

C. Quantitative Aspects

In Section I,A the generation of LDL from VLDL was considered mainly from a qualitative point of view. In this section the quantitative aspects of this process are considered in relation to human VLDL metabolism. As discussed below, there are considerable difficulties in measuring rates of flow through what is now known to be a complex system. Some of these difficulties have yet to be resolved. In particular, there is still disagreement about the quantitative significance of metabolic pathways for VLDL that must help to determine the rate of production of LDL.

1. How Can a VLDL Particle Be Labeled?

Although VLDL particles are well-defined physical entities, all their constituents except apoB either exchange continuously with constituents in other lipoproteins or are metabolized independently. Hence, the only meaningful sense in which the metabolism of a VLDL particle as a whole can be followed is by following the fate of its apoB component. In this context, it is helpful to think of a VLDL particle as consisting of a molecule of apoB associated with lipids and other apoproteins undergoing exchange or independent metabolism, the apoB molecule remaining in the particle throughout its intravascular life. Thus, VLDL-apoB is the appropriate marker for VLDL and its lipoprotein products, except when one is concerned specifically with the metabolism of some other VLDL component such as triglyceride.

2. Kinetics of ApoB Metabolism: Methodology

The intravascular phase of apoB metabolism may be investigated in man by following the fate of apoB in plasma lipoproteins that have been labeled with a radioactive isotope. The lipoproteins are labeled *in vivo* with a radioactive amino acid or *in vitro* with radioiodine.

Two approaches have been used in the analysis of the plasma radioactivity curves obtained from such experiments. In the one, it is assumed that each lipoprotein fraction is metabolically homogeneous, i.e., that all particles within each fraction have the same probability of following a given metabolic pathway. Rates of flow of apoB through VLDL and LDL are usually estimated by analyzing specific-radioactivity curves in terms of a single pool for VLDL and a two-pool system for LDL. In the other approach, the kinetics of radioactive apoB are analyzed in terms of models, each consisting of a system of compartments between which the tracer flows. In principle, the number of compartments in the model and the fractional rates of transfer between compartments are adjusted by trial and error to give optimum fit between the kinetics predicted from the model and the kinetics observed experimentally (plasma apoB specific-radioactivity curves in the present case). Workers using the two methods of analysis have obtained closely similar values for rates of turnover of apoB in VLDL and LDL and for precursor–product relationships between apoB in VLDL, IDL, and LDL. Nevertheless, multicompartmental analysis has revealed the presence of metabolic heterogeneity to an extent not previously thought to exist.

Since this method of analysis has become a standard procedure in lipoprotein methodology, a few remarks about its validity may be useful at this point. It has been of considerable value in providing more accurate estimates of the magnitude of known metabolic pathways and in suggesting the existence of previously unsuspected ones, especially in disorders of lipoprotein metabolism. However, conclusions derived from mathematical models must be regarded as tentative, at least until each element of the model has been shown to have a recognized physiological or biochemical counterpart and until independent evidence has been obtained for each of the pathways postulated. In this regard, see Phair (1982) for a discussion of the impossibility of proving that a particular one of several alternative models is the "correct" one for a given set of observations. It should also be noted that a model cannot be valid if artifacts or experimental inaccuracies in the observations on which the model is based are ignored; the likelihood of error due to uneven *in vitro* labeling of VLDL particles of different sizes is mentioned below. These limitations must be borne in mind in any discussion of estimates of lipoprotein kinetics based on multicompartmental analysis.

3. Precursor–Product Relationships

Bilheimer *et al.* (1972) showed some years ago that when VLDL, in which the apoproteins are labeled with radioiodine, is injected intravenously into human subjects, radioactivity appears in LDL-apoB. Since it was known that apoB does not undergo exchange between plasma lipoprotein fractions, this experiment showed that at least some of the apoB in LDL is derived from VLDL. Subsequent work has confirmed and extended this observation.

Figure 7.3 shows the results of a typical experiment in which radioiodine-labeled VLDL was injected into a normal human subject. The fall in specific radioactivity of apoB in VLDL, and its subsequent rise and fall in IDL and LDL, were then observed over several days. The specific radioactivity of VLDL-apoB fell monoexponentially until more than 90% of the radioactivity had left the VLDL fraction. Concurrently, radioactivity appeared in IDL-apoB and then in LDL-apoB. Within the limits of experimental error, the maximum of the IDL-apoB curve occurred at its intersection with the VLDL-apoB curve and the maximum of the LDL-apoB curve occurred at its intersection with the IDL-apoB curve. These kinetic relationships would be expected if, in normal subjects, all IDL-apoB is derived from VLDL and all LDL-apoB is derived from IDL. (See Zilversmit, 1960, for a discussion of the relationship between the specific-radioactivity curves given by a labeled precursor and its immediate product *in vivo*.) Berman *et al.* (1978), using multicompartmental analysis to interpret their observations, also concluded that all LDL-apoB is normally derived from VLDL. In keeping with this conclusion, estimates of the rates of turnover of apoB reported from many laboratories agree in showing that in normal subjects the rate of turnover of LDL-apoB (mg/day) never exceeds that of VLDL-apoB.

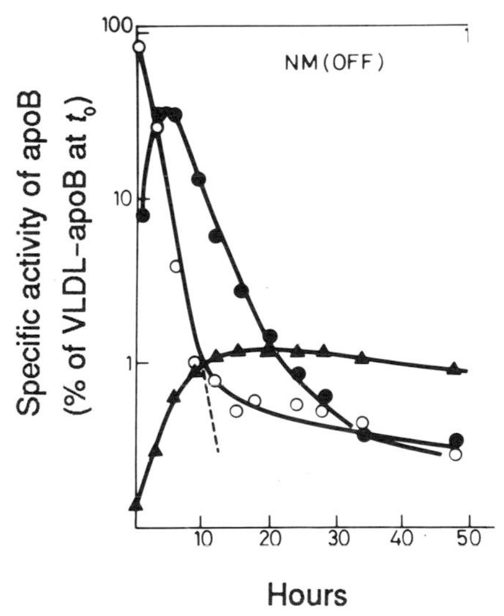

Figure 7.3. Specific radioactivity of apoB in VLDL (○), IDL (●), and LDL (▲) in the plasma of a normal subject after intravenous injection of ^{125}I-labeled VLDL. (From Soutar *et al.*, 1982.)

4. Metabolic Heterogeneity

As discussed in Chapter 5, VLDL and LDL are chemically and physically heterogeneous. It would be reasonable to suppose that this heterogeneity is reflected to some extent in metabolic heterogeneity. Indeed, Fig. 7.3 shows the presence of a prolonged "tail" following the initial exponential in the VLDL-apoB curve. This indicates the presence of subpopulations of VLDL particles that are metabolized more slowly than the bulk of VLDL, which usually leaves the circulation with a half-life of less than 6 hours. Metabolic heterogeneity within LDL is indicated by the presence of a tail that is often seen following the log-linear phase of the disappearance curve obtained after injections of radioiodine-labeled LDL. Fisher *et al.* (1980) also postulated the presence of more than one metabolically distinct subpopulation of LDL to explain the kinetics of LDL-apoB observed after endogenous labeling of VLDL-apoB in hypertriglyceridemic patients. In addition, Teng *et al.* (1986) have noted differences in the metabolic behavior of radiolabeled subfractions of LDL.

Phair and co-workers (Phair *et al.*, 1975; Berman *et al.*, 1978) and Fisher *et al.* (1980) have used multicompartmental models to analyze apoB specific-radioactivity curves obtained after endogenous or exogenous labeling of VLDL-apoB in normal and hypertriglyceridemic subjects (reviewed by Fisher, 1982). The kinetics of apoB metabolism could not be explained on the assumption that VLDL behaves metabolically as a homogeneous population of particles, but did conform to the model shown in Fig. 7.4. In this model, VLDL particles pass along a chain of delipidations, the particles generated at each step returning briefly to the plasma before reentering the delipidation chain; this process has been called a delipidation cascade (Berman, 1982). As shown in Fig. 7.4, the delipidation cascade continues into the IDL and LDL fractions.

The pathways shown for input of VLDL at several points along the chain were postulated by Fisher *et al.* (1980) to explain the rapid rise in apoB specific radioactivity in the denser VLDL subfractions after endogenous labeling of VLDL apoproteins with [^{3}H]leucine. These elements of the model imply that in a given individual the liver secretes a heterogeneous population of VLDL particles differing in size and lipid : protein ratio. This is consistent with reports that hepatic Golgi bodies contain nascent VLDL particles of different sizes (Alexander *et al.*, 1976; Gotto *et al.*, 1986). The essential features of the model shown in Fig. 7.4 are in agreement with independent evidence that subfractions of VLDL containing the larger, less dense particles are the precursors of smaller and more dense VLDL particles (Barter and Nestel, 1972; Eisenberg *et al.*, 1973). According to this model, in normal subjects apoB secreted in VLDL does not leave the circulation irreversibly until it has been transferred to LDL, whereas in primary hypertriglyceridemia some apoB leaves the circulation directly in IDL or in the more dense VLDL particles. The validity of these aspects of the model is discussed in the next section.

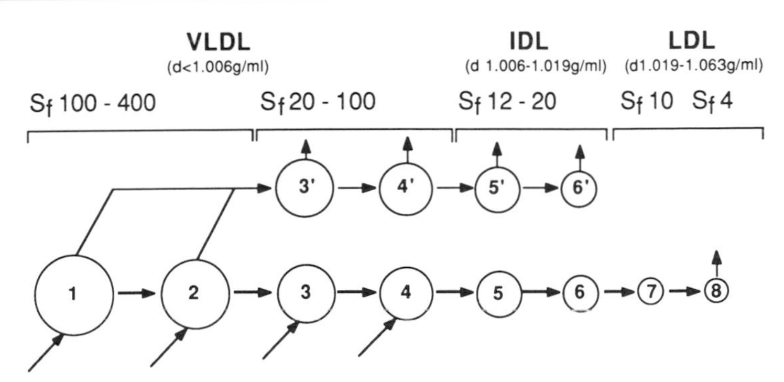

Figure 7.4. A model to show the possible routes by which hepatic VLDL-apoB is metabolized in normal man, combining features put forward by Berman *et al.* (1978), Fisher (1982), Packard *et al.* (1984), and Reardon *et al.* (1978). Each circle represents a population of particles with flotation rates (S_f) in the ranges at the top of the diagram. Diagonal arrows, secretion of particles into the circulation; vertical arrows, removal from the circulation. The larger particles (pools 1 and 2) are delipidated along two alternative pathways. In one pathway they are converted successively into small VLDL (3 and 4), IDL (5 and 6), and LDL (7 and 8). In the other they are converted into small VLDL (3′ and 4′) and IDL (5′ and 6′) that leave the circulation without conversion into LDL. Note that in this model the S_f 20–100 and S_f 12–20 fractions are metabolically heterogeneous, each fraction containing some particles that will become LDL and some that will be removed directly from the circulation. The two subfractions in LDL with precursor–product relationship (7 and 8) were postulated by Phair *et al.* (1975) to explain the kinetics of VLDL-apoB metabolism in subjects with polydisperse LDL.

5. The Fraction of the VLDL-ApoB Transferred to LDL

If VLDL-apoB is, in fact, the obligatory precursor of all LDL-apoB in normal man, then we need to know what proportion of the VLDL-apoB secreted into the circulation is transferred to LDL, since this must be one of the determinants of the rate of production of LDL. Despite the importance of this question, it cannot yet be answered with certainty.

Work on animals of several species has shown that up to 90% of an injected dose of VLDL leaves the circulation without conversion into LDL, mainly by hepatic uptake of VLDL or its remnants (Faergeman *et al.*, 1975; Barter *et al.*, 1977; Kita *et al.*, 1982). It would be surprising, therefore, if a significant fraction of the apoB secreted in VLDL in normal human subjects were not removed from the circulation at some stage before its transfer to LDL. However, Sigurdsson *et al.* (1975) and Reardon *et al.* (1978) concluded that at least 90% of the apoB in VLDL is transferred to LDL in normal man. Berman *et al.* (1978), using multicompartmental analysis, also concluded that "all VLDL goes to LDL" in human subjects with normal plasma lipid concentrations. (This is implied in the lower channel shown in Fig. 7.4, in which there is no exit pathway before the S_f 4 fraction of LDL.) Packard *et al.* (1980), also using multicompartmental analysis for the interpretation of their radioactivity curves, concluded that more than 95%

of the apoB in an injected dose of VLDL is transferred to LDL in normal subjects. In contrast, most workers who have used multicompartmental analysis have concluded that a substantial proportion of the VLDL-apoB in patients with primary hypertriglyceridemia is removed from the circulation without appearing in LDL. For example, Reardon *et al.* (1978) estimated that in one of their hypertriglyceridemic patients only a third of the apoB in an injected dose of VLDL appeared in LDL.

An alternative to multicompartmental analysis of plasma radioactivity curves is to estimate the proportion of VLDL-apoB that is transferred to LDL, using convolution–deconvolution analysis of LDL-apoB specific-radioactivity curves obtained after simultaneous injections of radioactive LDL and radioactive VLDL (Sigurdsson, 1982). An advantage of this technique is that no assumptions are made as to the steps involved in the conversion of VLDL into LDL. Applying this method to normal human subjects, Janus *et al.* (1980) estimated that, in contrast to estimates derived from multicompartmental analysis, less than 50% of VLDL-apoB is transferred to LDL.

More direct observations on the metabolism of VLDL have also called into question the view that essentially all the VLDL-apoB secreted into the circulation is transferred to LDL in normal man. Packard *et al.* (1984) isolated VLDL subfractions of defined flotation rate from human subjects, labeled the VLDL, and followed its metabolic fate after reinjection into the donor. Large labeled VLDL particles (S_f 100–400) were converted quantitatively into smaller particles (S_f 12–100), most of which were removed from the circulation without conversion into LDL. Small labeled VLDL particles (S_f 12–60), on the other hand, were converted into LDL. Packard *et al.* suggest that the liver secretes VLDL particles of different sizes, the larger ones generating remnants that are removed from the circulation (presumably via hepatic lipoprotein receptors), and the smaller ones acting as precursors for LDL. On this view, particles within the small-VLDL density range undergo metabolic channeling in the sense that, depending upon their origin, some are committed to direct removal, while others are committed to the pathway leading to LDL. In agreement with these observations, Stalenhoef *et al.* (1984) showed that large VLDL particles, isolated from the plasma of a patient with an inherited deficiency of lipoprotein lipase, were removed rapidly from the circulation without conversion into LDL when injected intravenously into normal subjects.

Some recent observations on rabbit lipoproteins may be relevant to the fate of VLDL-apoB in man, in a qualitative rather than a quantitative sense. Yamada *et al.* (1986) have shown that rabbit VLDL, IDL, and LDL each consist of particles that contain apoB and apoE (B,E particles) and those that contain only apoB (B particles). Experiments designed to provide information about the fate of apoB secreted in VLDL gave results that were consistent with the following sequence of events. About 90% of the apoB entering the blood circulation is secreted in VLDL B,E particles and about 10% is secreted in VLDL B particles. Most of the

VLDL B,E particles are removed rapidly from the circulation, as remnants of density <1.006 g/ml, without conversion into IDL or LDL. In contrast, about 27% of the VLDL B particles are converted into LDL. The net result of these alternative pathways for VLDL metabolism is that about 10% of the total apoB secreted in both classes of VLDL particles is transferred to LDL.

The marked differences between reported estimates of the proportion of VLDL converted into LDL in normal human subjects (<50% to more than 95%) are not easy to explain. Packard *et al.* (1984) have pointed out that small VLDL are more efficiently labeled by radioiodine than large VLDL. Hence, if small particles are preferentially converted into LDL, a labeled sample of total VLDL containing a higher proportion of small particles would give a higher VLDL-to-LDL conversion than one containing a lower proportion. Rapid lipolysis of newly secreted large VLDL particles would also lead to their underrepresentation in a labeled sample of total VLDL.

In evaluating the conclusions derived from mathematical modeling and those derived from more direct observation, it seems reasonable to give greater weight to the latter. In this case, we should conclude that in normal man a substantial proportion, perhaps 50%, of the VLDL-apoB secreted by the liver is removed from the circulation before it is transferred to LDL. Diversion of such a high proportion of VLDL-apoB from the pathway to LDL would give ample scope for regulation of the rate of production of LDL. The extent to which overproduction of LDL in familial hypecholesterolemia is due to decreased hepatic uptake of the precursors of LDL is discussed in Chapter 10. The possible effect of *increased* activity of hepatic LDL receptors on LDL production is dealt with in Section II below.

6. Determinants of the Uptake of Lipoproteins

As we have seen, mature VLDL particles are converted into partially delipidated remnants, some of which are removed rapidly from the plasma by the liver. Those not removed directly are converted into LDL, which are cleared much more slowly. Both VLDL remnants and LDL are thought to be removed from the circulation largely by the liver via the LDL-receptor pathway (see Section II below). Hence, the different rates of clearance of lipoprotein particles at successive stages in the chain from VLDL to LDL should be explicable, at least partly, by differences in the affinity of these particles for the LDL receptor.

Mature VLDL that has acquired its full complement of apoC from HDL (see Fig. 7.2) is taken up only at a slow rate by the perfused rat liver (Windler *et al.*, 1980), despite the fact that VLDL contains apoB and apoE, both of which are recognized by the LDL receptor. Moreover, VLDL (S_f >60) from normal human plasma is bound poorly by LDL receptors *in vitro* (Gianturco *et al.*, 1982). The low affinity of normal VLDL for LDL receptors is probably due to masking of the receptor-recognition sites on apoB and apoE by apoC and, possibly, to an unfavorable effect of the high lipid content of the particles on the conformation of apoB and apoE. As VLDL is delipidated by the action of lipoprotein lipase there is a progressive loss of apoC from the particles and an increase in the ratio of apoE to apoC, followed by a virtually complete loss of apoE when VLDL remnants are converted into LDL. In keeping with a relative enrichment with apoE as the delipidation cascade proceeds, Nestel *et al.* (1983) have shown that VLDL particles with a low apoE : apoC ratio are converted into VLDL particles with a high apoE : apoC ratio in the plasma of human subjects *in vivo*.

When VLDL is hydrolyzed by lipoprotein lipase *in vitro*, the reactivity of the particles toward monospecific anti-apoB antibodies increases progressively, and this increase is accompanied by a progressive increase in uptake of the lipolyzed particles by LDL receptors on fibroblasts (Schonfeld *et al.*, 1979). More detailed studies have been carried out with proteolytic enzymes that hydrolyze apoE in preference to apoB in lipoprotein particles (Bradley *et al.*, 1984) and with monoclonal antibodies reacting with receptor-binding domains on apoB and apoE (Krul *et al.*, 1985). These studies have shown that the ability of VLDL remnants in the IDL density range to react with LDL receptors is mediated largely by apoB, rather than by apoE. On the other hand, unmasking of receptor-recognition sites on apoE, combined perhaps with an increased number of apoE molecules per particle, appears to play a predominant role in the receptor binding of VLDL remnants in the S_f 20–60 range. In agreement with this conclusion, Hui *et al.* (1984) have shown that β-VLDL particles containing apoB-100 (isolated from type III patients) bind to LDL receptors entirely by virtue of their content of apoE. In this regard, it may be noted that the inability of apoB-100 to mediate receptor binding of these VLDL remnants explains why β-VLDL are removed abnormally slowly in type III patients, whose lipoproteins contain a variant of apoE with defective receptor-binding ability (see Section I,A).

In summary, then, it seems likely that the determinant for lipoprotein binding to LDL receptors on hepatocytes switches from apoE in VLDL remnants within the VLDL density range to apoB in LDL, and that in IDL both apoE and apoB contribute to binding, with apoB playing the predominant role. The contribution of apoE to the receptor-mediated uptake of VLDL remnants explains why they are cleared from the plasma more rapidly than is LDL, since the affinity of apoE-containing lipoproteins for the LDL receptor is, in general, greater than that of LDL (see Chapter 8, Section II,A). As noted above, large VLDL particles give rise to remnants that are cleared from the plasma without conversion into LDL. This may be because large VLDL, which in man contain more apoE molecules per particle than small VLDL (Kane *et al.*, 1975), generate apoE-rich remnants that are taken up by hepatic LDL receptors with high affinity owing to polyvalent binding, either to more than one receptor or to multiple sites on a single receptor.

7. The Rates of Production of VLDL and LDL

The rates of production of VLDL-apoB and LDL-apoB are estimated indirectly from their rates of catabolism in the steady state, on the assumption that production and catabolism are equal under these conditions. Rates of catabolism are usually determined by analyzing the kinetics of plasma radioactive apoB after intravenous injection of radioiodine-labeled VLDL or LDL. Different workers have based their analysis of apoB kinetics on models of varying degrees of complexity. However, estimates based on different models do not differ systematically from one laboratory to another. Since there are no direct methods for estimating rates of production of VLDL or LDL in man, the accuracy of an indirect method cannot be evaluated by comparison with an absolute standard. The best we can say is that some models are theoretically superior to others.

Estimates of the rate of production of VLDL-apoB in normal human subjects usually lie within the range from 10 to 15 mg/kg body weight/day. In several laboratories plasma radioactivity curves have been analyzed on the assumption that almost all VLDL is present in a single plasma compartment. For reasons discussed in Section I,C,4 above, this must be an oversimplification. Nevertheless, estimates based on multicompartmental analysis, using a two-pool model (Reardon *et al.*, 1978) or a model with a delipidation chain of four equal compartments and a branch pathway with slow turnover (Phair *et al.*, 1975), do not differ significantly from those based on the one-compartment model. (See Kesäniemi *et al.*, 1982, for a survey of reports from several sources.)

More recently, Beltz *et al.* (1985) have put forward an integrated model that accommodates all the known pathways taken by apoB secreted into the plasma by the liver. In this model, a proportion of the VLDL secreted is rapidly delipidated to give IDL or LDL particles without passing through the delipidation chain. When this additional pathway is included in the analysis of plasma radioactivity curves, higher values are obtained for the rate of production of VLDL-apoB in a given subject (Kesäniemi *et al.*, 1985). Beltz *et al.* (1985) point out that a VLDL subfraction with very rapid turnover would have a low steady-state plasma concentration and would therefore remain essentially unlabeled when the total mass of plasma VLDL is labeled *in vitro*. This may have a bearing on the increased rate of production of LDL observed under certain abnormal conditions (see Chapter 10). It should be noted that neither the model of Beltz *et al.* nor the one proposed by Phair *et al.* (1975) allows for multiple pathways for the entry of VLDL into the plasma, as shown in Fig. 7.4.

The rate of synthesis of VLDL-apoB is normal in most FH patients but is raised in some patients with other types of hyperlipidemia (see Kesäniemi *et al.*, 1982).

The rate of production of LDL-apoB is usually estimated on the basis of the model introduced by Matthews (1957). In this model it is assumed that LDL is present in a single, metabolically homogeneous, intravascular pool in exchange equilibrium with an extravascular pool, and that LDL can enter and leave the system only through the intravascular pool. The fractional catabolic rate (FCR) is estimated from the plasma radioactivity curve; the absolute rate of catabolism, and hence the rate of production, of LDL-apoB is estimated from the FCR and the total mass of LDL-apoB in the plasma. Reported values for the FCR of LDL-apoB in normal subjects have ranged from 0.3 to 0.6 pools per day, giving production rates of between 7 and 15 mg/kg/day.

The assumption that there is a single intravascular pool of LDL is clearly an oversimplification in the light of evidence, discussed above, that distinct subpopulations of LDL are present in many human subjects (Fisher, 1982). However, Beltz *et al.* (1985) and Kesäniemi *et al.* (1985), in a study of normal and hyperlipidemic subjects, could find no deviation of the kinetics of radioiodine-labeled LDL-apoB from that expected from the Matthews model. In particular, analysis of LDL-apoB kinetics in terms of a model that included the separate S_f 10 and S_f 4 pools shown in Fig. 7.4 gave values for total LDL-apoB turnover that were similar to those derived from the Matthews model. A different conclusion was reached by Teng *et al.* (1986), who measured the rates of production of apoB in each of two subfractions of LDL, a low-density (light) and a high-density (heavy) fraction, in two normal subjects. About 75% of the light fraction was converted into heavy LDL, but there was also some direct production of heavy LDL. When total LDL-apoB production was estimated as the sum of light LDL-apoB production and independent production of heavy LDL-apoB, the values obtained were 20–30% higher than the values obtained by the method of Matthews.

II. LDL Catabolism *in Vivo*

A. Preliminary Remarks

In this section the catabolism of LDL, other than that in FH (see Chapter 10), is considered at the level of the whole body and of organs and tissues. LDL catabolism at the cellular and subcellular levels is dealt with in the chapters on the LDL receptor (Chapters 8 and 9). However, a few preliminary remarks about LDL-receptor function will be helpful at this stage. Some of these points have already been referred to briefly in relation to HMG-CoA reductase and apoB.

Functional LDL receptors have been demonstrated *in vitro* in a wide variety of mammalian tissues, including liver, intestine, skin fibroblasts, and steroid-hormone-forming tissues. The LDL receptor binds specific domains in apoB and apoE when these apoproteins are incorporated into lipoproteins in an appropriate conformation. Chemical modification of the arginine or lysine residues in the receptor-recognition domains abolishes binding of apoB- or apoE-containing lipoproteins by LDL receptors. As expressed in most cells in culture, LDL receptors have high affinity for LDL (half saturation at 10 μg of LDL protein/ml at 37°C) and are saturated at about 50 μg of LDL protein/ml. In cells observed *in vitro*, the activity of HMG-CoA reductase and the number of LDL receptors on the surface of the cell are regulated coordinately in accordance with the cell's requirement for cholesterol (see Chapter 3, Section VIII). How far the behavior and mode of regulation of LDL receptors *in vitro* reflect what happens *in vivo* is discussed in Chapter 10.

B. Methodology

1. Catabolism in the Whole Body

The absolute rate of catabolism of LDL is estimated from the FCR of a tracer dose of radioiodinated LDL and the plasma LDL concentration (see Section I,C above). Most workers have analyzed plasma-radioactivity disappearance curves in terms of the Matthews model, although, as noted above, this is bound to be an oversimplification in so far as LDL is metabolically heterogeneous. The usual practice is to reinject the subject's own LDL (autologous LDL) after labeling it *in vitro*. This procedure may be safer than the use of labeled homologous LDL and it does, of course, provide a "best estimate" of the FCR of the subject's circulating LDL at the time of observation. However, the use of autologous LDL may have disadvantages in some circumstances, especially in relation to the measurement of LDL-receptor activity *in vivo*. For example, Witztum *et al.* (1985) have shown that stimulation of hepatic LDL-receptor activity brought about by treatment with cholestyramine is accompanied by a change in the composition of circulating LDL toward particles with reduced affinity for the receptor. Hence, labeled autologous LDL from a cholestyramine-treated animal or human subject may be catabolized with normal or near-normal FCR when reinjected into the treated donor. In this case, the FCR would not reflect the drug-induced change in receptor activity.

Measurements of the rate of disappearance of labeled native LDL from the circulation provide the basis for estimates of the *total* rate of catabolism of LDL by all routes in the whole body, including LDL-receptor-dependent and LDL-receptor-independent pathways. The FCR of LDL via the LDL-receptor pathway may be estimated as the difference between the FCR of labeled native LDL and that of LDL in which the receptor-recognition site in apoB has been blocked by chemical modification. The validity of methods based on this principle depends upon the assumptions that the chemical modification is stable *in vivo*, that it completely blocks catabolism of LDL by the LDL-receptor pathway, and that it has no effect on catabolism by all other pathways.

1,2-Cyclohexanedione (CHD) reacts with the guanido group of arginine, or of arginyl residues in proteins, to form an addition product that dissociates slowly in weakly alkaline solution (Patthy and Smith, 1975). Mahley *et al.* (1977) have shown that when about half the residues in apoB of LDL are blocked by *treatment with CHD*, high-affinity binding and degradation of the LDL by fibroblasts in culture are almost completely inhibited. CHD modification increases the net negative charge on LDL, with a consequent increase in electrophoretic mobility. *Reductive methylation* of the ε-amino groups of 15–20% of the lysine residues in LDL-apoB also abolishes binding of LDL by LDL receptors *in vitro* (Weisgraber *et al.*, 1978) but does not alter the charge on the amino groups of lysine. Hence, there is no change in net charge on the LDL particle. Incubation of LDL for several days with glucose in the presence of a suitable reducing agent results in the *glycosylation* of up to 45% of the apoB lysine residues, with the formation of glucitol lysines (Witztum *et al.*, 1982; Kesäniemi *et al.*, 1983). When about a third of the lysine residues are glycosylated, high-affinity uptake of the modified LDL by fibroblasts *in vitro* is abolished. Like CHD modification of arginine residues, glycosylation of lysine residues in LDL-apoB increases the negative charge on LDL.

Modification by treatment with CHD (Mahley *et al.*, 1980), reductive methylation (Mahley *et al.*, 1980), and nonenzymatic glycosylation (Witztum *et al.*, 1982) decreases the FCR of LDL in normal experimental animals. This is consistent with selective inhibition of LDL-receptor-mediated uptake of LDL *in vivo*. In keeping with this, the FCR of CHD-modified LDL in human homozygous FH (Thompson *et al.*, 1981) and of glycosylated LDL in WHHL rabbits (Steinbrecher *et al.*, 1983) (a strain with an inherited inability to express LDL receptors) is in each case the same as that of native LDL. However, the extent to which the FCR of LDL is decreased differs according to the method used for blocking the receptor recognition site on apoB, CHD treatment giving the smallest decrease and glycosylation the greatest. Nor is any direct method available for determining which of the three procedures gives the best estimate of the contribution of the LDL-receptor pathway to LDL catabolism in the whole body.

The increased electronegativity of CHD-treated and glycosylated LDL does not seem to lead to increased uptake by the acetyl-LDL receptor pathway (see chapter 8), since the rate of nonsaturable degradation of CHD-LDL by human monocyte-macrophages (Knight and Soutar, 1982), and of glycosylated LDL by mouse peritoneal macrophages (Witztum *et al.*, 1982), is no greater than that of native LDL. Slow reversibility of CHD modification might lead to partial dissociation of CHD-modified LDL *in vivo*, and thus to underestimation of the contribution of the LDL-receptor pathway to total LDL catabolism. The observations of Thompson *et al.* (1981) on monkeys, and of Slater *et al.* (1982) on rabbits, argue against this. However, it remains possible that inhibition of the catabolism of CHD-modified LDL by LDL receptors is incomplete *in vivo*.

Glycosylation of LDL is not reversible *in vivo* (Kesäniemi *et al.*, 1983), but there is no way of excluding the possibility that glycosylated LDL is catabolized by LDL-receptor-independent pathways more slowly than is native LDL. If this does occur, the glycosylation method would overestimate the contribution of the LDL-receptor pathway. Glycosylated homologous LDL is strongly immunogenic in guinea pigs (Witztum *et al.*, 1983) and is immunogenic in some human subjects (Kesäniemi *et al.*, 1983). For this reason it has not been widely used in the study of LDL metabolism in man.

Reductively methylated LDL, when injected intravenously into human subjects, is treated as a foreign protein and is rapidly removed from the circulation (Slater *et al.*, 1982). It is therefore unsuitable for use in human studies, though it has been used successfully in the study of LDL-receptor-independent catabolism of LDL in experimental animals (see Section II,D below). Because methylated LDL is catabolized *in vitro* at the same rate by fibroblasts from normal and homozygous FH human subjects, it is generally assumed to be an appropriate tracer for LDL-receptor-independent catabolism of LDL in whole animals. However, methylated LDL may fail to "detect" catabolism of LDL by specific mechanisms in the RE system, since it is not recognized by acetyl-LDL receptors on macrophages (Brown and Goldstein, 1983).

From a practical point of view, the safest and most convenient method for estimating the rate of LDL catabolism by the LDL-receptor pathway in man is to use CHD-modified LDL. Although the true value may be underestimated, it is reasonable to assume that changes in the contribution of the LDL-receptor pathway, brought about by diet, drugs, hormones, or metabolic disorders, are reflected in changes in the relative rates of catabolism of native and CHD-LDL. Clearly, there is a need for an improved method for measuring LDL catabolism by the two routes in man that is both theoretically acceptable and safe to use.

2. Catabolism in Organs and Tissues

The contribution of a particular tissue to total catabolism of LDL *in vivo* cannot be estimated from the amount of radioiodine that it accumulates after an intravenous injection of radioiodine-labeled LDL. When the labeled LDL enters lysosomes, the protein component is digested and the radioiodine rapidly leaves the cell, eventually appearing in the urine as iodide or iodinated amino acid. Hence, the radioiodine content of the tissue may bear no relation to the amount of LDL it has taken up and degraded.

What is needed is a method for labeling LDL such that the label is carried into lysosomes with the internalized LDL and is then retained in lysosomes without being degraded when the particles are catabolized. Pittman *et al.* (1979a,b) have shown that [^{14}C]sucrose linked covalently to the protein of LDL satisfies these requirements sufficiently well for practical purposes. When fibroblasts are incubated *in vitro* with [^{125}I]LDL labeled with [^{14}C]sucrose, the [^{14}C]sucrose accumulates within the cells at a rate equal to the rate of degradation of [^{125}I]LDL protein. Moreover, when [^{14}C]sucrose LDL is injected into pigs, the amount of [^{14}C]sucrose recovered in the major organs corresponds to about 75% of the labeled LDL catabolized in the whole body. Thus, the [^{14}C]sucrose method provides an adequate estimate of the cumulative amount of labeled LDL catabolized by a particular tissue *in vivo* over a specified interval. The method may also be used for the measurement of LDL-receptor-dependent catabolism of LDL in organs or tissues *in vivo*. The accumulation of [^{14}C]sucrose in the tissues is measured after an injection of [^{14}C]sucrose-labeled native or reductively methylated LDL; catabolism via the LDL-receptor pathway is then estimated as the difference between the uptakes of native and methylated LDL (Carew *et al.*, 1982). Pittman *et al.* (1983) have described a modification of the [^{14}C]sucrose procedure in which sucrose is replaced by tyramine cellobiose as the ligand. Tyramine cellobiose can be radioiodinated to high specific activity before it is coupled covalently to LDL and is therefore well suited to the measurement of LDL uptake and catabolism in very small samples of tissue.

Dietschy and his colleagues (Koelz *et al.*, 1982; Spady *et al.*, 1983) have adapted the [^{14}C]sucrose labeling procedure to the measurement of rates of clearance of the plasma LDL by particular organs or tissues in experimental animals *in vivo*. The animal is given a priming intravenous injection of [^{14}C]sucrose LDL, followed by continuous infusion of the same labeled LDL at a constant rate. Once the plasma radioactivity (dpm/ml) has reached a plateau, radioactivity accumulates in each tissue at a constant rate, from which the plasma clearance rate (milliliters of plasma cleared of LDL per hour per gram of tissue) may be calculated. Uptake of plasma LDL by a particular tissue, expressed in milligrams of LDL-apoB taken up per hour per gram, may then be determined from the clearance rate for that tissue and the plasma concentration of endogenous LDL-apoB. The contribution of the LDL-receptor pathway to total uptake of LDL by each tissue may be estimated from the clearance rates of [^{14}C]sucrose-labeled native and reductively methylated LDL.

C. Rates of Catabolism in the Whole Body

1. Total and Receptor-Mediated FCR

As noted in Section I,C above, the FCR of LDL-apoB in normal man ranges from 0.3 to 0.6 plasma pools per day, corresponding to the degradation of 7–15 mg of LDL-apoB/kg/day by all catabolic routes. Since the internalization of LDL by cells involves the uptake of whole particles, this corresponds to a total cellular uptake of 10–20 mg of LDL cholesterol/kg/day, or roughly 1 g of LDL cholesterol per day in the whole body of a 70-kg man. In hamsters (Spady *et al.*, 1983) and rats (Koelz *et al.*, 1982) the whole-body plasma clearance rates of homologous LDL, estimated by the primed continuous-infusion technique, range from 600 to 850 μl of plasma per hour per 100 g body weight. These values are equivalent to FCRs of between 0.2 and 0.3 plasma pools per hour.

Whole-body catabolism of LDL by the LDL-receptor pathway, as estimated by the CHD method in normal human subjects, ranges from 30 to 50% of total

LDL catabolism (Shepherd *et al.*, 1979). Thus, the CHD method gives values that suggest that LDL-receptor-independent pathways normally account for more than half the total LDL catabolized. However, the glycosylation method of labeling gives values of up to 80% for the fraction of LDL catabolized by the LDL-receptor pathway. As we have seen, neither method has been validated against an absolute standard. However, it is worth nothing that if the FCR via LDL-receptor-independent pathways is independent of the plasma LDL concentration and is equal to the FCR of LDL in FH homozygotes, then the LDL-receptor pathway in normal human subjects must be responsible for at least two-thirds of total catabolism (Goldstein and Brown, 1977).

2. Variability in Normal Populations

The effects of physiological factors on the FCR of LDL in man are of considerable potential interest because they may contribute to variability in plasma LDL concentration within the normal population.

It is well established that in industrialized countries the plasma LDL concentration increases with age in men and women. Miller (1984), after reviewing all relevant published reports, concluded that the age-related rise in plasma LDL level is due primarily to a fall in the FCR of LDL-apoB and that this, in turn, is due to a decline in LDL-receptor function. Since plasma LDL concentration does not increase after puberty in some populations, Miller suggested that in industrialized populations, environmental factors (including diet) lead to a decline in LDL-receptor function with age. An age-related decline in the activity of LDL receptors was also noted by Mahley *et al.* (1981) in the liver membranes of adult dogs. In contrast to the conclusion of Miller (1984), Grundy *et al.* (1985) found that the age-related increase in plasma LDL level that they observed in a group of normal U. S. men was due to a combination of increased LDL production rate and decreased FCR. However, it should be noted that a decrease in the activity of hepatic LDL receptors, in addition to decreasing the FCR of LDL, might also increase the LDL production rate by diminishing hepatic uptake of VLDL remnants. Thus, it is possible to reconcile the observations of Grundy *et al.*, that there is an age-related increase in LDL production rate, with Miller's conclusion that the major cause of the rise in plasma LDL concentration with increasing age is a fall in LDL-receptor activity.

Variability in the FCR of LDL also contributes to intrapopulation variability in plasma LDL concentration in normal people within the same age range. In a random sample of normal British men aged 35–49 examined by Turner *et al.* (1984), the FCR was higher in men with plasma LDL levels in the lowest decile than in those with levels in the modal decile. On the other hand, men with the highest plasma LDL levels were characterized by high LDL production rates. Results similar in most respects to these were obtained by Grundy *et al.* (1985) in a study of healthy young U. S. men eating the same diet. Over the whole range of plasma LDL concentrations, plasma LDL level was correlated negatively with FCR and positively with LDL production rate. Thus, the results of both studies are consistent with the conclusion that the FCR of LDL is a determinant of plasma LDL concentration in normal people, at least in those with LDL levels at the lower end of the normal range. In neither study was the contribution of the LDL-receptor pathway to total LDL catabolism estimated separately. However, if, as Grundy *et al.* suggest, the variable component of the FCR of LDL is that due to the LDL-receptor pathway, it would be reasonable to suppose that the high FCRs in normal men with low plasma LDL levels are due to high LDL-receptor activity.

Variability in LDL-receptor activity within normal populations could be due to genetic or environmental factors. With regard to genetic factors, Weight *et al.* (1982), in a study of normal twins, have demonstrated a high degree of heritability of LDL-receptor activity in lymphocytes in which LDL receptors are maximally induced *in vitro*. The maximal receptor activity expressed *in vitro* may not reflect activity normally expressed *in vivo,* but the maximal capacity for expressing receptors might help to determine the ability of an individual to respond to an environmental challenge such as an increase in the intake of dietary cholesterol. A genetic contribution to normal variability in LDL-receptor activity could be mediated by polymorphism in the receptor gene itself, leading to small interindividual differences in the rate of synthesis of receptors or in their affinity for LDL; the identification of RFLPs in or near the LDL-receptor gene (see, for example, Humphries *et al.*, 1985) should eventually make it possible to assess the contribution of polymorphism in this gene to variability in plasma LDL levels in the normal population. However, it should be borne in mind that heritable differences in receptor activity *in vivo* need not be due to differences in the receptor gene. They could be mediated by mutations in any of the genes coding for products concerned, directly or indirectly, in any of the steps in the LDL-receptor pathway.

Polymorphism in the coding region of the apoB gene could also be a source of variability in the FCR of autologous LDL, especially if the mutations affected the behavior of the receptor-recognition domain in apoB-100. Here, we are considering minor variants that have a small effect on the phenotype but are present at appreciable frequencies in the population, rather than rare mutations in the apoB gene that cause monogenic disorders such as familial defective apoB-100 or some of the apoB deficiency states (see Chapter 6). Associations between RFLPs in the apoB gene and plasma cholesterol concentration in groups of unrelated individuals were mentioned in the previous chapter. Some of these associations could reflect linkage disequilibrium between the RFLP and a polymorphic site in the apoB gene that affects the recognition of LDL by the LDL receptor. In support of this, preliminary observations of Houlston *et al.* (1988) point to a population association between the X^+ allele in the apoB gene and a reduced FCR of LDL in normolipidemic men.

Among the environmental factors likely to contribute to variability in LDL catabolism, diet is probably the most important. This is considered below.

3. Effects of Diet

The well-established association between plasma cholesterol concentration and intake of dietary cholesterol in different populations has been difficult to demonstrate within free-living populations (see Myant, 1981, Chapter 12). Under controlled metabolic conditions an increase in cholesterol intake leads to a significant increase in plasma cholesterol and LDL concentrations in some individuals (responders) but has no detectable effect in others (nonresponders). The ability of nonresponders to maintain a constant plasma LDL level when cholesterol intake is increased seems to be due to a combination of decreased synthesis of cholesterol in the liver and increased conversion of cholesterol into bile acids (see Myant, 1981, Chapter 10), with little or no effect on the FCR of LDL (Ginsberg *et al.*, 1981). Packard *et al.* (1983) have shown that the increase in plasma LDL level brought about by increased dietary cholesterol in responders is accompanied by a fall in the total FCR of LDL and a rise in LDL production rate. The fall in total FCR is due to a fall in FCR via the LDL-receptor pathway, without significant change in FCR via LDL-receptor-independent pathways. The net effect of these changes is an increase in the absolute catabolic rate (ACR) of LDL via receptor-independent pathways, with essentially no change in the ACR via the LDL-receptor pathway.

In view of the predominant rôle of the liver in LDL catabolism by the LDL-receptor pathway in animals (see Section II,D below), it might be supposed that the fall in total FCR of LDL in human responders fed cholesterol is due to suppression of hepatic LDL-receptor activity. This possibility has not been examined in human subjects, but Mahley *et al.* (1981) and Angelin *et al.* (1983) have shown that LDL receptors on liver membranes of young dogs are suppressed by cholesterol feeding or by intravenous infusions of chylomicrons. However, there are marked species differences in the effect of cholesterol feeding on LDL catabolism. Thus, Spady *et al.* (1985a) have shown that, whereas dietary cholesterol decreases the clearance of homologous LDL in male hamsters, the addition of up to 0.5% of cholesterol to the diet has no such effect in rats.

In man, the decrease in plasma LDL concentration brought about by dietary polyunsaturated fat is accompanied by an increase in the total FCR of LDL (Shepherd *et al.*, 1980b). In hamsters, the addition of saturated fat to a diet containing 0.12% of cholesterol raises the plasma LDL concentration and almost completely suppresses hepatic clearance of LDL via the LDL-receptor pathway (Spady and Dietschy, 1985).

4. Effects of Drugs

Cholestyramine and other anion exchangers that bind bile salts in the intestine lower the plasma LDL concentration in human subjects and in some, but not all, animal species. The immediate effect of these resins is to stimulate the conversion of hepatic cholesterol into bile acids, with a consequent increase in hepatic synthesis of cholesterol. In those species, such as the rat and the pig, in which the compensatory increase in cholesterol synthesis is sufficient to keep the cholesterol content of hepatocytes at the normal level, cholestyramine has no effect on plasma cholesterol concentration. In human subjects and young dogs, on the other hand, the net effect of interference with the reabsorption of bile salts from the small intestine seems to be to lower the cholesterol content of hepatocytes (see Myant, 1981, Chapter 5 for discussion). This effect is associated with an increase in the FCR of the plasma LDL, with no change in LDL production rate (Langer *et al.*, 1969). Shepherd *et al.* (1980a) have shown that the increased FCR in heterozygous FH patients treated with cholestyramine is due to an increased FCR of LDL via the LDL-receptor pathway, with no change in FCR by LDL-receptor-independent pathways. As noted above, the increased FCR of LDL during cholestyramine treatment leads to an increase in the protein : cholesterol ratio in LDL and a decrease in mean particle size. These changes may be due to a selective increase in the rate of degradation of a subpopulation of large LDL particles with a low protein : cholesterol ratio.

The stimulatory effect of cholesterol depletion on the expression of LDL receptors in cells in culture (see Chapter 8) suggests that the increased activity of the LDL-receptor pathway during cholestyramine treatment *in vivo* is mediated by increased LDL-receptor activity in liver cells. Receptor activity in hepatocytes has not been examined in cholestyramine-treated human subjects. However, Kovanen *et al.* (1981) have shown that LDL-receptor activity in membranes prepared from the livers of young dogs is increased by colestipol, an analog of cholestyramine. Thus, it seems likely that the increased total FCR of LDL in human subjects treated with bile-salt-binding resins is due predominantly to increased expression of LDL receptors on hepatocytes in response to a decreased cholesterol content of liver cells. As would be expected on this hypothesis, in patients with homozygous FH (who are incapable of expressing LDL receptors), cholestyramine, in doses large enough to produce a maximal increase in the conversion of hepatic cholesterol into bile salts, has no effect on the plasma cholesterol concentration (Moutafis and Myant, 1969).

Interruption of the reabsorption of bile salts from the intestine by partial bypass of the ileum, like treatment with cholestyramine, lowers the plasma LDL level. Again, this effect is due to a selective increase in the FCR of LDL via the LDL-receptor pathway (Spengel *et al.*, 1982).

Species differences in the effect of cholestyramine on plasma cholesterol concentration have already been mentioned. In this regard, Spady *et al.* (1985a) have investigated the effects *in vivo* of diet and cholestyramine on LDL-receptor activity in rats and female hamsters, both of which have high basal rates of hepatic cholesterol synthesis, and in male hamsters, in which the basal rate of hepatic synthesis is low. In rats and female hamsters, changes in the rate of hepatic synthesis of cholesterol compensate fully for the effects of dietary cholesterol and cholestyramine, neither of which produces any change in hepatic clearance of homologous LDL or in plasma LDL concentration. In male hamsters, compensatory changes in hepatic synthesis of cholesterol are not sufficient to maintain a constant level of cholesterol in the liver. Hence, cholesterol feeding leads to an increase in hepatic cholesterol content, decreased hepatic clearance of LDL, and a rise in plasma LDL concentration. In male hamsters, cholestyramine has effects on the liver opposite to those of cholesterol feeding, resulting in a fall in plasma LDL level.

These observations have a bearing on the design of therapeutic measures likely to increase the catabolism of LDL in hypercholesterolemic patients. Since, as we shall see below, the liver is probably the major site of LDL catabolism in man, the most effective strategy should be to stimulate LDL-receptor activity in the liver by depleting liver cells of cholesterol. As we have seen, cholestyramine stimulates the breakdown of hepatic cholesterol, but this effect induces a compensatory increase in cholesterol synthesis which, in some species, nullifies the increase in cholesterol catabolism. Hence, a rational approach would be to combine a bile-salt-binding resin with an inhibitor of cholesterol synthesis. Moutafis *et al.* (1971) tried to overcome the increase in cholesterol synthesis, induced by cholestyramine in a patient with FH, by combining cholestyramine with nicotinic acid. This combination had little effect on her plasma LDL level, probably because the patient, who had homozygous FH, had only a negligible capacity for expressing LDL receptors. Kovanen *et al.* (1981) have shown that a combination of colestipol and mevinolin (an analog of the HMG-CoA-reductase inhibitor, compactin) given to young dogs causes a marked fall in plasma LDL concentration, accompanied by an increase in LDL-receptor activity in liver membranes that is greater than the sum of the effects of either drug given alone. The major effect of this combination of drugs on LDL metabolism *in vivo* is to double the FCR, but there is also a significant decrease in the production rate of LDL. The effectiveness of cholestyramine and mevinolin, alone or in combination, in the treatment of FH is discussed in Chapter 10.

Probucol, a plasma-cholesterol-lowering agent, has the somewhat unusual property of bringing about a substantial fall in plasma LDL concentration in FH homozygotes. Since these patients have little or no capacity for expressing LDL receptors, the effect of this drug on LDL metabolism cannot be mediated by induction of LDL-receptor activity. Naruszewicz *et al.* (1984) have shown that when WHHL rabbits are given probucol, their LDL undergoes an intrinsic modification that alters its behavior *in vitro* and *in vivo*. LDL from probucol-treated animals is catabolized abnormally rapidly by normal rabbit fibroblasts in culture and by untreated normal and WHHL rabbits *in vivo*. These findings suggest that probucol-modified LDL is catabolized at an increased rate via LDL-receptor-dependent and LDL-receptor-independent pathways.

5. Effects of Hormones

The effects of hormones on LDL-receptor activity at the level of the cell are considered in the next chapter. Two hormonal effects on LDL metabolism *in vivo* are worth mentioning here.

Hypercholesterolemia, due mainly to increased plasma LDL concentration, is a well-recognized feature of reduced *thyroid function*. Walton *et al.* (1986) have shown that both the FCR and the rate of production of LDL are decreased in myxoedema, the changes in FCR having the dominant effect on plasma LDL concentration. The observations of Thompson *et al.* (1981) on a myxoedematous patient indicate that the low FCR in this condition is due mainly to a decrease in the FCR of LDL via the LDL-receptor pathway, but that there is also defective catabolism of LDL by LDL-receptor-independent pathways. Defective LDL-receptor activity *in vivo* in the cells of myxoedematous patients is consistent with the observation that thyroxine stimulates LDL-receptor activity in cells *in vitro* (see Chapter 8).

Pharmacological doses of natural or synthetic *estrogens* lower the plasma cholesterol concentration in men (Boyd, 1963) and male rats (Fewster *et al.*, 1967). In male rats, large doses of ethinyl estradiol bring about the almost complete disappearance of all lipoproteins from the plasma, including LDL (Davis and Roheim, 1978). Hay *et al.* (1971) have shown that the fall in plasma LDL level is accompanied by an increase in the FCR of a lipoprotein fraction containing IDL and LDL, both *in vivo* and in perfused livers from treated donor animals. The increased rate of LDL catabolism by the livers of estrogen-treated rats is paralleled by a 10-fold increase in the number of LDL receptors expressed on liver membranes *in vitro* (Kovanen *et al.*, 1979). In parallel with the rise in LDL-receptor activity in liver membranes, there is also a specific increase in the rate at which perfused livers from the treated rats catabolize lipoproteins that are recognized by the LDL receptor (Chao *et al.*, 1979). These and other observations are consistent with the conclusion that the increased FCR of LDL in estrogen-treated male rats is due partly to increased hepatic catabolism of LDL by the LDL-receptor pathway. In agreement with this conclusion, Kovanen *et al.* (1979) have shown that treatment with estrogen does not increase the rate of catabolism of CHD-modified LDL in rats *in vivo*.

Boyd (1963) showed that large doses of estrogen stimulate the activity of the reticuloendothelial (RE) system in rats. This led him to suggest that increased phagocytic activity is partly responsible for the effect of estrogens on plasma lipid levels. Some such general effect would certainly help to explain why pharmacological doses of estrogen lead to a fall in the plasma concentrations of all lipoproteins. However, it seems unlikely that estrogens stimulate the removal of LDL by the RE system, since they have no effect on the catabolism of acetylated LDL *in vivo* (Kovanen *et al.*, 1979), a modified form of LDL that is catabolized by specific receptors on cells of the RE system (see next chapter).

The induction of LDL receptors by large doses of estrogen probably has little relevance to the role of endogenous estrogens under physiological conditions, though there is some evidence to suggest that the FCR of LDL is greater in women than in men (Walton *et al.*, 1965). The importance of this pharmacological effect lies, rather, in the opportunity it affords for the study of normally expressed hepatic receptors and of the intracellular events that occur during LDL-receptor-mediated catabolism of LDL and VLDL remnants by the liver.

D. Catabolism in Organs and Tissues

1. Total Catabolism

The contribution of individual organs to LDL catabolism in the whole body has been estimated in several species of experimental animal by measuring tissue uptake of [^{14}C]sucrose-labeled human or homologous LDL *in vivo*. Pittman and his co-workers measured uptake after single injections of labeled LDL, whereas Dietschy and co-workers used the primed continuous-infusion procedure. With the latter technique it is possible to estimate absolute rates of clearance of endogenous LDL by individual tissues (μl of plasma cleared per hour) as the product of rate of uptake of labeled LDL and specific radioactivity of the plasma LDL at a time when the labeled LDL has equilibrated throughout the extracellular fluid space.

Results obtained by the two approaches agree in all essentials. The liver is responsible for degrading 50–70% of the total LDL catabolized in the whole body in pigs, rats, rabbits, and hamsters (see Pittman *et al.*, 1979a, 1982a,b; and Spady *et al.*, 1983, 1985b, 1987). This is several times greater than the contribution of any other organ. Pittman *et al.* (1982a,b) have shown that the parenchymal cells are responsible for more than 90% of the catabolism of LDL in whole liver. Other tissues that make a significant contribution to LDL catabolism in the whole body, partly by virtue of their mass, are small intestine, skeletal muscle, kidney, and spleen; those that make only a small or negligible contribution include steroid-hormone-forming glands, lung, and brain. Representative values, obtained from rats by Spady *et al.* (1985b), are shown in Table 7.1.

When expressed in terms of wet weight of tissue, clearance rates of LDL are highest in liver, adrenals, ovary, and spleen; in rats and rabbits the clearance rates per gram of tissue are higher in adrenals than in any other tissue. Other organs with comparatively high rates of clearance per gram are small intestine, lung, and kidney; those with low or negligible rates include skeletal muscle, adipose tissue, and brain. Note that skeletal muscle makes a significant contribution to whole-body catabolism of LDL because of its large mass. The adrenals and ovaries, on the other hand, make only a small contribution, despite their very high catabolic activity, because of their small mass.

2. Catabolism by the LDL-Receptor Pathway

The contributions of the LDL-receptor-dependent and LDL-receptor-independent pathways to total LDL catabolism in individual organs have been estimated by Pittman and co-workers, using the single-injection technique, and by Dietschy and co-workers, using the primed continuous-infusion procedure. In most of these studies, the tissue uptake of [^{14}C]sucrose-labeled human or homologous LDL was compared with that of reductively methylated labeled LDL. However, Pittman *et al.* (1982b) have also estimated the contribution of the receptor-independent pathway to LDL catabolism in the tissues of rabbits by comparing the uptake of labeled rabbit LDL in the tissues of normal rabbits with that in the tissues of LDL-receptor-deficient (WHHL) rabbits. For these estimates, Pittman *et al.* assumed that the clearance rate of LDL by the receptor-independent pathway is the same at normal plasma LDL concentrations as at the very high concentrations reached in WHHL rabbits.

The results obtained from the two laboratories are in essential agreement (see Table 7.1 for representative values obtained in rats). In general, the LDL-receptor pathway accounts for up to 90% of total LDL catabolism in the liver and for more than 90% in the adrenals and ovaries. The LDL-receptor pathway also makes a substantial contribution to LDL catabolism in small intestine, kidney, and lung. However, in several tissues, including muscle, skin, and brain, the contribution of this pathway is negligible.

In the livers of rats (Carew *et al.*, 1982) and rabbits (Pittman *et al.*, 1982b) the parenchymal cells, as distinct from Kupffer cells, are responsible for 90% of the catabolism of LDL occurring via both the LDL-receptor-dependent and LDL-receptor-independent routes.

3. Dominant Role of the Liver

Taken together, the above observations point to several important features of LDL metabolism in animals *in vivo*.

In the first place, the role of the liver in the catabolism of LDL in the whole body dominates that of all other tissues. Moreover, the contribution of the liver is

Table 7.1
Clearance Rates of Plasma LDL by the LDL-Receptor-Independent and LDL-Receptor-Dependent Pathways in Whole Organs of the Rat

	Clearance rate per organ[a] (μl of plasma/hr/100 g body weight)				
Organ	Total		Receptor-independent	Receptor-dependent	Receptor-dependent / Total (%)
Liver	352	(53)	29	323	91.8
Small intestine	64	(9.7)	27	37	57.8
Skeletal muscle	36	(5.5)	36	0	0
Kidney	21	(3.2)	3.6	17.4	82.9
Spleen	17	(2.6)	12	5	29.4
Skin	16	(2.4)	18	0	0
Adipose tissue	12	(1.8)	11	1	8.3
Adrenal	6.7	(1.0)	0.1	6.6	98.5
Lung	5.2	(0.8)	1.0	4.2	80.8
Ovary	4.7	(0.7)	0.2	4.5	95.7
Brain	<0.1	<0.1)	<0.1	0	0

[a]Clearance rates were measured by the primed continuous-infusion procedure (see Section II, A above). Total clearance of LDL by all pathways was estimated from the clearance rate of [^{14}C]sucrose-labeled rat LDL. Clearance by the receptor-independent pathway was estimated from the clearance rate of [^{14}C]sucrose-labeled methylated human LDL. Clearance by the receptor-dependent pathway was estimated as the difference between the total and receptor-independent rates. Values in parentheses in column 1 are the percentage of total clearance rate of LDL in whole rats (mean value, 659 μl/hr/100 g body weight). (Values taken from Spady *et al.*, 1985b.)

due largely to its high LDL-receptor activity. Since LDL receptors are regulated *in vivo,* the high level of receptor activity in the liver must enable this organ to make a major contribution to the regulation of LDL metabolism in the body as a whole, not only by modifying the rate of LDL catabolism, but also by determining the proportion of hepatic VLDL that is converted into LDL. Direct observations on hepatic catabolism of LDL have not been made in man. However, Spengel *et al.* (1982) have shown that, in the rhesus monkey with an ileal bypass, the liver makes a larger contribution than any other tissue to total LDL catabolism and that more than 80% of hepatic catabolism occurs by the LDL-receptor pathway. Furthermore, the effects of cholestyramine on LDL catabolism in man show that the human liver is capable of regulating its expression of LDL-receptor activity *in vivo*. For these reasons, it seems likely that the role of the liver in the regulation of LDL metabolism in man is as predominant as it seems to be in experimental animals.

The second point of interest is that in organs or tissues with high catabolic activity, such as liver, adrenals, ovaries, and kidney, catabolism occurs mainly via the LDL-receptor pathway. Finally, several tissues, including skeletal muscle and brain, catabolize LDL at very low rates. As noted in Chapter 3, these tissues satisfy their requirements for cholesterol mainly by endogenous synthesis.

4. LDL Metabolism in the Arterial Wall

Vascular endothelium and arterial smooth-muscle cells are both capable of catabolizing LDL. However, the overall metabolism of LDL in arteries is complicated by the fact that if LDL is to supply cholesterol to cells within the walls of a normal artery it must first cross a layer of endothelial cells. The ability of whole LDL particles to cross the vascular endothelium from luminal to abluminal surface is shown by the presence of native LDL in peripheral lymph and in the subendothelial space of normal arteries (see Myant, 1981, Chapter 13). Simionescu *et al.* (1976) have shown that molecules of the size of an LDL particle are transported through capillary endothelial cells in large vesicles of diameter 400–1000 Å. These vesicles are formed by invagination of the plasma membrane and are called plasmalemmal vesicles. Since they do not fuse with lysosomes their contents are not exposed to lysosomal enzymes. Plasmalemmal vesicles are also present in the endothelial cells of arteries (Schwartz *et al.*, 1977). Transport of LDL across arterial endothelium by this route probably explains the rapid penetration of plasma LDL into the normal artery wall *in vivo,* noted by Stender and Zilversmit (1981, 1982).

Steinberg and co-workers (see Pittman *et al.*, 1982b; Pittman and Steinberg, 1984; Carew *et al.*, 1984; Wiklund *et al.*, 1985) have estimated the amounts of LDL transported into and out of the arterial wall *in vivo,* compared with the amounts irreversibly degraded by endothelial and subendothelial cells. Flux into the artery was estimated from the uptake of [^{125}I]LDL at 30 minutes after an intravenous injection; catabolism was estimated using [^{125}I]tyramine cellobiose as the label for LDL.

In normal rabbit aorta the total rate of transport of LDL into the wall was equivalent to a plasma clearance of about 2000 nl/g/hr. Rates were similar for labeled native and methylated LDL, showing that transport across the endothelium is essentially independent of LDL-receptor activity. Since the rate of irreversible degradation by the whole artery was equivalent to about 700 nl of plasma per gram per hour, Wiklund *et al.* (1985) concluded that, in the steady state, at least two-thirds of the LDL entering the artery wall must return to the plasma without degradation. Catabolism of LDL by the intima accounted for about 40% of the total catabolized by the whole artery. However, when expressed in terms of cell protein, the rate of catabolism by the intima, most of which normally consists of endothelial cells, was more than 40 times that by the rest of the artery and was about one-fifth of that observed in rabbit liver. Comparisons of the rates of degradation of labeled native and methylated LDL showed that the LDL-receptor pathway accounted for about 40% of the LDL catabolized in the intima.

These results provide a picture of what may happen when LDL is taken up from the plasma by normal artery walls. LDL particles enter endothelial cells predominantly by the plasmalemmal route, but a small proportion also enters by LDL-receptor-mediated uptake. (The similar rate of entry of native and methylated LDL is explained by the small relative contribution of the receptor-mediated pathway.) About 90% of the particles that enter endothelial cells are transported into the subintimal space without degradation. The remaining 10% are catabolized in endothelial-cell lysosomes. A proportion of the particles that reach the subendothelial space, equivalent to about 20% of the total influx of LDL, is taken up and catabolized, presumably by smooth-muscle cells. The particles that escape degradation, equivalent to about 70% of total influx, return to the plasma, either via the lymphatic system or by transport in reverse across the endothelium.

5. Lipoproteins as a Source of Cell Cholesterol

LDL is clearly an important source of cholesterol for the cells of many tissues *in vivo*, particularly those that express high LDL-receptor activity. Indeed, in some tissues the uptake of LDL appears to satisfy all the cells' requirement for cholesterol. Pittman *et al.* (1979a), for example, estimate that the uptake of LDL by the adrenals of the pig could supply all the cholesterol needed for synthesis of corticosteroids, even under conditions of maximal stress. However, as discussed in Chapter 2, cell cholesterol may also be supplied by synthesis *in situ* and by uptake of lipoproteins other than LDL. Uptake of remnants of chylomicrons and VLDL by the liver is mentioned in the next chapter. Another example of a tissue that does not rely upon LDL as its major source of cholesterol is the rat adrenal. Pittman *et al.* (1982a) estimate that uptake of LDL by this tissue would not be sufficient to supply the cholesterol needed for hormone synthesis, even under basal conditions. Evidence from several sources suggests that the rat adrenal satisfies its requirement for cholesterol by a combination of receptor-mediated uptake of LDL, synthesis *in situ*, and uptake of HDL (Gwynne *et al.*, 1976; Andersen and Dietschy, 1977, 1981).

It may well be that HDL makes a larger contribution to cell cholesterol in general in those species (such as the rat) in which HDL is the major carrier of plasma cholesterol than in species (such as man and the pig) in which most of the plasma cholesterol is carried in LDL. A discussion of the mechanisms by which HDL interacts with animal cells is beyond the scope of this monograph. However, a brief mention of one or two points relevant to HDL as a source of cell cholesterol may help to put the role of LDL into perspective.

Cholesterol-enriched HDL containing apoE is recognized by LDL receptors and by the hepatic apoE receptors described by Mahley *et al.* (1981). Uptake and catabolism of apoE-containing HDL mediated by these two receptors on liver cells may play a major role in the transport of surplus cholesterol from extrahepatic cells to the liver. This reverse cholesterol transport would provide the body with a mechanism for disposing of cholesterol from cells that are unable to metabolize it; reverse transport may be of particular importance in the removal of cytoplasmic cholesterol from macrophages (see Reichl and Miller, 1986, for review).

Thus, HDL, in addition to acting as a source of cholesterol for some cells, also acts as an acceptor during the excretion of cellular cholesterol. In some cells, such as macrophages, the direction of flow of cholesterol is probably always from cell surface to HDL particle. In others, the direction may depend upon the relative concentrations of free cholesterol in the plasma membrane and the HDL particle.

Both uptake of HDL by cells and the transfer of cholesterol from cells to HDL are likely to involve a physical interaction between cell surface and HDL particle. However, the nature of these interactions is by no means clear. There is evidence for specific high-affinity binding of apoE-free HDL by a variety of animal cells, including fibroblasts, vascular endothelial cells, cells of the adrenals and gonads, hepatocytes, arterial smooth-muscle cells, and intestinal cells (for review, see Catapano, 1987). In some cases, the binding observed *in vitro* is consistent with the presence of discrete cell-surface receptors that recognize specifically the apoA-I component of HDL. Schmitz *et al.* (1985) have described an interaction between HDL and cholesterol-enriched macrophages that is of considerable interest in relation to reverse cholesterol transport. Their observations suggest that, in these cells, HDL particles are internalized via coated vesicles (see next chapter) and are then released into the external medium after taking up cholesterol from intracellular lipid droplets.

Pittman and his co-workers have described a mechanism by which the liver and steroid-hormone-forming tissues take up cholesteryl esters from HDL without an equivalent uptake of HDL apoprotein (reviewed by Pittman and Steinberg, 1984). For these studies, HDL particles were prepared in which the cholesteryl esters were replaced by cholesteryl ethers and the apoA-I was labeled with radioiodinated tyramine cellobiose. Since cholesteryl ethers and tyramine cellobiose are not degraded by lysosomal enzymes and cannot cross membranes, these labels acted as markers for the cumulative cellular uptake of HDL cholesteryl esters and HDL apoA-I. When doubly labeled HDL is injected intravenously into rats, cholesteryl ethers and apoA-I are taken up by most tissues at equal rates, consistent with internalization of whole HDL particles. In liver, adrenals, and ovaries, however, uptake of cholesteryl ethers is severalfold greater than that of apoA-I. Selective uptake of cholesteryl ethers is also observed when doubly labeled HDL is incubated with primary cultures of rat hepatocytes and adrenal cells.

Preferential uptake of HDL cholesteryl esters may facilitate the transfer of cholesterol to the liver in reverse cholesterol transport and may also mediate the entry of HDL cholesterol into the cells of the adrenals and gonads. The mechanism by which this process occurs is not understood. It may require specific binding of the HDL particle to the cell surface but probably does not involve internalization of whole HDL particles, with retroendocytosis of apoA-I. Nor is it likely that binding of apoE in the HDL particles to LDL receptors plays any part in selective uptake of cholesteryl esters, since reductive methylation of the reconstituted particles has no effect on this process.

E. LDL-Receptor-Independent Catabolism of LDL *in Vivo*

1. The Mechanisms Responsible

The ability of FH homozygotes and of WHHL rabbits to catabolize LDL shows that there must be one or more alternative pathways for LDL catabolism, in addition to that mediated by LDL receptors. Indeed, observations on the FCR of native and modified LDL show that this route(s) may account for 30–50% of the total LDL catabolized *in vivo* in the presence of normal LDL-receptor activity. Attempts to identify the mechanisms responsible for this process have not yet provided clear answers. However, several mechanisms that could contribute to LDL-receptor-independent catabolism of LDL *in vivo* are suggested by the behavior of cells *in vitro*. As discussed in the next chapter, these mechanisms include (1) adsorptive endocytosis by cell-surface receptors with low affinity for LDL; (2) pinocytosis (Gr. *pinein,* to drink; *kytos,* a cell), a process by which cells engulf droplets of extracellular fluid by invagination of the plasma membrane; (3) uptake of modified LDL by high-affinity receptors on cells of the RE system, of which the best understood is the acetyl-LDL receptor.

In the absence of LDL receptors, as in receptor-negative FH homozygotes, LDL-derived cholesterol accumulates selectively in cells of the RE system, especially in macrophages and other cells known collectively as scavenger cells. For this reason, Goldstein and Brown (1977) proposed that the LDL-receptor-independent catabolism of LDL is mediated *in vivo* by uptake and degradation via a pathway involving pinocytosis by scavenger cells, a class of cell in which this process is particularly active. The clearance of LDL by pinocytosis (milliliters of fluid cleared per hour) is independent of the concentration of LDL in the extracellular medium. Therefore the absolute rate of degradation of LDL by this route is directly proportional to plasma LDL concentration, even at the highest concentrations, i.e., degradation is nonsaturable. Moreover, degradation by pinocytosis must be unregulated, in the sense that it continues irrespective of the cholesterol content of the cell. Hence, uptake by pinocytosis could account for the massive deposition of cholesterol in macrophages in FH.

Implicit in the proposal of Goldstein and Brown (1977) is the assumption that the scavenger-cell pathway accounts not only for the localized tissue deposition of cholesterol that occurs at high plasma LDL concentrations, but also for LDL-receptor-independent catabolism of LDL occurring at normal LDL levels in the presence of normal LDL-receptor activity. However, it is possible that LDL is catabolized *in vivo* by more than one LDL-receptor-independent process. For example, there could be a generally distributed mechanism operating at normal plasma LDL levels, together with another mechanism restricted to cells of the macrophage class and responsible for the formation of foam cells in xanthomas and the atheromatous lesions of arteries. The generally distributed mechanism might involve nonspecific pinocytosis of unmodified LDL particles. The other might require the local conversion of LDL, perhaps by vascular endothelial cells (see Steinberg, 1983), into a form that is recognized by acetyl-LDL or other receptors on macrophages (see next chapter for details). It seems unlikely that acetyl-LDL receptors make more than a small contribution to total LDL-receptor-independent catabolism of LDL in the whole body, because CHD-modified LDL, which is not recognized by the acetyl-LDL receptor, is catabolized at the same rate as native LDL in FH homozygotes.

Attempts to assess the contribution of the RE system to total catabolism of LDL in the whole body have led to contradictory results. Shepherd *et al.* (1985) found that selective suppression of the RE system in rabbits, by intravenous injections of ethyl oleate, led to a significant decrease in the FCR of human LDL. Since the decrease in total FCR was due partly to decreased clearance of LDL by LDL-receptor-independent pathways, Shepherd *et al.* concluded that the RE system contributes significantly to total LDL catabolism *in vivo*. In keeping with these observations, Harkes and Van Berkel (1983) have shown that hepatic uptake of native human LDL by the intact rat occurs predominantly in hepatocytes, whereas methylated human LDL is taken up mainly in nonparenchymal cells. The latter include the Kupffer cells, comprising the largest single component of the RE system in the body. However, these results are difficult to interpret because of the inability of macrophages to recognize methylated LDL by a specific receptor (see Section II,B,1 above).

The suggestion that the RE system contributes specifically to LDL catabolism *in vivo* is in conflict with the observations of Pittman *et al.* (1982b), referred to above. Pittman and co-workers compared the uptake of [^{14}C]sucrose-labeled rabbit LDL into the organs of WHHL rabbits with that in normal rabbits. The pattern of distribution of ^{14}C in the different organs was broadly similar in the two groups of rabbits, although the relative contributions of liver and gut were lower and those of skin and skeletal muscle were higher in WHHL than in normal rabbits. In both groups about 90% of the ^{14}C in the liver was recovered in hepatocytes, with only negligible amounts in Kupffer cells. Pittman *et al.* concluded that the RE system does not make a major contribution to the LDL-receptor-independent pathway in WHHL rabbits. They suggested that this pathway operates in many, or perhaps all, tissues. Spady *et al.* (1985b), in a study of the LDL-receptor-independent catabolism of LDL in normal rats, also concluded that this catabolic pathway functions in the parenchymal cells of most tissues, including liver and small intestine, rather than in any specific set of cells. However, as noted above, methylated LDL may be taken up by RE cells only by nonspecific mechanisms, such as pinocytosis, and may not reveal specific uptake in these cells via receptors such as the acetyl-LDL receptor.

In favor of the view that LDL catabolism by alternative pathways is mediated mainly by nonselective mechanisms such as pinocytosis, Spady *et al.* (1985b) observed similar rates of clearance of albumin and methylated human LDL in most tissues of the rat. On the other hand, Pittman *et al.* (1982b) observed much higher FCRs for homologous LDL in WHHL rabbits (0.53 pools/day) than the FCR of albumin in normal rabbits (0.12 pools/day) (McFarlane and Koj, 1970). The FCR of homologous LDL in patients with homozygous FH (0.15 pools/day) is also much higher than the FCR of albumin in normal human subjects (0.04–0.05 pools/day) (Berson *et al.*, 1953). These discrepancies could be due either to failure of methylated human LDL to detect an LDL-receptor-independent pathway involving selective uptake of LDL, or to the presence of residual LDL-receptor activity in WHHL rabbits and in many FH homozygotes.

2. Metabolic Consequences of the Two Pathways

The coexistence of two pathways for LDL catabolism *in vivo* has several interesting consequences. These have been investigated by Dietschy and co-workers with their primed continuous-infusion procedure (summarized in Dietschy and Spady, 1985). Clearance rates of native and methylated LDL in organs of intact animals (μl of plasma cleared per hour per gram) were measured under conditions in which plasma-LDL concentration was raised over a 25-fold range by infusing LDL at increasing concentrations. Figure 7.5 shows what happens in the liver during short-term experiments in which the number of LDL receptors expressed may be assumed to remain constant throughout an infusion.

As the plasma LDL level is raised, total clearance of LDL (sum of clearance rates by the two routes) falls toward an asymptote, while the clearance by the alternative route remains constant even at the highest LDL levels tested. The fall in total clearance rate is due to a decrease in clearance by the LDL-receptor pathway as hepatic receptors approach saturation. The alternative pathway, on the other hand, is essentially nonsaturable (panel A, lower curve).

The lower panel of Fig. 7.5 shows the absolute rates of uptake of LDL (μg/hr/g). When the plasma LDL level is raised, the rate of total uptake rises steeply at low concentrations, but at concentrations higher than about 200 mg of LDL per 100 ml of plasma the total uptake rate becomes directly proportional to LDL concentration, i.e., the curve becomes linear. This curvilinear relationship reflects the summation of two processes: LDL-receptor-mediated uptake that is completely saturated at plasma LDL concentrations above about 200 mg/100 ml, and a nonsaturable mechanism by which a constant volume of plasma is cleared at all plasma LDL levels. The two curves shown in panel B are analogous to the

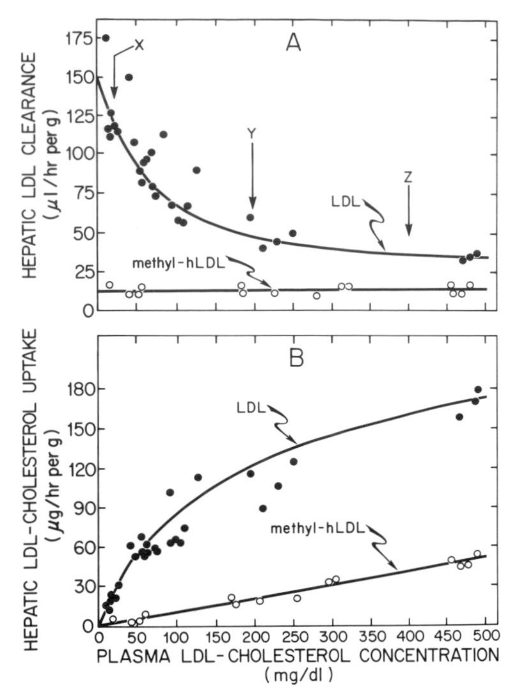

Figure 7.5. Kinetics of LDL transport by the liver of the intact hamster. The two panels show the effect of increasing the plasma LDL concentration on (A) The total and LDL-receptor-independent clearance rates of LDL (μl of plasma cleared/hr/g of liver), and (B) the absolute rates of total and LDL-receptor-independent uptake of LDL (μg of LDL cholesterol/hr/g of liver). For methods see Section II,B. Plasma LDL concentration was raised by adding increasing quantities of hamster LDL to the perfusion fluid. LDL-receptor-independent rates were measured with reductively methylated human LDL. (From Dietschy and Spady, 1985, with the permission of the authors.)

curves for total and nonsaturable uptake of LDL by normal human fibroblasts incubated in the presence of increasing concentrations of LDL in the medium (see Chapter 8).

It is possible to draw several inferences from the above relationship between uptake of LDL and plasma LDL concentration.

1. As the plasma LDL concentration is raised from a level below that required to saturate LDL receptors, the LDL-receptor pathway makes a diminishing relative contribution to total uptake of LDL in organs, such as liver and adrenals, in which LDL receptors are most active. This change, seen in the relationship between the two curves in Figure 7.5B, is a consequence of the saturability of

LDL receptors and will occur whether or not there is a decrease in receptor number.

2. In addition to the shift from LDL-receptor-dependent to LDL-receptor-independent catabolism in the liver and other organs, there is also a change in the relative contributions made by different organs to total LDL catabolism in the whole body. When the plasma LDL level rises, organs (such as skeletal muscle) that express few LDL receptors make an increasing relative contribution to total LDL catabolism. In the whole animal, this shift toward receptor-independent catabolism results in a decreased total FCR of LDL. Again, the change occurs irrespective of any change in receptor number. Hence, as Spady and Dietschy (1985) point out, a reduced FCR of LDL in the presence of a raised plasma LDL concentration does not necessarily mean that LDL-receptor activity in the whole animal is decreased.

3. A similar shift in LDL catabolism away from organs that normally express high LDL-receptor activity also occurs when there is a genetic or nongenetic reduction in receptor activity. For example, the liver accounts for only 30% of total LDL catabolism in WHHL rabbits, compared with 70% in normal rabbits (Spady *et al.*, 1987).

4. Since the LDL-receptor-independent pathway is nonsaturable, as much LDL can be catabolized by this route as by the LDL-receptor pathway, provided that the plasma LDL concentration rises to high enough levels. Indeed, at very high LDL levels, as in WHHL rabbits or FH homozygotes, the rate of LDL catabolism by the alternative route may be several times greater than the total rate at low LDL concentrations in the presence of normal LDL-receptor activity (see Chapter 8).

The presence of saturable and nonsaturable pathways for LDL catabolism bears on the interpretation of changes in the FCR of LDL in human subjects. Under physiological conditions the plasma LDL level is determined by the production rate of LDL and LDL-receptor activity. As noted above, an increase in production rate (the physiological equivalent of an infusion of LDL into an animal) will raise the plasma LDL concentration and lower the FCR in the presence of a constant level of LDL-receptor activity. Conversely, the plasma LDL level will rise and the FCR will fall if receptor activity decreases in the presence of a constant production rate. Thus, the relationship between production rate, receptor activity, and plasma LDL concentration is complex. Meddings and Dietschy (1986) have derived equations expressing the relationship between these three variables, from which it may be possible to deduce the level of LDL-receptor activity in human subjects. These equations have been applied to LDL kinetics in human subjects with primary hypercholsterolemia not due to FH. The results suggest that in some cases the rise in plasma cholesterol concentration is due to increased production of LDL, while in others it is due to decreased LDL-receptor activity.

References

Alexander, C. A., Hamilton, R. L., and Havel, R. S. (1976). Subcellular localization of B apoprotein of plasma lipoproteins in rat liver. *J. Cell Biol.* **69,** 241–263.

Andersen, J. M., and Dietschy, J. M. (1977). Regulation of sterol synthesis in 16 tissues of rat. 1. Effect of diurnal light cyclic, fasting, stress, manipulation of enterohepatic circulation, and administration of chylomicrons and triton. *J. Biol. Chem.* **252,** 3646–3651.

Andersen, J. M., and Dietschy, J. M. (1981). Kinetic parameters of the lipoprotein transport systems in the adrenal gland of the rat determined *in vivo.* Comparison of low and high density lipoproteins of human and rat origin. *J. Biol. Chem.* **256,** 7362–7370.

Angelin, B., Raviola, C. A., Innerarity, T. L., and Mahley, R. W. (1983). Regulation of hepatic lipoprotein receptors in the dog. Rapid regulation of apolipoprotein B,E receptors, but not of apolipoprotein E receptors, by intestinal lipoproteins and bile acids. *J. Clin. Invest.* **71,** 816–831.

Barter, P. J., and Nestel, P. J. (1972). Precursor–product relationship between pools of very low density lipoprotein triglyceride. *J. Clin. Invest.* **51,** 174–180.

Barter, P., Faegeman, O., and Havel, R. J. (1977). Metabolism of cholesteryl esters of very low density lipoproteins in the guinea pig. *Metabolism* **26,** 615–622.

Beltz, W. F., Kesäniemi, Y. A., Howard, B. V., and Grundy, S. M. (1985). Development of an integrated model for analysis of the kinetics of apolipoprotein B in plasma very low density lipoproteins, intermediate density lipoproteins, and low density lipoproteins. *J. Clin. Invest.* **76,** 575–585.

Berman, M. (1982). Kinetic analysis and modeling: Theory and applications to lipoproteins. *In* "Lipoprotein Kinetics and Modeling" (M. Berman, S. M. Grundy, and B. V. Howard, eds.), pp. 3–36. Academic Press, New York.

Berman, M., Hall, M., Levy, R. I., Eisenberg, S., Bilheimer, D. W., Phair, R. D., and Goebel, R. H. (1978). Metabolism of apoB and apoC lipoproteins in man: kinetic studies in normal and hyperlipoproteinemic subjects. *J. Lipid Res.* **19,** 38–56.

Berson, S. A., Yalow, R. S., Schreiber, S. S., and Post, J. (1953). Tracer experiments with ^{131}I labeled human serum albumin: distribution and degradation studies. *J. Clin. Invest.* **32,** 746–768.

Bilheimer, D. W., Eisenberg, S., and Levy, R. I. (1972). The metabolism of very low density lipoproteins. I. Preliminary *in vitro and in vivo* observations. *Biochim. Biophys. Acta* **260,** 212–221.

Boogaerts, J. R., Malone-McNeal, M., Archambault, J. L., and Davis, R. A. (1982). Expression in cultured hepatocytes of *in vivo* carbohydrate-induced changes in lipogenesis and apolipoprotein synthesis. *Fed. Proc., Fed. Am. Soc. Exp. Biol.* **41,** 1217.

Boogaerts, J. R., Malone-McNeal, M., Archambault-Schexnayder, J., and Davis, R. A. (1984). Dietary carbohydrate induces lipogenesis and very low density lipoprotein synthesis. *Am. J. Physiol.* **246,** E77–E83.

Boström, K., Wettesten, M., Borén, J., Bondjers, G., Wiklund, O., and Olofsson, S-O. (1986). Pulse-chase studies of the synthesis and intracellular transport of apolipoprotein B-100 in Hep G2 cells. *J. Biol. Chem.* **261,** 13800–13806.

Boyd, G. S. (1963). Hormones and cholesterol metabolism. *Biochem. Soc. Symp., No. 24,* 79–97.

Bradley, W. A., Hwang, S-L. C., Karlin, J. B., Lin, A. H. Y., Prasad, S. C., Gotto, A. M., and Gianturco, S. H. (1984). Low-density lipoprotein receptor binding determinants switch from apolipoprotein E to apolipoprotein B during conversion of hypertriglyceridemic very-low-density lipoprotein to low-density lipoproteins. *J. Biol. Chem.* **259,** 14728–14735.

Breckenridge, W. C., Little, J. A., Alaupovic, P., Wang, C. S., Kuksis, A., Kakis, G., Lindgren, F., and Gardiner, G. (1982). Lipoprotein abnormalities associated with a familial deficiency of hepatic lipase. *Atherosclerosis* **45,** 161–179.

Brown, M. S., and Goldstein, J. L. (1983). Lipoprotein metabolism in the macrophage. Implications for cholesterol deposition in atherosclerosis. *Ann. Rev. Biochem.* **52,** 223–261.

Carew, T. E., Pittman, R. C., and Steinberg, D. (1982). Tissue sites of degradation of native and reductively methylated [^{14}C]sucrose-labeled low density lipoprotein in rats. Contribution of receptor-dependent and receptor-independent pathways. *J. Biol. Chem.* **257,** 8001–8008.

Carew, T. E., Pittman, R. C., Marchand, E. R., and Steinberg, D. (1984). Measurement *in vivo* of irreversible degradation of low density lipoprotein in the rabbit aorta. Predominance of intimal degradation. *Atherosclerosis* **4,** 214–224.

Catapano, A. L. (1987). Binding sites for HDL and HDL catabolism. *Atherosclerosis Rev.* **16,** 109–123.

Chait, A., Brunzell, J. D., Albers, J. J., and Hazzard, W. R. (1977). Type-III hyperlipoproteinaemia ("remnant removal disease"). Insight into the pathogenic mechanism. *Lancet* **1,** 1176–1178.

Chao, Y-S., Windler, E. E., Chen, G. C., and Havel, R. J. (1979). Hepatic catabolism of rat and human lipoproteins in rats treated with 17α-ethinyl estradiol. *J. Biol. Chem.* **254,** 11360–11366.

Coleman, R. A., Haynes, E. B., Sand, T. M., and Davis, R. A. (1988). Developmental coordinate expression of triacylglycerol and small molecular weight apoB synthesis and secretion by rat hepatocytes. *J. Lipid Res.* **29,** 33–42.

Davidson, N. O., and Glickman, R. M. (1985). Apolipoprotein A-I synthesis in rat small intestine: regulation by dietary triglyceride and biliary lipid. *J. Lipid Res.* **26,** 368–379.

Davidson, N. O., Glickman, R. M., and Kollmer, M. G. (1986). Apolipoprotein B synthesis in rat small intestine: regulation by dietary triglyceride and biliary lipid. *J. Lipid Res.* **27,** 30–39.

Davidson, N. O., Nagun, A. M., Brasitus, T. A., and Glickman, R. M. (1987). Intestinal apolipoprotein A-I and B-48 metabolism: effects of sustained alterations in dietary triglyceride and mucosal cholesterol flux. *J. Lipid Res.* **28,** 388–402.

Davis, R. A., and Boogaerts, J. R. (1982). Intrahepatic assembly of very low density lipoproteins. Effect of fatty acids on triacylglycerol and apolipoprotein synthesis. *J. Biol. Chem.* **257,** 10908–10913.

Davis, R. A., and Malone-McNeal, M. (1985). Dietary cholesterol does not affect the synthesis of apolipoproteins B and E by rat hepatocytes. *Biochem. J.* **227,** 29–35.

Davis, R. A., and Roheim, P. S. (1978). Pharmacologically induced hypolipidemia. The ethinyl estradiol-treated rat. *Atherosclerosis* **30,** 293–299.

Deckelbaum, R. J., Eisenberg, S., Fainaru, M., Barenholz, Y., and Olivecrona, T. (1979). *In vitro* production of human plasma low density lipoprotein-like particles. A model for very low density lipoprotein catabolism. *J. Biol. Chem.* **254,** 6079–6087.

Demant, T., Carlson, L. A., Holmquist, L., Karpe, F., Nilsson-Ehle, P., Packard, C. J., and Shepherd, J. (1988). Lipoprotein metabolism in hepatic lipase deficiency: studies on the turnover of apolipoprotein B and on the effect of hepatic lipase on high density lipoprotein. *J. Lipid Res.* **29,** 1603–1611.

Dietschy, J. M., and Spady, D. K. (1985). Regulation of low density lipoprotein levels in animals and man with particular emphasis on the role of the liver. *In* "Receptor-Mediated Uptake in the Liver" (H. Greten, E. Windler, and U. Beisiegel, eds.), pp. 56–65. Springer-Verlag, Berlin.

Durrington, P. N., Newton, R. S., Weinstein, D. B., and Steinberg, D. (1982). The effect of insulin and glucose on very low density lipoprotein triglyceride secretion by cultured rat hepatocytes. *J. Clin. Invest.* **70,** 63–73.

Eisenberg, S., Bilheimer, D. W., Levy, R. I., and Lindgren, F. T. (1973). On the metabolic conversion of human plasma very low density lipoprotein to low density lipoprotein. *Biochim. Biophys. Acta* **326,** 361–377.

Elovson, J., Chatterton, J. E., Bell, G. T., Schumaker, V. N., Reuben, M. A., Puppione, D. L., Reeve, J. R., and Young, N. L. (1988). Plasma very low density lipoproteins contain a single molecule of apolipoprotein B. *J. Lipid Res.* **29,** 1461–1473.

Enholm, C., Mahley, R. W., Chappell, D. A., Weisgraber, K. H., Ludwig, E., and Witztum, J. L. (1984). The role of apolipoprotein E in the lipolytic conversion of β-very low density lipoproteins to low density lipoproteins in type III hyperlipoproteinemia. *Proc. Natl. Acad. Sci. U.S.A.* **81,** 5566–5570.

Faergeman, O., Sata, T., Kane, J. P., and Havel, R. J. (1975). Metabolism of apoprotein B of plasma very low density lipoproteins in the rat. *J. Clin. Invest.* **56,** 1396–1403.

Fainaru, M., Mahley, R. W., Hamilton, R. L., and Innerarity, T. L. (1982). Structural and metabolic heterogeneity of β-very low density lipoproteins from cholesterol-fed dogs and humans with type III hyperlipoproteinemia. *J. Lipid Res.* **23,** 702–714.

Fewster, M. E., Pirrie, R. E., and Turner, D. A. (1967). Effect of estradiol benzoate on lipid metabolism in the rat. *Endocrinology* **80,** 263–271.

Fisher, W. R. (1982). Apoprotein B kinetics in man: Concepts and questions. *In* ''Lipoprotein Kinetics and Modeling'' (M. Berman, S. M. Grundy, and B. V. Howard, eds.), pp. 44–68. Academic Press, New York.

Fisher, W. R., Zech, L. A., Bardalaye, P., Warmke, G., and Berman, M. (1980). The metabolism of apolipoprotein B in subjects with hypertriglyceridemia and polydisperse LDL. *J. Lipid Res.* **21,** 760–774.

Ghiselli, G., Schaefer, E. J., Gascon, P., and Brewer, H. B. (1981). Type III hyperlipoproteinemia associated with apolipoprotein E deficiency. *Science* **214,** 1239–1241.

Gianturco, S. H., Brown, F. B., Gotto, A. M., and Bradley, W. A. (1982). Receptor-mediated uptake of hypertriglyceridemic very low density lipoproteins by normal human fibroblasts. *J. Lipid Res.* **23,** 984–993.

Ginsberg, H., Le, N-A., Mays, C., Gibson, J., and Brown, W. V. (1981). Lipoprotein metabolism in nonresponders to increased dietary cholesterol. *Arteriosclerosis* **1,** 463–470.

Go, M. F., Schonfeld, G., Pfleger, B., Cole, T. G., Sussman, N. L., and Alpers, D. H. (1988). Regulation of intestinal and hepatic apoprotein synthesis after chronic fat and cholesterol feeding. *J. Clin. Invest.* **81,** 1615–1620.

Goldberg, I. J., Le, N-A., Paterniti, J. R., Ginsberg, H. N., Lindgren, F. T., and Brown, W. V. (1982). Lipoprotein metabolism during acute inhibition of hepatic triglyceride lipase in the cynomolgus monkey. *J. Clin. Invest.* **70,** 1184–1192.

Goldstein, J. L., and Brown, M. S. (1977). The low-density lipoprotein pathway and its relation to atherosclerosis. *Ann. Rev. Biochem.* **46,** 897–930.

Gotto, A. M. Pownall, H. J., and Havel, R. J. (1986). Introduction to the plasma lipoproteins. *Methods Enzymol.* **128,** 3–41.

Grundy, S. M., Vega, G. L., and Bilheimer, D. W. (1985). Kinetic mechanisms determining variability in low density lipoprotein levels and rise with age. *Arteriosclerosis* **5,** 623–630.

Gwynne, J. T., Mahaffee, D., Brewer, H. B., and Ney, R. L. (1976). Adrenal cholesterol uptake from plasma lipoproteins: regulation by corticotropin. *Proc. Natl. Acad. Sci. U.S.A.* **73,** 4329–4333.

Harkes, L., and Van Berkel, T. J. C. (1983). Cellular localization of the receptor-dependent and receptor-independent uptake of human LDL in the liver of normal and 17α-ethinyl estradiol-treated rats. *FEBS Lett.* **154,** 75–80.

Havel, R. J. (1984). The formation of LDL: mechanisms and regulation. *J. Lipid Res.* **25,** 1570–1576.

Hay, R. V., Pottenger, L. A., Reingold, A. L., Getz, G. S., and Wissler, R. W. (1971). Degradation of ^{125}I-labelled serum low density lipoprotein in normal and oestrogen-treated male rats. *Biochem. Biophys. Res. Comm.* **44,** 1471–1477.

Heimberg, M., and Wilcox, H. G. (1972). The effect of palmitic and oleic acids on the properties and composition of the very low density lipoprotein secreted by the liver. *J. Biol. Chem.* **247,** 875–880.

Houlston, R. S., Turner, P. R., Revill, J., Lewis, B., and Humphries, S. E. (1988). The fractional catabolic rate of low density lipoprotein in normal individuals is influenced by variation in the apolipoprotein B gene: a preliminary study. *Atherosclerosis* **71,** 81–85.

Hui, D. Y., Innerarity, T. L., and Mahley, R. W. (1984). Defective hepatic lipoprotein receptor binding of β-very low density lipoproteins from type III hyperlipoproteinemic patients. Importance of apolipoprotein E. *J. Biol. Chem.* **259,** 860–869.

Humphries, S. E., Kessling, A. M., Horsthemke, B., Donald, J. A., Seed, M., Jowett, M., Holm, M., Galton, D. J., Wynn, V., and Williamson, R. (1985). A common DNA polymorphism of the low-density lipoprotein (LDL) receptor gene and its use in diagnosis. *Lancet* **1,** 1003–1005.

Janus, E. D., Nicoll, A. M., Turner, P. R., Magill, P., and Lewis, B. (1980). Kinetic bases of the primary hyperlipidaemias: studies of apolipoprotein B turnover in genetically defined subjects. *Eur. J. Clin. Invest.* **10,** 161–172.

Kane, J. P., Sata, T., Hamilton, R. L., and Havel, R. J. (1975). Apoprotein composition of very low density lipoproteins of human serum. *J. Clin. Invest.* **56,** 1622–1634.

Kesäniemi, Y. A., Vega, G. L., and Grundy, S. M. (1982). Kinetics of apolipoprotein B in normal and hyperlipidemic man: Review of current data. *In* "Lipoprotein Kinetics and Modeling" (M. Berman, S. M. Grundy, and B. V. Howard, eds.), pp. 181–205. Academic Press, New York.

Kesäniemi, Y. A., Witztum, J. L., and Steinbrecher, U. P. (1983). Receptor-mediated catabolism of low density lipoprotein in man. Quantitation using glucosylated low density lipoprotein. *J. Clin. Invest.* **71,** 950–959.

Kesäniemi, Y. A., Beltz, W. F., and Grundy, S. M. (1985). Comparisons of metabolism of apolipoprotein B in normal subjects, obese patients, and patients with coronary heart disease. *J. Clin. Invest.* **76,** 586–595.

Kita, T., Brown, M. S., Bilheimer, D. W., and Goldstein, J. L. (1982). Delayed clearance of very low density and intermediate density lipoproteins with enhanced conversion to low density lipoprotein in WHHL rabbits. *Proc. Natl. Acad. Sci. U.S.A.* **79,** 5693–5697.

Knight, B. L., and Soutar, A. K. (1982). Changes in the metabolism of modified and unmodified low-density lipoproteins during the maturation of cultured blood monocyte-macrophages from normal and homozygous familial hypercholesterolaemic subjects. *Eur. J. Biochem.* **125,** 407–413.

Koelz, H. R., Sherrill, B. C., Turley, S. D., and Dietschy, J. M. (1982). Correlation of low and high density lipoprotein binding *in vivo* with rates of lipoprotein degradation in the rat. A comparison of lipoproteins of rat and human origin. *J. Biol. Chem.* **257,** 8061–8072.

Kovanen, P. T., Brown, M. S., and Goldstein, J. L. (1979). Increased binding of low density lipoprotein to liver membranes from rats treated with 17α-ethinylestradiol. *J. Biol. Chem.* **254,** 11367–11373.

Kovanen, P. T., Bilheimer, D. W., Goldstein, J. L., Jaramillo, J. J., and Brown, M. S. (1981). Regulatory role for hepatic low density lipoprotein receptors *in vivo* in the dog. *Proc. Natl. Acad. Sci. U.S.A.* **78,** 1194–1198.

Krul, E. S., Tikkanen, M. J., Cole, T. G., Davie, J. M., and Schonfeld, G. (1985). Roles of apolipoproteins B and E in the cellular binding of very low density lipoproteins. *J. Clin. Invest.* **76,** 361–369.

Langer, T., Levy, R. I., and Fredrickson, D. S. (1969). Dietary and pharmacologic perturbation of beta lipoprotein (BLP) turnover. *Circulation* **40,** Suppl. III, III-14.

McFarlane, A. S., and Koj, A. (1970). Short-term measurement of catabolic rates using iodine-labeled plasma proteins. *J. Clin. Invest.* **49,** 1903–1911.

Mahley, R. W. (1982). Atherogenic hyperlipoproteinemia: the cellular and molecular biology of plasma lipoproteins altered by dietary fat and cholesterol. *Med. Clin. N. Am.* **66,** 375–402.

Mahley, R. W. (1985). Atherogenic lipoproteins and coronary artery disease: concepts derived from recent advances in cellular and molecular biology. *Circulation* **72,** 943–948.

Mahley, R. W., Innerarity, T. L., Pitas, R. E., Weisgraber, K. H., Brown, J. H., and Gross, E. (1977). Inhibition of lipoprotein binding to cell surface receptors of fibroblasts following selective modification of arginyl residues in arginine-rich and B apoproteins. *J. Biol. Chem.* **252,** 7279–7287.

Mahley, R. W., Weisgraber, K. H., Melchior, G. W., Innerarity, T. L., and Holcombe, K. S. (1980). Inhibition of receptor-mediated clearance of lysine and arginine-modified lipoproteins from the plasma of rats and monkeys. *Proc. Natl. Acad. Sci. U.S.A.* **77,** 225–229.

Mahley, R. W., Hui, D. Y., Innerarity, T. L., and Weisgraber, K. H. (1981). Two independent lipoprotein receptors on hepatic membranes of dog, swine, and man. Apo-B,E and apoE receptors. *J. Clin. Invest.* **678,** 1197–1206.

Marsh, J. B., and Sparks, C. E. (1982). The effect of fasting on the secretion of lipoproteins and two forms of apoB by perfused rat liver. *Proc. Soc. Exp. Biol. Med.* **170,** 178–181.

Matthews, C. M. E. (1957). The theory of tracer experiments with ^{131}I-labelled plasma proteins. *Phys. Med. Biol.* **2,** 36–53.

Meddings, J. B., and Dietschy, J. M. (1986). Regulation of plasma levels of low-density lipoprotein cholesterol: interpretation of data on low-density lipoprotein turnover in man. *Circulation* **74,** 805–814.

Miller, N. E. (1984). Why does plasma low density lipoprotein concentration in adults increase with age? *Lancet* **1,** 263–267.

Mjøs, O. D., Faergeman, O., Hamilton, R. L., and Havel, R. J. (1975). Characterization of remnants produced during the metabolism of triglyceride-rich lipoproteins of blood plasma and intestinal lymph in the rat. *J. Clin. Invest.* **56,** 603–615.

Moutafis, C. D., and Myant, N. B. (1969). The metabolism of cholesterol in two hypercholesterolaemic patients treated with cholestyramine. *Clin. Sci.* **37,** 443–454.

Moutafis, C. D., Myant, N. B., Mancini, M., and Oriente, P. (1971). Cholestyramine and nicotinic acid in the treatment of familial hyperbetalipoproteinaemia in the homozygous form. *Atherosclerosis* **14,** 247–258.

Myant, N. B. (1981). "The Biology of Cholesterol and Related Steroids." Heinemann, London.

Naruszewicz, M., Carew, T. E., Pittman, R. C., Witztum, J. L., and Steinberg, D. (1984). A novel mechanism by which probucol lowers low density lipoprotein levels demonstrated in the LDL receptor-deficient rabbit. *J. Lipid Res.* **25,** 1206–1213.

Nestel, P., Billington, T., Tada, N., Nugent, P., and Fidge, N. (1983). Heterogeneity of very-low-density lipoprotein metabolism in hyperlipidemic subjects. *Metabolism* **32,** 8110–8117.

Packard, C. J., Shepherd, J., Joerns, S., Gotto, A. M., and Taunton, O. D. (1980). Apolipoprotein B metabolism in normal, type IV, and type V hyperlipoproteinemic subjects. *Metabolism* **29,** 213–222.

Packard, C. J. McKinney, L., Carr, K., and Shepherd, J. (1983). Cholesterol feeding increases low density lipoprotein synthesis. *J. Clin. Invest.* **72,** 45–51.

Packard, C. J., Munro, A., Lorimer, A. R., Gotto, A. M., and Shepherd, J. (1984). Metabolism of apolipoprotein B in large triglyceride-rich very low density lipoproteins of normal and hypertriglyceridemic subjects. *J. Clin. Invest.* **74,** 2178–2192.

Packard, C. J., Boag, D. E., Clegg, R., Bedford, D., and Shepherd, J. (1985). Effects of 1,2 cyclohexanedione modification on the metabolism of very low density lipoprotein apolipoprotein B: potential role of receptors in intermediate density lipoprotein catabolism. *J. Lipid Res.* **26,** 1058–1067.

Pagnan, A., Havel, R. J., Kane, J. P., and Kotite, L. (1977). Characterization of human very low density lipoproteins containing two electrophoretic populations: double pre-beta lipoproteinemia and primary dysbetalipoproteinemia. *J. Lipid Res.* **18,** 613–622.

Patsch, W., Franz, S., and Schonfeld, G. (1983a). Role of insulin in lipoprotein secretion by cultured rat hepatocytes. *J. Clin. Invest.* **71,** 1161–1174.

Patsch, W., Tamai, T., and Schonfeld, G. (1983b). Effect of fatty acids on lipid and apoprotein secretion and association in hepatocyte cultures. *J. Clin. Invest.* **72,** 371–378.

Patthy, L., and Smith, E. L. (1975). Identification of functional arginine residues in ribonuclease A and lysozyme. *J. Biol. Chem.* **250,** 565–569.

Phair, R. D. (1982). The role of kinetic modeling in physiology. *In* "Lipoprotein Kinetics and Modeling" (M. Berman, S. M. Grundy, and B. V. Howard, eds.), pp. 37–40. Academic Press, New York.

Phair, R. D., Hammond, M. G., Bowden, J. A., Fried, M., Fisher, W. R., and Berman, M. (1975). A preliminary model of human lipoprotein metabolism in hyperlipoproteinemia. *Fed. Proc., Fed Am. Soc. Exp. Biol.* **34,** 2263–2270.

Pittman, R. C., and Steinberg, D. (1984). Sites and mechanisms of uptake and degradation of high density and low density lipoproteins. *J. Lipid Res.* **25,** 1577–1585.

Pittman, R. C., Attie, A. D., Carew, T. E., and Steinberg, D. (1979a). Tissue sites of degradation of low density lipoprotein: application of a method for determining the fate of plasma proteins. *Proc. Natl. Acad. Sci. U.S.A.* **76,** 5345–5349.

Pittman, R. C., Green, S. R., Attie, A. D., and Steinberg, D. (1979b). Radiolabeled sucrose covalently linked to protein. A device for quantifying degradation of plasma proteins catabolized by lysosomal mechanisms. *J. Biol. Chem.* **254,** 6876–6879.

Pittman, R. C., Attie, A. D., Carew, T. E., and Steinberg, D. (1982a). Tissue sites of catabolism of rat and human low density lipoproteins in rats. *Biochim. Biophys. Acta* **710,** 7–14.

Pittman, R. C., Carew, T. E., Attie, A. D., Witztum, J. L., Watanabe, Y., and Steinberg, D. (1982b). Receptor-dependent and receptor-independent degradation of low density lipoprotein in normal rabbits and in receptor-deficient mutant rabbits. *J. Biol. Chem.* **257,** 7994–8000.

Pittman, R. C., Carew, T. E., Glass, C. K., Green, S. R., Taylor, C. A., and Attie, A. D. (1983). A radioiodinated, intracellularly trapped ligand for determining the sites of plasma protein degradation *in vivo*. *Biochem. J.* **212,** 791–800.

Reardon, M. F., Fidge, N. H., and Nestel, P. J. (1978). Catabolism of very low density lipoprotein B apoprotein in man. *J. Clin. Invest.* **61,** 850–860.

Reichl, D., and Miller, N. E. (1986). The anatomy and physiology of reverse cholesterol transport. *Clin. Sci.* **70,** 221–231.

Schmitz, G., Robenek, H., Lohmann, U., and Assman, G. (1985). Interaction of high density lipoproteins with cholesteryl ester-laden macrophages: biochemical and morphological characterization of cell surface receptor binding, endocytosis and resecretion of high density lipoproteins by macrophages. *EMBO J.* **4,** 613–622.

Schneider, W. J., Kovanen, P. T., Brown, M. S., Goldstein, J. L., Utermann, G., Weber, W., Havel, R. J., Kotite, L., Kane, J. P., Innerarity, T. L., and Mahley, R. W. (1981). Familial dysbetalipoproteinemia. Abnormal binding of mutant apoprotein E to low density lipoprotein receptors of human fibroblasts and membranes from liver and adrenals of rats, rabbits and cows. *J. Clin. Invest.* **68,** 1075–1085.

Schonfeld, G., Patsch, W., Pfleger, B., Witztum, J. L., and Weidman, S. W. (1979). Lipolysis produces changes in the immunoreactivity and cell reactivity of very low density lipoproteins. *J. Clin. Invest.* **64,** 1288–1297.

Schwartz, C. J., Gerrity, R. J., Lewis, L. J., Chisolm, G. M., and Bretherton, K. N. (1977). Arterial endothelial permeability to macromolecules. *In* "Atherosclerosis IV." *Proc. Fourth Int. Symp.* (G. Schettler, Y. Goto, Y. Hata, and G. Klose, eds.), pp. 1–11. Springer-Verlag, Berlin.

Shepherd, J., Bicker, S., Lorimer, A. R., and Packard, C. J. (1979). Receptor-mediated low density lipoprotein catabolism in man. *J. Lipid Res.* **20,** 999–1006.

Shepherd, J., Packard, C. J., Bicker, S., Lawrie, T. D., and Morgan, H. G. (1980a). Cholestyramine promotes receptor-mediated low-density- lipoprotein catabolism. *N. Engl. J. Med.* **302,** 1219–1222.

Shepherd, J., Packard, C. J., Grundy, S. M., Yeshurun, D., Gotto, A. M., and Taunton, O. D. (1980b). Effects of saturated and polyunsaturated fat diets on the chemical composition and metabolism of low density lipoproteins in man. *J. Lipid Res.* **21,** 91–99.

Shepherd, J., Packard, C. J., and Gibson, S. A. W. (1985). The reticuloendothelial system and low density lipoprotein metabolism. *In* "Receptor-mediated Uptake in the Liver" (H. Greten, E. Windler, and U. Beisiegel, eds.), pp. 87–90. Springer-Verlag, Berlin.

Sigurdsson, G. (1982). Deconvolution analysis of the conversion of VLDL apoprotein B to LDL apoprotein B. *In* "Lipoprotein Kinetics and Modeling" (M. Berman, S. M. Grundy, and B. V. Howard, eds.), pp. 113–120. Academic Press, New York.

Sigurdsson, G., Nicoll, A., and Lewis, B. (1975). Conversion of very low density lipoprotein to low density lipoprotein. *J. Clin. Invest.* **56,** 1481–1490.

Simionescu, N., Simionescu, M., and Palade, G. E. (1976). Recent studies on vascular endothelium. *Ann. N.Y. Acad. Sci.* **275,** 64–75.

Siuta-Mangano, P., Howard, S. C., Lennarz, W. J., and Lane, M. D. (1982). Synthesis, processing, and secretion of apolipoprotein B by the chick liver cell. *J. Biol. Chem.* **257,** 4292–4300.

Slater, H. R., Packard, C. J., and Shepherd, J. (1982). Measurement of receptor-independent lipoprotein catabolism using 1,2 cyclohexanedione-modified low density lipoprotein. *J. Lipid Res.* **23,** 92–96.

Soutar, A. K., Myant, N. B., and Thompson, G. R. (1982). The metabolism of very-low-density and intermediate-density lipoproteins in patients with familial hypercholesterolaemia. *Atherosclerosis* **43,** 217–231.

Spady, D. K., and Dietschy, J. M. (1985). Dietary saturated triacylglycerols suppress hepatic low density lipoprotein receptor activity in the hamster. *Proc. Natl. Acad. Sci. U.S.A.* **82,** 4526–4530.

Spady, D. K., Turley, S. D., and Dietschy, J. M. (1983). Receptor-independent LDL transport in the rat. *Arteriosclerosis* **3,** 506a.

Spady, D. K., Turley, S. D., and Dietschy, J. M., (1985a). Rates of low density lipoprotein uptake and cholesterol synthesis are regulated independently in the liver. *J. Lipid Res.* **26,** 465–472.

Spady, D. K., Turley, S. D., and Dietschy, J. M. (1985b). Receptor-independent low density lipoprotein transport in the rat *in vivo.* Quantitation, characterization and metabolic consequences. *J. Clin. Invest.* **76,** 1113–1122.

Spady, D. K. Huettinger, M. M., Bilheimer, D. W., and Dietschy, J. M. (1987). Role of receptor-independent low density lipoprotein transport in the maintenance of tissue cholesterol balance in the normal and WHHL rabbit. *J. Lipid Res.* **28,** 32–41.

Sparks, J. D., and Sparks, C. E. (1985). Apolipoprotein B and lipoprotein metabolism. *Adv. Lipid Res.* **21,** 1–46.

Sparks, C. E., Sparks, J. D., Bolognino, M., Salhanick, A., Strumph, P. S., and Amatruda, J. M. (1986). Insulin effects on apolipoprotein B synthesis and secretion by primary cultures of rat hepatocytes. *Metabolism* **35,** 1128–1136.

Spengel, F. A., Harders-Spengel, K., Duffield, R., Wood, C., Myant, N. B., and Thompson, G. R.

(1982). The effect of partial ileal bypass on receptor-mediated uptake and catabolism of low density lipoprotein in the rhesus monkey. *Res. Exp. Med. (Berl.)* **180,** 263–270.

Stalenhoef, A. F., Malloy, M. J., Kane, J. P., and Havel, R. J. (1984). Metabolism of apolipoproteins B-48 and B-100 of triglyceride-rich lipoproteins in normal and lipoprotein lipase-deficient humans. *Proc. Natl. Sci. U.S.A.* **81,** 1839–1843.

Steinberg, D. (1983). Lipoproteins and atherosclerosis. A look back and a look ahead. *Arteriosclerosis* **3,** 283–301.

Steinbrecher, U. P., Witztum, J. L., Kesäniemi, Y. A., and Elam, R. L. (1983). Comparison of glucosylated low density lipoproteins with methylated or cyclohexanedione-treated low density lipoprotein in the measurement of receptor-independent low density lipoprotein catabolism. *J. Clin. Invest.* **71,** 960–964.

Stender, S., and Zilversmit, D. B. (1981). Transfer of plasma lipoprotein components and of plasma proteins into aortas of cholesterol-fed rabbits. Molecular size as a determinant of plasma lipoprotein influx. *Arteriosclerosis* **1,** 38–49.

Stender, S., and Zilversmit, D. B. (1982). Comparison of cholesteryl ester transfer from chylomicrons and other plasma lipoproteins to aortic intima media of cholesterol-fed rabbits. *Arteriosclerosis* **2,** 493–499.

Teng, B., Sniderman, A. D., Soutar, A. K., and Thompson, G. R. (1986). Metabolic basis of hyperapobetalipoproteinemia. *J. Clin. Invest.* **77,** 663–672.

Thompson, G. R., Soutar, A. K., Spengel, F. A., Jadhav, A., Gavigan, S. J. P., and Myant, N. B. (1981). Defects of receptor-mediated low density lipoprotein catabolism in homozygous familial hypercholesterolemia and hypothyroidism *in vivo*. *Proc. Natl. Acad. Sci. U.S.A.* **78,** 2591–2595.

Turner, P. R., Miller, N. E., Cortese, C., Hazzard, W., Coltart, J., and Lewis, B. (1981). Splanchnic metabolism of plasma apolipoprotein B. Studies of artery-hepatic vein differences of mass and radiolabel in fasted human subjects. *J. Clin. Invest.* **67,** 1678–1686.

Turner, P. R., Konarska, R., Revill, J., Masana, L., LaVille, A., Jackson, P., Cortese, C., Swan, A. V., and Lewis, B. (1984). Metabolic study of variation in plasma cholesterol level in normal men. *Lancet* **2,** 663–665.

Walton, K. W., Scott, P. J., Dykes, P. W., and Davies, J. W. L. (1965). The significance of alterations in serum lipids in thyroid dysfunction. II. Alterations of the metabolism and turnover of ^{131}I-low-density lipoproteins in hypothyroidism and thyrotoxicosis. *Clin. Sci.* **29,** 217–238.

Weight, M., Cortese, C., Şule, U., Miller, N. E., and Lewis, B. (1982). Heritability of the low density lipoprotein receptor activity of human blood mononuclear cells: studies in normolipidaemic adult male twins. *Clin. Sci.* **62,** 397–401.

Weisgraber, K. H., Innerarity, T. L., and Mahley, R. W. (1978). Role of the lysine residues of plasma lipoproteins in high affinity binding to cell surface receptors on human fibroblasts. *J. Biol. Chem.* **253,** 9053–9062.

Wiklund, O., Carew, T. E., and Steinberg, D. (1985). Role of the low density lipoprotein receptor in penetration of low density lipoprotein into rabbit aortic wall. *Arteriosclerosis* **5,** 135–141.

Windler, E., Chao, Y-S., and Havel, R. J. (1980). Regulation of the hepatic uptake of triglyceride-rich lipoproteins in the rat. Opposing effects of homologous apolipoprotein E and individual C apoproteins. *J. Biol. Chem.* **255,** 8303–8307.

Witztum, J. L., Mahoney, E. M., Branks, M. J., Fisher, M., Elam, R., and Steinberg, D. (1982). Nonenzymatic glucosylation of low-density lipoprotein alters its biologic activity. *Diabetes* **31,** 283–291.

Witztum, J. L., Steinbrecher, U. P., Fisher, M., and Kesäniemi, Y. A. (1983). Nonenzymatic glucosylation of homologous low-density lipoprotein and albumin renders them immunogenic in the guinea pig. *Proc. Natl. Acad. Sci. U.S.A.* **80,** 2757–2761.

Witztum, J. L., Young, S. G., Elam, R. L., Carew, T. E., and Fisher, M. (1985). Cholestyramine-induced changes in low density lipoprotein composition and metabolism. I. Studies in the guinea pig. *J. Lipid Res.* **26,** 92–103.

Yamada, N., Shames, D. M., Stoudemire, J. B., and Havel, R. J. (1986). Metabolism of lipoproteins containing apoprotein B-100 in blood plasma of rabbits: heterogeneity related to the presence of apolipoprotein E. *Proc. Natl. Acad. Sci. U.S.A.* **83,** 3479–3483.

Zilversmit, D. B. (1960). The design and analysis of isotope experiments. *Am. J. Med.* **29,** 832–848.

Chapter 8

The LDL Receptor: Biochemistry and Cell Biology

I. Historical Background

Brown and Goldstein discovered the LDL receptor during the course of their attempt to identify the mutation responsible for the raised plasma LDL concentration that characterizes FH. Work on mammalian cells in culture, carried out mainly in the 1960s, had already shown that lymphoblasts, fibroblasts, and L-cells (a line of transformed mouse fibroblast) could satisfy essentially all their requirements for sterol by taking up cholesterol from a growth medium containing complete serum. In the presence of serum, sterol synthesis in these cells was almost completely suppressed by cholesterol acquired from the medium, and when the serum was replaced by lipoprotein-deficient serum (LPDS) containing no cholesterol, sterol synthesis rose to a maximum within 24 hours. (For references, see Rothblat, 1972; Williams and Avigan, 1972; Bailey, 1973; Bates and Rothblat, 1974).

Extending these observations, Brown and Goldstein and their co-workers investigated the effect of adding specific lipoproteins to a culture medium in which human skin fibroblasts were growing. When HMG-CoA reductase was maximally induced by preincubating the cells in the presence of LPDS, addition of LDL or, to a lesser extent, VLDL to the medium led to a concentration-dependent suppression of reductase activity. HDL, on the other hand, had no such effect. These observations suggested that apoB-containing lipoproteins exert a specific regulatory effect on HMG-CoA reductase by delivering cholesterol to the cells.

In contrast to normal cells, fibroblasts from patients with homozygous FH[1]

[1]Throughout this chapter, unless otherwise specified, homozygous FH refers to the condition in which LDL receptors are absent or markedly deficient in their capacity to bind LDL.

failed to show any regulatory response to lipoproteins in the growth medium. In the presence of complete serum, reductase activity in FH cells was 40–60 times that in normal cells (Fig. 8.1). Moreover, the replacement of serum by LPDS in the medium, or the addition of LDL, had no effect on reductase activity in FH cells. However, cholesterol added to the medium in ethanolic solution suppressed HMG-CoA reductase by more than 90% in both normal and FH cells. This showed that the failure of LDL to suppress reductase activity in FH cells could not be due to a genetic abnormality in the structure of the enzyme, such that it failed to respond to a normal regulatory signal. From these and other observations, Brown *et al.* (1974) deduced that the mutation in FH is in a gene whose product is necessary for the delivery of cholesterol from LDL in the external medium to a regulatory site within the cell.

One obvious possibility was that the normal gene product is a cell-surface receptor that mediates a specific interaction between LDL and the plasma mem-

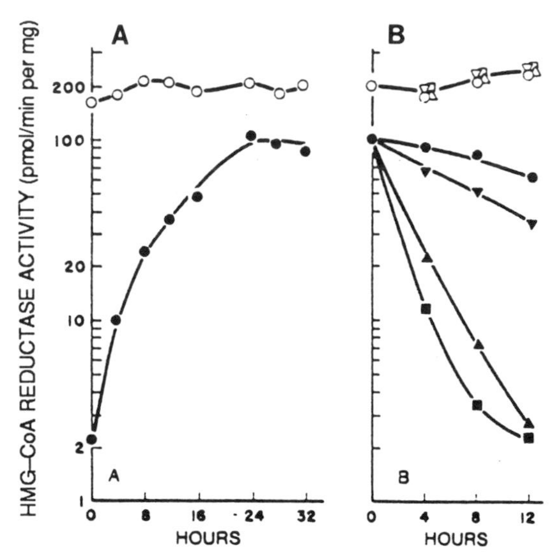

Figure 8.1. HMG-CoA reductase activity in fibroblasts from a normal subject (closed symbols) and a patient with homozygous FH (open symbols). Skin fibroblasts were grown in culture medium containing 10% fetal calf serum. In (A) the medium was replaced at zero time by medium containing lipoprotein-deficient plasma and HMG-CoA reductase activity was then measured at the indicated times in extracts prepared from the cells in each dish. In (B) after incubation for 24 hours in medium containing lipoprotein-deficient plasma, LDL was added to the medium to give the indicated concentrations and the incubation was continued for 12 hours. HMG-CoA reductase activity was measured in the cells at the indicated times. LDL concentrations are (●,○) none; (▼,▽) 2 μg/ml; (▲,△) 10 μg/ml; (■,□) 20 μg/ml. Enzyme activity is expressed as pmol of mevalonate formed from HMG-CoA/min/mg of protein. (From Goldstein and Brown, 1973, with the permission of the authors.)

brane. In support of this, Brown and Goldstein (summarized in Brown and Goldstein, 1974) showed that normal fibroblasts in culture exhibit saturable, specific, and high-affinity binding of LDL, and that binding is followed by internalization and intracellular degradation of the bound LDL and by suppression of HMG-CoA reductase. With cells from FH homozygotes or with normal cells pretreated with pronase (a proteolytic enzyme), no high-affinity binding of LDL could be detected, and LDL at low concentrations was not degraded and did not suppress reductase. Thus, it appeared that high-affinity binding of LDL to the surface of the cell is a necessary step in the intracellular degradation of LDL and in the suppression of reductase by LDL at low concentrations in the medium.

At this point, it was difficult to avoid the conclusion that a specific cell-surface receptor, probably a protein, was responsible for the binding of LDL by normal cells in culture. However, there were some who remained unconvinced until the partial isolation, from normal cells, of a membrane protein with all the functional properties of the postulated LDL receptor and the demonstration that this protein was absent from the cells of an FH homozygote (Schneider *et al.*, 1979, 1980). In the remaining sections of this chapter, I deal with the functional properties of the LDL receptor in cells *in vitro,* as expressed at the biochemical and subcellular levels.

II. Surface Binding and Intracellular Degradation of LDL

A. Binding

1. Heparin-Releasable LDL

References for Sections II,A and II,B of this chapter will be found in Goldstein and Brown (1976, 1977) and Goldstein *et al.* (1979a, 1983). The binding of LDL by LDL receptors is readily demonstrated in monolayer cultures of normal human skin fibroblasts in which the receptors have been induced by preincubating the cells in medium containing LPDS. Pretreated cells are incubated in the presence of [^{125}I]LDL and are then washed to remove unbound LDL. The amount of surface-bound LDL is determined from the radioactivity released from the washed cells by heparin, a sulfated glycosaminoglycan that competes with the LDL receptor for binding to LDL. At 4°C all the [^{125}I]LDL associated with the washed cells is surface bound because endocytosis is completely inhibited at this temperature. At 37°C, total cell-associated radioactivity includes surface-bound LDL plus LDL that has been internalized during the incubation; only the surface-bound component is released by heparin.

Figure 8.2 shows the amount of heparin-releasable LDL bound by normal and FH homozygote fibroblasts at 4°C as a function of LDL concentration in the medium. The curve for normal cells shows an initial steep rise merging, at a

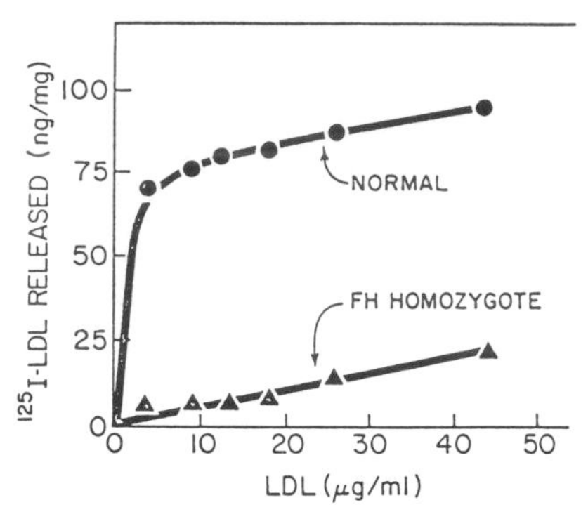

Figure 8.2. Saturation curves for heparin-releasable ^{125}I-LDL binding at 4°C in normal (●) and FH homozygote (▲) fibroblasts. Skin fibroblasts from a normal subject and from a patient with the receptor-negative form of homozygous FH were grown in culture. The cells were preincubated at 37°C for 48 hours in the presence of lipoprotein-deficient serum (LPDS). They were then incubated at 4°C for 2 hours in the presence of LPDS and the indicated concentrations of ^{125}I-LDL. At the end of the incubation the cells were washed repeatedly at 4°C and were then incubated at 4°C for 60 minutes in medium containing heparin (10 mg/ml). Heparin-releasable ^{125}I-IDL was determined from the amount of ^{125}I released into the medium. (From Goldstein *et al.*, 1976, with the permission of the authors.)

concentration of 5–10 μg of LDL protein/ml, with a much less steep linear increase. The curve for FH cells shows only the linear component. When the incubations are carried out at 37°C curves of a similar shape are obtained, but the inflection occurs at a concentration of about 50 μg/ml. Curves such as these were originally interpreted in terms of two processes: (1) high-affinity saturable binding of LDL to a limited number of specific binding sites (LDL receptors) present only in normal cells, and (2) low-affinity, nonsaturable binding present in both normal and FH homozygote cells. Additional evidence for low-affinity binding of LDL by human fibroblasts has also been adduced by Miller *et al.* (1978) and it has been suggested that this process contributes to LDL-receptor-independent catabolism of LDL *in vivo* (see Chapter 7). However, it seems more likely that the low-affinity binding exhibited by normal and FH homozygote cells in culture is due to nonspecific attachment of LDL particles to the intercellular matrix and to exposed areas of the culture dish. As discussed below, ultrastructural studies with the electron microscope fail to show any binding of LDL particles to the surfaces of fibroblasts lacking LDL receptors.

High-affinity binding at each LDL concentration may be calculated by subtracting the low-affinity component from the total amount of LDL bound by the cells (see Goldstein and Brown, 1974). Curves derived in this way show half-saturation of LDL receptors at about 2 μg of LDL protein/ml at 4°C and at 10–15 μg/ml at 37°C. The amount of LDL bound to maximally induced fibroblasts in culture at LDL concentrations high enough to saturate all the receptors corre-

Table 8.1
Some Functional Properties of the LDL Receptor on Skin Fibroblasts in Culture

Binding affinity.	Half-saturation for LDL: 2 μg of protein/ml at 4°C; 10–15 μg of protein/ml at 37°C.
Binding capacity.	In fully induced cells: 7,500–10,000 LDL particles per cell at 4°C; 15,000–70,000 at 37°C.
Specificity.	Binds lipoproteins containing apoB-100 or apoE in suitable conformation. Does not bind apoB-48. Affinity for LDL 200 times that for apoE-free HDL; also binds HDL_c, VLDL from hypertriglyceridemic patients, chylomicron remnants, VLDL remnants, and β-VLDL. (For details see text.)
Optimum pH.	7.5
Requires Ca^{2+} (binding inhibited by EDTA).	
Agents that displace LDL from receptors	Heparin, dextran sulfate, polyphosphates with chain length greater than 4.

sponds to about 10,000 LDL particles per cell at 4°C and to 15,000–70,000 particles per cell at 37°C. These and other functional properties of the LDL receptor on fibroblasts in culture are summarized in Table 8.1. Several of these properties are considered below in more detail.

2. Binding Affinity

The higher binding affinity for LDL at 4°C than at 37°C may be due to rapid internalization of bound particles at the higher temperature. This could result in the presence of an unstirred layer of medium at the cell surface containing LDL at a lower concentration than that determined in the medium as a whole. Another possibility, suggested by the observations of Gavigan *et al.* (1988), is that LDL receptors, which may reach the cell surface as dimers (Schneider *et al.*, 1983; Van Driel *et al.*, 1987), are unable to dissociate at 4°C.

3. Specificity

As discussed in Chapter 6, apoE, like apoB-100, has a domain that is recognized by binding sites on the LDL receptor. Binding of apoE-containing lipoproteins by LDL receptors was first noted by Mahley and Innerarity (1977), who showed that HDL_c [a lipoprotein containing apoE but no apoB (see p. 275)] competes with LDL for binding to LDL receptors on fibroblasts. Subsequent work in Mahley's laboratory extended this observation in considerable detail.

Binding of HDL_c by LDL receptors is followed by internalization and intracellular degradation, and by the biochemical consequences that follow LDL-receptor-mediated uptake of LDL (see Section I,E,2 below). As with LDL, binding to the receptor is inhibited by chemical modification of a limited number of lysine or arginine residues in the protein ligand. At 37°C the affinity of HDL_c

for the LDL receptor is 10–20 times that of LDL, and about 4 times as many LDL particles as HDL_c particles are required to saturate LDL receptors on fibroblasts at this temperature. The observations of Pitas *et al.* (1979, 1980) on the binding of HDL_c and of artificial phospholipid–apoE complexes to LDL receptors are consistent with the hypothesis that each HDL_c particle binds to four receptors and that each LDL particle binds to one receptor. Since each HDL_c particle has about 16 molecules of apoE, each with a single receptor-recognition domain, one HDL_c particle could bind to four receptors via four of its apoE molecules. An alternative possibility, suggested by the radiation-inactivation[2] experiments of Innerarity *et al.* (1981), is that one receptor binds either one HDL_c particle via four binding sites, or four LDL particles, each via a single binding site (see Mahley *et al.*, 1986, for review). Either hypothesis explains the higher affinity of the receptor for HDL_c and the larger number of LDL particles required for saturation. If each LDL receptor can, in fact, interact with more than one molecule of apoB or apoE, this would provide a functional explanation for the presence of multiple repeats in the binding domain of the receptor (see Chapter 9). However, the observations of Knight *et al.* (1987), in which the amount of LDL-receptor protein in fibroblasts was determined directly by radioimmunoassay, indicate that one molecule of receptor binds only one particle of LDL in the presence of saturating concentrations of LDL.

Although the presence of apoB-100 or apoE as an integral component of a lipoprotein is necessary for binding to the LDL receptor, this is not a sufficient condition. As we saw in Chapter 7 (Section I,C), if binding of a lipoprotein is to occur, it must not only contain a specific apoprotein; the apoprotein must also be present in a conformation such that its binding domain is exposed at the surface of the particle. This is well exemplified by the S_f 60–400 VLDL from human subjects with hypertriglyceridemia (HTG-VLDL). Normal S_f 60–400 VLDL binds to the LDL receptor only to a small or negligible extent; according to Gianturco *et al.* (1982b) the receptor binding of normal VLDL reported from some laboratories is due to nonspecific adsorption. HTG-VLDL, on the other hand, binds to fibroblast receptors with an affinity several times that of LDL (Gianturco *et al.*, 1982b). The binding of HTG-VLDL to LDL receptors is mediated entirely by apoE (Gianturco *et al.*, 1983; Bradley and Gianturco, 1986). However, since the apoE content of normal VLDL is similar to that of HTG-VLDL, the ability of HTG-VLDL to bind to LDL receptors with high affinity must be due to the presence of some apoE molecules in a conformation that favors multivalent binding, similar to the binding of HDL_c discussed above.

The binding of chylomicron remnants (Florén *et al.*, 1981) and of VLDL remnants (Gianturco *et al.*, 1982b) to LDL receptors *in vitro* must also depend to

[2]Radiation inactivation is a method for estimating the size of the unit whose function is abolished when any part of the unit is hit by a single ionizing radiation.

some degree upon the conformation, rather than the quantity, of apoE and apoB-100 in the particles, since intact chylomicrons (which contain apoE) and VLDL (which contain apoE and apoB-100) are bound weakly, if at all, by LDL receptors. The binding of human β-VLDL by LDL receptors is mediated by the apoE present in these particles, the apoB-100 present in fraction II (see p. 190) playing no part in binding unless the particles are partially delipidated (Hui *et al.*, 1984). ApoB-48, which lacks the LDL-receptor-binding domain present in apoB-100 (see Chapter 6), makes no contribution to the binding of β-VLDL fraction I by LDL receptors (Hui *et al.*, 1984).

The LDL receptor exhibits a high degree of specificity toward lipoproteins containing apoB-100 or apoE. For example, the affinity of human LDL for human LDL receptors is about 200 times that for human HDL from which the apoE-containing subfraction has been removed. However, as mentioned in Chapter 6 (Section VI,D), LDL receptors from a given species recognize apoB-100-containing lipoproteins from a wide range of different species (Young *et al.*, 1986). Likewise, LDL receptors from widely different species cross-react immunochemically and also exhibit extensive homology at the DNA level (see Section III below). This lack of interspecies specificity reflects conservation of functional domains during the evolution of the LDL receptor and of the apoproteins responsible for the binding of lipoproteins. Incomplete specificity has also been of considerable help in the isolation of the LDL receptor and its gene.

4. Interaction with Ligands

When skin fibroblasts in culture are incubated with [^{125}I]LDL at 4°C, binding reaches completion in about 2 hours (Brown and Goldstein, 1974). Binding of [^{125}I]LDL is inhibited competitively by unlabeled LDL added initially to the incubation medium. However, once binding has occurred, [^{125}I]LDL dissociates only very slowly from its receptors (30% dissociation at 4°C in about 3 hours), even in the presence of high concentrations of unlabeled LDL. In contrast to this, bound LDL is displaced very rapidly from receptors by heparin and other sulfated glycosaminoglycans (Goldstein *et al.*, 1976). Heparin, for example, displaces more than 80% of ^{125}I-labeled LDL bound to fibroblasts within 20 minutes at 4°C. Since heparin is known to bind to LDL by ionic interaction between negative charges on heparin and positively charged regions on LDL (Iverius, 1972), Goldstein *et al.* (1976) suggested that sulfated glycosaminoglycans displace LDL from its receptor by competing with receptors for positive charges on LDL. This led to the suggestion that the same positively charged regions on LDL interact with negatively charged regions on the LDL receptor. These ideas have been fully substantiated by what is now known about the binding sites on the LDL receptor and about the receptor-recognition domains on apoB-100 and apoE (see Chapter 6 and Section III of this chapter).

In view of the probability that binding of LDL to its receptors is mediated by charge interaction, the very slow reversibility of binding in the presence of high concentrations of LDL is surprising. It is likely to remain unexplained until more is known about the details of this interaction. Goldstein and Brown (1976) have suggested that several sites on each LDL particle bind to multiple receptor sites on the cell surface and that the bound particle is released only when all receptor sites are occupied concurrently by small competitor molecules such as heparin. However, this is not consistent with the presence of one apoB molecule, each with one receptor-recognition domain, in each LDL particle.

B. Internalization and Degradation

1. Time Course of Uptake

When fibroblasts in culture that have bound [^{125}I]LDL at 4°C are washed to remove unbound LDL and are then warmed to 37°C in the absence of [^{125}I]LDL, heparin-releasable LDL disappears rapidly from the cells. At the same time, [^{125}I]LDL, equal in amount to that lost from the cell surfaces, appears within the cells. Within 15–30 minutes, acid-soluble radioactivity (mainly as iodotyrosines) begins to appear in the medium, reflecting the degradation of internalized [^{125}I]LDL. If the incubation is carried out at 37°C in the continued presence of [^{125}I]LDL, the amount of heparin-releasable LDL bound to the cells reaches a plateau within 15 minutes. At this time, the rate at which new LDL particles are bound by LDL receptors is exactly balanced by the rate of internalization of bound particles. The amount of [^{125}I]LDL within the cells also increases, reaching a plateau when the rate of internalization is balanced by the rate of degradation. As the internalized LDL is degraded, [^{125}I]iodotyrosines appear in the medium after an initial lag period of several minutes, the amount increasing at a constant rate throughout the incubation. These events, reflecting the receptor-mediated uptake of whole LDL particles and the intracellular degradation of their protein component, are shown schematically in Fig. 8.3. See also Goldstein and Brown (1976).

From time-course experiments at 37°C, Goldstein *et al.* (1976) have estimated that about half the LDL particles bound at the cell surface are internalized every 3 minutes and that an amount of LDL equal to the total within the cell is degraded every 120 minutes. Experiments with LDL containing [^{3}H]cholesteryl linoleate have shown that intracellular degradation of LDL-receptor-bound LDL also involves the hydrolysis of the cholesteryl esters present in the core of the particle. However, as discussed below, the free cholesterol resulting from this hydrolysis is retained within the cell.

2. The Role of Lysosomes

Two lines of evidence from biochemical studies have established that the degradation of LDL that enters cells through the LDL-receptor pathway is mediated by

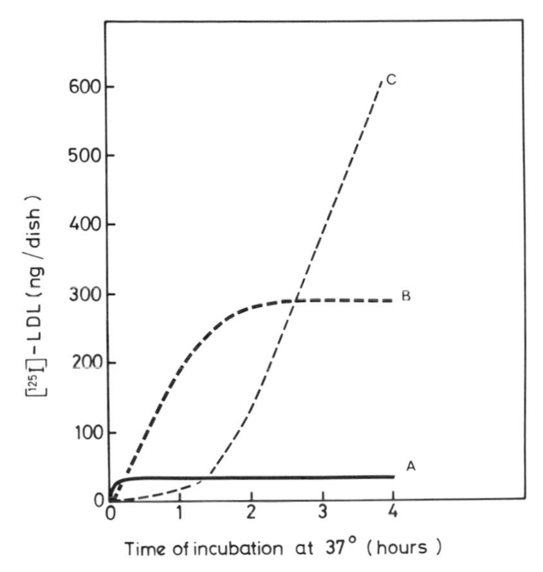

Figure 8.3. Schematic representation of the time course of surface binding, internalization, and proteolytic degradation of ^{125}I-labeled LDL by cultured human fibroblasts preincubated in medium containing LPDS for 24 hours at 37°C. After preincubation, [^{125}I]LDL (925 μg of LDL protein/ml) was added to the medium and the incubation was continued at 37°C. Surface-bound [^{125}I]LDL was determined from the amount of radioactivity released from the cells by heparin; internalized [^{125}I]LDL was estimated as total cell-associated radioactivity minus the amount released by heparin; degradation of labeled LDL protein was determined as the amount of acid-soluble radioactivity in the medium. A, surface-bound; B, internalized; C, acid-soluble. (From Goldstein and Brown, 1976, with the permission of the authors.)

lysosomal hydrolysis. First, chloroquine, a drug that inhibits lysosomal hydrolases, blocks the hydrolysis of both the protein and the cholesteryl esters of LDL by fibroblasts in culture, but has no effect on the surface binding or internalization of LDL. Hence, when fibroblasts are incubated with [^{125}I]LDL in the presence of chloroquine, [^{125}I]LDL enters the cells by the receptor pathway but is not degraded. Instead, it accumulates progressively as intact LDL particles within secondary lysosomes (see Goldstein and Brown, 1977). The second line of evidence for lysosomal hydrolysis is based on the study of fibroblasts from patients with monogenic disorders characterized by a specific deficiency of lysosomal acid lipase (Wolman's disease and cholesteryl-ester storage disease). Cultured fibroblasts from these patients bind and internalize [^{125}I]LDL at the normal rate and they degrade the protein component of LDL normally. However, hydrolysis of the cholesteryl esters is defective, resulting in the intralysosomal accumulation of esterified cholesterol with a fatty-acid pattern characteristic of the cholesteryl esters of LDL. The ultrastructural studies of Anderson *et al.*

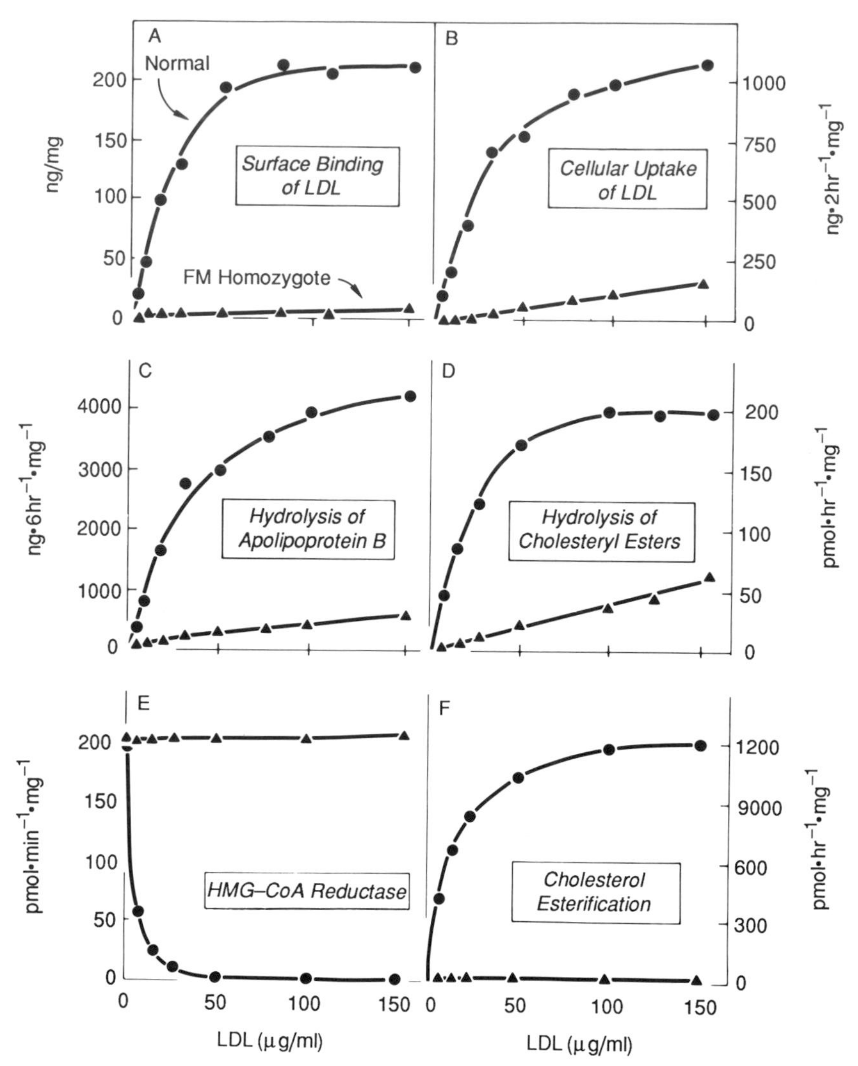

Figure 8.4. Interactions between LDL and LDL receptors on cultured fibroblasts from a normal subject (●) and a patient with familial hypercholesterolemia in the homozygous form (▲). After culture for 6 days in a medium containing fetal calf serum, the medium was replaced by medium containing 10% human lipoprotein-deficient serum to induce receptor formation. On day 8 the medium was replaced with 2 ml of fresh medium containing (A)–(C) ^{125}I-labeled LDL, (D) [^{3}H]cholesteryl linoleate labeled LDL, or (E)–(F) unlabeled LDL. After incubation with LDL at 37°C for 2 hours (A) and (B) or for 6 hours (C)–(F), the indicated measurements were made.

(A) and (B) Surface binding and cellular uptake of ^{125}I-labeled LDL. Each monolayer was washed six times at 4°C with an albumin-containing buffer, and a solution containing heparin (10 mg/ml) was added to each dish. The dishes were then incubated at 4°C for 1 hour. The

(1977a), demonstrating the participation of lysosomes in the LDL-receptor pathway, are described in Section II,C below.

3. Effects of Free Cholesterol

As we have seen, the esterified cholesterol in LDL taken up by fibroblasts via the LDL-receptor pathway is rapidly hydrolyzed in lysosomes. The free cholesterol so released is used partly for membrane synthesis, especially in nonconfluent cells. It also has three well-defined effects on cell metabolism.

As discussed in Chapters 3 and 4, HMG-CoA reductase activity is suppressed and ACAT activity is stimulated, with the net result that the free cholesterol content of the cells increases slightly while that of the cytoplasmic pool of esterified cholesterol increases markedly (Brown *et al.*, 1975). The effect of these two regulatory events is to minimize changes in the cell content of free cholesterol. LDL that enters fibroblasts other than by the LDL-receptor pathway, as in FH homozygote fibroblasts, does not increase the cholesterol content of the cells and has no effect on HMG-CoA reductase or ACAT (see Fig. 8.4). A similar lack of any effect of receptor-independent uptake of lipoproteins on reductase activity in rat hepatocytes is mentioned below.

The failure of FH cells in culture to respond normally to LDL in the medium may be due in part to the slow rate at which LDL at low concentrations enters cells by pinocytosis. However, this may not be the whole explanation because reductase activity is not suppressed in FH cells even at LDL concentrations high enough to give rates of uptake and degradation by pinocytosis that are comparable with those achieved by receptor-mediated uptake at much lower concentrations (Goldstein and Brown, 1976). One possibility is that at the very high concentrations of LDL required to support these rates of uptake, free cholesterol released by lysosomal hydrolysis of esterified cholesterol is withdrawn from the cells by LDL in the medium acting as acceptor for cholesterol.

Degradation of LDL taken up by nonsaturable pathways in fibroblasts in

---

heparin-containing medium was removed and the amount of ^{125}I-labeled LDL bound to cell surfaces (and hence removable by heparin) was determined. The cells were dissolved in 0.1 *N* NaOH and the amount of ^{125}I-labeled LDL that had entered the cells was determined. (C) Proteolytic hydrolysis of ^{125}I-labeled LDL. The medium was assayed for ^{125}I-labeled trichloroacetic-acid-soluble degradative products formed during the 6-hour incubation. (D) Hydrolysis of LDL-cholesteryl esters. The cellular content of unesterified [^{3}H]cholesterol formed by the hydrolysis of [^{3}H]cholesteryl linoleate-labeled LDL was measured after separation of free from esterified cellular cholesterol. (E) Suppression of HMG-CoA reductase activity. Cells were harvested and enzyme activity determined in detergent-solubilized extracts. (F) Stimulation of cholesteryl [^{14}C]oleate formation, determined from the incorporation of [1-^{14}C]oleate during a 1-hour incubation at 37°C. The horizontal axes show the concentration of LDL (μg of LDL protein/ml of incubation medium) during the experimental incubations on day 8. (From Brown and Goldstein, 1976, with the permission of the authors.)

culture is inhibited by chloroquine (Goldstein and Brown, 1977). Hence, the inability of LDL to suppress HMG-CoA reductase, when the particles are taken up in this manner, cannot be due to failure of the internalized LDL to enter lysosomes. However, the observations of Ostlund *et al.* (1979) on cultured human fibroblasts are consistent with the possibility that LDL particles internalized other than via coated pits (see Section II,C below) are delivered to a subpopulation of lysosomes which release free cholesterol that is not immediately accessible to the regulatory pool. Another possibility, suggested by the observations of Attie *et al.* (1980) on rat hepatocytes in culture, is that LDL internalized by nonreceptor pathways takes longer to reach lysosomes than LDL entering cells by the LDL-receptor pathway. Either mechanism would help to resolve the apparent contradiction between (1) the failure of receptor-independent uptake of LDL to suppress reductase during comparatively short-term experiments *in vitro,* and (2) the normal activity of reductase in the cells of FH homozygotes *in vivo* (see Chapter 10).

Figure 8.4 shows the rates of hydrolysis of LDL apoprotein and esterified cholesterol by normal and FH fibroblasts in culture, each expressed as a function of LDL concentration in the medium. With normal fibroblasts, both curves are parallel to the curve for surface binding of LDL. That is, they reach a maximum value at about 50 μg of LDL protein/ml, with half maximum at 10–15 μg/ml. Similar saturation curves are shown for suppression of HMG-CoA reductase activity and for stimulation of cholesterol esterification, both reaching a maximum at LDL concentrations of about 50 μg/ml. In FH cells, surface binding of LDL is negligible at all concentrations tested, hydrolysis of LDL apoB-100 and of esterified cholesterol occurs only by nonsaturable mechanisms, and LDL has no effect on the activities of reductase and ACAT. As already mentioned (see Chapters 3 and 4 for details), the effects of LDL on the activities of reductase and ACAT in normal fibroblasts in culture may be brought about in FH cells by adding ethanolic solutions of free cholesterol, or of an oxygenated analog such as 7-ketocholesterol, to the culture medium.

The third regulatory effect of LDL uptake by the LDL-receptor pathway is suppression of the synthesis of LDL receptors. When fibroblast monolayers in which LDL receptors have been maximally induced are incubated in the presence of LDL, there is a progressive and concentration-dependent decrease in the number of receptors expressed by the cells (Brown and Goldstein, 1975). As shown in Fig. 8.5, in the presence of LDL at a concentration of 10 μg of protein/ml, LDL-binding capacity, measured in the presence of saturating concentrations of LDL, declines to less than 30% of the initial value within 48 hours. A similar effect is seen when an ethanolic solution of free cholesterol or 25-hydroxycholesterol is added to the incubation medium.

When LDL is removed from the medium, receptor number is restored at a rate depending upon the concentration of LDL to which the cells were previously

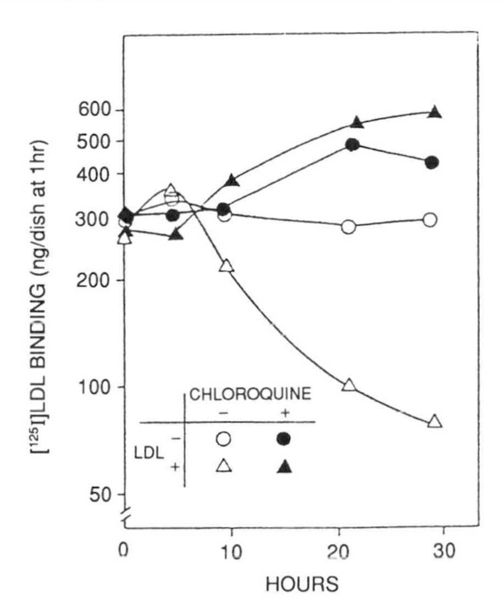

Figure 8.5. Feedback suppression of [^{125}I]LDL binding in fibroblast monolayers by unlabeled LDL. (This figure also shows that no suppression occurs when lysosomal degradation of internalized [^{125}I]LDL is inhibited by chloroquine.) On day 7 (zero time) the culture dishes were divided into four groups and each received 2 ml of incubation medium containing (○) no LDL, no chloroquine; (●) no LDL, 75 μ*M* chloroquine; (△) 10 μg/ml unlabeled LDL, no chloroquine; (▲) 10 μg/ml unlabeled LDL, 75 μ*M* chloroquine. At the indicated times during incubation at 37°C, duplicate dishes containing 10 μg/ml of unlabeled LDL (△,▲) received an additional 10 μg/ml of [^{125}I]LDL, and duplicate dishes containing no LDL (○,●) received 20 μg/ml of [^{125}I]LDL. Thus, during the labeling period all dishes contained 20 μg/ml of [^{125}I]LDL. The cell content of ^{125}I was determined in each dish after 1 hour of incubation. (From Brown and Goldstein, 1975, with the permission of the authors.)

exposed. Since the restoration of receptor number could be prevented by cycloheximide, Brown and Goldstein (1975) suggested that the regulation of LDL receptors by LDL is mediated by control of the rate of synthesis of receptor protein. Additional observations on the rate of turnover of LDL receptors, under conditions in which they were partially suppressed by LDL, were consistent with the assumption that suppression is mediated entirely by a decrease in the rate of synthesis without any change in the rate of catabolism of receptors. The half-life of LDL receptors on fibroblasts in culture, estimated from the rate at which receptor number declines in the presence of cycloheximide, is about 25 hours. However, a more rapid turnover ($t_{1/2}$ 10–12 hours) is suggested by the observations of Knight *et al.* (1987) on the rate at which radioactivity is lost from endogenously labeled receptors in fibroblasts *in vitro*. See also Davis *et al.* (1986) and Esser and Russell (1988).

Because suppression of LDL receptors by LDL is prevented by chloroquine (an inhibitor of lysosomal hydrolases), Brown and Goldstein concluded that receptor number is regulated in accordance with the amount of free cholesterol released from lysosomes during the degradation of LDL that enters cells by the LDL-receptor pathway. This led them to propose that HMG-CoA reductase activity, ACAT activity, and synthesis of LDL receptors are regulated coordinately in response to the cell's requirement for free cholesterol. This proposal is still broadly acceptable. However, it is now recognized that HMG-CoA reductase is under dual control by cholesterol itself and by nonsterol products of mevalonic acid (for details see Chapter 3, Section VIII,D). Regulation of LDL receptors, on the other hand, seems to be mediated largely by LDL-derived free cholesterol, since LDL-receptor activity in fibroblast monolayers is reduced by LDL to the same extent whether or not compactin is present at a concentration high enough to inhibit the synthesis of mevalonic acid (Brown *et al.*, 1978). This would not be the case if the nonsterol repressor of HMG-CoA reductase also suppresses LDL receptors. The mechanisms by which cholesterol regulates LDL-receptor synthesis are discussed in Chapter 9.

C. Ultrastructural Basis

1. Endocytosis via Coated Pits

The ultrastructural basis of the biochemical events described in the previous section has been investigated by Anderson and co-workers (reviewed in Goldstein *et al.*, 1979a; Anderson *et al.*, 1982; Goldstein *et al.*, 1985).

The distribution of LDL receptors on cells in culture may be examined by conventional electron microscopy, using LDL linked covalently to ferritin as a visual marker for receptors. When monolayers of normal fibroblasts are incubated at 4°C with LDL-ferritin and are then washed to remove unbound particles, the bound LDL-ferritin is seen to be localized preferentially over specialized regions of the cell surface called *coated pits*. Seen in cross section, these are shallow depressions, 2000–5000 Å (0.2–0.5 μm) in diameter, with a fuzzy coating on the cytoplasmic face due to the presence of a network of protein attached noncovalently to the plasma membrane (see Fig. 8.6). Coated pits on fibroblasts occupy a total of about 2% of the whole surface. As noted below, coated pits on human fibroblasts are arranged in lines along the cell surface. Using the technique of freeze-fracture/deep-etching, with which the two leaflets of the plasma membrane are viewed *en face*, Orci *et al.* (1978) have shown that coated pits are irregular, craterlike depressions with a higher concentration of intramembrane particles than that present in the remainder of the plasma membrane (Fig. 8.6). Montesano *et al.* (1979) have shown that filipin, an antibiotic that forms complexes with free cholesterol in cell membranes, fails to label the coated pits of fibroblasts. Montesano *et al.* pointed out that a deficiency of free cholesterol in coated pits would be expected to increase the fluidity of the pit membrane and thus to facilitate its invagination during the formation of coated

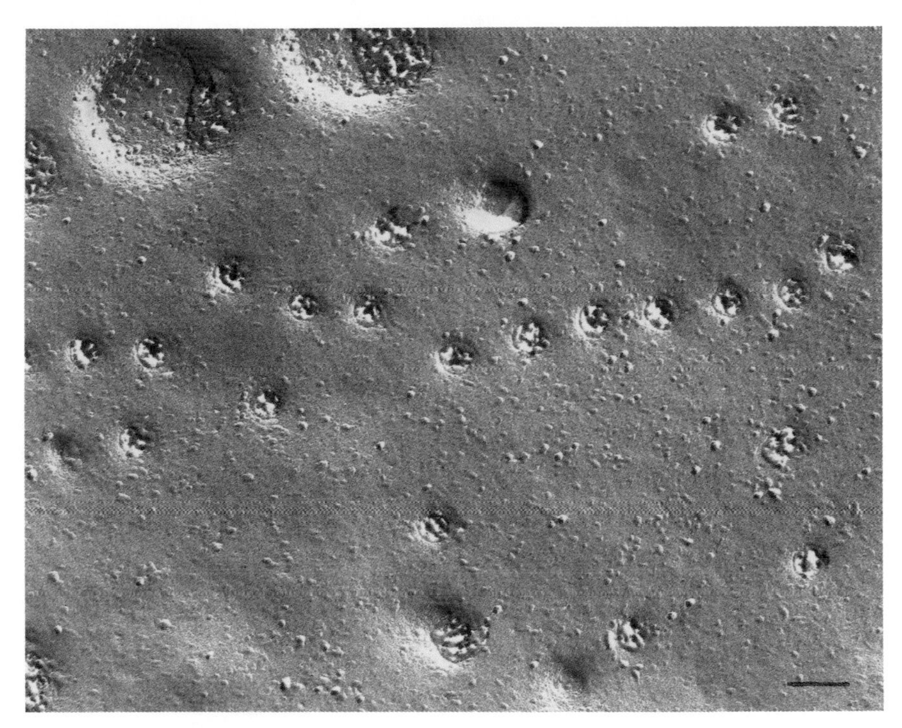

Figure 8.6. Freeze-fracture replicas of the outer face of the inner leaflet (the P face) of a fibroblast plasma membrane. Two coated pits in the upper left corner appear as craterlike depressions about 0.2 μ*M* in diameter. The more numerous smaller depressions (diameter ~ 0.05 μm) correspond to the small flask-shaped invaginations observed by Anderson *et al.* (1976) in transmission electron micrographs of fibroblasts. Bar = 0.1 μm (1000 Å); ×101,000. (From Orci *et al.*, 1978, with the permission of the authors.)

vesicles, as described below. However, McGookey *et al.* (1983) have shown that, although filipin does not react with coated vesicles derived from coated pits, when the coat dissociates to form uncoated endocytic vesicles their membranes bind filipin. This suggests either that the coat prevents filipin from reacting with free cholesterol present in normal amounts in the coated-vesicle membrane or that the membrane acquires cholesterol as soon as the coat is lost.

On normal fibroblasts, 50–80% of the LDL-ferritin particles bound at 4°C are present in coated pits, the other 20–50% occurring at random over the remaining 98% of the cell surface. Particles bound on noncoated regions of the surface of normal cells are presumably bound to LDL receptors that are not clustered in pits since, as noted below, receptor-negative homozygous FH fibroblasts bind no LDL-ferritin particles on coated or noncoated regions. A few particles are also attached to the proteinaceous matrix between the cells. This spurious binding

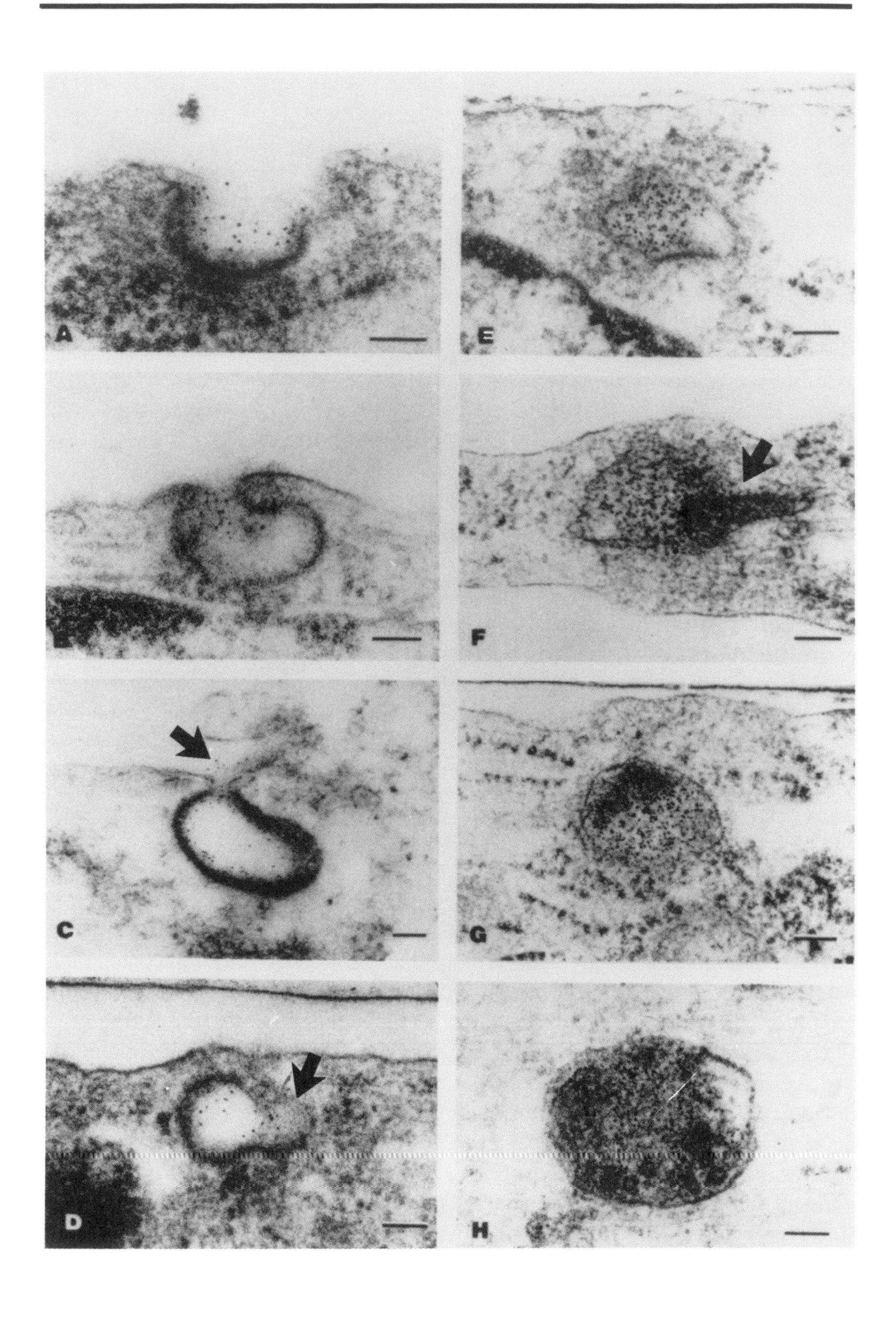
A
B
C
D
E
F
G
H

may account for the low-affinity, nonsaturable binding of LDL mentioned in the previous section.

Fibroblasts from FH homozygotes have an electron-microscopic appearance identical to that of normal cells. In particular, they contain coated pits similar in number, distribution, and ultrastructure to those on normal fibroblasts. However, when receptor-negative FH homozygote cells are incubated at 4°C with LDL-ferritin, no particles are bound to the cell surface, either in coated pits or on noncoated regions. Binding of particles to intercellular proteinaceous material is similar to that in cultures of normal cells.

If fibroblast monolayers that have bound LDL-ferritin at 4°C are warmed to 37°C, the coated pits invaginate into the cytoplasm within 1–3 minutes and become pinched off from the plasma membrane to form coated vesicles enclosing surface-bound particles. The coat is removed from the cytoplasmic face of the vesicle, either during invagination of the pits (Willingham and Pastan, 1980) or shortly thereafter (Anderson *et al.*, 1977a), to form noncoated vesicles called endocytic vesicles or receptosomes. Endocytic vesicles rapidly increase in size, possibly by fusing with each other, and migrate through the cytoplasm, eventually fusing with primary or secondary lysosomes. Within 10–15 minutes of warming to 37°C, LDL-ferritin particles appear within lysosomes. These events are shown sequentially in Fig. 8.7. It will be noted that the time taken for coated pits to invaginate and for endocytic vesicles to fuse with lysosomes on warming cells to 37°C coincides closely with the time taken for heparin-releasable [^{125}I]LDL to become internalized and for its lysosomal degradation to begin (see Section II,B above).

In their experiment on normal fibroblasts, Anderson *et al.* (1977a) noted that particles which bound to noncoated regions at 4°C disappeared from the cell surface within 10 minutes of warming to 37°C. Since LDL-ferritin was never seen to enter cells other than through coated pits, Anderson *et al.* suggested that

Figure 8.7. Electron micrograph showing representative stages in the endocytosis of ferritin-labeled LDL by cultured normal human fibroblasts, with subsequent appearance in a lysosome. The fibroblasts were incubated with the labeled LDL at 4°C to permit surface binding, washed to remove unbound labeled LDL, and then warmed at 37°C for various times. (A) A typical coated pit (seen after 1 minute at 37°C; ×77,500). (B) A coated pit being transformed into a vesicle enclosing labeled LDL particles (seen after 1 minute at 37°C; ×65,000). (C) Formation of a coated vesicle by complete invagination of the coated pit (seen after 1 minute at 37°C; ×43,500). (D) A fully formed coated vesicle that appears to be losing its cytoplasmic coat on one side (arrow) (seen after 2 minutes at 37°C; ×60,000). (E) An endocytic vesicle that has completely lost its cytoplasmic coat (seen after 2 minutes at 37°C; ×60,000). (F) An irregularly shaped endocytic vesicle that contains more labeled LDL than a typical coated vesicle and also has a region of increased electron density within the lumen (arrow) (seen after 6 minutes at 37°C; ×60,000). (G) An endocytic vesicle similar to that in (F) (seen after 6 minutes at 37°C; ×55,000). (H) A secondary lysosome that contains labeled ferritin (seen after 8 minutes at 37°C; ×60,000). (From Anderson *et al.*, 1977a, with the permission of the authors.)

LDL receptors on noncoated regions move laterally, together with their bound ligands, into coated pits and are then internalized when the pits invaginate. The suggestion that LDL receptors reach coated pits by lateral diffusion from non-coated regions was also in agreement with electron-microscopic observations on the fibroblasts of a patient (J.D.) carrying a very rare allele specifying a receptor that binds LDL but does not internalize it (the internalization-defective receptor). When fibroblasts from this patient are incubated at 4°C with LDL-ferritin, the particles are bound at random over the whole cell surface, with no clustering in coated pits. These findings were interpreted to mean that normal LDL receptors make their first appearance on the cell surface at random sites before moving into coated pits, whereas internalization-defective receptors, having reached the cell surface normally, are unable to carry their bound LDL into coated pits by lateral migration (Anderson *et al.*, 1977b). These suggestions are in line with the manner in which LDL receptors are now thought to cycle through the cell (see below).

Observations on fibroblasts from receptor-negative FH homozygotes at 37°C show that, despite their failure to bind any LDL-ferritin particles, the coated pits of these cells invaginate and form coated vesicles at about the same rate as normal cells. Moreover, coated pits on normal fibroblasts invaginate at the same rate whether or not LDL is present in the medium. Thus, it seems that at 37°C coated pits invaginate and are replaced, continuously and at a constant rate, irrespective of the binding of LDL to LDL receptors clustered within the pits. When fibroblasts in culture are treated with formaldehyde or (as noted above) are cooled to 4°C and are then exposed to LDL-ferritin, the particles are bound selectively in coated pits. Since lateral movement of receptors through the plasma membrane is abolished or markedly reduced in formaldehyde-fixed cells and in cells at 4°C, this shows that binding to LDL particles is not necessary for the movement of LDL receptors into a coated pit.

2. Localization with Antibodies

Immunohistochemical studies with antibodies to coat protein have provided independent evidence for the conclusions drawn from experiments with LDL-ferritin, in which, of course, only the location of the ferritin label is revealed directly by the electron microscope.

For the subcellular localization of structures associated with coat protein, Anderson *et al.* (1978) raised a rabbit antibody against the coat protein of ox-brain coated vesicles. For light microscopy, human fibroblasts made permeable with detergent were incubated with the antibody and were then treated with fluorescein-labeled goat anti-rabbit IgG. For electron microscopy, the anti-IgG antibody was coupled to horseradish peroxidase instead of the fluorescein label.

Fluorescence microscopy showed that the coated pits were arranged linearly in parallel with intracellular stress fibers lying just beneath the cell surface. A similar pattern was observed by fluorescence microscopy of fibroblasts that had been incubated at 4°C with LDL-ferritin, followed by treatment with a fluorescence-labeled antibody to ferritin. This observation confirmed the selective localization of LDL receptors within coated pits. [Note that the linear arrangement of coated pits and LDL receptors is not seen in all types of cells (Goldstein *et al.*, 1979a).]

Immunoperoxidase electron microscopy showed the presence of peroxidase-positive regions on the cytoplasmic faces of coated pits and on the outer aspect of vesicles similar in size and shape to coated endocytic vesicles (Fig. 8.8). These peroxidase-positive vesicles were shown to contain LDL-ferritin when the cells

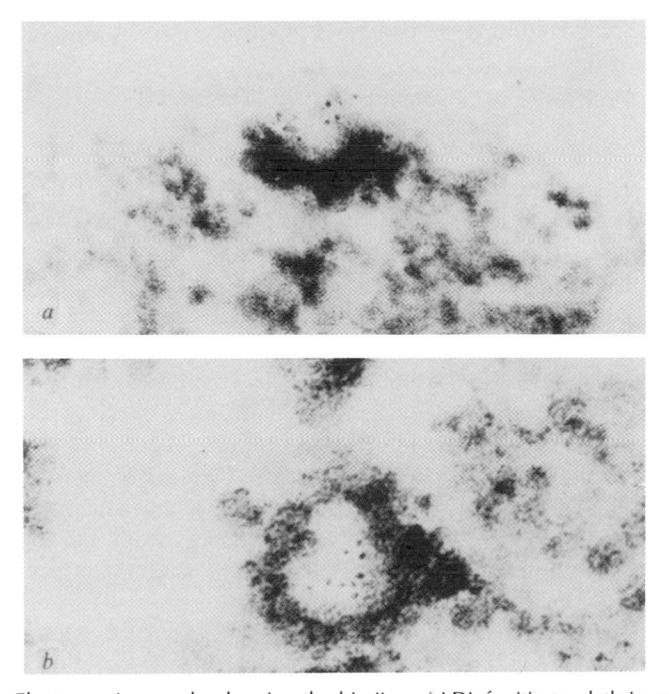

Figure 8.8. Electron micrographs showing the binding of LDL-ferritin to clathrin-containing regions of cell membrane. Monolayers of human fibroblasts were chilled to 4°C and incubated with LDL-ferritin for 30 minutes. The cells were then warmed to 37°C for 8 minutes, washed at 4°C to remove excess LDL-ferritin, fixed with 3% formaldehyde, and made permeable with 0.05% Triton X-100. The cells were then incubated first with rabbit anti-coat protein γ-globulin and then with goat anti-rabbit IgG coupled to horseradish peroxidase, followed by staining for peroxidase. (A) Ferritin was associated with an indented, peroxidase-positive segment of membrane that has the typical morphology of a coated pit, indicating that LDL-ferritin and the antibody to coat protein bind to the same region of surface membrane. (B) Ferritin cores contained within endocytic vesicles rimmed by peroxidase reaction product that delineates the clathrin coat. (×100,000) (From Goldstein *et al.*, 1979a, with the permission of the authors. Reprinted by permission from *Nature*, Vol. 279, pp. 679–685. Copyright © 1979 Macmillan Magazines Ltd.)

were incubated with LDL-ferritin before treatment with detergent (Goldstein *et al.*, 1979a).

Beisiegel *et al.* (1981b) have raised a monoclonal antibody (IgG-C7) to bovine LDL receptors that reacts with receptors on human fibroblasts. At 37°C this antibody is taken up and degraded by normal fibroblasts via the LDL-receptor pathway at a rate comparable with that for LDL. When IgG-C7 was incubated at 37°C for 1 hour with normal fibroblasts, the monoclonal antibody, localized by indirect immunofluorescence, was found to be distributed within the cell in vacuoles corresponding to lysosomes. No intracellular fluorescence was observed after incubating the antibody with FH homozygote cells. These observations provide additional evidence that ligands internalized by the LDL-receptor pathway are delivered to lysosomes.

D. Coated Vesicles, Coat Proteins, and the Coated Pit

1. General Significance

Roth and Porter, in a seminal paper published in 1964, have described how mosquito oocytes take up yolk protein from the external medium. When they examined oocytes with the electron microscope, the plasma membrane was seen to contain shallow pits with a bristle coat (about 200 Å thick) on the cytoplasmic surface and a layer of protein on the external face. Vesicles with bristle coats similar to those on coated pits were also seen immediately beneath the plasma membrane. When the mosquito took a blood meal there was a 15-fold increase in the number of coated pits on the oocyte surface, many of which appeared to be in various stages of invagination into the cytoplasm. There was also a marked increase in the number of vesicles, those nearest the plasma membrane retaining their coat, while those deeper into the cell had no coat and contained material of density similar to that in intracellular yolk granules. Roth and Porter (1964) concluded that yolk protein in the external medium is adsorbed selectively onto the external face of a coated pit and is then carried into the cell by invagination of the pit to form a coated vesicle. They proposed that coated vesicles lose their coat and then fuse to form protein-filled yolk bodies. They were unable to suggest any obvious function for the bristle coat in the formation of coated vesicles.

As we shall see in Section II,D,3 below, subsequent work has shown that the coated-pit/coated-vesicle pathway for the selective uptake of extracellular materials is used by a wide variety of eukaryotic cells, though the final stages of this pathway are not always the same. For example, the protein taken up by the mosquito oocyte is stored for future use, whereas LDL taken up via LDL receptors is delivered to lysosomes and degraded.

Coated vesicles, some much smaller than those described by Roth and Porter, have been detected in virtually all animal cells in which their presence has been sought and they have also been observed in plant cells (Mersey *et al.*, 1985). In many cases they arise by invagination of coated pits in the plasma membrane, as in the uptake of LDL by fibroblasts and of yolk protein by oocytes. But they have also been shown to carry out other functions, including the transport of enzymes from the Golgi apparatus to lysosomes, the secretion of protein by certain cells, and the recycling of surplus plasma membrane from presynaptic neurones (see Goldstein *et al.*, 1985). The widespread occurrence of coated pits and coated vesicles raises many questions concerning the coat. "What is its structure and chemical composition?" "How is it assembled and taken apart?" "What part, if any, does it play in the invagination of coated pits?" Considerable progress has been made toward answering these questions during the past few years.

2. Structure of the Coat

Among the earliest attempts to determine the structure of the coated vesicle was that of Kanaseki and Kadota (1969), who studied the electron-microscopic appearance of Golgi-derived coated vesicles from guinea-pig brain. They concluded that the vesicle coat is a polyhedral[3] lattice of regular hexagons and pentagons, forming a cage or basket that encloses the vesicle. One of the structures they proposed had 12 pentagons and 20 hexagons, with a total of 60 corners. Since a plane surface cannot be constructed from any combination of regular hexagons and pentagons, they suggested that while the coat is attached to the parent membrane it is formed entirely from regular hexagons and that as the membrane invaginates into the cytoplasm some hexagons are converted into pentagons to produce a convex surface. In their view, the driving force for the invagination of a coated membrane is the conversion of a limited number of hexagons into pentagons.

Crowther *et al.* (1976), by examining coated vesicles tilted at various angles in the electron microscope, were able to obtain a more accurate picture of the construction of the vesicle coat. The basic structure is an icosahedral[3] barrel constructed from polygons with side length 186 Å (Crowther and Pearse, 1981) (Fig. 8.9A). At each pole there is a regular hexagon surrounded by a ring of 6 regular pentagons and the two polar assemblies are separated by a ring of 6 puckered hexagons. (From geometric considerations the 6 hexagons could not be regular.) This gives a total of 20 polygons with 36 corners or vertices and 3 sides radiating from each corner. Larger cages are formed by the addition of more hexagons, as in the 32-faced cage shown in Fig. 8.9B. The molecular details of the coat are discussed below.

Heuser and Evans (1980) examined the cytoplasmic surfaces of coated pits of fibroblasts by deep etching after removal of the cytoplasm. Each coated region was seen to consist of a polygonal lattice formed predominantly from hexagons (Fig. 8.10), with an occasional pentagon adjacent to a heptagon (a "5-7 disloca-

[3]A polyhedron is a solid figure bounded by many faces. An icosahedron has 20 faces.

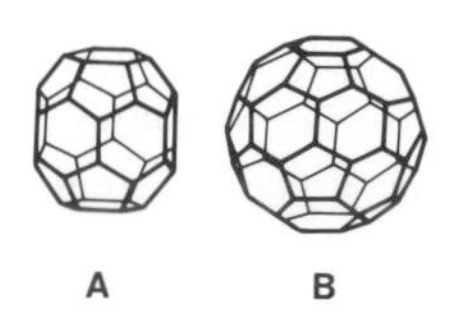

Figure 8.9. Structure of the cages forming the outer shells of coated-vesicle coats. The models are based on electron micrographs of tilted specimens of coated vesicles (Crowther *et al.*, 1976). (A) A barrel-shaped polyhedron constructed from 12 pentagons and 8 hexagons, with 1 hexagon at the top of the cage and one at the bottom. This structure has 36 corners and could be built from 108 clathrin molecules (36 triskelions). (B) A polyhedron constructed from 12 pentagons and 20 hexagons. This structure has 60 corners and would require 180 clathrin molecules. (From Pearse and Crowther, 1987. Reproduced, with permission, from the authors and from the *Annual Review of Biophysics and Biophysical Chemistry*, Volume 16, © 1987 by Annual Reviews Inc.)

tion'') (Fig. 8.11). The diameters of the coated regions ranged continuously from less than 100 Å to more than 3000 Å, suggesting that the coat is formed by progressive addition of coat material to a few polygons, rather than by deposition of a complete coat from a recycling coated vesicle. In agreement with this, the margins of the coats were composed of incomplete polygons, as shown in Fig. 8.11. Heuser and Evans noted that in three-dimensional views of coats the ratio of pentagons to hexagons increased as the degree of curvature increased, and that 5-7 dislocations tended to occur at the margins of convexities. They suggested that the rearrangement of two adjacent hexagons into a 5-7 pair is a transient stage in the process of coat curvature. This suggestion supports the earlier ideas of Kanaseki and Kadota but is not consistent with the mechanism proposed by Pearse and Crowther (1987), as discussed below. A circular pit of diameter 2500 Å could hold a maximum of not more than 100 close-packed LDL particles. This is consistent with the observation of Heuser and Evans (1980) showing 20–40 particles of diameter 250 Å per coated pit on the surfaces of fibroblasts.

3. Coat Proteins and the Triskelion

Pearse (1975, 1976) showed that the major constituent of the coat of coated vesicles from a variety of sources is a protein with apparent M_r 180K. She called this protein *clathrin* (Gr. *kleethra*, cage or lattice). Coated vesicles also contain smaller amounts of other proteins with apparent molecular weights of approximately 100K, 50–55K, 36K, and 33K (Pearse, 1978). In current terminology, the 180K protein is called clathrin heavy chain and the two low-M_r proteins (36K and 33K) are called clathrin light chains (LCa or Lα and LCb or Lβ). For reasons mentioned in Section E,3, the 100K proteins are called *adaptins*.

When coated vesicles are exposed to 2 *M* urea or to high concentrations of Tris at neutral pH, the coats dissociate into 8.4 S subunits which can be separated from the 100K and 50K–55K ''accessory'' proteins by gel filtration. Examina-

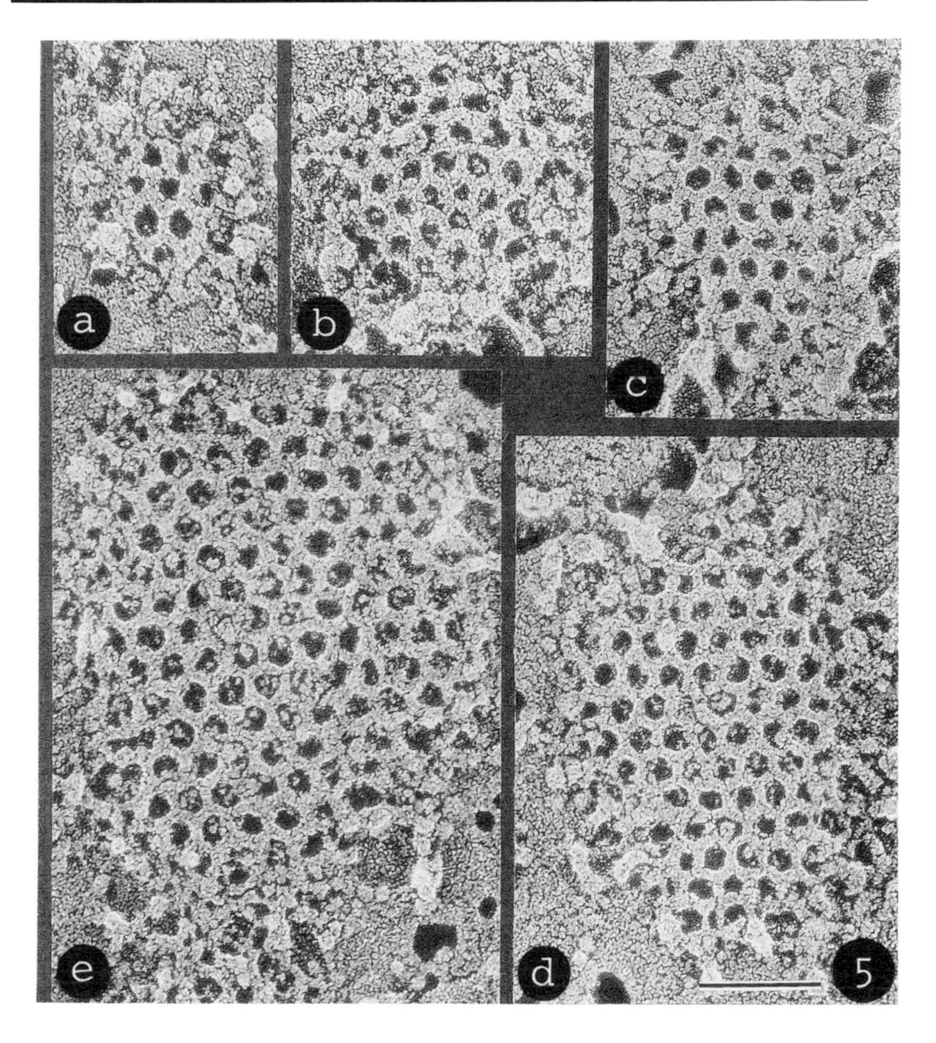

Figure 8.10. Electron micrographs of coated areas of fibroblast plasma membranes seen from the cytoplasmic surface. Hexagonal lattices of various sizes (increasing from *a* to *e*) are shown to suggest continuous, progressive growth. Note the incomplete polygons at the periphery of each lattice. The pictures were obtained by the quick-freeze, deep-etch, rotary-replication method. Bar = 1000 Å. (From Heuser and Evans, 1980. Reproduced from the *Journal of Cell Biology,* 1980, Vol. 84, pp. 560–583 by copyright permission of the Rockefeller University Press and with the permission of the authors.)

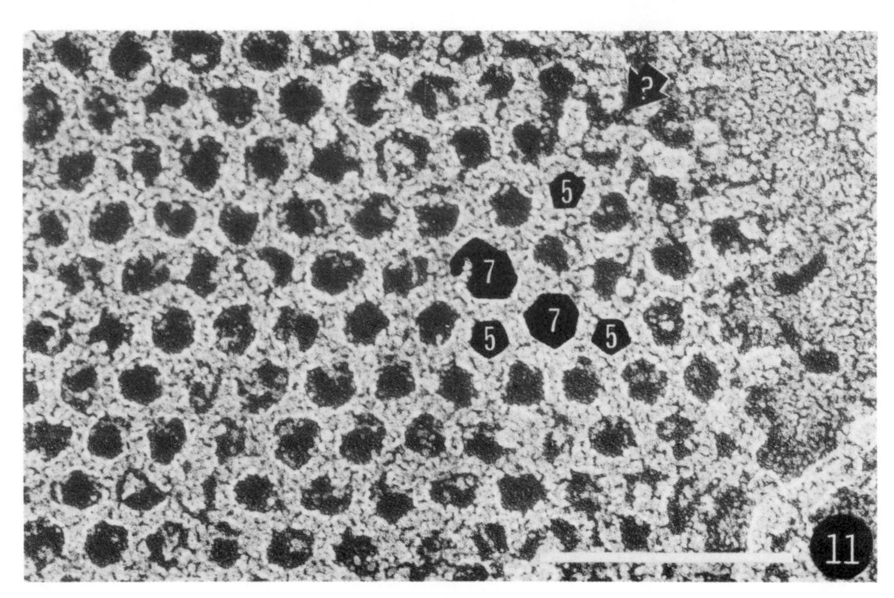

Figure 8.11. Electron micrograph of the edge of a coated pit seen from the cytoplasmic surface of a fibroblast. Most of the polygons in the lattice are hexagons, but three pentagons are also seen, two of which are adjacent to heptagons, forming 5–7 dislocations. Bar = 1000 Å. (From Heuser and Evans, 1980. Reproduced from the *Journal of Cell Biology,* 1980, Vol. 84, pp. 560–583 by copyright permission of the Rockefeller University Press and with the permission of the authors.)

tion of the 8.4 S sediment in the electron microscope shows that the subunit is a triskelion (Gr. *skelos,* leg) consisting of three legs radiating symmetrically from a common center (Ungewickell and Branton, 1981). Each leg is about 450 Å long and the three legs are bent in the same direction at about 160 Å from the center of the triskelion, as shown in Fig. 8.12. The composition and structure of the triskelion are now known in considerable detail (for references see Pearse and Crowther, 1987). Each triskelion is a trimer of three clathrin chains, with their C-terminal ends joined at the center, and a light chain bound tightly to the proximal (central) segment of each heavy chain. Both species of light chain are present in any population of triskelions and the distribution of LCa and LCb proteins within a given triskelion is random.

The amino acid sequence of rat clathrin heavy chains has been determined from a series of overlapping cDNAs corresponding to the complete message (Kirchhausen *et al.,* 1987a). The peptide chain has 1675 amino acids. Kirchhausen *et al.* have discussed the positions of functional domains in the clathrin heavy chain in the light of its amino acid sequence and of the fragments obtained by limited proteolytic digestion of triskelions. They suggest that a

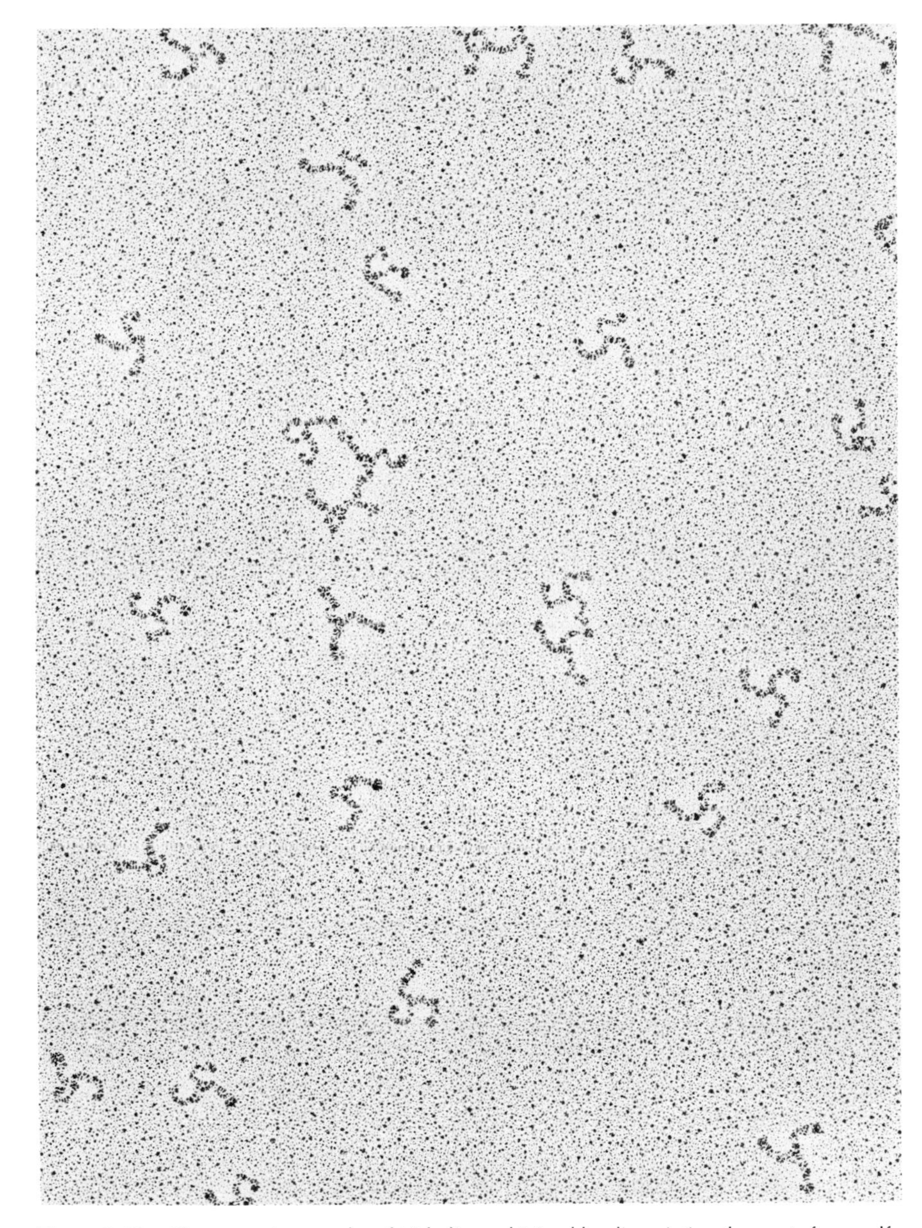

Figure 8.12. Electron micrographs of triskelions obtained by dissociating the coats from calf-brain coated vesicles in 2 *M* urea at pH 7.5. Note that the legs are bent in the same direction in all triskelions (clockwise in the view shown here). The molecular weight of triskelions determined by sedimentation equilibrium was 630K, consistent with the summed molecular weights of three 180K clathrin molecules and three 34K light chains. ×65,000 (From Ungewickell and Branton, 1981, with the permission of the authors. Reprinted by permission from *Nature*, Vol. 289, pp. 420–422. Copyright © 1981 Macmillan Magazines Ltd.)

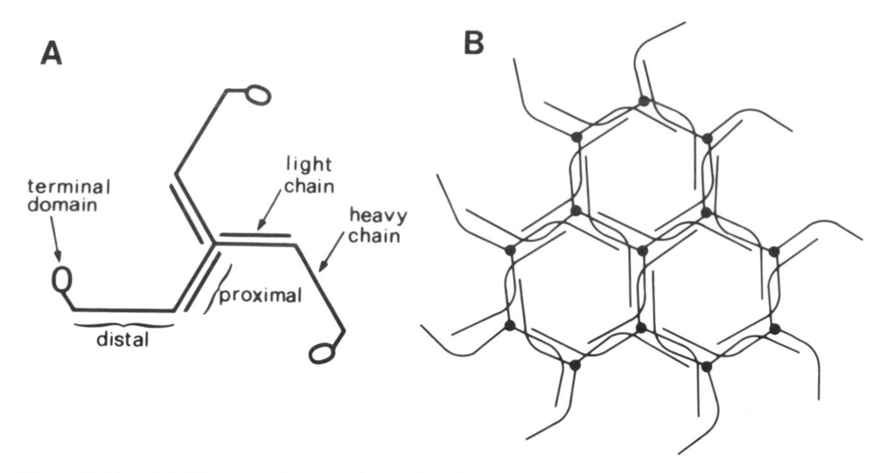

Figure 8.13. (A) Diagram showing the molecular structure of a triskelion. The total length of each arm is about 450 Å and the length of the light chain is about 160 Å. (B) Probable mode of packing of triskelion legs in the formation of a hexagonal cage. In the fully closed lattice each side of a hexagon is formed from the proximal segments of two legs and the distal segments of two other legs. In the packing diagram shown here, the proximal segments of legs from adjacent vertices cross over and lie between the distal segments of two other legs. This mode of packing is consistent with the electron-microscopic evidence of Crowther and Pearse (1981). The terminal domains of the triskelion legs have been omitted. (From Pearse and Crowther, 1987. Reproduced, with permission, from the authors and from the *Annual Review of Biophysics and Biophysical Chemistry,* Volume 16, © 1987 by Annual Reviews Inc.)

globular amino-terminal domain is joined to the distal segment (see Fig. 8.13) by a flexible linker sequence of 45 amino acids. They also suggest that a repeated sequence, Tyr-Gly-Gln-Pro-Gln, near the C terminus is involved in holding together the three legs at the vertex of a triskelion by noncovalent bonds.

The amino acid sequences of the two clathrin light chains (LCa and LCb) have also been determined from full-length cDNAs isolated from bovine and rat libraries (Jackson *et al.*, 1987; Kirchhausen *et al.*, 1987b). The two light chains have different, but homologous, sequences and are encoded in different genes, which probably arose by duplication from an ancestral gene. Jackson *et al.* (1987) have compared the LCs from bovine brain and lymphocytes. Each brain LC is identical to its nonbrain counterpart, except for the insertion of an additional amino acid sequence near the C-terminal end of the brain LC. LCa from brain, for example, has a 30-residue segment not present in nonbrain LCa. Examination of the nucleotide sequences of brain and nonbrain mRNAs for each LC shows that the differences in amino acid lengths are due to differential splicing of the primary RNA transcripts, resulting in the expression of exons in brain LCs that are spliced out in other tissues. Brodsky *et al.* (1987) have shown that monoclonal antibodies reacting with a segment of clathrin light chains comprising residues 93–157 block the binding of light chains by heavy chains. They conclude that this segment mediates the interaction between light and heavy chains.

4. Assembly of Triskelions and Accessory Proteins

Purified triskelions, when exposed to acid pH under appropriate conditions, associate rapidly to form cages with the characteristic lattice of vesicle coats (Crowther and Pearse, 1981). The cages formed under these conditions are heterogeneous in diameter. However, in the presence of the 100K and 50–55K proteins, triskelions associate at physiological pH and salt concentration to form cages of uniform diameter containing these proteins in definite molar ratios (Zaremba and Keen, 1983). This suggests that the accessory proteins participate in the assembly and stabilization of the vesicle coat in the intact cell. The role of the light chains may be to facilitate enzymic disassembly of coats (see below). Winkler and Stanley (1983) showed that the light chains could be removed from triskelions by exposure to NaSCN. Under standard reassembly conditions these heavy-chain trimers associated to form cage structures that were indistinguishable from those formed by the association of native triskelions. This suggests that the lattice of the vesicle coat is held together largely by interactions between the heavy chains, as in Fig. 8.13.

Crowther and Pearse (1981) have shown how the cage lattice of hexagons and pentagons with 186-Å sides is constructed from triskelions with legs of length about 450 Å. Each vertex of the polyhedral cage is the center of a triskelion and each leg of the triskelion extends along two adjacent polyhedral sides. As shown in Fig. 8.13, each side of a polygon in a completed polyhedron is formed by the *proximal* halves of two legs from adjacent vertices and the *distal* halves of two legs from vertices one step away. This remarkable arrangement, by which triskelions are held together in multiple noncovalent embraces, would give the coat its necessary combination of strength and flexibility. If each vertex in a polyhedral cage corresponds to one triskelion ($M_r \sim$ 642K), the icosahedral cage shown in Fig. 8.8A would have 36 triskelions and should have $M_r \sim 2.3 \times 10^7$. This is in close agreement with the value determined in the ultracentrifuge (2.2×10^7) by Crowther *et al.* (1976).

Vigers *et al.* (1986) have constructed three-dimensional maps from multiple electron micrographs of coated vesicles. Their findings indicate that the coat is a three-layered shell. The outer layer is formed by the lattice of heavy and light clathrin chains; the middle layer is formed by the terminal domains of the heavy chains projecting inward; the innermost layer is formed by the 100–50K accessory proteins in contact with the vesicle membrane. Pearse and Crowther (1987) suggest that the terminal domains of the heavy chains mediate the binding of the outer lattice to the accessory proteins.

5. Disassembly of Coats

As discussed in Section II,D,3, the vesicles formed by invagination of coated pits in plasma membranes lose their coats almost as soon as they are pinched off. The functional significance of such rapid uncoating is not clear. Presumably, loss of the coat must occur at some stage if the vesicle is to fuse with other intracellular membranes. Uncoating may also be necessary for the exposure of proteins in the vesicle membrane that enable the vesicle to be recognized by specific elements within the cell. As we have seen, the coats of coated vesicles can be dissociated nonenzymically into triskelions *in vitro,* and these can reassemble spontaneously under appropriate conditions. Hence, it would be reasonable to suppose that uncoating under physiological conditions leads to the release of triskelions into a cytoplasmic pool from which they can be recruited during the formation of a new membrane coat. However, very little is known about the mechanism of uncoating in intact cells. One suggestion is that disassembly is promoted by a rapid fall in pH within the vesicle (Forgac *et al.*, 1983). Rothman and co-workers (see Schmid and Rothman, 1985, for references) have described a cytosolic clathrin-dependent ATPase that catalyzes the release of triskelions from coated vesicles *in vitro.* Binding of the uncoating enzyme to the coat requires the presence of clathrin light chains in the triskelions of the cage lattice. The product of the uncoating reaction is an enzyme–triskelion complex that cannot reassemble to form cages. However, if this enzyme acts in the intact cell on the lattice covering a spherical vesicle, why should it not also act to dissociate the lattice covering the more planar surface of a coated pit and thus hinder the formation of coated vesicles?

6. The Generation of Coated Pits *in Vivo*

The function of coated pits on the plasma membrane is to mediate adsorptive endocytosis of macromolecules bound to specific receptors in the bilayer membrane of the pit. Each pit contains receptors of several types, each type characterized by high binding affinity for a particular class of macromolecules; most animal cells express LDL receptors, but some types of receptors predominate only in certain classes of cells. The small coated vesicles arising from Golgi membranes are also thought to originate in coated pits containing a characteristic set of receptors, including the mannose-6-phosphate (M6P) receptor that recognizes the M6P label on a lysosomal enzyme. All the receptors in plasma-membrane pits whose structure and orientation have been determined are transmembrane proteins, with an external ligand-binding domain and an internal domain extending into the cytoplasm (Goldstein *et al.*, 1985). Thus, the complete assembly of a coated pit must require, in addition to the formation of a coat, the clustering together of a heterogeneous set of receptors in the membrane.

In view of the tendency of triskelions to associate spontaneously into lattices it is natural to assume that the immediate precursor molecule of the coat lattice is a triskelion dissolved in the cytoplasm. Indeed, Brodsky (1985) has shown that immunoprecipitable triskelions are present in the cytosol of intact cells and that clathrin heavy chains associate to form triskelions as, or before, they are incorporated into coats. However, there is more to the generation of a complete coated pit than the deposition of a clathrin lattice on the plasma membrane. We also need to explain how the other elements of the pit are assembled, how the rate of assembly is regulated to maintain a more or less constant number of pits on the cell surface, and how the pattern of receptors in the pit is adjusted to the needs of the cell for particular macromolecules in the extracellular medium. Answers to some of these questions are now beginning to emerge, especially with regard to the mode of assembly of pits (see Pearse (1987) and Pearse and Crowther (1987) for references and discussion).

The electron-microscopic appearance of structures formed *in vitro* from purified coat proteins and M6P receptors suggests that the 100K proteins in the coats of native coated vesicles interact both with the clathrin lattice and with the cytoplasmic tails of receptors embedded in the bilayer of the membrane; the composition of reconstituted structures is consistent with a ratio of one receptor per molecule of 100K protein in coated vesicles in the intact cell (see Pearse and Crowther, 1987, for references). From these experiments *in vitro,* and from other evidence, Pearse and Robinson (1984) and Pearse and Crowther (1987) have put forward a model for the assembly of coated pits *in vivo*. They suggest that a coated pit begins with the attachment of 100K protein molecules to the cytoplasmic surface of the cell membrane, followed by interaction of the 100K proteins with the cytoplasmic extensions of receptors already diffusing at random in the plane of the membrane, and that capture of receptors is accompanied by deposition of a clathrin lattice on the membrane-bound 100K proteins. Presumably, growth of the pit by continuous extension from the first unit to be assembled would be favored by interaction between neighboring 100K molecules. On the basis of this model, Pearse and Crowther (1987) have discussed the factors that may determine the size, rate of formation, and receptor composition of coated pits. With regard to receptor composition, the 100K proteins are very heterogeneous (Pearse and Robinson, 1984), raising the possibility that each species of molecule binds to a specific receptor. As we shall see below, analysis of the J.D. mutation in the LDL-receptor gene has provided independent evidence for an interaction between the cytoplasmic domain of the LDL receptor and the coated-pit coat in intact fibroblasts.

Pearse and Crowther (1987) suggest that the inward convexity of pit membranes is induced by interaction between receptors and accessory proteins. Binding of clathrin would tend to stabilize the curvature of the membrane, since clathrin trimers associate spontaneously into closed baskets rather than into flat sheets. They suggest that the proportion of hexagons to pentagons in the clathrin lattice is determined by the degree of curvature of the membrane on which clathrin is deposited. This view contrasts with the scheme put forward by Kanaseki and Kadota (1969), in which the coat starts with a flat array of hexagons and is then changed into a convex structure by conversion of some hexagons into pentagons.

In the model proposed by Pearse and Crowther (1987), the cytoplasmic domains of receptors participate actively in the lateral extension of a growing coated pit. In support of this idea, Iacopetta *et al.* (1988) have demonstrated a positive correlation between the number of transferrin receptors expressed at the surface of a mouse L-cell and the density of coated pits on the cell surface. No such correlation has been demonstrated for LDL receptors on fibroblasts. Moreover, the observations of Anderson *et al.* (1977b) on J.D.'s receptors indicate that an LDL receptor with a normal cytoplasmic domain moves into a coated pit that has already been assembled.

E. Recycling of LDL Receptors and the Sorting Problem

1. Evidence for Recycling

The rate at which fibroblasts internalize receptor-bound LDL at 37°C is equivalent to the endocytosis of all the LDL receptors on the cell surface about once every 12 minutes. Despite this rapid rate of entry into the cell, the number of receptors on the fibroblast surface declines by only 50% in 20–25 hours when the synthesis of new receptors is inhibited by cycloheximide (for references see Goldstein *et al.*, 1979a). Since there is no reservoir of LDL receptors in the fibroblast (Basu *et al.*, 1978) this means that each LDL receptor must recycle to the cell surface many times before it is degraded. Basu *et al.* (1981) estimate that each receptor recycles from cell surface to cell interior and back again about 100 times every 20 hours.

Evidence of a similar kind indicates that certain other receptors that enter cells via coated pits (including the receptors for asialoglycoproteins and peptide hormones) also recycle back to the cell surface, usually after dissociating from their bound ligands before the ligand is delivered to lysosomes. Other receptors, such as that for epidermal growth factor (EGF), do not dissociate from their ligand and are degraded, together with the ligand, in lysosomes. These two routes, one leading to return of the receptor to the cell surface and the other to its degradation in lysosomes, are exemplified by the behavior of LDL receptors bound to anti-receptor antibodies.

As noted in Section II,C above, monoclonal anti-receptor antibody IgG-C7 is taken up and degraded by fibroblasts via the LDL-receptor pathway. Beisiegel *et al.* (1981b) have shown that when fibroblasts are incubated with this antibody at 37°C, the antibody continues to be degraded for several hours without any decrease in the number of LDL receptors on the cell surface. This indicates that internalized LDL receptors dissociate from the monoclonal antibody and return to the surface. When fibroblasts are incubated with a polyclonal antireceptor antibody the results are quite different (Anderson *et al.*, 1982). The antibody is rapidly internalized, but LDL receptors disappear from the cell surface within 15 minutes. If the antibody is removed from the incubation medium, receptors return to the surface over a period of several hours. This must be due to synthesis of new receptor protein, since the reappearance of receptors on the cell surface is inhibited by cycloheximide. Evidently, the polyclonal antibody prevents recycling of LDL receptors. Anderson *et al.* (1982) suggest that this is because the antibody induces extensive cross linking of receptors.

2. The Site of Receptor–Ligand Dissociation

In those cases in which the internalized receptor returns to the cell surface without its ligand, the question arises as to where dissociation of the receptor–ligand complex takes place within the cell. The intracellular route taken by LDL receptors in fibroblasts has been investigated with monensin, a drug that inhibits the transport of Golgi-derived vesicles to the plasma membrane (see Brown *et al.*, 1982, for references). When fibroblasts in culture are incubated in the presence of monensin and LDL, the number of receptors on the cell surface declines by about 75% in 15 minutes owing to trapping of receptors within the cell. Immunofluorescence studies in which both LDL receptors and LDL particles are visualized within cells have shown that LDL receptors and their LDL ligands are trapped within the same vesicles in monensin-treated fibroblasts. These vesicles have not been identified unequivocally, but there is a considerable body of evidence (reviewed in Brown *et al.*, 1983) to suggest that dissociation of receptor–ligand complexes, including that of the receptor–LDL complex, normally takes place in endocytic vesicles and that dissociation is induced by a rapid fall in pH within the vesicles. Davis *et al.* (1987) have produced a line of hamster ovary cells that synthesizes, as its sole LDL receptor, a protein lacking the whole of the region homologous to a sequence of 400 amino acids in the human EGF precursor (see Chapter 9 for details). When this receptor internalizes a lipoprotein ligand, the receptor fails to return to the cell surface because it cannot dissociate at the normal rate from its bound ligand at acid pH.

Since intact LDL receptors return to the plasma membrane, they must leave the endocytic vesicle before it fuses with a lysosome. Geuze *et al.* (1983) have investigated the recycling of asialoglycoprotein receptors in rat hepatocytes *in vivo*. Using a double immunolabeling technique, they have shown that the receptors and their ligands dissociate in vesicles with long tubular extensions. They called this vesicle the *compartment* of *uncoupling* of *receptor* and *ligand* (CURL). In some CURLs the receptors were confined to the membrane of the extensions, and the ligand was present mainly in the lumen of the vesicle. Geuze *et al.* suggest that the tubular extensions, carrying membrane-bound receptors, bud off and are transported to the plasma membrane, possibly after acquiring a clathrin coat, and that the vesicle then fuses with a lysosome (see Fig. 8.14). Pathak *et al.* (1988) have shown that LDL receptors internalized from the surface of a normal fibroblast enter multivesicular bodies with irregular projections. If these projections correspond to the recycling vesicles postulated by Geuze *et al.*,

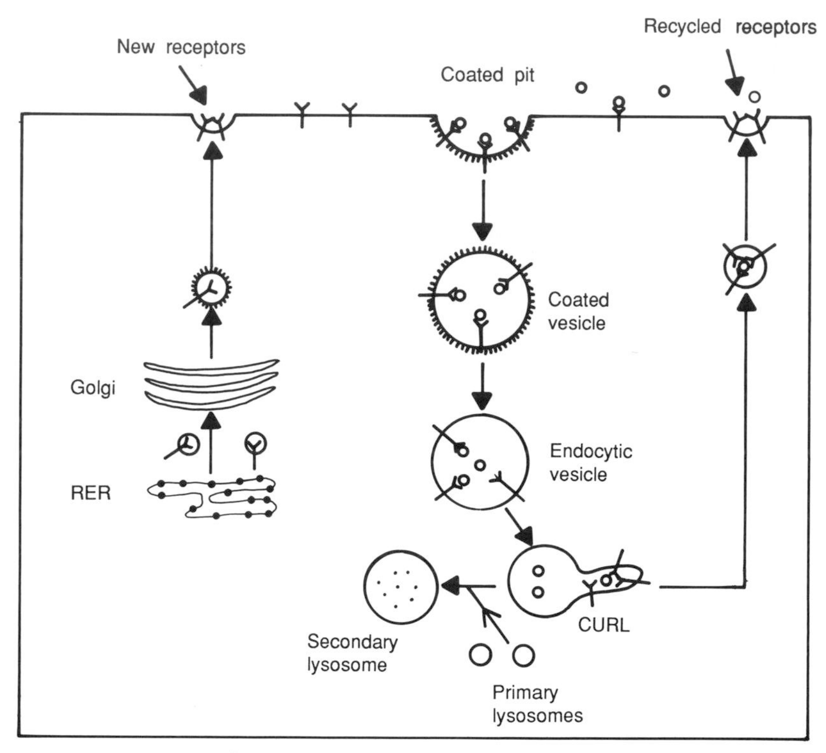

Figure 8.14. Diagram to show the probable sequence of events in the LDL-receptor pathway. The right-hand side shows the endocytosis of LDL (○) bound to LDL receptors (**Y**) that have clustered in a coated pit, the recycling of internalized LDL receptors via CURL (see text), and the delivery of LDL particles to a lysosome in which the particles are degraded. The left-hand side shows the synthesis of LDL receptors in the rough endoplasmic reticulum (RER), their transport to Golgi cisternae (possibly in vesicles) for processing, and the transport of a mature LDL receptor in a small coated vesicle for insertion into the plasma membrane. Most of the receptors in noncoated regions of the plasma membrane will eventually cluster in coated pits. Goldstein *et al.* (1979a) estimate that the time taken for one LDL receptor to undergo a complete cycle is about 12 minutes, of which 6 minutes are spent in a coated pit, 3 minutes within the cell, and 3 minutes in diffuse distribution on the plasma membrane. Note that in the scheme shown here, newly synthesized receptors are transported in vesicles from the RER to the Golgi and from the *trans* face of the Golgi to the plasma membrane. Note also that receptors remain membrane bound during endocytosis and recycling and that their orientation with respect to the cytoplasm is maintained throughout the cycle.

then this would explain why LDL receptors spend so little time within fibroblasts after endocytosis via coated pits (see legend to Fig. 8.14).

As noted above, when an LDL receptor fails to dissociate from its ligand (as with LDL receptors bound to polyclonal antibodies) both receptor and ligand usually undergo lysosomal degradation. However, it does not follow that dissociation of ligand from receptor is a necessary condition for recycling of receptors via CURL tubules. The transferrin receptor recycles without dissociating from its ligand (Octave *et al.*, 1981), and Gavigan *et al.* (1988) have shown that a monoclonal antibody to the LDL receptor (10A2) returns to the cell surface, presumably bound to receptor, after endocytosis by fibroblasts.

Aulinskas *et al.* (1981, 1985) have shown that up to 10% of the LDL endocytosed via the LDL-receptor pathway in fibroblasts and other cells is regurgitated rapidly into the external medium, essentially without physical or immunochemical change. They suggest that "retroendocytosis" of LDL is due to trapping of LDL particles in the tubular extensions of CURL, so that some internalized LDL that has not been exposed to lysosomal enzymes returns to the plasma membrane and is discharged into the external medium by exocytosis.

Figure 8.14 shows a working model for the LDL-receptor pathway in fibroblasts, including the recycling and synthesis of receptors, based on the evidence discussed above and in Chapter 9.

3. Sorting of Receptors and Other Membrane Proteins

The clustering of cell-surface receptors of different classes in coated pits must involve some mechanism for separating receptor proteins from those "resident" proteins that never leave the plasma membrane. At various stages after endocytosis there must also be separation of different classes of receptors and, in some cases, of receptors from their ligands. This is so because, as noted above, not all receptors in a given coated vesicle are destined to follow the same intracellular route. Some examples of the different routes followed by cell-surface receptors have already been mentioned. They include the routes taken by (1) receptors for LDL and asialoglycoproteins, both of which leave their ligands, probably in CURL vesicles, and return to the plasma membrane; (2) the EGF receptor, which is delivered to lysosomes and degraded with its ligand; (3) the maternal-antibody receptor in intestinal cells of newborn rats, which carries ingested antibodies from the luminal to the abluminal face of the cell in vesicles that do not fuse with lysosomes; and (4) the yolk-protein receptor in oocytes, which is carried in vesicles that deliver yolk protein to storage inclusions.

The question as to how surface receptors are separated initially from other membrane proteins, and are then separated into different classes wherever there is a divergence of pathways, is part of the general problem of how each protein is directed sequentially to its specific targets in the cell. This, one of the more challenging questions in current cell biology, is sometimes referred to as the *sorting problem.* If it were not for the efficient sorting of proteins during their transport from one organelle or membrane to another, the distribution of membrane proteins would eventually become randomized throughout the cell.

Pearse and Bretscher (1981) and Goldstein *et al.* (1985) have discussed what is required of an efficient system for guiding proteins through a sequence of diverging intracellular pathways to their final destinations. One requirement is a mechanism for concentrating specific membrane proteins into a restricted region of the membrane that can then bud off to form a transport vesicle. In this regard, the coated pit seems to play a general role as a molecular filter, retaining those membrane proteins whose cytoplasmic extensions it recognizes and ignoring those it does not. Coated pits are now thought to act in this way in Golgi membranes as well as in the plasma membrane. Thus, it seems likely that newly synthesized receptors for LDL and for other ligands, after being processed in the Golgi apparatus, are concentrated in coated pits which bud off to form vesicles that carry the mature receptors to the plasma membrane.

Since receptors of many different classes cluster in coated pits, it would be reasonable to expect them to bear a common label recognized by some element in the protein coats of coated pits (see Section II,D,6 above). With this in mind, Goldstein *et al.* (1985) have compared the amino acid sequences of the cytoplasmic extensions of the LDL receptor with those of five other receptors. The only common feature they could discern was the presence of one or more cysteine residues in the cytoplasmic region, suggesting that interchain disulfide bonds play a role in the clustering of receptors in coated pits. However, Davis *et al.* (1986) have shown that a tyrosine or other aromatic residue at position 807 in the cytoplasmic domain of the LDL receptor is required for efficient clustering in coated pits (see Chapter 9, Section II,D for details). Lazarovits and Roth (1988) have also shown that replacement of serine by tyrosine (but not by phenylalanine) at position 543 in the cytoplasmic domain of influenza virus hemagglutinin changes the hemagglutinin from a protein that is largely excluded from coated pits to one that is endocytosed rapidly via coated pits. According to Lazarovits and Roth (1988), tyrosine at a specific position in the cytoplasmic domain is also required for clustering of poly-Ig receptors in coated pits. Taken together, these observations suggest that tyrosine (or, in the case of the LDL receptor, another aromatic residue) at a specific position in the cytoplasmic domain is an essential component of a recognition signal that facilitates endocytosis of many transmembrane proteins via coated pits.

As noted below, Herz *et al.* (1988) have identified an LDL-receptor-related protein in liver. The cytoplasmic domain of this protein has two copies of the sequence N-P-X-Y, where X is either T or V (see Abbreviations list). Tyrosine 807 in the cytoplasmic tail of the LDL receptor is also the last amino acid in the sequence N-P-V-Y. If it turns out that the protein described by Herz *et al.* is endocytosed via coated pits, then this would suggest that tyrosine is a component of a conserved tetrapeptide sequence that is recognized by one or more proteins in the coated pit and, hence, that the sequence acts as a clustering signal. In support of this suggestion, Pearse (1988) has shown that LDL receptors and other transmembrane proteins containing the N-P-X-Y consensus sequence bind specifically to 100/50 kDa complexes present in the coated pits of plasma membranes. Thus, the 100/50 kDa proteins appear to act as adaptors that mediate the binding of receptors to the clathrin lattice of coated pits. Presumably, adaptor proteins in plasma-membrane coated pits contain an amino-acid sequence that recognizes the N-P-X-Y clustering signal in the cytoplasmic tails of certain transmembrane proteins. Because the ≃100 kDa proteins form the bulk of the adaptor unit, Pearse (1988) suggests calling them *adaptins*. It should be noted that the 100/50 kDa proteins present in coated pits of Golgi membranes do not bind LDL receptors but do bind receptors that recognize the M6P label on lysosomal enzymes (see above). This could provide the basis for the sorting of Golgi proteins (such as the newly-formed mature LDL receptor) that are destined for the plasma membrane from those that are to be transported to lysosomes.

While it seems clear that the coated-pit coat is essential for the initial sorting of receptors in the plasma membrane, the sorting that takes place after uncoating must be mediated by other recognition signals in the receptors and in the vesicles in which they become segregated.

III. LDL Receptors in Nonfibroblast Cells

A. Methods of Detection and Measurement

LDL receptors are detectable in most animal tissues, although their activity differs widely from one cell type to another. Various methods have been used to establish the presence of LDL-receptor activity in specific tissues *in vitro* and *in vivo*. These methods, some of which have already been referred to, are summarized below.

Intact cells in culture or suspension are assumed to express LDL-receptor activity if they bind LDL with all the characteristics exhibited by LDL binding by human skin fibroblasts in culture. An additional test for LDL-receptor activity in normal human cells is the absence of characteristic LDL binding by the corresponding cells obtained from a receptor-negative FH homozygote. By these criteria, LDL-receptor activity has been demonstrated in smooth-muscle cells and endothelial cells of arterial wall, monocyte-macrophages, cells of the adrenal cortex and ovary, lymphocytes, adipocytes, and hepatocytes, as well as in skin fibroblasts from many species. Using the same criteria for the presence of LDL-receptor activity, Kovanen *et al.* (1979a) have identified receptors in subcellular membranes prepared from fresh bovine tissues, including adrenal cortex, ovary, liver, small intestine, and kidney. Polyclonal (Beisiegel *et al.*, 1981a) and monoclonal (Beisiegel *et al.*, 1981b) anti-receptor antibodies have been used to provide independent evidence for LDL-receptor activity in isolated intact cells or subcellular membranes by showing that the antibodies compete specifically with LDL for high-affinity binding. Immunoblot analysis[4] with a monoclonal antibody to the receptor has also been used to detect LDL-receptor protein in tissue extracts (Beisiegel *et al.*, 1982). In addition, the amount of LDL-receptor mRNA in a particular tissue may be estimated *in vitro* by Northern blot hybridization with a radioactive cDNA probe. (See, for example, Russell *et al.*, 1983; Hofmann *et al.*, 1987.) The use of ligand blotting in the study of the LDL-receptor protein is mentioned in the next chapter.

As we have seen (Chapter 7), in experimental animals LDL-receptor activity may be detected and quantitated in particular tissues *in vivo* by comparing the uptake of native LDL with that of chemically modified LDL. The distribution of active receptors *in vivo* has also been investigated with anti-receptor antibodies, using two different approaches. In the first, tissues are identified in which uptake of LDL is inhibited by pretreating the animal with a polyclonal antibody to its LDL receptors (Kita *et al.*, 1981); in the second approach, tissue uptake of a labeled monoclonal anti-receptor antibody is measured. Using the latter method, Huettinger *et al.* (1984) have demonstrated high levels of LDL-receptor activity per gram of tissue in the adrenals and liver of the rabbit.

Taken as a whole, the results obtained with the methods described above have provided clear evidence that LDL-receptor activity is expressed by the cells of many tissues *in vivo*. There is also good agreement between *in vivo* and *in vitro* estimates of the relative levels of LDL-receptor activity in different tissues. In particular, the results from several independent methods for measuring receptor activity *in vivo* and *in vitro* agree in showing relatively high levels in the liver and adrenal cortex or whole adrenal.

The LDL receptors synthesized in different tissues are encoded in the same gene. This is shown by the fact that in the presence of the mutation responsible for FH, LDL receptors are defective in all tissues in which the receptor gene is normally expressed. Moreover, from a consideration of the way in which the domain structure of the LDL receptor is adapted to clustering in coating pits (see next chapter), it seems certain that LDL receptors, in whatever cells they are expressed, undergo the cycle of events observed in skin fibroblasts. Nevertheless, there are aspects of the way in which LDL receptors function in certain specialized tissues, especially with regard to regulation, that merit discussion. Accordingly, LDL receptors in hepatocytes, lymphocytes, macrophages, and adrenal-cortex cells are discussed below. This will also provide a suitable opportunity for dealing with other lipoprotein receptors whose function is relevant to that of the LDL receptor.

[4]For analysis by immunoblotting (Western blotting) a tissue extract containing the proteins of interest is submitted to electrophoresis on polyacrylamide-gel slabs. The protein bands are then blotted onto nitrocellulose filters and are localized by incubating the filters with the first antibody, followed by incubation with a labeled second antibody to the first antibody.

B. Hepatic Receptors for LDL and Other Lipoproteins

1. Evidence for LDL Receptors in the Liver

For reviews and discussion, see Myant (1982), Brown and Goldstein (1983a), and Havel (1986). As we saw in Chapter 7, observations on the tissue uptake of labeled LDL *in vivo* have shown that in many species the liver makes a larger contribution to total LDL catabolism than any other organ and that hepatic catabolism of LDL occurs predominantly via the LDL-receptor pathway. Attempts to determine which cells are responsible for hepatic catabolism of LDL *in vivo* have led to inconsistent findings (see Chapter 7). However, the electron-microscopic studies described below indicate that LDL is taken up in the intact liver mainly by hepatocytes, with little contribution from nonparenchymal cells. Observations on perfused livers, liver membranes, and hepatocytes in culture have also shown that liver cells from animals, in both the stimulated and unstimulated state, are capable of expressing LDL-receptor activity. The LDL receptors in liver membranes from rats in which receptor activity is enhanced by estrogen treatment show high affinity for HDL_c, VLDL remnants, apoE-phospholipid vesicles, and VLDL secreted by perfused livers, as well as for LDL. LDL receptors have also been detected in liver proteins by immunoblotting (Huettinger *et al.*, 1984) and ligand blotting (Wade *et al.*, 1985). Failure to demonstrate LDL receptors on freshly prepared hepatocytes, reported from several laboratories, may have been due to removal of surface receptors during preparation of the cells.

Mahley *et al.* (1981) were unable to detect LDL-receptor activity in liver membranes prepared from adult dogs or human subjects, though receptor activity was present in liver membranes from immature dogs and from mature dogs fed cholestyramine. However, Harders-Spengel *et al.* (1982) and Hoeg *et al.* (1984) have reported LDL-receptor activity in liver membranes from adult human subjects. Moreover, LDL receptor protein has been identified in liver samples from adult human subjects (Soutar *et al.*, 1986) and adult dogs (Wade *et al.*, 1986) by ligand blotting with biotin-modified LDL as ligand.

2. Endocytosis of Lipoproteins in the Liver: Ultrastructural Aspects

The coated-pit/coated-vesicle pathway for adsorptive endocytosis of specific macromolecules functions in hepatocytes in a manner similar to that described in other cells. (For review, see Wileman *et al.*, 1985.) The ultrastructural basis of this process may be seen in the eye-catching electron micrographs of Hirokawa and Heuser (1982) (Fig. 8.15). Other workers have studied the time-course of endocytosis of labeled lipoprotein particles by hepatocytes *in vivo* or in perfused livers.

Handley *et al.* (1981) investigated the uptake of gold-labeled LDL by perfused livers from normal and estrogen-treated rats. After perfusion for 15 minutes,

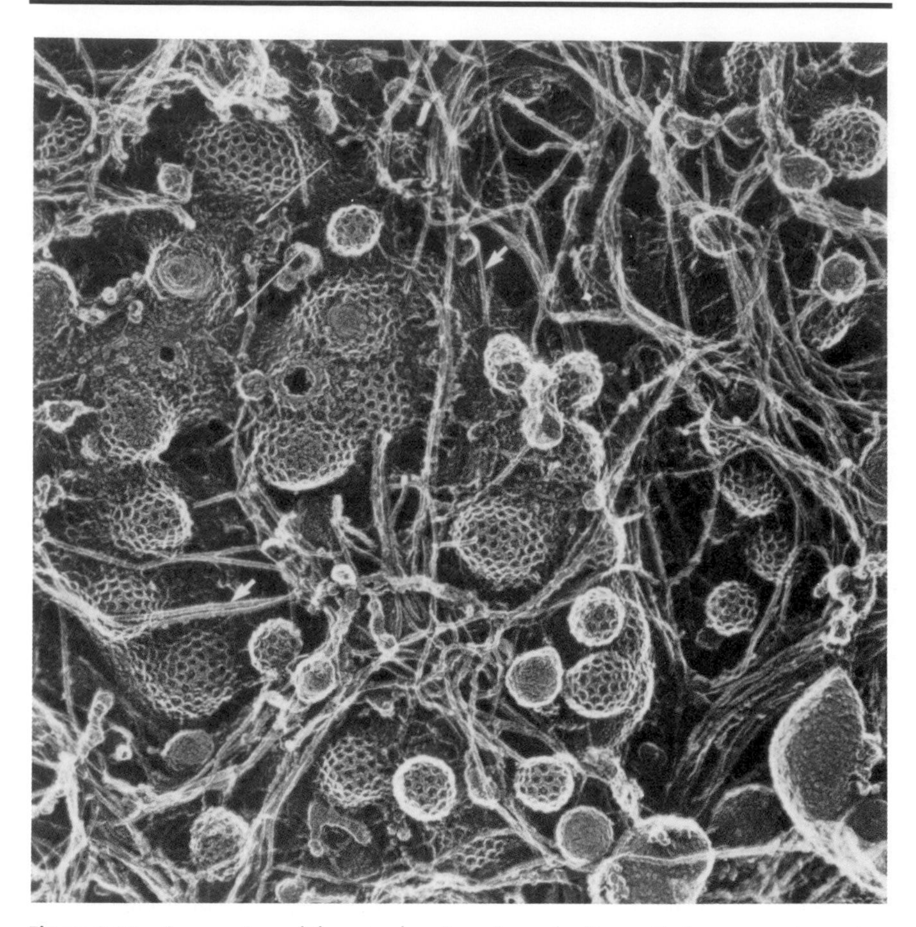

Figure 8.15. Survey view of the cytoplasmic surface of a liver-cell plasma membrane in a nonjunctional area. Numerous coated pits and intermediate filaments (short thick arrows) are seen, as well as a few thinner filaments that may be active (long thin arrows). ×65,000. (From Hirokawa and Heuser, 1982, with the permission of the authors.)

electron microscopy showed that most of the labeled LDL in the liver was bound to the surfaces of the microvilli of parenchymal cells and that the number of particles bound in treated livers was about 10 times that in normal livers. Bound particles appeared to be distributed more or less evenly, with no clustering in the coated pits, most of which were confined to the bases of the microvilli. At later intervals, labeled LDL particles were seen in coated vesicles and multivesicular bodies[5] (MVB). Internalization of labeled LDL particles by Kupffer cells was observed only in livers from estrogen-treated rats.

[5]Multivesicular bodies are large vesicles (5,000–10,000 Å in diameter) present in many types of cells in the region of the Golgi apparatus. They contain numerous small vesicles, usually with

Chao *et al.* (1981) followed the fate of ^{125}I-labeled LDL taken up by the liver *in vivo,* or by perfused livers, from estrogen-treated and normal rats. The LDL particles were localized by autoradiography. Again, LDL was bound initially to the surfaces of the microvilli, with no clustering in coated pits. At later intervals, photographic blackening appeared sequentially over endocytic vesicles and then over MVB, within which numerous LDL particles (diameter 200–230 Å) could be seen. Very few grains were seen over Kupffer cells at any time after exposing the liver to labeled LDL. Subsequently, Hornick *et al.* (1985) were able to isolate and characterize MVB from the livers of estrogen-treated rats. MVB could be distinguished chemically and morphologically from the smaller, Golgi-derived vesicles containing newly synthesized VLDL particles on their way to the plasma membrane for secretion into the space of Disse. Figure 8.16 shows an isolated MVB packed with bilayer vesicles and with spherical particles 200–850 Å in diameter representing LDL and VLDL-and chylomicron-remnants.

Absence of clustering of gold-labeled and ^{125}I-labeled LDL particles in coated pits on the hepatocyte plasma membrane may reflect a genuine difference in the initial stage of adsorptive endocytosis in liver cells and fibroblasts. It is possible, for instance, that LDL receptors on hepatocytes do not move into pits until they have bound their ligand. On the other hand, the number of pits on the cell surface at 37°C would be far fewer than that at 4°C (the temperature at which clustering has usually been demonstrated in fibroblasts) owing to continuous invagination of pits at the higher temperature.

Jones *et al.* (1984) have shown that chylomicron remnants and VLDL remnants taken up by the liver follow a route similar to that followed by LDL. Remnants of both triglyceride-rich lipoproteins are bound to the plasma membranes of hepatocytes, with little or no binding to Kupffer cells. Subsequently, the surface-bound remnant particles appear in endocytic vesicles and then in MVB. Thus, although chylomicron remnants and LDL are thought to be bound by different receptors on the hepatocyte plasma membrane (see below), the two lipoproteins are endocytosed by the same pathway because they are both internalized by invagination of coated pits. The finding that remnants are taken up predominantly by hepatocytes, with little or no uptake by Kupffer cells, is in agreement with the earlier electron microscopic observations of Stein *et al.* (1969, 1980) on the uptake of labeled chylomicrons by the liver in the intact rat.

3. Hepatic Receptors for Remnant Particles

The remnants of triglyceride-rich lipoproteins are taken up and catabolized in the liver by high-affinity, saturable processes. This has been established by observa-

---

bilaminar limiting membranes (see Fig. 8.16), which probably arise by invaginations from the membrane of the parent vesicle (McKanna *et al.*, 1979). MVB are thought to arise by fusion of several endocytic vesicles and to represent the last step in the endocytic pathway before fusion with a primary lysosome.

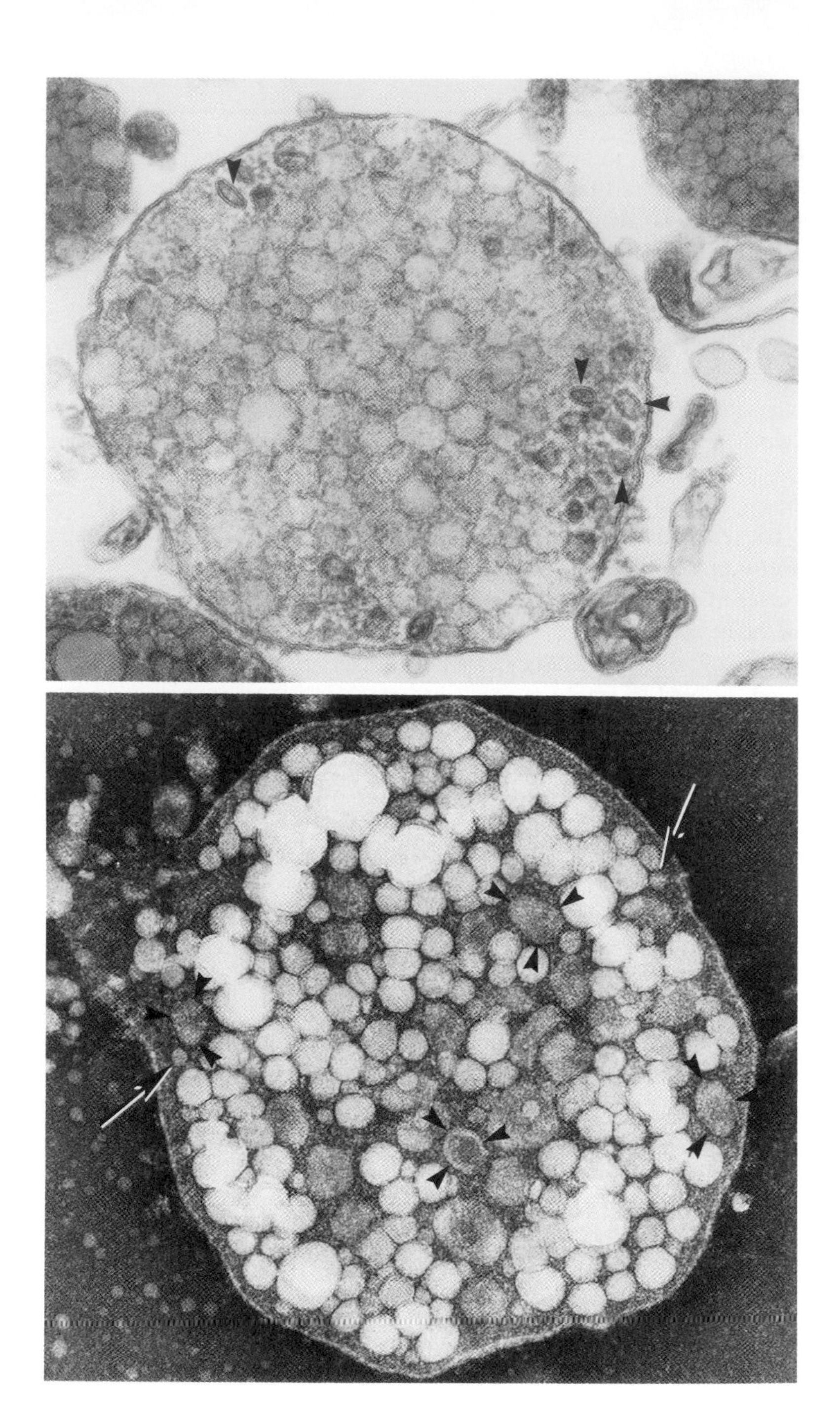

tions on intact rats (Andersen *et al.*, 1977), isolated perfused liver (Sherrill and Dietschy, 1978; Windler *et al.*, 1980; Arbeeny and Rifici, 1984), rat-liver cells in culture (Florén and Nilsson, 1977), and liver membranes (Carella and Cooper, 1979).

Both chylomicron remnants (Florén *et al.*, 1981) and VLDL remnants (references in Chapter 7) are recognized by the LDL receptor. However, there is a good deal of evidence to indicate that in intact animals chylomicron remnants are catabolized via hepatic lipoprotein receptors that are distinct from the LDL receptor. The existence of a separate chylomicron receptor in the liver was first suggested by the observation that in FH homozygotes and in homozygous WHHL rabbits neither chylomicrons nor chylomicron remnants accumulate in the plasma, whereas the clearance rates of LDL and of VLDL remnants are defective. In agreement with this, cholesterol feeding suppresses LDL-receptor activity in the liver (Kovanen *et al.*, 1981) but has little effect on the hepatic clearance of chylomicron remnants (Arbeeny and Rifici, 1984). Further evidence for a specific chylomicron-remnant receptor has been provided by competition studies. Carella and Cooper (1979) found that LDL competed much less effectively than unlabeled chylomicron remnants for the binding of labeled chylomicron remnants to rat-liver membranes. Comparable results have been reported by Krempler *et al.* (1987), using human Hep G2 cells in culture. Arbeeny and Rifici (1984) also found that the saturable uptake of ^{125}I-labeled chylomicron remnants by perfused liver was inhibited to a greater extent by unlabeled chylomicron remnants than by VLDL remnants (known to be taken up predominantly by the LDL-receptor pathway).

Differences between the regulation of hepatic catabolism of LDL and that of chylomicron remnants also point to the existence of a separate remnant receptor (see Mahley and Innerarity, 1983, and Brown and Goldstein, 1983a, for references). In general, hepatic LDL-receptor activity is increased by factors that tend to deplete the liver of cholesterol and is decreased in conditions that lead to an increase in the cholesterol content of the liver. Thus, hepatic LDL receptors are induced by fasting or treatment with cholestyramine or mevinolin and are suppressed by cholesterol feeding or by infusions of intestinal lymph or bile salts. In dogs and pigs there is also a marked decrease in LDL-receptor activity in liver membranes as the animals mature (Hui *et al.*, 1981). None of these factors has more than a small effect on the binding or metabolism of chylomicron remnants.

Figure 8.16. Electron micrograph of two multivesicular bodies (MVB) isolated from the liver of an estrogen-treated rat 10–15 minutes after an intravenous injection of [^{125}I]LDL. Top: Thin section of an intensely stained MVB, showing bilayer vesicles (arrowheads) among remnant lipoprotein particles. Bottom: Negative stain of a single MVB containing numerous remnants together with collapsed internal bilayer vesicles (arrowheads) and some injected [^{125}I]LDL particles (arrows). ×65,000. (From Hamilton, 1985, with the permission of the author.)

As we saw in Chapter 7 (Section II,C), pharmacological doses of estrogens increase the rate of catabolism of LDL by the rat's liver, owing to a 10-fold rise in hepatic LDL-receptor activity. Arbeeny and Rifici (1984), using perfused livers from estrogen-treated rats, have shown that the rise in LDL-receptor activity is not accompanied by a comparable rise in uptake and catabolism of chylomicron remnants. This experiment provides further evidence that chylomicron remnants are catabolized, at least in part, by a route other than the LDL-receptor pathway.

A receptor, other than the LDL receptor, that could be responsible for binding and uptake of chylomicron remnants by the liver has not yet been identified. Sherrill *et al.* (1980), from a study of the competition of HDL_c and chylomicron remnants for uptake by perfused livers, concluded that hepatic uptake of remnants is mediated solely by the apoE present in these particles. Following up this observation, Hui *et al.* (1981) and Mahley *et al.* (1981) reported the presence of a receptor on liver membranes from dogs, pigs, and human subjects that binds certain lipoproteins containing apoE (HDL_c) or apoE plus apoB (chylomicron remnants) but does not bind LDL. They called this the apoE receptor and they suggested that it is responsible for the hepatic uptake of chylomicron remnants *in vivo*. Expression of apoE-receptor activity on liver membranes resembles the saturable uptake of chylomicron remnants by the intact liver in showing little or no response to cholesterol feeding, infusions of bile salts, or treatment with cholestyramine.

Although the apoE receptor postulated by Mahley and his co-workers differs markedly from the hepatic LDL receptor with regard to regulation, the two receptors resemble each other in several respects. The binding of VLDL remnants by the LDL receptor and of chylomicron remnants by the apoE receptor is mediated by the apoE component of the particles. In each case, binding of the lipoprotein is diminished when wild-type apoE is replaced by apoE2—hence the accumulation of both VLDL remnants and chylomicron remnants in patients with type III hyperlipoproteinemia (see Chapter 7, Section I,A). Finally, lipoprotein binding by both receptors is Ca^{2+}-dependent and is abolished by reductive methylation of apoE in the particles or digestion of the receptor with pronase.

The probable existence of two hepatic receptors with overlapping specificities raises the question as to how the two species of remnant find their receptor targets *in vivo*. Both remnants must have apoE in a conformation that makes it recognizable by the LDL receptor *in vitro*, so why, in the presence of the two receptors *in vivo*, are chylomicron remnants taken up by the hepatic remnant receptor while VLDL remnants are taken up by the hepatic LDL receptor? It does not seem likely that the difference in targeting is due to the exclusive presence of apoB-48 in chylomicron remnants, since the clearance of chylomicron remnants from plasma is unaffected by removal of apoB-48 from the particles (Borensztajn *et al.*, 1982). Possibly, the presence of apoB-100 in VLDL remnants interferes in some way with their uptake by remnant receptors. But this would not explain the preferential uptake of chylomicron remnants by the remnant receptor. The very high affinity of HDL_c for LDL receptors makes it unlikely that the presence of apoB-100 is required for uptake of lipoprotein particles by the hepatic LDL receptor *in vivo,* though the observations of Arbeeny *et al.* (1987) suggest that apoB-100 makes a minor contribution to uptake of VLDL remnants by the perfused rat liver.

Attempts to isolate the apoE receptor from liver membranes have not, so far, been successful. Hui *et al.* (1986), using HDL_c-affinity chromatography, isolated a 56kDa protein from dog liver that bound apoE-containing lipoproteins but did not bind LDL. This protein cross-reacted with a polyclonal antibody to the LDL receptor. However, subsequent work has shown that the 56kDa protein is a mitochondrial ATPase (Beisiegel *et al.,* 1988). Another apoE-binding protein (59kDa) has been isolated from dog and human liver membranes by Beisiegel *et al.* (1988). Monoclonal antibodies against this protein do not react with the 56kDa protein. The 59kDa protein cannot be a product of the LDL-receptor gene since it has been detected in the liver of a homozygous FH patient whose cells produce no immunologically reactive LDL receptor. The relation between the 59kDa apoE-binding protein and the proposed hepatic remnant receptor is uncertain. According to Beisiegel *et al.* (1988), the 59kDa protein is localized predominantly in the endoplasmic reticulum. Moreover, it binds apoE2 and apoE3 with similar affinities. As discussed in Chapter 7 (Section I), the accumulation of apoE2-containing chylomicron remnants in type III hyperlipoproteinemia indicates that the receptor that is responsible for normal clearance of chylomicron remnants has lower affinity for E2 than for E3.

Although the balance of all the evidence favors the view that chylomicron remnants are catabolized via a receptor that is genetically and functionally distinct from the LDL receptor, an element of doubt must remain until the postulated remnant receptor has been identified and has been shown to function on the surface of hepatocytes, both *in vitro* and *in vivo.* In most of the published work on competition between chylomicron remnants and other lipoproteins for binding or uptake by receptors, some displacement of remnants is seen at high concentrations of the competing lipoprotein. This would be expected if chylomicron remnants bind to the same receptor as the competitor, but with much higher affinity. The apparently normal clearance of chylomicrons in FH homozygotes and WHHL rabbits also needs further study. In many FH homozygotes, and in all WHHL rabbits, the liver and other tissues express some LDL-receptor activity. This residual level of activity could conceivably maintain a near-normal clearance of chylomicron remnants if their affinity for LDL receptors is sufficiently high. What is required is a careful investigation of the clearance rate of chylomicron remnants in human subjects homozygous for the null allele at the LDL-receptor locus, compared with that in normal subjects. Better still would be the identification of an inherited disorder of chylomicron-remnant clearance analogous to FH.

Whatever the final answer to this question, it is not difficult to see the biological advantage of two independent pathways for receptor-mediated uptake of lipoproteins, one to extract chylomicron remnants from the plasma whenever they are presented to the liver, the other (a regulated pathway) to satisfy the variable needs of the liver for cholesterol. A common recent ancestry for the two receptor genes would explain the similarities between the LDL receptor and the apoE receptor, considered by Mahley and co-workers to function as a chylomicron-remnant receptor *in vivo*.

Herz *et al.* (1988) have described a high-molecular-weight protein [LDL-receptor-related protein (LRP)], present in liver and other tissues, that has a remarkable structural similarity to the LDL receptor. The complete amino acid sequence determined from overlapping cDNAs has 4544 residues including a cleavable signal sequence. Analysis of the primary structure shows that the protein has a single membrane-spanning segment separating a long N-terminal segment from a shorter C-terminal sequence of 100 amino acids. Immunochemical studies show that LRP is expressed on the surface of Hep G2 cells, with the N-terminal segment outside and the C-terminal segment inside the cell. The extracellular portion of the mature protein consists essentially of 31 imperfect copies of the cysteine-rich sequence that is repeated seven times in the binding domain of the LDL receptor, plus 22 imperfect copies of the "EGF-precursor" sequence present in domain 2 of the LDL receptor (see Chapter 9, Section II). The cytoplasmic extension, which is twice the length of the cytoplasmic tail of the LDL receptor, has two copies of the 4-residue consensus sequence N-P-X-Y, one of which is identical to the sequence ending with tyrosine 807 in the LDL receptor. Herz *et al.* (1988) point out that the extracellular domain of LRP has four clusters of negatively charged cysteine-rich sequences similar to sequences in the LDL receptor that are responsible for binding lipoproteins containing apoB or apoE. This strongly suggests that the protein is a surface receptor capable of binding lipoproteins containing positively charged apoproteins. The relationship, if any, between LRP and a possible chylomicron-remnant receptor remains to be determined. See Soutar (1989) for a discussion of recent evidence on this question.

4. Low-Affinity Binding of LDL by Liver *in Vitro*

Liver membranes and intact hepatocytes *in vitro* express low-affinity binding sites for LDL and other lipoproteins, in addition to the LDL receptor. Kita *et al.* (1982) observed low-affinity, EDTA-resistant binding of HDL_c and chylomicron remnants by liver membranes prepared from normal and WHHL rabbits. Low-affinity binding of LDL to human liver membranes, in addition to binding by the LDL receptor, has also been reported by Harders-Spengel *et al.* (1982) and by Hoeg *et al.* (1984).

Low-affinity binding of lipoproteins by heterogeneous membrane fractions *in vitro* may have no physiological significance. However, binding, uptake, and catabolism by low-affinity processes with only partial Ca^{2+}-dependence has been observed in suspensions of freshly prepared liver cells (Ose *et al.*, 1980), hepatocytes in culture (Edge *et al.*, 1986; Salter *et al.*, 1986), and cultured human hepatocytes of the Hep G2 cell line (Havekes *et al.*, 1983). Edge *et al.* (1986) have identified a low-affinity, nonsaturable pathway for binding and catabolism of LDL in primary cultures of hepatocytes from normal and homozygous FH human subjects. This pathway, which is equally active in normal and FH cells, is responsible for the degradation of about 25% of the total LDL degraded by normal cells in the presence of low concentrations of LDL in the medium. Reductively methylated LDL is catabolized at the same rate as native LDL by the low-affinity pathway. This pathway was not detectable in cultured fibroblasts tested under identical conditions. The low-affinity process for LDL catabolism expressed by nontransformed human hepatocytes *in vitro* could be the mechanism responsible for the high rate of catabolism of LDL by the liver at high plasma LDL concentrations in WHHL rabbits and, presumably, in FH homozygotes. As we saw in Chapter 7, it is not known whether hepatic catabolism of LDL by LDL-receptor-independent pathways *in vivo* is due to pinocytosis or to endocytosis mediated by low-affinity receptors. According to Edge *et al.* (1986) the rate of catabolism of LDL by the low-affinity pathway in cultured hepatocytes is higher than would be expected by pinocytosis alone.

Receptor-mediated uptake of modified LDL by Kupffer cells is dealt with in Section III,C below.

5. Feedback Regulation of LDL-Receptor Activity *in Vitro*

Owing to the presence of fenestrae in the walls of hepatic sinusoids, the LDL concentration in the fluid in the space of Disse in contact with hepatocytes must be close to the plasma LDL concentration (500–1000 μg of protein/ml in normal human subjects). Hence, if the response of hepatocytes *in vivo* to LDL in the external medium were similar to that of fibroblasts in culture (see Fig. 8.17), LDL-receptor activity in the human liver would be almost completely suppressed at normal plasma LDL concentrations. In fact, several studies of the behavior of normal hepatocytes in culture (Pangburn *et al.*, 1981; Edge *et al.*, 1986; Salter *et al.*, 1987a) and of Hep G2 cells (Havekes *et al.*, 1986) have shown that hepatocytes are much less responsive to feedback regulation by LDL than are fibroblasts when tested under the same conditions. In general, preincubation in the presence of LDL concentrations high enough to bring about almost complete suppression of LDL-receptor activity in fibroblasts decreases receptor activity in hepatocytes by only 25–50%. In addition, Havekes *et al.* (1986) have shown that HDL_3 has a much greater stimulatory effect on LDL-receptor activity in Hep G2 cells than in fibroblasts. (With both cell types the effect of HDL_3 is presumably due to withdrawal of free cholesterol from the cells.) Figure 8.17 shows the

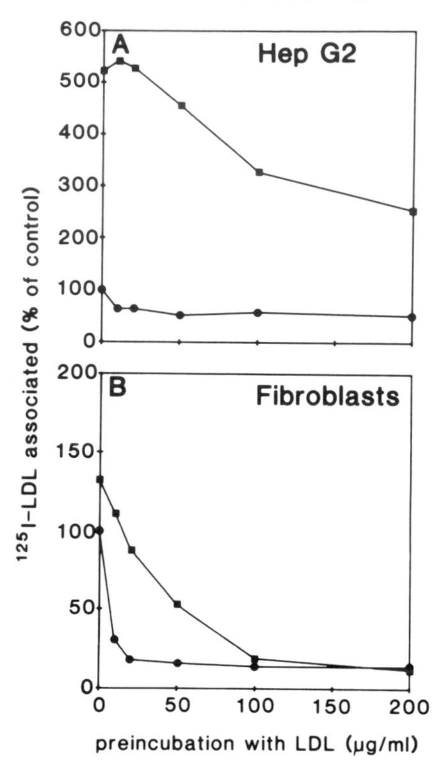

Figure 8.17. Suppression of LDL-receptor-mediated uptake of [^{125}I]LDL by (A) Hep G2 cells and (B) fibroblasts by preincubation with increasing amounts of unlabeled LDL in the absence (●) and in the presence (■) of HDL (200 μg of apoA-I/ml). After preincubation in the absence of LDL and HDL, receptor-mediated uptake was 123 and 398 μg of [^{125}I]LDL per mg cell protein for Hep G2 cells and fibroblasts, respectively. These values were taken as 100%. (From Havekes *et al.*, 1986, with the permission of the authors.)

results of an experiment in which LDL-receptor activity was measured in Hep G2 cells and fibroblasts after preincubation in the presence of increasing concentrations of LDL in the presence or absence of HDL. Note that in the presence of the highest concentrations of LDL plus HDL tested, receptor activity in fibroblasts is almost completely suppressed, whereas receptor activity in Hep G2 cells is more than 50% of that observed in the absence of either lipoprotein.

Havekes *et al.* (1986) suggest that decreased sensitivity to LDL and increased sensitivity to cholesterol depletion combine to enable hepatocytes to express high levels of LDL-receptor activity *in vivo.*

C. Lymphocytes

Much attention has been paid to the measurement of LDL-receptor activity in circulating blood cells in the expectation that this would provide information about LDL-receptor status in the intact animal or human subject under different conditions. However, the results that have been obtained in man are disappointing because LDL-receptor activity in blood cells capable of expressing receptors is almost completely suppressed by LDL and other lipoproteins present at high concentration in human plasma. Freshly isolated human lymphocytes that have not been induced by preincubation in LPDS exhibit little or no saturable binding and degradation of LDL; the low level of LDL-receptor activity observed in mixed populations of freshly isolated, uninduced blood mononuclear cells is due largely to the presence of monocytes (Chait *et al.*, 1982). As we shall see, both lymphocytes and monocytes express LDL-receptor activity when maximally induced by LPDS. Measurement of the maximal capacity of leukocytes for binding and degradation of LDL has been useful in studies of the heritability of LDL-receptor capacity in the general population and may also be helpful in the diagnosis of FH in the heterozygous form (see Chapter 10).

1. LDL Receptors

LDL-receptor activity has been demonstrated in established lines of lymphoid cells in which activity has been induced by preincubation in LPDS (Ho *et al.*, 1976b). The functional properties of the LDL receptor expressed by these cells are identical with the properties of fibroblast receptors. In particular, binding affinity of lymphoid-cell receptors for LDL is similar to that of fibroblast receptors, and receptors on cells of both types are saturated at LDL concentrations of 25–50 μg of protein/ml. However, in the induced state, lymphoid cells express far fewer LDL receptors than do fibroblasts (<3,600 particles of LDL bound per lymphoid cell versus >15,000 per fibroblast). As expected, lymphoid cells in culture derived from FH homozygotes express no LDL receptors.

Lymphocytes freshly isolated from normal human subjects express little or no LDL-receptor activity (Reichl *et al.*, 1976; Ho *et al.*, 1976a; Shechter *et al.*, 1981). However, after preincubation in LPDS for at least 36 hours they develop the capacity to bind and degrade LDL by the LDL-receptor pathway (Ho *et al.*, 1976a). In agreement with these findings, LDL-receptor protein is barely detectable by ligand-blotting or immunoblotting of extracts of freshly isolated human lymphocytes but is readily demonstrated in extracts of cells that have been incubated for 48 hours in LPDS (Semenkovich and Ostlund, 1986). The addition of LDL to the preincubation medium prevents induction of receptors in normal lymphocytes, indicating that the low number expressed by lymphocytes before induction *in vitro* is due to suppression by plasma LDL *in vivo*. As with lymphoid cells in culture, lymphocytes from FH homozygotes fail to develop LDL receptors after preincubation in LPDS.

Ho *et al.* (1977) have investigated the role of LDL receptors in the regulation of cholesterol synthesis in lymphocytes. Cholesterol synthesis is almost completely suppressed in freshly isolated lymphocytes from both normal subjects and FH homozygotes. If freshly isolated normal and FH cells are depleted of cholesterol by incubation in LPDS the rate of synthesis of cholesterol increases markedly in cells of both types. If LDL is added to the medium after incubation in LPDS, cholesterol synthesis is suppressed in the normal cells but is unaffected in the FH cells. These findings may be interpreted as follows. In normal lymphocytes present in the circulation, cholesterol synthesis is suppressed by LDL taken up via the LDL-receptor pathway. In the presence of LPDS *in vitro,* cholesterol is withdrawn from the regulatory pool of free cholesterol in the cells and HMG-CoA reductase is induced. When LDL is added back to the medium, reductase is suppressed by rapid uptake of LDL via the LDL-receptor pathway. In FH lymphocytes *in vivo,* cholesterol synthesis is suppressed in the long term partly by cholesterol synthesized within the cells and partly by LDL, present at very high concentration in the plasma, taken up by receptor-independent pathways. In relatively short-term experiments *in vitro,* LDL does not suppress reductase in FH cells that have been incubated in LPDS because the LDL-receptor pathway is not available for rapid delivery of cholesterol to the regulatory pool. On this interpretation, induction of HMG-CoA reductase in freshly isolated lymphocytes by LPDS *in vitro* and its suppression by the subsequent addition of LDL to the medium is analogous to the induction of reductase in certain tissues by 4-APP *in vivo* and the resuppression of reductase by infusion of plasma lipoprotein into 4-APP-treated animals (Balasubramaniam *et al.*, 1976; and see Chapter 2).

Although suppression of LDL-receptor activity in circulating lymphocytes of normal human subjects may well be due predominantly to receptor-mediated uptake of plasma LDL, other mechanisms must be capable of bringing about suppression in the absence of LDL. In patients with abetalipoproteinemia, the rate of cholesterol synthesis in freshly isolated lymphocytes is normal (Reichl *et al.*, 1978) or only slightly increased (Ho *et al.*, 1977) despite the complete absence of LDL from the plasma. Serum from abetalipoproteinemic patients does not suppress cholesterol synthesis in skin fibroblasts *in vitro* (Brown and Goldstein, 1974). However, it is possible that continuous exposure of lymphocytes to apoE-containing lipoproteins in the plasma results in sufficient receptor-mediated uptake of lipoprotein cholesterol to suppress HMG-CoA reductase in abetalipoproteinemic plasma *in vivo*. An additional possibility is that endogenously synthesized cholesterol contributes to suppression of reductase in the absence of LDL.

Cuthbert and Lipsky (1984a) have shown that mitogen-stimulated proliferation of normal and FH lymphocytes grown in the absence of LDL is inhibited by the addition to the medium of mevinolin in doses sufficient to suppress cholesterol synthesis but not high enough to prevent the synthesis of nonsterol metabolites of mevalonic acid. LDL at low concentrations reverses the inhibitory effect of mevinolin on the response to mitogens in normal lymphocytes but has little or no effect in lymphocytes from FH homozygotes. Cuthbert and Lipsky conclude that LDL-receptor-mediated uptake of LDL is necessary for the normal response of lymphocytes to mitogens *in vitro* when endogenous synthesis of cholesterol is limited. This effect of LDL at low concentrations, it should be noted, is quite distinct from the inhibitory effect of LDL and other lipoproteins at high concentrations on the mitogen-stimulated proliferation of lymphocytes (see Section III,C,3 below).

2. Mononuclear Cells in the Measurement of LDL-Receptor Activity

Skin fibroblasts in culture are not a suitable system for the rapid assessment of LDL-receptor function in the diagnosis of FH or in population studies. Nor, for obvious reasons, can any genetically normal cells in culture provide information about LDL-receptor activity as expressed *in vivo*. Several attempts have been made to overcome these limitations by measuring LDL receptors in circulating lymphocytes or mixed populations of blood mononuclear cells.

As a diagnostic test for FH, Bilheimer *et al.* (1978) measured the rate of saturable degradation of LDL by lymphocytes in which LDL receptors had been maximally induced by preicubation in LPDS for 67 hours at 37°C. Using this test, Bilheimer *et al.* investigated a large family with inheritance of the FH gene through five generations. There was no overlap between the values given by the known FH heterozygotes and the known normal individuals in this family. An additional point of interest is that there was no correlation between maximal LDL-receptor activity expressed *in vitro* and plasma LDL-cholesterol concentration within either group of FH heterozygotes and normal subjects, though the mean plasma LDL concentration in the FH heterozygotes was about twice that in the normals. Bilheimer *et al.* were also able to demonstrate a significant difference between the mean values for LDL degradation in freshly isolated (uninduced) mononuclear cells from the normals and FH heterozygotes in this family. The rates of degradation of LDL by the cells from both groups were very low and there was overlap between the values from the two groups. However, the results of this study suggest that the number of LDL receptors expressed *in vivo* by mononuclear cells in FH heterozygotes is about half the number expressed in normal subjects. Cuthbert *et al.* (1986) have suggested that the ability of mevinolin-inhibited lymphocytes to respond to a mitogenic stimulus in the presence of LDL could be used as the basis of a rapid diagnostic test for FH. Preliminary results suggest that this test may be useful in the diagnosis of FH in the heterozygous state.

Mistry *et al.* (1981) have measured the rate of saturable degradation of LDL by blood mononuclear cells, either freshly isolated or fully induced by LPDS, in normal human subjects before and after the ingestion of eggs in amounts sufficient to cause a rise in mean plasma LDL concentration. The egg diet led to an increase in the cholesterol content of the cells and a decrease in HMG-CoA reductase activity. These changes were accompanied by a decrease in LDL-receptor activity in freshly isolated cells, but there was no effect on receptor activity in fully induced cells. Thus, the response of mononuclear cells *in vivo* to a rise in plasma LDL level is similar to that of cultured fibroblasts to LDL added to the culture medium. Weight *et al.* (1982), using similar methods, have demonstrated a high degree of heritability of LDL-receptor activity (0.7 to 1.0) in maximally induced mononuclear cells from a group of normal twins living in the London area. They conclude that in normal people the maximal capacity for expressing LDL receptor activity is determined largely by genetic factors. Note that all normal individuals have the same number of LDL-receptor genes per diploid cell. Heritability of LDL-receptor-mediated degradation of LDL in the maximally induced state may, therefore, reflect inheritance of polymorphic variants of the normal receptor gene that affect the function of its product. There could also be inheritance of other genes that help to determine the efficiency of the LDL-receptor pathway as a whole. These could include genes encoding many proteins, such as those concerned in the internalization and recycling of LDL receptors.

It will be clear from the above discussion that useful information has been obtained by measurement of the binding and degradation of LDL by blood mononuclear cells. Nevertheless, there is a need for improvements in methodology. The development of a sensitive and accurate noninvasive method for determining LDL-receptor activity in cells *in vivo* would open the way to investigation of a wide range of important problems, especially those relating to cholesterol homeostasis in man. Wojciechowski *et al.* (1987) have described a method for the rapid determination of the uptake of fluorescent LDL by lymphocytes *in vitro,* using a cell sorter. Other approaches that might be worth pursuing are determination of the amount of LDL-receptor mRNA in freshly isolated cells and measurement of the uptake of a radioactive monoclonal antibody to the LDL receptor by tissues *in vivo* (see ''Discussion'' in Huettinger *et al.*, 1984). In any consideration of this problem it should be borne in mind that determinations of LDL-receptor activity in circulating mononuclear cells may fail to provide valid information relevant to receptor activity in other tissues that are not in contact with LDL at the high concentration existing in plasma.

3. LDL and the Immune Response of Lymphocytes

LDL and other plasma lipoproteins that contain apoB or apoE suppress early biochemical events in the response of T-lymphocytes to specific antigens or nonspecific mitogens *in vitro* (for references, see Cuthbert and Lipsky, 1983). Curtiss and Edgington (1976, 1978) have described experiments from which they conclude that human LDL contains a minor subfraction (LDL-In) that inhibits DNA synthesis in mitogen-stimulated lymphocytes, and that inhibition is mediated by binding of LDL-In particles to specific saturable receptors on lymphocytes (LDL-In receptors) that are distinct from the LDL receptor. Observations reported by Hui *et al.* (1980) and by Harmony and Hui (1981) indicate that LDL-In is an apoE-rich lipoprotein, probably IDL, and that binding to the postulated receptor, without internalization, is sufficient for suppression of the response of lymphocytes to mitogens.

It seems clear that LDL and other lipoproteins, particularly those enriched with apoE, are capable of suppressing the response of lymphocytes to an immunological challenge under appropriate conditions *in vitro* and that lymphocytes from normal subjects and homozygous FH patients (Cuthbert and Lipsky, 1984a) are equally sensitive to this inhibition. However, evidence for the presence of a specific ''immunoregulatory'' receptor for lipoproteins on lymphocytes is controversial. If such a receptor does exist, its affinity for ligand must be unusually low [about 100-fold less than that of the LDL receptor for LDL at 4°C (Hui *et al.*, 1980)]. It must also have low specificity, since all lipoproteins are capable of suppression to some degree (Morse *et al.*, 1977).

The mechanism by which LDL suppresses the response of T-lymphocytes to mitogenic stimuli is not understood. Curtiss and Edgington (1977) concluded that LDL inhibits lymphocyte activation by a direct action on T-lymphocytes. However, Nakayasu *et al.* (1986) suggest that the primary action of LDL is to suppress the activity of accessory cells (present in lymphocyte preparations) that facilitate proliferation of mitogen-stimulated lymphocytes. Cuthbert and Lipsky (1984b), on the other hand, have shown that transferrin reverses the inhibitory effect of LDL on lymphocytes, suggesting that LDL inhibits lymphocyte activity by interfering with the metabolism of transferrin. Inhibition of transferrin metabolism by LDL is not mediated by an effect on the expression of transferrin receptors or on their ability to bind transferrin. Nor is inhibition dependent upon the presence of LDL receptors on lymphocytes, since transferrin reverses the inhibitory effect of LDL on mitogen-stimulated DNA synthesis in lymphocytes from patients with homozygous FH. One possibility, suggested by Cuthbert and Lipsky (1984b), is that LDL affects the recycling of transferrin or its receptor by transferring cholesterol to the lymphocyte plasma membrane and thus altering its fluidity.

It is not easy to see how LDL could act as a regulator of lymphocyte responses to immunological stimuli *in vivo,* since the plasma LDL concentration remains essentially constant in the short term. However, it is possible that apoE-containing lipoproteins, which are known to have a marked inhibitory effect on mitogen-activated lymphocytes (Hui *et al.*, 1980), play some part in the regulation of the lymphocyte response *in vivo*. Macrophages normally secrete apoE (see Brown and Goldstein, 1983b, for review) and this secretion is suppressed in cells obtained from mice treated with various inflammatory agents (Werb and Chin, 1983). Thus, it is possible that diminished secretion of apoE by macrophages in an inflammatory lesion helps to promote an immune response in lymphocytes present in the lesion by releasing them from inhibition.

D. Macrophages and Monocytes

1. The Acetyl-LDL Receptor

Peritoneal macrophages and other cells of the RE system express saturatable, high-affinity surface receptors for acetylated LDL (acetyl-LDL) *in vitro*. (See Table 8.2 for some properties of the acetyl-LDL receptor, and see Brown and Goldstein, 1983b, for review.) At 37°C, surface binding of acetyl-LDL is followed by internalization of the lipoprotein particles, their delivery to lysosomes, and the digestion of their lipid and protein components by lysosomal enzymes. The free cholesterol released by lysosomal hydrolysis of cholesteryl esters in acetyl-LDL enters the cytoplasm and is esterified by ACAT. The esterified cholesterol enters the cytoplasmic pool of cholesteryl ester in the macrophage, where it participates in the cycle of hydrolysis and reesterification described in Chapter 4, Section III. Uptake and hydrolysis of acetyl-LDL by macrophages does not lead to suppression of acetyl-LDL receptors. Hence (as discussed in Chapter 4), when macrophages are incubated for several days in the presence of acetyl-LDL, they continue to take up acetyl-LDL particles, eventually accumulating massive amounts of esterified cholesterol in lipid droplets in the cytoplasm. Macrophages that have been treated in this way resemble the foam cells seen in atherosclerotic lesions and in the xanthomas of FH patients. Activation of peritoneal macrophages by inflammatory agents that increase the activity of receptors for mannose-conjugated proteins has no effect on acetyl-LDL-receptor activity (Imber *et al.*, 1982). Nor is acetyl-LDL-receptor activity on macrophages increased by preincubating the cells in the presence of LPDS.

Acetyl-LDL receptors are expressed by macrophages of all classes from all species that have been examined. In addition to their presence on peritoneal macrophages, they have been demonstrated on Kupffer cells, monocyte-derived

Table 8.2
Some Functional Properties of the Acetyl-LDL Receptor on Mouse Peritoneal Macrophages *in Vitro*

Binding affinity	Half-saturation for acetyl-LDL: 5 μg of protein/ml at 4°C; 25 μg of protein/ml at 37°C.
Binding capacity	Binds 20,000–40,000 particles per cell at 4°C.
Specificity	Binds LDL modified by agents that block ε-amino groups of lysine residues in apoB-100, thus increasing the net negative charge on LDL. Modified LDLs that act as ligands include acetyl-LDL, succinyl-LDL, malondialdehyde-treated LDL and LDL incubated in the presence of endothelial cells. Polyanionic compounds that also bind to the receptor include fucoidin, dextran sulfate, and polyvinyl sulphate. Does not bind LDL or reductively methylated LDL or CHD-LDL.
No requirement for Ca^{2+}	

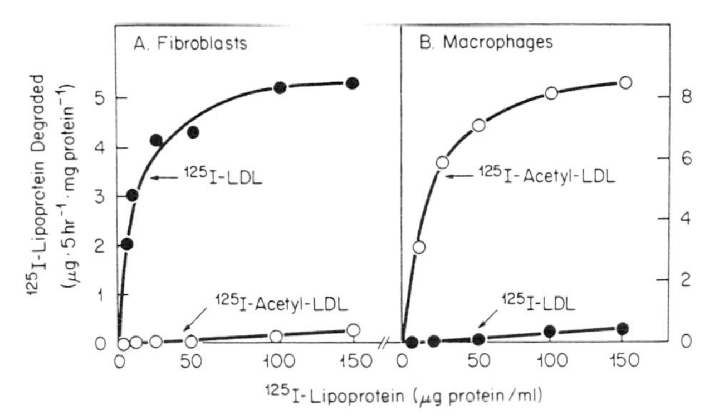

Figure 8.18. Differences in the uptake and degradation of [^{125}I]LDL (●) and [^{125}I]acetyl-LDL (○) by human fibroblasts (A) and mouse peritoneal macrophages (B). Monolayers of growing fibroblasts were incubated in lipoprotein-deficient serum for 48 hours before the experiments to induce maximal LDL-receptor activity. Monolayers of freshly isolated macrophages were studied without preincubation in lipoprotein-deficient serum. For measurement of LDL degradation, each monolayer received medium containing 10% lipoprotein-deficient serum and the indicated concentration of either human [^{125}I]LDL (●) or [^{125}I]acetyl-LDL (○). After incubation for 5 hours at 37°C the amount of ^{125}I-labeled acid-soluble material released into the medium was measured. (From Brown and Goldstein, 1983b. Reproduced, with permission, from the authors and from the *Annual Review of Biochemistry,* Volume 52, © 1983 by Annual Reviews Inc.)

macrophages, and mouse macrophages of the J774 line. Acetyl-LDL-receptor activity is also expressed by bovine vascular endothelial cells in culture (Stein and Stein, 1980) and by endothelial cells of rat-liver sinusoids *in vivo* and *in vitro* (Nagelkerke *et al.*, 1983). In general, cells that express acetyl-LDL receptors do not express LDL receptors and those that express LDL receptors express few, if any, acetyl-LDL receptors. This is well exemplified by the difference between the expression of the two receptors on fibroblasts and macrophages (Fig. 8.18). An interesting exception is the human blood monocyte, which expresses both receptors in about equal numbers at one stage of its conversion into a macrophage *in vitro* (see below). Macrophage-derived foam cells in explants of atherosclerotic lesions express acetyl-LDL receptors, as shown by the ability of these cells to take up fluorescent acetoacetyl-LDL *in situ* (Pitas *et al.*, 1983).

Acetylation of LDL by acetic anhydride blocks the ε-amino groups of lysine residues in apoB-100 in the lipoprotein. This removes positive charges and thus increases the net negative charge on LDL. Hence, acetyl-LDL has greater electrophoretic mobility than native LDL. Modification of LDL by other reagents that increase its electronegativity by blocking lysine residues also results in the formation of ligands for the acetyl-LDL receptor. These ligands include acetoacetyl-LDL, maleyl-LDL, succinyl-LDL, and malondialdehyde-treated LDL; LDL modified by incubation with endothelial cells is discussed below. Blocking lysine residues in LDL cannot, by itself, be sufficient to convert it into a ligand because reductively methylated LDL, in which lysine residues are blocked but the net negative charge is unchanged, does not bind to the receptor. Haberland *et al.* (1982) have investigated the effect of blocking lysine residues in native LDL on its affinity for LDL receptors and acetyl-LDL receptors. As the number of residues blocked by malondialdehyde is increased progressively, there is a parallel decline in affinity for the LDL receptor. Affinity falls to zero when 20 mol of malondialdehyde are incorporated per mol of LDL. In contrast, the ability to bind to the acetyl-LDL receptor does not appear until 30 mol of malondialdehyde are incorporated per mol of LDL.

Most proteins, including albumin, are not converted into ligands for the acetyl-LDL receptor by acetylation, presumably because the effect on net negative charge is not great enough. However, maleylated albumin binds to the receptor. In this case, maleic acid (a dicarboxylic acid) not only blocks lysine residues in the albumin molecule but also adds negatively charged carboxyl groups. The acetyl-LDL receptor also binds several high-molecular-weight polyanions, including fucoidin, dextran sulfate, polyinosinic acid, and polyvinyl sulfate. However, many polyanions with strong negative charge do not act as ligands.

From a comparison of the affinities of various compounds for the acetyl-LDL receptor, Brown *et al.* (1980) concluded that binding requires the presence of a large number of negatively charged residues, not necessarily carboxyl groups, within a specific region of the ligand.

Some progress has been made toward the isolation and characterization of the acetyl-LDL receptor. Via *et al.* (1982) reported the partial purification of a solubilized mouse-macrophage protein that bound acetyl-LDL and other ligands for the acetyl-LDL receptor expressed on intact cells. The apparent molecular weight of the protein-detergent complex was 283K. Via *et al.* (1985) have also isolated a protein (MW 260K) from solubilized mouse-macrophage membranes by ligand blotting with acetyl-LDL, maleylated albumin, and malondialdehyde-LDL as ligands. This protein does not bind native LDL. [See also Kodoma *et al.*, *Nature* (1990), **343,** 531–535.]

Observations on the metabolism *in vivo* of radioiodine-labeled ligands for the acetyl-LDL receptor have shown that it is expressed in the living animal. Thus, labeled acetoacetylated LDL (Mahley *et al.*, 1979a,b) and labeled succinylated LDL (Chao *et al.*, 1979a) are taken up very rapidly by nonparenchymal cells of the liver after intravenous injection into intact animals. Nagelkerke *et al.* (1983), using a low-temperature collagenase procedure for separating liver cells, have shown that ^{125}I-labeled human acetyl-LDL, when injected intravenously into rats, is taken up predominantly by the endothelial cells of liver sinusoids.

2. LDL Modified by Endothelial Cells

Henriksen and co-workers have shown that when LDL is incubated in the presence of vascular endothelial cells or aortic smooth-muscle cells in culture, the

LDL undergoes several modifications (see Henriksen *et al.*, 1983, for review of initial studies). LDL modified by endothelial cells (EC-modified LDL) has increased electrophoretic mobility, and increased density due to selective loss of lipid. Moreover, EC-modified LDL is recognized by the acetyl-LDL receptor, so that when the modified LDL is incubated with macrophages it is taken up and degraded by the acetyl LDL-receptor pathway, resulting in the intracellular accumulation of cholesterol. None of these changes occurs when LDL is incubated in the presence of fibroblasts.

Subsequent work (Morel *et al.*, 1983, 1984; Steinbrecher *et al.*, 1984; Parthasarathy *et al.*, 1985, 1986) has gone some way toward explaining the changes undergone by EC-modified LDL. The initial step in the whole sequence of events is probably the generation of H_2O_2 and the free radical O_2^- (superoxide anion) by endothelial cells and arterial smooth-muscle cells and macrophages (but not, apparently, by fibroblasts). In keeping with this, EC modification can be prevented by adding EDTA or free-radical scavengers to the incubation medium. The presence of free radicals in the medium leads to the peroxidation of LDL lipids, including the fatty acids of phospholipids, and the extensive breakdown of apoB. Other mechanisms that may contribute to the peroxidation of the phospholipid fatty acids of LDL are reviewed in Steinberg *et al.* (1989).

Peroxidation of the fatty acids in the 2 position of LDL lecithin increases their susceptibility to hydrolysis by a phospholipase A_2 that is intimately associated with the LDL particle. This results in the formation of lysolecithin and of peroxidized free fatty acids which, in turn, may be responsible for the breakdown of apoB adjacent to the phospholipids. Parthasarathy *et al.* (1985) suggest that changes in apoB, perhaps brought about by peroxidized fatty acids, are responsible for the increased electronegativity of EC-modified LDL and its recognition by the acetyl-LDL receptor. In confirmation of this, Steinbrecher (1987) has shown that fragments of oxidized LDL fatty acids combine covalently with the ϵ-amino groups of lysine residues in apoB. This is analogous to the covalent modification of LDL by acetic anhydride and other ligands (see above) and would explain the increased electronegativity of EC-modified LDL and its recognition by acetyl-LDL receptors. It is not known how this complex interaction between free radicals and phospholipase A_2 leads to loss of cholesterol from the LDL particle and the consequent increase in hydrated density.

Quinn *et al.* (1987) have described an interesting property of EC-modified LDL that may be relevant to the formation of foam cells in the arterial wall. EC-modified LDL acts as a chemotactic agent for human blood monocytes *in vitro*. If this activity is exhibited *in vivo* it could promote the movement of monocytes from the blood circulation into the subendothelial space of arteries.

It is unlikely that oxidized LDL ever reaches a significant concentration in the circulation *in vivo* because any that entered the plasma would be removed very rapidly by acetyl-LDL receptors in the liver (Nagelkerke *et al.*, 1983). However, within the arterial wall oxidized LDL could reach a local concentration high enough to contribute to the formation of foam cells by uptake via the acetyl-LDL-receptor pathway. In agreement with this possibility, Palinski *et al.* (1989) have demonstrated the presence of oxidized LDL in rabbit atherosclerotic lesions by immunostaining with antibodies that react specifically with oxidized forms of LDL.

3. Other Receptors for Lipoproteins

It has been suggested that macrophages express separate high-affinity receptors for several lipoproteins, in addition to the receptor for acetyl-LDL. Additional receptors on macrophages for which there is some evidence, based largely on the results of competition studies, include those for β-VLDL, LDL–dextran sulfate complexes and cholesteryl ester–protein complexes present in extracts of human atherosclerotic plaques (see Brown and Goldstein, 1983b, for references). None of these receptors has been characterized other than by its functional activity. Macrophages also bind LDL–anti-LDL complexes, probably by the Fc receptors that bind all antigen–antibody complexes.

The presence of a separate class of β-VLDL receptors on macrophages, distinct from the LDL receptor and the acetyl-LDL receptor, was first postulated by Goldstein *et al.* (1980). (For a detailed account of receptor-mediated uptake of β-VLDL by macrophages, see Brown and Goldstein, 1983b.)

Mouse peritoneal macrophages express high-affinity saturable receptors that bind β-VLDL from cholesterol-fed animals and from patients with type III hyperlipoproteinemia. At 37°C, binding is followed by internalization and lysosomal degradation of the particles and the deposition of cholesteryl esters in the cytoplasm. Uptake of β-VLDL by macrophages *in vitro* leads to partial suppression of β-VLDL receptors, but the extent of their regulation is not sufficient to prevent the formation of lipid droplets in the cytoplasm of cells incubated continuously in the presence of β-VLDL. Both fraction I (chylomicron derived) and fraction II (hepatic-VLDL derived) of the β-VLDL of cholesterol-fed animals and type III patients are taken up by macrophages via the β-VLDL receptor.

When macrophages are incubated under conditions in which LDL receptors are barely expressed they take up β-VLDL via a saturable, high-affinity process. Moreover, receptor-mediated uptake of β-VLDL by macrophages is not inhibited by acetyl-LDL or other ligands for the acetyl-LDL receptor. For these reasons, Goldstein *et al.* (1980) concluded that the receptor responsible for the high-affinity binding of β-VLDL by macrophages is different from the receptors for LDL and acetyl-LDL. In support of this conclusion, Van Lenten *et al.* (1985) have shown that alveolar macrophages obtained from WHHL rabbits degrade β-VLDL by a saturable pathway. Other observations reported by Gianturco *et al.* (1982a) may also be relevant to the possible presence of β-VLDL receptors on macrophages. VLDL from hypertriglyceridemic human subjects (but not normal VLDL) is taken up by mouse peritoneal macrophages *in vitro*, resulting in the massive intracellular accumulation of triglyceride. Uptake of "hypertriglyceridemic VLDL" is saturable and is inhibited by β-VLDL but not by acetyl-LDL. This is consistent with the presence of a β-VLDL receptor that recognizes the abnormal apoE-rich VLDL in the plasma of hypertriglyceridemic subjects. Intracellular accumulation of triglyceride, rather than esterified cholesterol, is presumably due to the preponderance of triglyceride in the lipid component of these abnormal VLDL particles.

β-VLDL is also taken up and degraded by human monocyte-derived macrophages by a saturable pathway, degradation leading to marked stimulation of cholesteryl-ester synthesis in the cytoplasm (Mahley *et al.*, 1980; Van Lenten *et al.*, 1983). The rate of degradation of β-VLDL is much greater than that of LDL and is unaffected by the presence of ligands for acetyl-LDL, again suggesting the presence of a specific receptor for β-VLDL on cells of the macrophage class. However, experiments on the binding and uptake of β-VLDL by monocyte-macrophages are difficult to interpret because these cells express LDL receptors (see below), for which β-VLDL is an efficient ligand (Mahley *et al.*, 1980). In favor of the presence of a separate β-VLDL receptor on monocyte-macrophages, Soutar and Knight (1984) have shown that cultured monocytes from receptor-defective FH homozygotes degrade β-VLDL by a saturable mechanism.

Despite the evidence that the saturable uptake and degradation of β-VLDL by macrophages is mediated by a separate β-VLDL receptor, no such receptor has yet been isolated and characterized. In several respects, the question as to whether or not there is a separate β-VLDL receptor on macrophages is similar to the problem of the existence of a separate receptor for chylomicron remnants in the liver; in both cases ligands for the postulated specific receptor are also recognized by the LDL receptor.

The possibility that high-affinity binding of β-VLDL by macrophages is, in fact, mediated by LDL receptors cannot yet be ruled out. Innerarity *et al.* (1986) point out that the receptor responsible for high-affinity binding of β-VLDL by macrophages has several properties in common with the LDL receptor on skin fibroblasts. In particular, both receptors have high affinity for ligands containing apoE, both are Ca^{2+} dependent, and the binding exhibited by both receptors is inhibited by chemical modification of lysine residues in their ligands. Koo *et al.* (1986) used ligand blotting, with ^{125}I-labeled β-VLDL as ligand, to examine the proteins in extracts of mouse macrophages. They identified a single protein which, by immunoblotting with antibody to the bovine LDL receptor, was shown to be identical to the LDL receptor in mouse fibroblasts. No other proteins that bound β-VLDL could be detected in the macrophage extracts. From the results of competitive binding studies with β-VLDL, LDL, and HDL_c, Koo *et al.* (1986) concluded that binding of β-VLDL by mouse macrophages is due to the presence of an LDL receptor with atypical properties. The LDL receptor on macrophages has the same high affinity for apoE-containing lipoproteins (including β-VLDL) as the LDL receptor on fibroblasts. However, the LDL receptor on macrophages has an 18-fold lower affinity for LDL than has the fibroblast receptor. LDL receptors on macrophages are also relatively resistant to suppression by uptake of lipoprotein cholesterol. In support of their conclusion that macrophages bind β-VLDL via the LDL receptor, Koo *et al.* showed that antibody to the LDL receptor inhibited binding of β-VLDL to mouse macrophages. If receptor-mediated uptake of β-VLDL by macrophages is due to the presence of LDL receptors with low affinity for LDL and high affinity for β-VLDL, the saturable uptake of β-VLDL by macrophages from WHHL rabbits and by monocyte-macrophages from receptor-defective FH homozygotes may be explained by residual LDL-receptor activity expressed by these cells.

4. Monocyte-Macrophages

When human blood monocytes are incubated in a suitable medium, they grow without dividing and eventually acquire many of the morphological and functional characteristics of resident macrophages. When freshly isolated, monocytes express LDL receptors and acetyl-LDL receptors in about equal numbers (Fogelman *et al.*, 1981; Knight and Soutar, 1982). If the cells are maintained in culture in the presence of 20–30% serum, the activities of the two receptors change in the manner shown in Fig. 8.19. LDL-receptor activity per cell (or per milligram of DNA) increases to a maximum by about the end of the first week, when the growth rate is maximal, and then declines slowly or remains constant. Acetyl-LDL-receptor activity, on the other hand, increases progressively to more than 10 times the initial value by the end of the second week. It should be noted that these changes in LDL-receptor activity take place while the cells are growing in the presence of LDL at a concentration much greater than that required to saturate the receptors expressed on the cell surface. Under these conditions, LDL receptors on skin fibroblasts in culture would be almost completely suppressed (see Section II,B above).

In monocytes growing in 20% serum, the affinity of the LDL receptor for LDL (half saturation at 20 μg of LDL protein/ml) is much lower than that in fibroblasts (Soutar and Knight, 1982) or monocytes (Knight and Soutar, 1986) grown in LPDS. Knight and Soutar (1986) suggest that the increased affinity of the receptor for LDL after preincubation of monocytes in LPDS is accompanied by an increase in the number of receptors that bind to each particle.

Despite these striking differences between monocyte-macrophages and fibroblasts, LDL-receptor activity in growing monocyte-macrophages is suppressed by cholesterol entering through receptor-mediated pathways. Thus, LDL receptors are suppressed when the cells are incubated with malondialdehyde-treated LDL (a ligand for the acetyl-LDL receptor) (Fogelman *et al.*, 1981). Conversely, LDL-receptor activity increases when the cells are incubated in LPDS (Shechter *et al.*, 1981). Hence, although LDL receptors are expressed at a relatively high

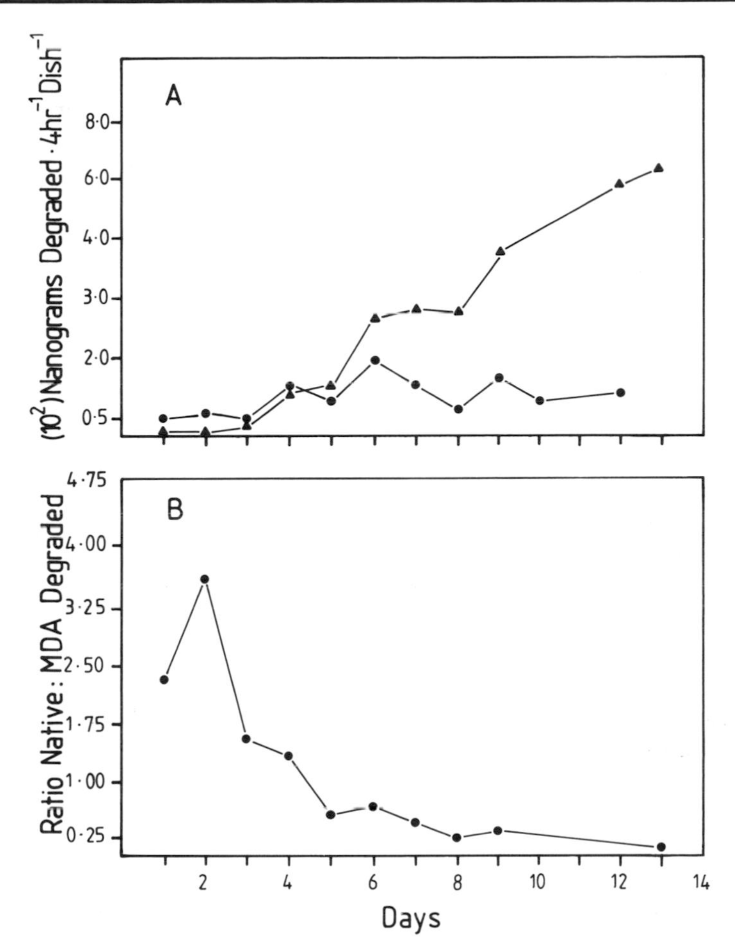

Figure 8.19. The activity of LDL receptors and of acetyl-LDL receptors in human monocytes as a function of time in culture. Freshly isolated blood monocytes were cultured in the presence of 30% autologous serum in plastic Petri dishes. At the times shown on the horizontal axis the medium was replaced by 1 ml of medium containing either 20 μg of ^{125}I-labeled native LDL (●) or 10 μg of ^{125}I-labeled malondialdehyde-LDL (▲) (a ligand for the acetyl-LDL receptor). After incubation for 4 hours at 37°C the amount of ^{125}I-labeled acid-soluble material in the medium was determined. (A) shows the rates of degradation of native and malondialdehyde-LDL. (B) shows the ratio of the rates of degradation of the two LDLs. [From Fogelman *et al.*, 1981, (slightly modified) with the permission of the authors.]

level by monocytes growing in the presence of serum, receptor activity is not maximally induced under these conditions.

The regulation of HMG-CoA reductase activity in growing monocyte-macrophages also differs from that in fibroblasts in culture (see Soutar and Knight, 1985, for review). In contrast to fibroblasts in culture, growing monocyte-macrophages maintain HMG-CoA reductase activity at a high level in the presence of LDL at concentrations sufficient to saturate their LDL receptors. When the concentration of LDL in the medium is raised above the level required for saturation, LDL enters the cells through nonsaturable pathways, and this leads to suppression of reductase activity (Knight *et al.*, 1983). Nonsaturable uptake of LDL also leads to suppression of reductase in monocyte-macrophages derived from FH homozygotes. Hence, reductase activity in FH monocyte-macrophages is not significantly higher than in normal cells when both are grown in the presence of 20% serum. This is in marked contrast to the behavior of fibroblasts. As we saw in Section II,B, nonsaturable uptake of LDL by fibroblasts in culture has no effect on reductase activity. Consequently reductase activity in FH fibroblasts is many times higher than that in normal fibroblasts when both are grown in the presence of high concentrations of LDL.

Patel *et al.* (1984) have investigated the sources of the cholesterol used for membrane formation by monocyte-macrophages growing in culture. They conclude that, during the period of maximal growth, cells maintained in the presence of 20% serum satisfy about 50% of their requirement for cholesterol by endogenous synthesis and most of the remaining 50% by nonsaturable uptake of LDL. Despite the fact that growing monocyte-macrophages are capable of expressing enough LDL receptors to satisfy all their requirement for cholesterol, LDL-receptor-mediated uptake of LDL appears to make only a small contribution when the cells are grown in 20% serum. Patel and Knight (1985) suggest that growing monocyte-macrophages express high reductase activity, even in the presence of 20% serum, because they require large amounts of a nonsterol product of mevalonate metabolism that they cannot obtain from the medium.

5. The Role of Lipoprotein Receptors on Macrophages *in Vivo*

As we saw in Section III,D,1 above, the acetyl-LDL receptor is expressed *in vivo*. However, acetyl-LDL itself is unlikely to be a natural ligand for this receptor because there is no known mechanism by which LDL could be acetylated in the extracellular fluids of animal tissues. The question of the natural ligand for the acetyl-LDL receptor is important for two reasons. First, the foam cells that are an essential component of atherosclerotic lesions are thought to consist predominantly of macrophages that have entered the arterial wall as blood monocytes and have then become filled with esterified cholesterol derived from the plasma LDL (Ross and Glomset, 1976; Gerrity, 1981; Aqel *et al.*, 1984). However, fully developed macrophages express few LDL receptors and do not accumulate cholesteryl esters when incubated in the presence of LDL *in vitro*. Moreover, in receptor-negative FH homozygotes (who express no LDL receptors at all), macrophages and other cells of the RE system accumulate massive amounts of esterified cholesterol, much of which must be derived ultimately from the plasma LDL. Thus it is reasonable to postulate the existence of a mechanism for the unregulated uptake of a modified form of LDL by macrophages *in vivo* by a pathway independent of the LDL receptor. Indeed, it was this line of thought that led to the identification of the acetyl-LDL receptor on mouse resident macrophages (Goldstein *et al.*, 1979b). The second reason for considering the natural ligand for acetyl-LDL receptors is its possible role in the LDL-receptor-independent catabolism of LDL in the whole body (see Chapter 7, Section III).

When human monocyte-macrophages are incubated with malondialdehyde-treated LDL, the lipoprotein is taken up and degraded via the acetyl-LDL-receptor pathway, leading to accumulation of esterified cholesterol in the cytoplasm of the cells (Shechter *et al.*, 1981). Fogelman *et al.* (1980) have pointed out that aggregated platelets secrete malondialdehyde, formed as a by-product of the oxidative conversion of arachidonic acid into prostaglandins. They suggested that malondialdehyde produced in this way could bring about the local formation of malondialdehyde-LDL, which might then be taken up by macrophages at the site of an atherosclerotic lesion. In support of this suggestion, they have shown that LDL that has been exposed to platelet aggregates is capable of bringing about the deposition of esterified cholesterol in monocyte-macrophages *in vitro*.

Another possible natural ligand for the acetyl-LDL receptor is the EC-modified LDL described by Steinberg and co-workers (see Section III,D,2 above). Although there is, as yet, no direct evidence that EC modification of LDL occurs *in vivo*, electronegative forms of LDL have been detected in various body fluids. Thus, Reichl *et al.* (1975) showed that the LDL-like lipoprotein of human peripheral lymph is more electronegative than the plasma LDL. A lipoprotein with electrophoretic mobility greater than that of plasma LDL has also been detected in the fraction of density 1.019–1.063 g/ml obtained from rabbit inflammatory fluid (Raymond and Reynolds, 1983). More recently, Shaikh *et al.* (1988) have shown that LDL extracted from human atherosclerotic plaques is more electronegative than plasma LDL, and that plaque LDL is recognized by the acetyl-LDL receptor on monocyte-macrophages.

It should be noted that if malondialdehyde-LDL and EC-modified LDL are, in fact, naturally occurring ligands for the acetyl-LDL receptor on macrophages, it is unlikely that they would be present at appreciable concentrations in the circulation. Their formation in or near the arterial wall would make these lipoproteins readily accessible to macrophages in the subendothelium, and any that escaped into the circulation would be removed rapidly by hepatic uptake mediated by acetyl-LDL receptors.

It is unlikely that macrophages are responsible for the 30–50% of total LDL catabolism that must occur by LDL-receptor-independent pathways under normal conditions (the "scavenger-cell pathway" postulated by Goldstein and Brown, 1977). If up to half the LDL in the plasma is degraded in macrophages, either before or after conversion into a modified form in the extracellular fluids, this should be apparent from experiments in which the tissue distribution of a radioactive label is examined after intravenous injections of radiolabeled native LDL. As discussed in Chapter 7 (Section II,E), the observations of Pittman *et al.* (1982) and of Spady *et al.* (1985) with [^{14}C]-sucrose-labeled LDL show that catabolism of LDL by LDL-receptor-independent pathways occurs mainly in parenchymatous cells. Thus, acetyl-LDL receptors on macrophages and other RE cells are unlikely to make more than a small contribution to LDL catabolism in the body as a whole. On the other hand, they may well make a major contribution to the local uptake of modified LDLs by macrophages in the arterial wall and in xanthomas.

Lipoprotein ligands for macrophage receptors other than the acetyl-LDL receptor may also contribute to the formation of foam cells. These ligands include LDL complexed with sulfated proteoglycans, which might be taken up by the macrophage receptors that recognize LDL–dextran sulfate complexes and β-VLDL. (For a discussion of the possible role of β-VLDL in the formation of foam cells *in vivo,* see Mahley, 1982). β-VLDL are very similar in apoprotein and lipid composition to the remnants derived from chylomicrons under normal conditions. It might, therefore, be expected that macrophages would be capable of taking up and degrading chylomicron remnants by the receptor through which they take up β-VLDL (either the LDL receptor or a separate β-VLDL receptor). Ostlund-Lindqvist *et al.* (1983) have, in fact, shown that when macrophages are incubated with human chylomicrons, the lipoprotein lipase secreted by the cells converts the chylomicrons into remnants that are taken up and degraded by the cells by a saturable mechanism. Uptake of chylomicron remnants by macrophages is accompanied by the intracellular accumulation of esterified cholesterol and is therefore a potential mechanism for the formation of foam cells *in vivo*.

Macrophages also participate in the receptor-mediated redistribution of lipids in nerve tissue during the regeneration of injured nerves. Ignatius *et al.* (1987) have shown that the macrophages that enter a damaged nerve secrete large amounts of apoE that become associated with lipid. This apoE-rich lipoprotein is taken up and internalized by LDL receptors on the growth cones at the ends of neurites. Ignatius *et al.* suggest that this mechanism provides the growing tip of the regenerating nerve with cholesterol and phospholipid required for the formation of new axonal membrane. They point out that there is little accumulation of apoE in injured CNS nerves, which do not regenerate. Nevertheless, apoE is secreted by astrocytes throughout the brain (Boyles *et al.*, 1985) and is present in cerebrospinal fluid (Pitas *et al.*, 1987). Moreover, Hofmann *et al.* (1987) have demonstrated the presence of mRNA for the LDL receptor in the CNS of immature and mature rabbits, and Pitas *et al.* (1987) have shown that LDL receptors are expressed by astrocytes associated with the pial membranes of rat and monkey brain. The presence of apoE and a receptor for apoE-containing lipoproteins in brain raises fascinating questions for future investigation (see Mahley, 1988).

E. Adrenal-Cortex Cells

The adrenal cortex requires a regulated supply of free cholesterol as precursor for the synthesis of steroid hormones, in addition to the cholesterol needed for membrane formation. As we saw in Section II of the Chapter 7, uptake of LDL from the plasma by the LDL-receptor pathway is a major source of cholesterol for the adrenals in many species. Pittman *et al.* (1979), for example, have calculated that uptake of LDL by this pathway could supply all the cholesterol needed by the pig's adrenal cortex, even when steroid-hormone synthesis is maximally stimulated by ACTH. In keeping with this functional role for LDL receptors, in most species the uptake of LDL by the LDL-receptor pathway *in vivo,* expressed per gram of tissue, is greater in the adrenals than in any other tissue. Moreover, in membranes prepared from human fetal and adult bovine tissues, LDL-receptor activity is highest in adrenal membranes (Fig. 8.20, and see Brown *et al.*, 1979, for review). In fact, membranes from bovine adrenals were used for the first successful isolation of the LDL receptor and for the preparation of the first anti-LDL-receptor antibody (see Chapter 9).

The regulation of LDL-receptor activity in the adrenal cortex has been investigated in mouse adrenal-tumor cells (Y-1 cells) in culture, in primary cultures of bovine adrenocortical cells, and in human fetal adrenal cells in culture.

Faust *et al.* (1977) have shown that in Y-1 cells maintained in medium containing no LDL, HMG-CoA reductase activity rises to a level high enough to sustain normal growth, but not high enough to permit a normal response to ACTH. When mouse or human LDL is added to the medium, the cells take up LDL by the LDL-receptor pathway, reductase activity falls, and cholesteryl esters accumulate in the cytoplasm. Under these conditions the cells are able to respond fully to ACTH by a maximal increase in the rate of formation of steroid hormones. The response to ACTH is accompanied by a marked increase in the number of LDL receptors expressed on the cell surface. Faust *et al.* (1977) have estimated that when Y-1 cells in culture are maximally stimulated by ACTH in the presence of LDL in the medium, about 75% of the cholesterol used for hormone synthesis in the steady state is supplied by LDL-receptor-mediated uptake of LDL. The behavior of bovine adrenal-cortex cells in culture is similar to that of mouse Y-1 cells, except that in bovine cells grown in the presence of LDL, the cholesterol used for hormone synthesis in the fully stimulated state appears to be derived entirely from LDL taken up by the LDL-receptor pathway (Kovanen *et al.*, 1979b).

The human fetal adrenal cortex converts large quantities of cholesterol into steroids that are used for the formation of estrogens in the placenta (see Myant,

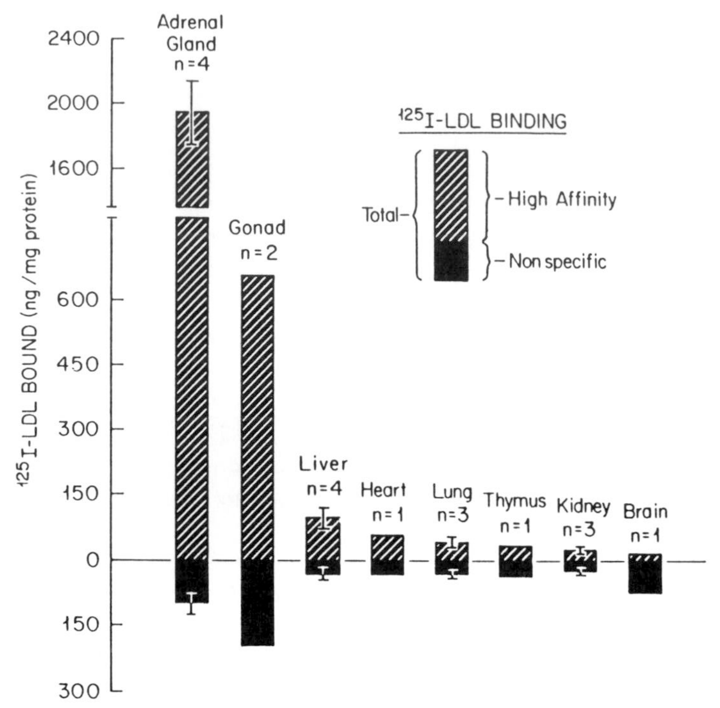

Figure 8.20. Comparison of ^{125}I-labeled low-density lipoprotein ([^{125}I]LDL) binding activity in human fetal membranes prepared from various organs. Membranes were prepared from 20-week-old human fetuses and incubated with human [^{125}I]LDL. High-affinity and nonspecific binding were determined separately. All the gonads were testes. (From Brown *et al.*, 1979, with the permission of the authors.)

1981, Chapter 6). The very high activity of LDL receptors in membranes prepared from human fetal adrenal glands (Fig. 8.20) suggests that a major source of fetal adrenal-cell cholesterol is the uptake of LDL by the LDL-receptor pathway. This is supported by the observations of Carr *et al.* (1980a,b) on steroid-hormone synthesis in human fetal adrenal cells in culture. These cells take up LDL (but not HDL) from the medium by a saturable process, presumably mediated by LDL receptors, and when hormone synthesis is stimulated by addition of ACTH to the medium there is a marked increase in the rate of degradation of LDL by the cells.

Most of the cholesterol used for synthesis of corticosteroids by the adrenals of adult human subjects in the steady state enters the adrenal cortex by uptake from the plasma (Borkowski *et al.*, 1972). However, the contribution of the LDL-receptor pathway to this process is difficult to evaluate. Illingworth *et al.* (1982) have shown that in abetalipoproteinemic patients the rate of production of corticosteroids is normal in the basal state. However, when challenged by a continuous infusion of ACTH for 36 hours, these patients are unable to maintain a normal increase in steroid-hormone output. This suggests that uptake of LDL from the plasma is required for a normal response of the adrenal cortex to prolonged stress in adult man. Homozygous FH patients appear to have a normal capacity for responding to a severe physical challenge, such as a heart operation. Moreover, Allen *et al.* (1983) observed a normal increase in plasma cortisol concentration in FH homozygotes after single injections of ACTH, suggesting that the human adrenal cortex does not depend upon the LDL-receptor pathway when there is a need for increased hormone production. However, it is possible that at the very high plasma LDL concentrations present in these patients, uptake of LDL by nonsaturable pathways compensates for the absence of LDL receptors in the adrenal cortex.

IV. Effects of Growth Factors and Hormones *in Vitro*

A. Growth Factors

Platelet-derived growth factor (PDGF) stimulates LDL-receptor activity in aortic smooth-muscle cells (Chait *et al.*, 1980a) and skin fibroblasts (Witte and Cornicelli, 1980) in culture. The effect of PDGF on receptor activity is due to an increase in the number of receptors expressed on the cell surface, with no change in their affinity for LDL. Other mitogenic proteins that increase LDL-receptor activity in cells in culture include EGF and fibroblast growth factor (Chait *et al.*, 1980b), endothelial-cell-derived growth factor (Witte *et al.*, 1982), and a protein released by human monocyte-macrophages (Chait and Mazzone, 1982). The macrophage protein, which may be the macrophage-derived growth factor described by Leibovich and Ross (1976), enhances LDL-receptor activity in fibroblasts and smooth-muscle cells in culture.

Stimulation of LDL-receptor activity by PDGF and endothelial-cell-derived growth factor is preceded by a rise in HMG-CoA reductase activity and is followed by increased DNA synthesis. This sequence of events suggests that uptake of cholesterol from the medium and endogenous synthesis of cholesterol are both increased in preparation for the increased membrane formation that occurs during cell division in response to the mitogenic stimulus. In a sense, the effect of the mitogen is to raise the level of cholesterol in the regulatory pool at which HMG-CoA reductase and LDL-receptor activity are down-regulated.

B. Hormones

Several hormonal effects on LDL-receptor activity, including the effects of estrogens and thyroid hormones were mentioned in previous chapters in the context of lipoprotein metabolism in the whole body. In this section I describe some of the effects of hormones on LDL receptors that have been observed in cells in culture. (The cell-specific effects of ACTH on adrenocortical cells *in vitro* have already been dealt with in the previous section.)

Semenkovich and Ostlund (1987) have shown that estrogens at pharmacological concentration increase binding and degradation of LDL by the LDL-receptor pathway in human Hep G2 cells grown in the absence of LDL but have no effect on LDL-receptor activity in cultured fibroblasts. This effect observed in Hep G2 cells was accompanied by a decrease in the cell content of free cholesterol, suggesting that stimulation of LDL-receptor activity by large doses of estrogen is secondary to depletion of the regulatory pool of cholesterol in the hepatocyte. These observations may help to explain the marked stimulation of hepatic LDL-receptor activity by large doses of estrogen given to intact rats (Chao *et al.*, 1979b). However, there is no evidence to suggest that estrogens play any part in the regulation of hepatic LDL receptors under physiological conditions. Wade *et al.* (1988) were unable to detect any change in LDL-receptor activity in cultured Hep G2 cells when estrogens were added to the medium at physiological concentrations.

Insulin stimulates high-affinity binding of human LDL by cultured human fibroblasts (Chait *et al.*, 1978) and rat hepatocytes (Salter *et al.*, 1987b). Wade *et al.* (1988) have investigated the effect of insulin on LDL-receptor activity in Hep G2 cells in culture, using ligand blotting to determine the amount of LDL-receptor protein expressed by the cells. Addition of insulin to the medium increased high-affinity binding and degradation of human LDL by cells preincubated in the presence of LPDS. This effect was concentration dependent, reaching a maximum at about 100 mU of insulin/ml. As shown in Fig. 8.21, the increase in receptor activity determined in intact cells was accompanied by a parallel increase in the amount of receptor protein that could be extracted from the cells. Stimulation of LDL-receptor activity by insulin was observed even when activity was suppressed by addition of LDL to the medium. Wade *et al.* (1988) point out that the ability of insulin to stimulate LDL-receptor activity in the presence of LDL may explain why hepatocytes express LDL receptors *in vivo* despite the high concentration of LDL that must exist in the space of Disse. A relationship between the effect of insulin on LDL-receptor activity in hepatocytes *in vitro* and the hyperlipidemia of diabetes has not been established. However, in this regard it is worth noting that Howard *et al.* (1986) have shown that the FCR of plasma LDL via the LDL-receptor pathway is directly related to the plasma insulin concentration in normal human subjects.

Triiodothyronine at physiological concentrations increases the LDL-receptor-mediated binding and degradation of LDL by human fibroblasts in culture, due to an increase in receptor number without significant change in the affinity of LDL receptors for LDL (Chait *et al.*, 1979). The effects of insulin and thyroid hormone on LDL-receptor activity in human fibroblasts are additive. Thyroid hor-

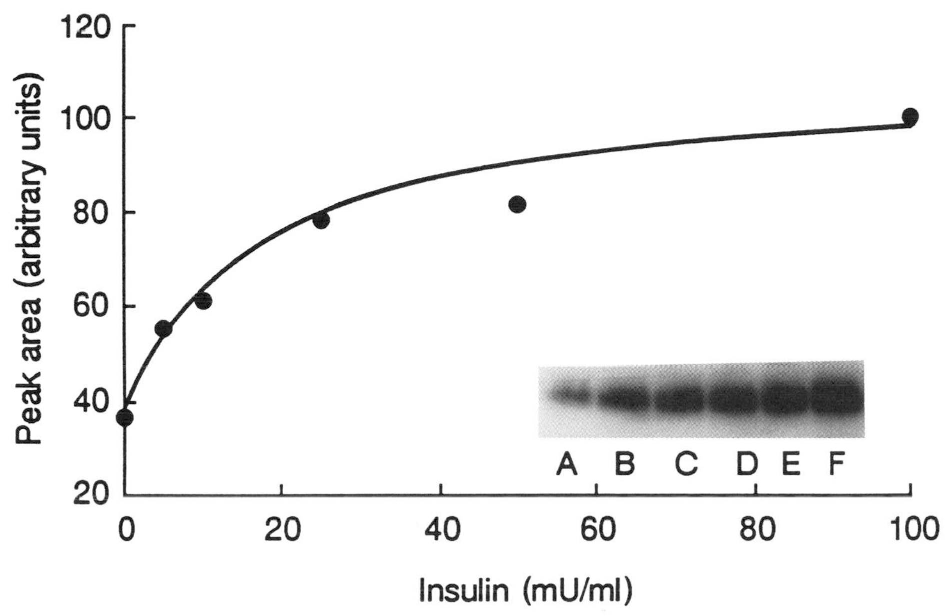

Figure 8.21. The effect of insulin on the amount of LDL-receptor protein in Hep G2 cells, as determined by ligand blotting. Hep G2 cells were incubated for 48 hours in medium containing 10% lipoprotein-deficient serum and insulin at the concentrations shown on the horizontal axis. Ligand blots were performed on solubilized extracts of the cells with ^{125}I-labeled β-VLDL obtained from cholesterol-fed rabbits as ligand. After incubation with the ligand the nitrocellulose blots were dried and exposed to autoradiography for 2–7 days. Radioactivity in the radioactive bands, present at a position corresponding to M_r 130,000, was determined by densitometric scanning. The area under the peak was expressed in arbitrary units. The inset shows an autoradiograph of a ligand blot from one experiment of extracts of cells preincubated with (A) 0, (B) 5 mU, (C) 10 mU, (D) 25 mU, (E) 50 mU, and (F) 100 mU insulin/ml. (From Wade *et al.*, 1988, with the permission of the authors.)

mone also increases the binding of human LDL by rat hepatocytes in culture (Salter *et al.*, 1988). Thompson *et al.* (1981) showed that the raised plasma LDL concentration in hypothyroid patients is due partly to defective LDL-receptor catabolism of LDL. In the study shown in Fig. 8.22 a myxoedematous patient, whose plasma LDL-cholesterol concentration was 319 mg/100 ml, was given an intravenous injection of a mixture of ^{131}I-labeled native LDL and cyclohexanedione-modified ^{125}I-labeled LDL. Plasma radioactivity was then measured daily for 14 days and the FCRs of native and CHD-LDL were estimated by the methods described in Chapter 7. The FCR of LDL via the LDL-receptor pathway was calculated as the difference between the FCRs for native and CHD-LDL. The observations were repeated after the patient had been given 2 mg of L-thyroxine daily for 6 weeks. Before treatment, the FCR of native LDL was 0.12

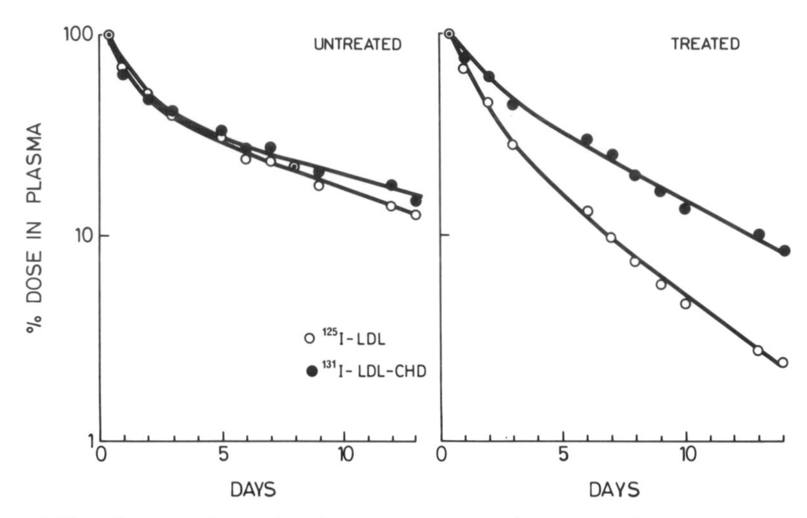

Figure 8.22. Plasma radioactivity disappearance curves in a myxoedematous patient after intravenous injection of ^{125}I-labeled native LDL (○) and cyclohexanedione-modified ^{131}I-labeled LDL (●). (A) before treatment; (B) after treatment with L-thyroxine (2 mg/day) for 6 weeks. (From Thompson *et al.*, 1981.)

pools/day, compared with 0.32 pools/day in normal control subjects. The low FCR in the patient was due to a fall in FCR by both the LDL receptor and LDL-receptor-independent pathways. After treatment, the plasma LDL-cholesterol concentration fell to 137 mg/100 ml and the FCR of native LDL increased to 0.29 pools/day, the rise being due to increased FCRs by both pathways (see Table 10.3). Presumably, diminished LDL-receptor activity in myxoedema is due to the absence of a permissive or stimulatory action of thyroid hormone on LDL receptors in the liver and other tissues. In agreement with this, Scarabottolo *et al.* (1986) have shown that surgical thyroidectomy of rats diminishes high-affinity binding of β-VLDL by membranes prepared from their livers. The specificity of this effect was confirmed by ligand blotting of solubilized liver proteins with β-VLDL as ligand.

Saturable binding of human LDL by rat hepatocytes in culture is decreased by dexamethasone (Salter *et al.*, 1987b). Henze *et al.* (1983) have shown that cortisol decreases the saturable degradation of human LDL by human fibroblasts in culture without affecting high-affinity binding.

Human chorionic gonadotropin stimulates LDL receptor activity in human ovarian granulosa cells in culture by increasing the cell content of receptor protein detectable by immunoblotting (Golos *et al.*, 1986). Golos *et al.* have shown that the rise in receptor number is due to increased synthesis of receptors rather than to a fall in their rate of degradation. Stimulation of receptor synthesis by gonadotropin is not mediated by increased utilization of cellular cholesterol for steroid-hormone formation, since stimulation occurs when hormone production is blocked. Thus, it seems likely that gonadotropin stimulates receptor synthesis by an effect that is independent of the regulatory pool of free cholesterol in the cell.

References

Allen, J. M., Thompson, G. R., and Myant, N. B. (1983). Normal adrenocortical response to adrenocorticotrophic hormone in patients with homozygous familial hypercholesterolaemia. *Clin. Sci.* **65,** 99–101.

Andersen, J. M., Nervi, F. O., and Dietschy, J. M. (1977). Rate constants for the uptake of cholesterol from various intestinal and serum lipoprotein fractions by the liver of the rat *in vivo*. *Biochim. Biophys. Acta* **486,** 298–307.

Anderson, R. G. W., Goldstein, J. L., and Brown, M. S. (1976). Localization of low density lipoprotcin rcccptors on plasma membrane of normal human fibroblasts and their absence in cells from a familial hypercholesterolemia homozygote. *Proc. Natl. Acad. Sci. U.S.A.* **73,** 2434–2438.

Andcrson, R. G. W., Brown, M. S., and Goldstein, J. L. (1977a). Role of the coated endocytic vesicle in the uptake of receptor-bound low density lipoprotein in human fibroblasts. *Cell* **10,** 351–364.

Anderson, R. G. W., Goldstein, J. L., and Brown, M. S. (1977b). A mutation that impairs the ability of lipoprotein receptors to localise in coated pits on the cell surface of human fibroblasts. *Nature (London)* **270,** 695–699.

Anderson, R. G. W., Vasilc, E., Mcllo, R. J., Brown, M. S., and Goldstein, J. L. (1978). Immunocytochemical visualization of coated pits and vesicles in human fibroblasts: relation to low density lipoprotein receptor distribution. *Cell* **15,** 919–933.

Anderson, R. G. W., Brown, M. S., Beisiegel, U., and Goldstein, J. L. (1982). Surface distribution and recycling of the low density lipoprotein as visualized with antireceptor antibodies. *J. Cell Biol.* **93,** 523–531.

Aqel, N. M., Ball, R. Y., Waldmann, H., and Mitchinson, M. J. (1984). Monocytic origin of foam cells in human atherosclerotic plaques. *Atherosclerosis* **53,** 265–271.

Arbeeny, C. M., and Rifici, V. A. (1984). The uptake of chylomicron remnants and very low density lipoprotein remnants by the perfused rat liver. *J. Biol. Chem.* **259,** 9662–9666.

Arbeeny, C. M., Rifici, V. A., Handley, D. A., and Eder, H. A. (1987). Determinants of the uptake of very low density lipoprotein remnants by the perfused rat liver. *Metabolism* **36,** 1106–1113.

Attie, A. D., Pittman, R. C., and Steinberg, D. (1980). Metabolism of native and of lactosylated human low density lipoprotein: evidence for two pathways for catabolism of exogenous proteins in rat hepatocytes. *Proc. Natl. Acad. Sci. U.S.A.* **77,** 5923–5927.

Aulinskas, T. H., van der Westhuyzen, D. R., Bierman, E. L., Gevers, W., and Coetzee, G. A. (1981). Retro-endocytosis of low density lipoprotein by cultured bovine smooth muscle cells. *Biochim. Biophys. Acta* **664,** 255–265.

Aulinskas, T. H., Oram, J. F., Bierman, E. L., Coetzee, G. A., Gevers, W., and van der Westhuyzen, D. R. (1985). Retro-endocytosis of low density lipoprotein by cultured human skin fibroblasts. *Arteriosclerosis* **5,** 45–54.

Bailey, J. M. (1973). Regulation of cell cholesterol content. *In* "Atherogenesis: Initiating Factors" *Ciba Found. Symp. 12 (New Ser.)* pp. 63–92. Elsevier, Amsterdam.

Balasubramaniam, S., Goldstein, J. L., Faust, J. R., and Brown, M. S. (1976). Evidence for regulation of 3-hydroxy-3-methylglutaryl coenzyme A reductase activity and cholesterol synthesis in nonhepatic tissues of rat. *Proc. Natl. Acad. Sci. U.S.A.* **73,** 2564–2568.

Basu, S. K., Goldstein, J. L., and Brown, M. S. (1978). Characterization of the low density lipoprotein receptor in membranes prepared from human fibroblasts. *J. Biol. Chem.* **253,** 2852–3856.

Basu, S. K., Goldstein, J. L., Anderson, R. G. W., and Brown, M. S. (1981). Monensin interrupts the recycling of low density lipoprotein receptors in human fibroblasts. *Cell* **24,** 493–502.

Bates, S. R., and Rothblat, G. H. (1974). Regulation of cellular sterol flux and synthesis of human serum lipoproteins. *Biochim. Biophys. Acta* **360,** 38–55.

Beisiegel, U., Kita, T., Anderson, R. G. W., Schneider, W. J., Brown, M. S., and Goldstein, J. L. (1981a). Immunologic cross-reactivity of the low density lipoprotein receptor from bovine adrenal cortex, human fibroblasts, canine liver and adrenal gland, and rat liver. *J. Biol. Chem.* **256,** 4071–4078.

Beisiegel, U., Schneider, W. J., Goldstein, J. L., Anderson, R. G. W., and Brown, M. S. (1981b). Monoclonal antibodies to the low density lipoprotein receptor as probes for study of receptor-mediated endocytosis and the genetics of familial hypercholesterolemia. *J. Biol. Chem.* **256,** 11923–11931.

Beisiegel, U., Schneider, W. J., Brown, M. S., and Goldstein, J. L. (1982). Immunoblot analysis of low density lipoprotein receptors in fibroblasts from subjects with familial hypercholesterolemia. *J. Biol. Chem.* **257,** 13150–13156.

Beisiegel, U., Weber, W., Havinga, J. R., Ihrke, G., Hui, D. Y., Wernette-Hammond, M. E., Turck, C. W., Innerarity, T. L., and Mahley, R. W. (1988). Apolipoprotein E-binding proteins isolated from dog and human liver. *Arteriosclerosis* **8,** 288–297.

Bilheimer, D. W., Ho, Y. K., Brown, M. S., Anderson, R. G. W., and Goldstein, J. L. (1978). Genetics of the low density lipoprotein receptor: diminished receptor activity in lymphocytes from heterozygotes with familial hypercholesterolemia. *J. Clin. Invest.* **61,** 678–696.

Borensztajn, J., Getz, G. S., Padley, R. J., and Kotlar, T. J. (1982). The apoprotein B-independent hepatic uptake of chylomicron remnants. *Biochem. J.* **204,** 609–612.

Borkowski, A., Delcroix, C., and Levin, S. (1972). Metabolism of adrenal cholesterol in man. I. *In vivo* studies. *J. Clin. Invest.* **51,** 1664–1678.

Boyles, J. K., Pitas,R. E., Wilson, E., Mahley, R. W., and Taylor, J. M. (1985). Apolipoprotein E associated with astrocytic glia of the central nervous system and with nonmyelinating glia of the peripheral nervous system. *J. Clin. Invest.* **76,** 1501–1513.

Bradley, W. A., and Gianturco, S. H. (1986). ApoE is necessary and sufficient for the binding of large triglyceride-rich lipoproteins to the LDL receptor; apoB is unnecessary. *J. Lipid Res.* **27,** 40–48.

Brodsky, F. M. (1985). Clathrin structure characterized with monoclonal antibodies. II. Identification of *in vivo* forms of clathrin. *J. Cell Biol.* **101,** 2055–2062.

Brodsky, F. M., Galloway, C. J., Blank, G. S., Jackson, A. P., Seow, H-F., Drickamer, K., and Parham, P. (1987). Localization of clathrin light-chain sequences mediating heavy-chain binding and coated vesicle diversity. *Nature (London)* **326,** 203–205.

Brown, M. S., and Goldstein, J. L. (1974). Familial hypercholesterolemia: defective binding of lipoproteins to cultured fibroblasts associated with impaired regulation of 3-hydroxy-3-methylglutaryl coenzyme A reductase activity. *Proc. Natl. Acad. Sci. U.S.A.* **71,** 788–792.

Brown, M. S., and Goldstein, J. L. (1975). Regulation of the activity of the low density lipoprotein receptor in human fibroblasts. *Cell* **6,** 307–316.

Brown, M. S. and Goldstein, J. L. (1976). Receptor-mediated control of cholesterol metabolism. *Science,* **191,** 150–154.

Brown, M. S., and Goldstein, J. L. (1983a). Lipoprotein receptors in the liver. Control signals for plasma cholesterol traffic. *J. Clin. Invest.* **72,** 743–747.

Brown, M. S., and Goldstein, J. L. (1983b). Lipoprotein metabolism in the macrophage. Implications for cholesterol deposition in atherosclerosis. *Ann. Rev. Biochem.* **52,** 223–261.

Brown, M. S., Dana, S. E., and Goldstein, J. L. (1974). Regulation of 3-hydroxy-3-methylglutaryl coenzyme A reductase activity in cultured human fibroblasts. Comparison of cells from a normal subject and from a patient with homozygous familial hypercholesterolemia. *J. Biol. Chem.* **249,** 789–796.

Brown, M. S., Faust, J. R., and Goldstein, J. L. (1975). Role of the low density lipoprotein receptor in regulating the content of free and esterified cholesterol in human fibroblasts. *J. Clin. Invest.* **55,** 783–793.

Brown, M. S., Faust, J. R., Goldstein, J. L., Kaneko, I., and Endo, A. (1978). Induction of 3-hydroxy-3-methylglutaryl coenzyme A reductase activity in human fibroblasts incubated with compactin (ML-236B), a competitive inhibitor of the reductase. *J. Biol. Chem.* **253,** 1121–1128.

Brown, M. S., Kovanen, P. T., and Goldstein, J. L. (1979). Receptor-mediated uptake of lipoprotein-cholesterol and its utilization for steroid synthesis in the adrenal cortex. *Recent Prog. Horm. Res.* **35,** 215–257.

Brown, M. S., Basu, S. K., Falck, J. R., Ho, Y. K., and Goldstein, J. L. (1980). The scavenger cell pathway for lipoprotein degradation: specificity of the binding site that mediates the uptake of negatively-charged LDL by macrophages. *J. Supramol. Struc.* **13,** 67–81.

Brown, M. S., Anderson, R. G. W., Basu, S. K., and Goldstein, J. L. (1982). Recycling of cell-surface receptors: observations from the LDL receptor system. *Cold Spring Harbor Symp. Quant. Biol.* **46,** 713–721.

Brown, M. S., Anderson, R. G. W., and Goldstein, J. L. (1983). Recycling receptors: the normal-trip itinerary of migrant membrane proteins. *Cell* **32,** 663–667.

Carella, M., and Cooper, A. D. (1979). High affinity binding of chylomicron remnants to rat liver plasma membranes. *Proc. Natl. Acad. Sci. U.S.A.* **76,** 338–342.

Carr, B. R., Parker, C. R., MacDonald, P. C., and Simpson, E. R. (1980a). Metabolism of high density lipoprotein by human fetal adrenal tissue. *Endocrinology* **107,** 1849–1854.

Carr, B. R., Porter, J. C., MacDonald, P. C., and Simpson, E. R. (1980b). Metabolism of low density lipoprotein by human fetal adrenal tissue. *Endocrinology* **107,** 1034–1040.

Chait, A., and Mazzone, T. (1982). A secretory product of human monocyte-derived macrophages stimulates low density lipoprotein receptor activity in arterial smooth muscle cells and skin fibroblasts. *Arteriosclerosis* **2,** 134–141.

Chait, A., Bierman, E. L., and Albers, J. J. (1978). Regulatory role of insulin in the degradation of low density lipoprotein by cultured human skin fibroblasts. *Biochim. Biophys. Acta* **529,** 292–299.

Chait, A., Bierman, E. L., and Albers, J. J. (1979). Low-density lipoprotein receptor activity in cultured human skin fibroblasts. Mechanism of insulin-induced stimulation. *J. Clin. Invest.* **64,** 1309–1319.

Chait, A., Ross, R., Albers, J. J., and Bierman, E. L. (1980a). Platelet-derived growth factor stimulates activity of low density lipoprotein receptors. *Proc. Natl. Acad. Sci. U.S.A.* **77,** 4084–4088.

Chait, A., Ross, R., Albers, J. J., and Bierman, E. L. (1980b). Stimulation of low density lipoprotein receptor activity by platelet-derived growth factor. *Clin. Res.* **28,** 517A.

Chait, A., Henze, K., Mazzone, T., Jensen, M., and Hammond, W. (1982). Low density lipoprotein receptor activity in freshly isolated human blood monocytes and lymphocytes. *Metabolism* **31,** 721–727.

Chao, Y. S., Chen, G. C., Windler, E., Kane, J. P., and Havel, R. J. (1979a). Catabolism of covalently modified human low density lipoprotein in the rat. *Fed. Proc., Fed. Am. Soc. Exp. Biol.* **38,** 896.

Chao, Y-S., Windler, E. E., Chen, G. C., and Havel, R. J. (1979b). Hepatic catabolism of rat and human lipoproteins in rats treated with 17α-ethinyl estradiol. *J. Biol. Chem.* **254,** 11360–11366.

Chao, Y-S., Jones, A. L., Hrader, G. T., Windler, E. E. T., and Havel, R. J. (1981). Autoradiographic localization of the sites of uptake, cellular transport, and catabolism of low density lipoproteins in the liver of normal and estrogen-treated rats. *Proc. Natl. Acad. Sci. U.S.A.* **78,** 597–601.

Crowther, R. A., and Pearse, B. M. F. (1981). Assembly and packing of clathrin into coats. *J. Cell Biol.* **91,** 790–797.

Crowther, R. A., Finch, J. T., and Pearse, B. M. F. (1976). On the structure of coated vesicles. *J. Mol. Biol.* **103,** 785–798.

Curtiss, L. K., and Edgington, T. S. (1976). Regulatory serum lipoproteins: regulation of lymphocyte stimulation by a species of low density lipoprotein. *J. Immunol.* **116,** 1452–1458.

Curtiss, L. K., and Edgington, T. S. (1977). Effect of LDL-In, a normal immunoregulatory human serum low density lipoprotein, on the interaction of macrophages with lymphocytes proliferating in response to mitogen and allogeneic stimulation. *J. Immunol.* **118,** 1966–1970.

Curtiss, L. K., and Edgington, T. S. (1978). Identification of a lymphocyte surface receptor for low density lipoprotein inhibitor, an immunoregulatory species of normal human serum low density lipoprotein. *J. Clin. Invest.* **61,** 1298–1308.

Cuthbert, J. A., and Lipsky, P. E. (1983). Immunoregulation by low density lipoproteins in man: low density lipoprotein inhibits mitogen-stimulated human lymphocyte proliferation after initial activation. *J. Lipid Res.* **24,** 1512–1524.

Cuthbert, J. A., and Lipsky, P. E. (1984a). Modulation of human lymphocyte responses by low density lipoprotein (LDL): enhancement but not immunosuppression is mediated by LDL receptors. *Proc. Natl. Acad. Sci. U.S.A.* **81,** 4539–4543.

Cuthbert, J. A., and Lipsky, P. E. (1984b). Immunoregulation by low density lipoproteins in man. Inhibition of mitogen-induced T lymphocyte proliferation by interference with transferrin metabolism. *J. Clin. Invest.* **73,** 992–1003.

Cuthbert, J. A., East, C. A., Bilheimer, D. W., and Lipsky, P. E. (1986). Detection of familial hypercholesterolemia by assaying functional low-density-lipoprotein receptors on lymphocytes. *N. Engl. J. Med.* **314,** 879–883.

Davis, C. G., Elhammer, A., Russell, D. W., Schneider, W. J., Kornfeld, S., Brown, M. S., and Goldstein, J. L. (1986). Deletion of clustered O-linked carbohydrates does not impair function of low density lipoprotein receptor in transfected fibroblasts. *J. Biol. Chem.* **261,** 2828–2838.

Davis, C. G., Goldstein, J. L., Südhof, T. C., Anderson, R. G. W., Russell, D. W., and Brown, M. S. (1987). Acid-dependent ligand dissociation and recycling of LDL receptor by growth factor homology region. *Nature (London)* **326,** 760–765.

Edge, S. B., Hoeg, J. M., Triche, T., Schneider, P. D., and Brewer, H. B. (1986). Cultured human hepatocytes. Evidence for metabolism of low density lipoproteins by a pathway independent of the classical low density lipoprotein receptor. *J. Biol. Chem.* **261,** 3800–3806.

Esser, V., and Russell, D. W. (1988). Transport-deficient mutations in the low density lipoprotein receptor. Alterations in the cysteine-rich and cysteine-poor regions of the protein block intracellular transport. *J. Biol. Chem.* **263,** 13276–13281.

Faust, J. R., Goldstein, J. L., and Brown, M. S. (1977). Receptor-mediated uptake of low density lipoprotein and utilization of its cholesterol for steroid synthesis in cultured mouse adrenal cells. *J. Biol. Chem.* **252,** 4861–4871.

Florén, C-H., and Nilsson, A. (1977). Binding, interiorization and degradation of cholesteryl ester-labelled chylomicron-remnant particles by rat hepatocyte monolayers. *Biochem. J.* **168,** 483–494.

Florén, C-H., Albers, J. J., Kudchodkar, B. J., and Bierman, E. L. (1981). Receptor-dependent uptake of human chylomicron remnants by cultured skin fibroblasts. *J. Biol. Chem.* **256,** 425–433.

Fogelman, A. M., Shechter, I., Seager, J., Hokom, M., Child, J. S., and Edwards, P. A. (1980). Malondialdehyde alteration of low density lipoproteins leads to cholesteryl ester accumulation in human monocyte-macrophages. *Proc. Natl. Acad. Sci. U.S.A.* **77,** 2214–2218.

Fogelman, A. M., Haberland, M. E., Seager, J., Hokom, M., and Edwards, P. A. (1981). Factors regulating the activities of the low density lipoprotein receptor and the scavenger receptor on human monocyte-macrophages. *J. Lipid Res.* **22,** 1131–1141.

Forgac, M., Cantley, L., Wiedenmann, B., Altstiel, L., and Branton, D. (1983). Clathrin-coated vesicles contain an ATP-dependent proton pump. *Proc. Natl. Acad. Sci. U.S.A.* **80,** 1300–1303.

Gavigan, S. J. P., Patel, D. D., Soutar, A. K., and Knight, B. L. (1988). An antibody to the low-density lipoprotein (LDL) receptor that partially inhibits the binding of LDL to cultured human fibroblasts. *Eur. J. Biochem.* **171,** 355–361.

Gerrity, R. G. (1981). The role of the monocyte in atherogenesis: 1. Transition of blood-borne monocytes into foam cells in fatty lesions. *Am. J. Pathol.* **103,** 181–190.

Geuze, J. H., Slot, J. W., Gerja Strous, A. M., and Schwartz, A. L. (1983). The pathway of the asialoglycoprotein-ligand during receptor-mediated endocytosis: a morphological study with colloidal gold/ligand in the human hepatoma cell line Hep G2. *Eur. J. Cell Biol.* **32,** 38–44.

Gianturco, S. H., Bradley, W. A., Gotto, A. M., Jr., Morrisett, J. D., and Peavy, D. L. (1982a). Hypertriglyceridemic very low density lipoproteins induce triglyceride synthesis and accumulation of mouse peritoneal macrophages. *J. Clin. Invest.* **70,** 168–178.

Gianturco, S. H., Brown, F. B., Gotto, A. M., and Bradley, W. A. (1982b). Receptor-mediated uptake of hypertriglyceridemic very low density lipoproteins by normal human fibroblasts. *J. Lipid Res.* **23,** 984–993.

Gianturco, S. H., Gotto, A. M., Hwang, S-L. C., Karlin, J. B., Lin, A. H. Y., Prasad, S. C., and Bradley, W. A. (1983). Apolipoprotein E mediates uptake of S_f 100–400 hypertriglyceridemic very low density lipoproteins by the low density lipoprotein receptor pathway in normal human fibroblasts. *J. Biol. Chem.* **258,** 4526–4533.

Goldstein, J. L., and Brown, M. S. (1974). Binding and degradation of low density lipoproteins by cultured human fibroblasts. Comparison of cells from a normal subject and from a patient with homozygous familial hypercholesterolemia. *J. Biol. Chem.* **249,** 5153–5162.

Goldstein, J. L., and Brown, M. S. (1976). The LDL pathway in human fibroblasts: a receptor-mediated mechanism for the regulation of cholesterol metabolism. *In* "Current Topics in Cellular Regulation" (B. Horecker and E. R. Stadtman, eds.), Vol. 11, pp. 147–181. Academic Press, New York.

Goldstein, J. L., and Brown, M. S. (1977). The low-density lipoprotein pathway and its relation to atherosclerosis. *Ann. Rev. Biochem.* **46,** 897–930.

Goldstein, J. L., and Brown, M. S. (1973). Familial hypercholesterolemia: identification of a defect in the regulation of 3-hydroxy-3-methylglutaryl coenzyme A reductase activity associated with overproduction of cholesterol. *Proc. Natl. Acad. Sci. U.S.A.* **70,** 2804–2808.

Goldstein, J. L., Basu, S. K., Brunschede, G. Y., and Brown, M. S. (1976). Release of low density lipoprotein from its cell surface receptor by sulfated glycosaminoglycans. *Cell* **7,** 85–95.

Goldstein, J. L., Anderson, R. G. W., and Brown, M. S. (1979a). Coated pits, coated vesicles, and receptor-mediated endocytosis. *Nature (London)* **279,** 679–685.

Goldstein, J. L., Ho, Y. K., Basu, S. K., and Brown, M. S. (1979b). Binding site on macrophages that mediates uptake and degradation of acetylated low density lipoprotein, producing massive cholesterol deposition. *Proc. Natl. Acad. Sci. U.S.A.* **76,** 333–337.

Goldstein, J. L., Ho, Y. K., Brown, M. S., Innerarity, T. L., and Mahley, R. W. (1980). Cholesteryl ester accumulation in macrophages resulting from receptor-mediated uptake and degradation of hypercholesterolemic canine β-very low density lipoproteins. *J. Biol. Chem.* **255,** 1839–1848.

Goldstein, J. L., Basu, S. K., and Brown, M. S. (1983). Receptor-mediated endocytosis of low-density lipoprotein in cultured cells. *Methods Enzymol.* **98,** 241–260.

Goldstein, J. L., Brown, M. S., Anderson, R. G. W., Russell, D. W., and Schneider, W. J. (1985). Receptor-mediated endocytosis: concepts emerging from the LDL receptor system. *Ann. Rev. Cell Biol.* **1,** 1–39.

Golos, T. G., August, A. M., and Strauss, J. F. (1986). Expression of low density lipoprotein receptor in cultured human granulosa cells: regulation by human chorionic gonadotropin, cyclic AMP, and sterol. *J. Lipid Res.* **27,** 1089–1096.

Haberland, M. E., Fogelman, A. M., and Edwards, P. A. (1982). Specificity of receptor-mediated recognition of malondialdehyde-modified low density lipoproteins. *Proc. Natl. Acad. Sci. U.S.A.* **79,** 1712–1716.

Hamilton, R. L. (1985). Subcellular dissection and characterization of plasma lipoprotein secretory (Golgi) and endocytic (multivesicular bodies) compartments of rat hepatocytes. *In* ''Receptor-mediated Uptake in the Liver'' (H. Greten, E. Windler, and U. Beisiegel, eds.), pp. 125–133. Springer-Verlag, Berlin.

Handley, D. A., Arbeeny, C. M., Eder, H. A., and Chien, S. (1981). Hepatic binding and internalization of low density lipoprotein-gold conjugates in rats treated with 17α-ethinylestradiol. *J. Cell Biol.* **90,** 778–787.

Harders-Spengel, K., Wood, C. B., Thompson, G. R., Myant, N. B., and Soutar, A. K. (1982). Differences in saturable binding of low density lipoprotein to liver membranes from normocholesterolemic subjects and patients with heterozygous familial hypercholesterolemia. *Proc. Natl. Acad. Sci. U.S.A.* **79,** 6355–6359.

Harmony, J. A. K., and Hui, D. Y. (1981). Inhibition by membrane-bound low-density lipoproteins of the primary inductive events of mitogen-stimulated lymphocyte activation. *Cancer Res.* **41,** 3799–3802.

Havekes, L., Van Hinsbergh, V., Kempen, H. J., and Emeis, J. (1983). The metabolism *in vitro* of human low-density lipoprotein by the human hepatoma cell line Hep G2. *Biochem. J.* **214,** 951–958.

Havekes, L. M., Schouten, D., DeWit, E. C. M., Cohen, L. H., Griffioen, M., Van Hinsberg, V. W. M., and Princen, H. M. G. (1986). Stimulation of the LDL receptor in the human hepatoma cell line Hep G2 by high-density serum fractions. *Biochim. Biophys. Acta* **875,** 236–246.

Havel, R. J. (1986). Functional activities of hepatic lipoprotein receptors. *Ann. Rev. Physiol.* **48,** 119–134.

Henriksen, T., Mahoney, E. M., and Steinberg, D. (1983). Interactions of plasma lipoproteins with endothelial cells. *Ann. N.Y. Acad. Sci.* **401,** 102–116.

Henze, K., Chait, A., Albers, J. J., and Bierman, E. L. (1983). Hydrocortisone decreases the internalization of low density lipoprotein in cultured human fibroblasts and arterial smooth muscle cells. *Eur. J. Clin. Invest.* **13,** 171–177.

Herz, J., Hamann, U., Rogne, S., Myklebost, O., Gausepohl, H., and Stanley, K. K. (1988). Surface location and high affinity for calcium of a 500 kDa liver membrane protein closely related to the LDL-receptor suggest a physiological role as lipoprotein receptor. *EMBO J.* **7,** 4119–4127.

Heuser, J., and Evans, L. (1980). Three-dimensional visualization of coated vesicle formation in fibroblasts. *J. Cell Biol.* **84,** 560–583.

Hirokawa, N., and Heuser, J. (1982). The inside and outside of gap-junction membranes visualized by deep etching. *Cell* **30,** 395–406.

Ho, Y. K., Brown, M. S., Bilheimer, D. W., and Goldstein, J. L. (1976a). Regulation of low density lipoprotein receptor activity in freshly isolated human lymphocytes. *J. Clin. Invest.* **58,** 1465–1474.

Ho, Y. K., Brown, M. S., Kayden, H. J., and Goldstein, J. L. (1976b). Binding, internalization and hydrolysis of low density lipoprotein in long-term lymphoid cell lines from a normal subject and a patient with homozygous familial hypercholesterolemia. *J. Exp. Med.* **144,** 444–455.

Ho, Y. K., Faust, J. R., Bilheimer, D. W., Brown, M. S., and Goldstein, J. L. (1977). Regulation of cholesterol synthesis by low density lipoprotein in isolated human lymphocytes. Comparison of cells from normal subjects and patients with homozygous familial hypercholesterolemia and abetalipoproteinemia. *J. Exp. Med.* **145,** 1531–1549.

Hoeg, J. M., Demosky, S. J., Schaefer, E. J., Starzl, T. E., and Brewer, H. B. (1984). Characterization of hepatic low density lipoprotein binding and cholesterol metabolism in normal and homozygous familial hypercholesterolemic subjects. *J. Clin. Invest.* **73,** 429–436.

Hofmann, S. L., Russell, D. W., Goldstein, J. L., and Brown, M. S. (1987). mRNA for low density lipoprotein receptor in brain and spinal cord of immature and mature rabbits. *Proc. Natl. Acad. Sci. U.S.A.* **84,** 6312–6316.

Hornick, C. A., Hamilton, R. L., Spaziani, E., Enders, G. H., and Havel, R. J. (1985). Isolation and characterization of multivesicular bodies from rat hepatocytes: an organelle distinct from secretory vesicles of the Golgi apparatus. *J. Cell Biol.* **100,** 1558–1569.

Howard, B. V., Egusa, G., Beltz, W. F., Kesaniemi, Y. A., and Grundy, S. M. (1986). Compensatory mechanisms governing the concentration of plasma low density lipoprotein. *J. Lipid Res.* **27,** 11–20.

Huettinger, M., Schneider, W. J., Ho, Y. K., Goldstein, J. L., and Brown, M. S. (1984). Use of monoclonal anti-receptor antibodies to probe the expression of the low density lipoprotein receptor in tissues of normal and Watanabe heritable hyperlipidemic rabbits. *J. Clin. Invest.* **74,** 1017–1026.

Hui, D. Y., Harmony, J. A. K., Innerarity, T. L., and Mahley, R. W. (1980). Immunoregulatory plasma lipoproteins. Role of apoprotein E and apoprotein B. *J. Biol. Chem.* **255,** 11775–11781.

Hui, D. Y., Innerarity, T. L., and Mahley, R. W. (1981). Lipoprotein binding to canine hepatic membranes. Metabolically distinct apo-E and apo-B,E receptors. *J. Biol. Chem.* **256,** 5646–5655.

Hui, D. Y., Innerarity, T. L., and Mahley, R. W. (1984). Defective hepatic lipoprotein receptor binding of β-very low density lipoproteins from type III hyperlipoproteinemic patients. Importance of apolipoprotein E. *J. Biol. Chem.* **259,** 860–869.

Hui, D. Y., Brecht, W. J., Hall, E. A., Friedman, G., Innerarity, T. L., and Mahley, R. W. (1986). Isolation and characterization of the apolipoprotein E receptor from canine and human liver. *J. Biol. Chem.* **261,** 4256–4267.

Iacopetta, B. J., Rothenberger, S., and Kühn, L. C. (1988). A role for the cytoplasmic domain in transferrin receptor sorting and coated pit formation during endocytosis. *Cell* **54,** 485–489.

Ignatius, M. J., Shooter, E. M., Pitas, R. E., and Mahley, R. W. (1987). Lipoprotein uptake by neuronal growth cells *in vitro*. *Science* **236,** 959–962.

Illingworth, D. R., Kenny, T. A., and Orwoll, E. S. (1982). Adrenal function in heterozygous and homozygous hypobetalipoproteinemia. *J. Clin. Endocrinol. Metab.* **54,** 27–33.

Imber, M. J., Pizzo, S. V., Johnson, W. J., and Adams, D. O. (1982). Selective diminution of the binding of mannose by murine macrophages in the late stages of activation. *J. Biol. Chem.* **257,** 5129–5135.

Innerarity, T. L., Kempner, E. S., Hui, D. Y., and Mahley, R. W. (1981). Functional unit of the low density lipoprotein receptor of fibroblasts: A 100,000-dalton structure with multiple binding sites. *Proc. Natl. Acad. Sci. U.S.A.* **78,** 4378–4382.

Innerarity, T. L., Arnold, K. S., Weisgraber, K. H., and Mahley, R. W. (1986). Apolipoprotein E is the determinant that mediates the receptor uptake of β-very low density lipoproteins by mouse macrophages. *Arteriosclerosis* **6,** 114–122.

Iverius, P-H. (1972). The interaction between human plasma lipoproteins and connective tissue glycosaminoglycans. *J. Biol. Chem.* **247,** 2607–2613.

Jackson, A. P., Seow, H-F., Holmes, N., Drickamer, K., and Parham, P. (1987). Clathrin light chains contain brain-specific insertion sequences and a region of homology with intermediate filaments. *Nature (London)* **326,** 154–159.

Jones, A. L., Hradek, G., Hornick, C. A., Renaud, G., Windler, E. E. T., and Havel, R. J. (1984). Uptake and processing of remnants of chylomicrons and very low density lipoproteins by rat liver. *J. Lipid Res.* **25,** 1151–1158.

Kanaseki, T., and Kadota, K. (1969). The "vesicle in a basket." A morphological study of the coated vesicle isolated from the nerve endings of the guinea pig brain, with special reference to the mechanism of membrane movements. *J. Cell Biol.* **42,** 202–220.

Kirchhausen, T., Harrison, S. C., Chow, E. P., Mattaliano, R. J., Ramachandran, K. L., Smart, J., and Brosius, J. (1987a). Clathrin heavy chain: molecular cloning and complete primary structure. *Proc. Natl. Acad. Sci. U.S.A.* **84,** 8805–8809.

Kirchhausen, T., Scarmato, P., Harrison, S. C., Monroe, J. J., Chow, E. P., Mattaliano, R. J., Ramachandran, K. L., Smart, J. E., Ahn, A. H., and Brosius, J. (1987b). Clathrin light chains LCA and LCB are similar, polymorphic, and share repeated heptad motifs. *Science* **236,** 320–324.

Kita, T., Beisiegel, U., Goldstein, J. L., Schneider, W. J., and Brown, M. S. (1981). Antibody against low density lipoprotein receptor blocks uptake of low density lipoprotein (but not high density lipoprotein) by the adrenal gland of the mouse *in vivo*. *J. Biol. Chem.* **256,** 4701–4703.

Kita, T., Goldstein, J. L., Brown, M. S., Watanabe, Y., Hornick, C. A., and Havel, R. J. (1982). Hepatic uptake of chylomicron remnants in WHHL rabbits: A mechanism genetically distinct from the low density lipoprotein receptor. *Proc. Natl. Acad. Sci. U.S.A.* **79,** 3623–3627.

Knight, B. L., and Soutar, A. K. (1982). Changes in the metabolism of modified and unmodified low-density lipoproteins during the maturation of cultured blood monocyte-macrophages from normal and homozygous familial hypercholesterolaemic subjects. *Eur. J. Biochem.* **125,** 407–413.

Knight, B. L., and Soutar, A. K. (1986). Low apparent affinity for low density lipoprotein of receptors expressed by human macrophages maintained with whole serum. *Eur. J. Biochem.* **156,** 205–210.

Knight, B. L., Patel, D. D., and Soutar, A. K. (1983). The regulation of 3-hydroxy-3-methylglutaryl-CoA reductase activity, cholesterol esterification and the expression of low-density lipoprotein receptors in cultured monocyte-derived macrophages. *Biochem. J.* **210,** 523–532.

Knight, B. L., Patel, D. D., and Soutar, A. K. (1987). Regulation of synthesis and cell content of the low-density-lipoprotein receptor protein in cultured fibroblasts from normal and familial hypercholesterolaemic subjects. *Eur. J. Biochem.* **163,** 189–196.

Koo, C., Wernette-Hammond, M. E., and Innerarity, T. L. (1986). Uptake of canine β-very low density lipoproteins by mouse peritoneal macrophages is mediated by a low density lipoprotein receptor. *J. Biol. Chem.* **261,** 11194–11201.

Kovanen, P. T., Basu, S. K., Goldstein, J. L., and Brown, M. S. (1979a). Low density lipoprotein receptors in bovine adrenal cortex. II. Low density lipoprotein binding to membranes prepared from fresh tissue. *Endocrinology* **104,** 610–616.

Kovanen, P. T., Faust, J. R., Brown, M. S., and Goldstein, J. L. (1979b). Low density lipoprotein receptors in bovine adrenal cortex. I. Receptor-mediated uptake of low density lipoprotein and utilization of its cholesterol for steroid synthesis in cultured adrenocortical cells. *Endocrinology* **104,** 599–609.

Kovanen, P. T., Brown, M. S., Basu, S. K., Bilheimer, D. W., and Goldstein, J. L. (1981). Saturation and suppression of hepatic lipoprotein receptors: A mechanism for the hypercholesterolemia of cholesterol-fed rabbits. *Proc. Natl. Acad. Sci. U.S.A.* **78,** 1396–1400.

Krempler, F., Kostner, G. M., Friedl, W., Paulweber, B., Bauer, H., and Sandhofer, F. (1987). Lipoprotein binding to cultured human hepatoma cells. *J. Clin. Invest.* **80,** 401–408.

Krieger, M., Goldstein, J. L., and Brown, M. S. (1978). Receptor-mediated uptake of low-density-lipoprotein reconstituted with 25-hydroxycholesteryl oleate suppresses 3-hydroxy-3-methylglutaryl coenzyme A reductase and inhibits growth of normal fibroblasts. *Proc. Natl. Acad. Sci. U.S.A.* **75,** 5052–5056.

Lazarovits, J., and Roth, M. (1988). A single amino acid change in the cytoplasmic domain allows the influenza virus hemagglutinin to be endocytosed through coated pits. *Cell* **53,** 743–752.

Leibovich, S. J., and Ross, R. A. (1976). A macrophage-dependent factor that stimulates the proliferation of fibroblasts *in vitro. Am. J. Pathol.* **84,** 501–504.

McGookey, D. J., Fagerberg, K., and Anderson, R. G. W. (1983). Filipin-cholesterol complexes form in uncoated vesicle membrane derived from coated vesicles during receptor-mediated endocytosis of low density lipoprotein. *J. Cell Biol.* **96,** 1273–1278.

McKanna, J. A., Haigler, H. T., and Cohen, S. (1979). Hormone receptor topology and dynamics: morphological analysis using ferritin-labeled epidermal growth factor. *Proc. Natl. Acad. Sci. U.S.A.* **76,** 5689–5693.

Mahley, R. W. (1982). Atherogenic hyperlipoproteinemia: the cellular and molecular biology of plasma lipoproteins altered by dietary fat and cholesterol. *In* "Medical Clinics in North America: Lipid Disorders" (R. J. Havel, ed.) Vol. 66, pp. 375–402. Saunders, Philadelphia.

Mahley, R. W. (1988). Apolipoprotein E: cholesterol transport protein with expanding role in cell biology. *Science* **240,** 622–630.

Mahley, R. W., and Innerarity, T. L. (1977). Interaction of canine and swine lipoproteins with the low density lipoprotein receptor of fibroblasts as correlated with heparin/manganese precipitability. *J. Biol. Chem.* **252,** 3980–3986.

Mahley, R. W., and Innerarity, T. L. (1983). Lipoprotein receptors and cholesterol homeostasis. *Biochim. Biophys. Acta* **737,** 197–222.

Mahley, R. W., Innerarity, T. L., Pitas, R. E., Weisgraber, K. H., Brown, J. H., and Gross, E. (1977). Inhibition of lipoprotein binding to cell surface receptors of fibroblasts following selective modification of arginyl residues in arginine-rich and B apoproteins. *J. Biol. Chem.* **252,** 7279–7287.

Mahley, R. W., Innerarity, T. L., Weisgraber, K. H., and Oh, S. Y. (1979a). Altered metabolism (*in vivo* and *in vitro*) of plasma lipoproteins after selective chemical modification of lysine residues of the apoproteins. *J. Clin. Invest.* **64,** 743–750.

Mahley, R. W., Weisgraber, K. H., Innerarity, T. L., and Windmueller, H. G. (1979b). Accelerated clearance of low-density and high-density lipoproteins and retarded clearance of E apoprotein-containing lipoproteins from the plasma of rats after modification of lysine residues. *Proc. Natl. Acad. Sci. U.S.A.* **76,** 1746–1750.

Mahley, R. W., Innerarity, T. L., Brown, M. S., Ho, Y. K., and Goldstein, J. L. (1980). Cholesteryl ester synthesis in macrophages: stimulation by β-very low density lipoproteins from cholesterol-fed animals of several species. *J. Lipid Res.* **21,** 970–980.

Mahley, R. W., Hui, D. Y., Innerarity, T. L., and Weisgraber, K. H. (1981). Two independent lipoprotein receptors on hepatic membranes of dog, swine and man. Apo-B,E and apo-E receptors. *J. Clin. Invest.* **68,** 1197–1206.

Mahley, R. W., Innerarity, T. L., Weisgraber, K. H., Boyles, J. K., Taylor, J. M., and Levy-Wilson, B. (1986). Cellular and molecular biology of lipoprotein metabolism: characterization of lipoprotein receptor–ligand interactions. *Cold Spring Harbor Symp. Quant. Biol.* **51,** 821–828.

Mersey, B. J., Griffing, L. R., Rennie, P. J., and Fowke, L. C. (1985). The isolation of coated vesicles from protoplasts of soybean. *Planta* **163,** 317–327.

Miller, N. E., Weinstein, D. B., and Steinberg, D. (1978). Uptake and degradation of high density lipoprotein: comparison of fibroblasts from normal subjects and from homozygous familial hypercholesterolemic subjects. *J. Lipid Res.* **19,** 644–653.

Mistry, P., Miller, N. E., Laker, M., Hazzard, W. R., and Lewis, B. (1981). Individual variation in the effects of dietary cholesterol on plasma lipoproteins and cellular cholesterol homeostasis in man. *J. Clin. Invest.* **67,** 493–502.

Montesano, R., Perrelet, A., Vassalli, P., and Orci, L. (1979). Absence of filipin–sterol complexes from large coated pits on the surface of cultured cells. *Proc. Natl. Acad. Sci. U.S.A.* **76,** 6391–6395.

Morel, D. W., Hessler, J. R., and Chisolm, G. M. (1983). Low density lipoprotein cytotoxicity induced by free radical peroxidation of lipid. *J. Lipid Res.* **24,** 1070–1076.

Morel, D. W., DiCorleto, P. E., and Chisolm, G. M. (1984). Endothelial and smooth muscle cells alter low density lipoprotein *in vitro* by free radical oxidation. *Arteriosclerosis* **4,** 357–364.

Morse, J. H., Witte, L. D., and Goodman, D. S. (1977). Inhibition of lymphocyte proliferation stimulated by lectins and allogeneic cells by normal plasma lipoproteins. *J. Exp. Med.* **146,** 1791–1803.

Myant, N. B. (1981). "The Biology of Cholesterol and Related Steroids." Heinemann, London.

Myant, N. B. (1982). Cholesterol transport through the plasma. *Clin. Sci.* **62,** 261–271.

Nagelkerke, J. F., Barto, K. P., and van Berkel, T. J. C. (1983). *In vivo* and *in vitro* uptake and degradation of acetylated low density lipoprotein by rat liver endothelial, Kupffer, and parenchymal cells. *J. Biol. Chem.* **258,** 12221–12227.

Nakayasu, T., Macy, M., Okano, Y., McCarthy, B. M., and Harmony, J. A. K. (1986). Plasma lipoproteins can suppress accessory cell function and consequently suppress lymphocyte activation. *Exp. Cell Res.* **163,** 103–116.

Octave, J-N., Schneider, Y. J., Crichton, R. R., and Trouet, A. (1981). Transferrin uptake by cultured rat embryo fibroblasts. The influence of temperature and incubation time, subcellular distribution and short-term kinetic studies. *Eur. J. Biochem.* **115,** 611–618.

Orci, L., Carpentier, J-L., Perrelet, A., Anderson, R. G. W., Goldstein, J. L., and Brown, M. S. (1978). Occurrence of low density lipoprotein receptors within large pits on the surface of human fibroblasts as demonstrated by freeze-etching. *Exp. Cell Res.* **113,** 1–13.

Ose, T., Berg, T., Norum, K. R., and Ose, L. (1980). Catabolism of [^{125}I] low density lipoproteins in isolated rat liver cells. *Biochem. Biophys. Res. Comm.* **97,** 192–199.

Ostlund, R. E., Pfleger, B., and Schonfeld, G. (1979). Role of microtubules in low density lipoprotein processing by cultured cells. *J. Clin. Invest.* **63,** 75–84.

Ostlund-Lindqvist, A-M., Gustafson, S., Lindqvist, P., Witztum, J. L., and Little, J. A. (1983). Uptake and degradation of human chylomicrons by macrophages in culture. Role of lipoprotein lipase. *Arteriosclerosis* **3,** 433–440.

Palinski, W., Rosenfeld, M. E., Ylä-Herttuala, S., Gurtner, G. C., Socher, S. S., Butler, S. W., Parthasarathy, S., Carew, T. E., Steinberg, D., and Witztum, J. L. (1989). Low density lipoprotein undergoes oxidative modification *in vivo*. *Proc. Natl. Acad. Sci. U.S.A.* **86,** 1372–1376.

Pangburn, S. H., Newton, R. S., Chang, C-M., Weinstein, D. B., and Steinberg, D. (1981). Receptor-mediated catabolism of homologous low density lipoproteins in cultured pig hepatocytes. *J. Biol. Chem.* **256,** 3340–3347.

Parthasarathy, S., Steinbrecher, U. P., Barnett, J., Witztum, J. L., and Steinberg, D. (1985). Essential role of phospholipase A_2 activity in endothelial cell-induced modification of low density lipoprotein. *Proc. Natl. Acad. Sci. U.S.A.* **82,** 3000–3004.

Parthasarathy, S., Printz, D. J., Boyd, D., Joy, L., and Steinberg, D. (1986). Macrophage oxidation of low density lipoprotein generates a modified form recognized by the scavenger receptor. *Arteriosclerosis* **6,** 505–510.

Patel, D. D., and Knight, B. L. (1985). The effect of mevalonate on 3-hydroxy-3-methylglutaryl-CoA reductase activity and the absolute rate of cholesterol biosynthesis in human monocyte-derived macrophages. *Eur. J. Biochem.* **153,** 117–123.

Patel, D. D., Pullinger, C. R., and Knight, B. L. (1984). The absolute rate of cholesterol biosynthesis in monocyte-macrophages from normal and familial hypercholesterolaemic subjects. *Biochem. J.* **219,** 461–470.

Pathak, R. K., Merkle, R. K., Cummings, R. D., Goldstein, J. L., Brown, M. S., and Richardson, R. G. W. (1988). Immunocytochemical localization of mutant low density lipoprotein receptors that fail to reach the Golgi complex. *J. Cell Biol.* **106,** 1831–1841.

Pearse, B. M. F. (1975). Coated vesicles from pig brain: purification and biochemical characterization. *J. Mol. Biol.* **97,** 93–98.

Pearse, B. M. F. (1976). Clathrin: a unique protein associated with intracellular transfer of membrane by coated vesicles. *Proc. Natl. Acad. Sci. U.S.A.* **73,** 1255–1259.

Pearse, B. M. F. (1978). On the structural and functional components of coated vesicles. *J. Mol. Biol.* **126,** 803–812.

Pearse, B. M. F. (1987). Clathrin and coated vesicles. *EMBO Journal,* **6,** 2507–2512.

Pearse, B. M. F. (1988). Receptors compete for adaptors found in plasma membrane coated pits. *EMBO Journal,* **7,** 3331–3336.

Pearse, B. M. F., and Bretscher, M. S. (1981). Membrane recycling by coated vesicles. *Ann. Rev. Biochem.* **50,** 85–101.

Pearse, B. M. F., and Crowther, R. A. (1987). Structure and assembly of coated vesicles. *Ann. Rev. Biophys. Biophys. Chem.* **16,** 49–68.

Pearse, B. M. F., and Robinson, M. F. (1984). Purification and properties of 100-kd proteins from coated vesicles and their reconstitution with clathrin. *EMBO J.* **3,** 1951–1957.

Pitas, R. E., Innerarity, T. L., Arnold, K. S., and Mahley, R. W. (1979). Rate and equilibrium constants for binding of apoE HDL_c (a cholesterol-induced lipoprotein) and low density lipoproteins to human fibroblasts: Evidence for multiple receptor binding of apo-E HDL_c. *Proc. Natl. Acad. Sci. U.S.A.* **76,** 2311–2315.

Pitas, R. E., Innerarity, T. L., and Mahley, T. L. (1980). Cell surface receptor binding of phospholipid protein complexes containing different ratios of receptor-active and -inactive E apoprotein. *J. Biol. Chem.* **255,** 5454–5460.

Pitas, R. E., Innerarity, T. L., and Mahley, R. W. (1983). Foam cells in explants of atherosclerotic rabbit aortas have receptors for β-very low density lipoproteins and modified low density lipoproteins. *Arteriosclerosis* **3,** 2–12.

Pitas, R. E., Boyles, J. K., Lee, S. H., Hui, D., and Weisgraber, K. W. (1987). Lipoproteins and their receptors in the central nervous system. Characterization of the lipoproteins in cerebrospinal fluid and identification of apolipoprotein B,E (LDL) receptors in the brain. *J. Biol. Chem.* **262,** 14352–14360.

Pittman, R. C., Attie, A. D., Carew, T. E., and Steinberg, D. (1979). Tissue sites of degradation of low density lipoprotein: application of a method for determining the fate of plasma proteins. *Proc. Natl. Acad. Sci. U.S.A.* **76,** 5345–5349.

Pittman, R. C., Carew, T. E., Attie, A. D., Witztum, J. L., Watanabe, Y., and Steinberg, D. (1982). Receptor-dependent and receptor-independent degradation of low density lipoprotein in normal rabbits and in receptor-deficient mutant rabbits. *J. Biol. Chem.* **257,** 7994–8000.

Quinn, M. T., Parthasarathy, S., Fong, L. G., and Steinberg, D. (1987). Oxidatively modified low density lipoproteins: a potential role in recruitment and retention of monocyte/macrophages during atherogenesis. *Proc. Natl. Acad. Sci. U.S.A.* **84,** 2995–2998.

Raymond, T. L., and Reynolds, S. A. (1983). Lipoproteins of the extravascular space: alterations in low density lipoproteins of interstitial inflammatory fluid. *J. Lipid Res.* **24,** 113–119.

Reichl, D., Postiglione, A., Myant, N. B., Pflug, J. J., and Press, M. (1975). Observations on the passage of apoproteins from plasma lipoproteins into peripheral lymph in two men. *Clin. Sci. Mol. Med.* **49,** 419–426.

Reichl, D., Postiglione, A., and Myant, N. B. (1976). Uptake and catabolism of low density lipoprotein by human lymphocytes. *Nature (London)* **260,** 634–635.

Reichl, D., Myant, N. B., and Lloyd, J. K. (1978). Surface binding and catabolism of low-density lipoprotein by circulating lymphocytes from patients with abetalipoproteinaemia, with observations on sterol synthesis in lymphocytes from one patient. *Biochim. Biophys. Acta* **530,** 124–131.

Ross, R., and Glomset, J. A. (1976). The pathogenesis of atherosclerosis. *N. Engl. J. Med.* **295,** 369–377, 420–425.

Roth, T. F., and Porter, K. R. (1964). Yolk protein uptake in the oocyte of the mosquito *Aedes aegypti. L. J. Cell Biol.* **20,** 313–332.

Rothblat, G. H. (1972). Cellular sterol metabolism. *In* "Growth Nutrition and Metabolism of Cells in Culture," (G. H. Rothblat and V. J. Cristofalo, eds.), Vol. 1, pp. 297–325. Academic Press, New York.

Russell, D. W., Yamamoto, T., Schneider, W. J., Slaughter, C. J., Brown, M. S., and Goldstein, J. L. (1983). cDNA cloning of the bovine low density lipoprotein receptor: Feedback regulation of receptor mRNA. *Proc. Natl. Acad. Sci. U.S.A.* **80,** 7501–7505.

Salter, A. M., Saxton, J., and Brindley, D. N. (1986). Characterization of the binding of human low-density lipoprotein to primary monolayer cultures of rat hepatocytes. *Biochem. J.* **240,** 549–557.

Salter, A. M., Bugaut, M., Saxton, J., Fisher, S. C., and Brindley, D. N. (1987a). Effects of preincubation of primary monolayer cultures of rat hepatocytes with low- and high-density lipoproteins on the subsequent binding and metabolism of human low-density lipoprotein. *Biochem. J.* **247,** 79–84.

Salter, A. M., Fisher, S. C., and Brindley, D. N. (1987b). Binding of low-density lipoprotein to monolayer cultures of rat hepatocytes is increased by insulin and decreased by dexamethasone. *FEBS Lett.* **220,** 159–162.

Salter, A. M., Fisher, S. C., and Brindley, D. N. (1988). Interactions of tri-iodothyronine, insulin and dexamethasone on the binding of human LDL to rat hepatocytes in monolayer culture. *Atherosclerosis* **71,** 77–80.

Scarabottolo, L., Trezzi, E., Roma, P., and Catapano, A. L. (1986). Experimental hypothyroidism modulates the expression of the low density lipoprotein receptor by the liver. *Atherosclerosis* **59,** 329–333.

Schmid, S. L., and Rothman, J. E. (1985). Enzymatic dissociation of clathrin cages in a two-stage process. *J. Biol. Chem.* **260,** 10044–10049.

Schneider, W. J., Basu, S. K., McPhaul, M. J., Goldstein, J. L., and Brown, M. S. (1979). Solubilization of the low density lipoprotein receptor. *Proc. Natl. Acad. Sci. U.S.A.* **76,** 5577–5581.

Schneider, W. J., Goldstein, J. L., and Brown, M. S. (1980). Partial purification and characterization of the low density lipoprotein receptor from bovine adrenal cortex. *J. Biol. Chem.* **255,** 11442–11447.

Schneider, W. J., Brown, M. S., and Goldstein, J. L. (1983). Kinetic defects in the processing of the low density lipoprotein receptor in fibroblasts from WHHL rabbits and a family with familial hypercholesterolemia. *Mol. Biol. Med.* **1,** 353–367.

Semenkovich, C. F., and Ostlund, R. E. (1986). The low density lipoprotein receptor on human peripheral blood monocytes and lymphocytes. visualization by ligand blotting and immunoblotting techniques. *J. Clin. Endocr. Metab.* **62,** 1279–1287.

Semenkovich, C. F., and Ostlund, R. E. (1987). Estrogens induce low-density lipoprotein receptor activity and decrease intracellular cholesterol in human hepatoma cell line Hep G2. *Biochemistry* **26,** 4987–4992.

Shaikh, M., Martini, S., Quiney, J. R., Baskerville, P., LaVille, A. E., Browse, N. L., Duffield, R., Turner, P. R., and Lewis, B. (1988). Modified plasma-derived lipoproteins in human atherosclerotic plaques. *Atherosclerosis* **69,** 165–172.

Shechter, I., Fogelman, A. M., Haberland, M. E., Seager, J., Hokom, M., and Edwards, P. A. (1981). The metabolism of native and malondialdehyde-altered low density lipoproteins by human monocyte-macrophages. *J. Lipid Res.* **22,** 63–71.

Sherrill, B. C., and Dietschy, J. M. (1978). Characterization of the sinusoidal transport process responsible for uptake of chylomicrons by the liver. *J. Biol. Chem.* **253,** 1859–1867.

Sherrill, B. C., Innerarity, T. L., and Mahley, R. W. (1980). Rapid hepatic clearance of the canine lipoproteins containing only the E apoprotein by a high affinity receptor. *J. Biol. Chem.* **255,** 1804–1807.

Soutar, A. K. (1989). Second receptor verified? *Nature (London),* **341,** 106–107.

Soutar, A. K., and Knight, B. L. (1982). Degradation by cultured monocyte-derived macrophages from normal and familial hypercholesterolaemic subjects of modified and unmodified low-density lipoproteins. *Biochem. J.* **204,** 549–556.

Soutar, A. K., and Knight, B. L. (1984). Degradation of lipoproteins by human monocyte-derived macrophages. Evidence for two distinct processes for the degradation of abnormal very-low-density lipoprotein from subjects with type III hyperlipidaemia. *Biochem. J.* **218,** 101–111.

Soutar, A. K., and Knight, B. L. (1985). Regulation of macrophage functions by lipoproteins. *In* "Mononuclear Phagocytes: Physiology and Pathology" (R. T. Dean and W. Jessup, eds.), pp. 265–282. Elsevier, Amsterdam.

Soutar, A. K., Harders-Spengel, K., Wade, D. P., and Knight, B. L. (1986). Detection and quantitation of low-density lipoprotein (LDL) receptors in human liver by ligand blotting, immunoblotting and radioimmunoassay. LDL receptor protein content is correlated with plasma LDL cholesterol concentration. *J. Biol. Chem.* **261,** 17127–17133.

Spady, D. K., Turley, S. D., and Dietschy, J. M. (1985). Receptor-independent low density lipoprotein transport in the rat *in vivo.* Quantitation, characterization, and metabolic consequences. *J. Clin. Invest.* **76,** 1113–1122.

Stein, O., and Stein, Y. (1980). Bovine aortic endothelial cells display macrophage-like properties towards acetylated ^{125}I-labelled low density lipoprotein. *Biochim. Biophys. Acta* **620,** 631–635.

Stein, O., Stein, Y., Goodman, D. S., and Fidge, N. H. (1969). The metabolism of chylomicron cholesteryl ester in rat liver. A combined radioautographic-electron microscopic and biochemical study. *J. Cell. Biol.* **43,** 410–431.

Stein, O., Halperin, G., and Stein, Y. (1980). Biological labeling of low density lipoproteins with cholesteryl linoleyl ether and its fate in the intact rat. *Biochim. Biophys. Acta* **620,** 247–260.

Steinberg, D., Parthasarathy, S., Carew, T. E., Khoo, J. C., and Witztum, J. L. (1989). Beyond cholesterol. Modifications of low-density lipoprotein that increase its atherogenicity. *N. Engl. J. Med.* **320,** 915–924.

Steinbrecher, U. P. (1987). Oxidation of human low density lipoprotein results in derivatization of lysine residues of apolipoprotein B by lipid peroxide decomposition products. *J. Biol. Chem.* **262,** 3603–3608.

Steinbrecher, U. P., Parthasarathy, S., Leake, D. S., Witztum, J. L., and Steinberg, D. (1984). Modification of low density lipoprotein by endothelial cells involves lipid peroxidation and degradation of low density lipoprotein phospholipids. *Proc. Natl. Acad. Sci. U.S.A.* **81,** 3883–3887.

Thompson, G. R., Soutar, A. K., Spengel, F. A., Jadhav, A., Gavigan, S. J. P., and Myant, N. B. (1981). Defects of receptor-mediated low density lipoprotein catabolism in homozygous familial hypercholesterolemia and hypothyroidism *in vivo*. *Proc. Natl. Acad. Sci. U.S.A.* **78,** 2591–2595.

Ungewickell, E., and Branton, D. (1981). Assembly units of clathrin coats. *Nature (London)* **289,** 420–422.

Van Driel, I. R., Davis, C. G., Goldstein, J. L., and Brown, M. S. (1987). Self-association of the low density lipoprotein receptor mediated by the cytoplasmic domain. *J. Biol. Chem.* **262,** 16127–16134.

Van Lenten, B. J., Fogelman, A. M., Hokom, M. M., Benson, L., Haberland, M. E., and Edwards, P. A. (1983). Regulation of the uptake and degradation of β-very low density lipoprotein in human monocyte macrophages. *J. Biol. Chem.* **258,** 5151–5157.

Van Lenten, B. J., Fogelman, A. M., Jackson, R. L., Shapiro, S., Haberland, M. E., and Edwards, P. A. (1985). Receptor-mediated uptake of remnant lipoproteins by cholesterol-loaded human monocyte-macrophages. *J. Biol. Chem.* **260,** 8783–8788.

Via, D. P., Dresel, H. A., and Gotto, A. M. (1982). Isolation and characterization of the murine macrophage acetyl LDL receptor. *Circulation* **66,** II–37.

Via, D. P., Dresel, H. A., Cheng, S-L., and Gotto, A. M. (1985). Murine macrophage tumors are a source of a 260,000-dalton acetyl-low density lipoprotein receptor. *J. Biol. Chem.* **260,** 7379–7386.

Vigers, G. P. A., Crowther, R. A., and Pearse, B. M. F. (1986). Location of the 100 kd–50 kd accessory proteins in clathrin coats. *EMBO J.* **5,** 2079–2085.

Wade, D. P., Knight, B. L., and Soutar, A. K. (1985). Detection of the low-density-lipoprotein receptor with biotin-low-density lipoprotein. A rapid new method for ligand blotting. *Biochem. J.* **229,** 785–790.

Wade, D. P., Knight, B. L., and Soutar, A. K. (1986). Binding of low-density lipoprotein and chylomicron remnants to the hepatic low-density lipoprotein receptor of dogs, rats and rabbits demonstrated by ligand blotting. Failure to detect a distinct chylomicron-remnant-binding protein by ligand blotting. *Eur. J. Biochem.* **159,** 333–340.

Wade, D. P., Knight, B. L., and Soutar, A. K. (1988). Hormonal regulation of low-density lipoprotein (LDL) receptor activity in human hepatoma Hep G2 cells. Insulin increases LDL receptor activity and diminishes its suppression by exogenous LDL. *Eur. J. Biochem.* **174,** 213–218.

Weight, M., Cortese, C., Sule, U., Miller, N. E., and Lewis, B. (1982). Heritability of the low density lipoprotein receptor activity of human blood mononuclear cells: studies in normolipidaemic adult male twins. *Clin. Sci.* **62,** 397–401.

Werb, Z., and Chin, J. R. (1983). Endotoxin suppresses expression of apoprotein E by mouse macrophages *in vivo* and in culture. *J. Biol. Chem.* **258,** 10642–10648.

Wileman, T., Harding, C., and Stahl, P. (1985). Receptor-mediated endocytosis. *Biochem. J.* **232,** 1–14.

Williams, C. D., and Avigan, J. (1972). *In vitro* effects of serum proteins and lipids on lipid synthesis in human skin fibroblasts and leukocytes grown in culture. *Biochim. Biophys. Acta* **260,** 413–423.

Willingham, M. C., and Pastan, I. (1980). The receptosome: an intermediate organelle of receptor-mediated endocytosis in cultured fibroblasts. *Cell* **21,** 67–77.

Windler, E., Chao, Y-S., and Havel, R. J. (1980). Determinants of hepatic uptake of triglyceride-rich lipoproteins and their remnants in the rat. *J. Biol. Chem.* **255,** 5475–5480.

Winkler, F. K., and Stanley, K. K. (1983). Clathrin heavy chain, light chain interactions. *EMBO J.* **2,** 1397–1400.

Witte, L. D., and Cornicelli, J. A. (1980). Platelet-derived growth factor stimulates low density lipoprotein receptor activity in cultured human fibroblasts. *Proc. Natl. Acad. Sci. U.S.A.* **77,** 5962–5966.

Witte, L. D., Cornicelli, J. A., Miller, R. W., and Goodman, D. S. (1982). Effects of platelet-derived and endothelial cell-derived growth factors on the low density lipoprotein receptor pathway in cultured human fibroblasts. *J. Biol. Chem.* **257,** 5392–5401.

Wojciechowski, A. P., Winder, A. F., and Campbell, A. C. (1987). Receptor-mediated endocytosis of fluorescent-probe-labelled low-density lipoprotein using human lymphocytes, fluorimetry and flow cytometry. *Biochem. Soc. Trans.* **15,** 251–252.

Young, S. G., Witztum, J. L., Casal, D. C., Curtiss, L. K., and Bernstein, S. (1986). Conservation of the low density lipoprotein receptor-binding domain of apoprotein B. *Arteriosclerosis* **6,** 178–188.

Zaremba, S., and Keen, J. H. (1983). Assembly polypeptides from coated vesicles mediate reassembly of unique clathrin coats. *J. Cell Biol.* **97,** 1339–1347.

Chapter 9

The LDL Receptor: Structure, Biosynthesis, and Molecular Genetics

I. The Route to Isolation of the Human Receptor and Its Gene

The first step toward isolation of the human receptor was the solubilization of the receptor from bovine adrenal cortex, a very rich source of LDL receptors. Human LDL was used as ligand for assaying receptor activity at each stage of the procedure (Schneider *et al.*, 1980). This relatively crude preparation was used to raise a rabbit antiserum (Beisiegel *et al.*, 1981a) and a mouse monoclonal antibody (IgG-C7) (Beisiegel *et al.*, 1981b), both of which cross-reacted with the human receptor. (The use of these antibodies in the study of LDL-receptor function in intact animals and cells in culture has already been mentioned in Chapters 7 and 8.) The availability of anti-receptor antibodies opened the way to the isolation and analysis of the normal and genetically abnormal human receptor by specific immunoprecipitation combined with polyacrylamide-gel electrophoresis.

It is worth noting that the success of this experimental approach depended upon the ability of LDL receptors from one species to recognize LDL from another species and to react with antibodies raised against receptors from other species. Both these properties of the LDL receptor are consequences of the considerable degree to which amino acid sequences in functional regions of the receptor and of apoB-100 have been conserved during evolution. Another factor that helped in the isolation of human LDL receptors was the existence of cells specifically lacking the receptor. This has provided a means of confirming unequivocally the presence of anti-receptor activity in antibody preparations and of LDL-receptor activity in proteins isolated from human tissues (see, for example, Fig. 9.1). The problem of the possible existence of other lipoprotein receptors,

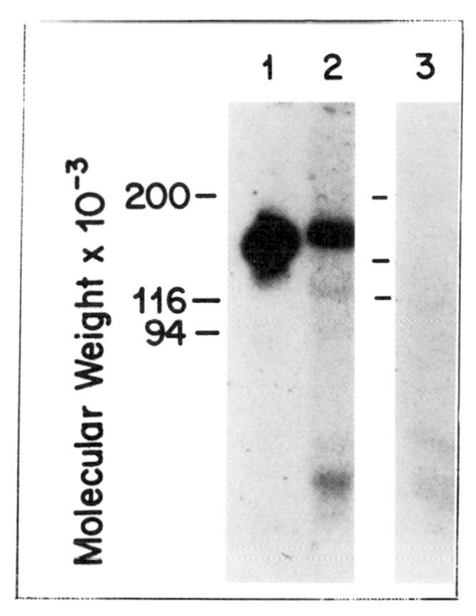

Figure 9.1. Autoradiographs of an immunoblot of LDL receptors from bovine adrenal cortex and human fibroblasts. LDL receptors, partially purified by DEAE-cellulose chromatography, were submitted to SDS–polyacrylamide-gel electrophoresis in the absence of reducing buffer. The proteins were transferred electrophoretically from the gel to nitrocellulose paper and the paper was incubated with mouse monoclonal anti-LDL-receptor antibody IgG-C7. The bound monoclonal antibody was detected by a second incubation with ^{125}I-labeled goat anti-mouse IgG. The dried paper was exposed to X-ray film for autoradiography. The positions of molecular weight markers are shown to the left of the autoradiograph. The samples submitted to electrophoresis were derived from 1, bovine adrenal cortex; 2, normal human fibroblasts; 3, fibroblasts from an FH homozygote whose cells bound no LDL and no IgG-C7 (receptor-negative, cross-reacting material negative). (From Beisiegel *et al.*, 1982, with the permission of the authors.)

discussed in the previous chapter (Section III), would be easier to settle if cells specifically lacking these receptors were also available.

An essential step in the isolation of the human LDL-receptor gene was the purification to homogeneity of the bovine receptor in milligram amounts (Schneider *et al.*, 1982). Oligonucleotides, encoding short amino acid sequences determined in fragments of the purified bovine receptor, were synthesized chemically and labeled with ^{32}P. These were used as hybridization probes to isolate cDNA clones for the LDL receptor from a bovine adrenal cDNA library (Russell *et al.*, 1983). A partial cDNA clone for the bovine receptor was then used as a probe in the initial stage in the isolation of a cDNA clone for the receptor from a human liver cDNA library (Yamamoto *et al.*, 1984), as discussed below. Here again, the success of the strategy depended upon the presence of homologous sequences in the coding regions of the LDL-receptor genes from different species.

II. The Normal LDL Receptor

A. Composition and Molecular Weight

The mature human LDL receptor is a transmembrane glycoprotein with a single polypeptide chain. The protein has 839 amino acids, whose complete sequence was deduced from a cDNA for the human receptor (Yamamoto *et al.*, 1984). The carbohydrate composition of the LDL receptor is consistent with the presence of 2 complex N-linked oligosaccharide chains and up to 18 O-linked chains (Cummings *et al.*, 1983). The calculated molecular weight of the protein of the human receptor is 93,102. With the addition of the oligosaccharide chains (MW ~22K), the true molecular weight of the whole receptor should therefore be about 115K. The apparent molecular weight of the LDL receptor, determined by SDS–polyacrylamide-gel electrophoresis with standards of known molecular weight, is considerably higher than this. Nevertheless, measurement of the apparent molecular weight of the mature receptor and of its immunoreactive precursor has been used extensively in studies of the biosynthesis of the normal receptor and of its mutant forms.

Schneider *et al.* (1982) obtained a value of ~164K for the apparent molecular weight of the bovine receptor, determined by gel electrophoresis in the presence of reducing buffer. Beisiegel *et al.* (1982) obtained similar values for the receptor in normal human fibroblasts by immunoblotting, with the antibody IgG-C7 as probe for detecting the protein bands. Solubilized extracts of fibroblasts from a receptor-negative FH homozygote gave no immunoreactive band on blots prepared under identical conditions, confirming that the bands detected in the samples from normal fibroblasts were due to the presence of LDL receptors. Figure 9.1 shows an autoradiograph of an immunoblot of the proteins solubilized from bovine adrenal cortex, normal human fibroblasts, and fibroblasts from an FH homozygote whose intact cells expressed no receptors detectable with IgG-C7 (for experimental details, see the figure legend). Note that the bovine and normal human LDL receptors have migrated to the same position between the 116K and 200K MW markers and that no band is visible in the lane with the sample from the FH homozygote cells.

LDL receptors separated by gel electrophoresis can also be visualized by blotting onto nitrocellulose paper, followed by incubation with LDL (Daniel *et al.*, 1983) or biotin-LDL (Wade *et al.*, 1985) as ligand (''ligand blotting''). The ligands bind specifically to the protein band containing the receptor. Bound native LDL is detected with an anti-LDL antibody and bound biotin-LDL is detected with a streptavidin–biotinylated-peroxidase complex. For efficient binding of LDL to the receptor on nitrocellulose blots, the electrophoresis, blotting, and incubation with ligand must be carried out in the absence of sulfydryl reducing agents. Under these conditions the binding domain of the LDL receptor is maintained in the appropriately folded conformation because the intrachain disulfide bonds in this region of the receptor are not disrupted (see Section II,C below). In the presence of nonreducing buffer, the apparent molecular weight of the receptor is decreased to ~130K. This, presumably, is a consequence of the tightly folded conformation of the native protein.

B. Biosynthesis

The posttranslational stages of the biosynthesis of the LDL receptor may be followed in cells incubated with a radioactive amino acid or with radioactive sugars.

Tolleshaug *et al.* (1982) showed that when human fibroblasts are incubated in the presence of [^{35}S]methionine, the label is rapidly incorporated into a protein, precipitable with IgG-C7, that has an apparent M_r ~120K in the presence of reducing buffer. Experiments in which pulse-labeling with [^{35}S]methionine was "chased" by a further incubation with nonradioactive methionine showed that conversion of the 120K protein into the mature receptor begins within 15–30 minutes of its synthesis and is essentially complete within 1 hour. Figure 9.2 shows an experiment in which normal fibroblasts were incubated for various intervals with [^{35}S]methionine, with or without a subsequent chase with nonradioactive methionine, and were then solubilized with detergent. The precursor and its mature product in the solubilized extracts were precipitated with IgG-C7 and the precipitated receptor proteins submitted to gel electrophoresis. Autoradiographs of the gel showed that after 30 minutes of incubation with [^{35}S]methionine all the newly synthesized immunoreactive protein was in the 120K form (lane B). After 1 hour, radioactivity had begun to appear in the 160K protein (lane C) and by 2 hours radioactivity was about equally distributed between the two immunoprecipitable proteins (lane D). When the pulse incubation was followed by a chase for 1 hour all the radioactive 120K protein synthesized during the pulse was converted into the 160K form (lane E).

The above sequence of events shows that the LDL receptor is first detectable immunochemically as a precursor with apparent M_r ~120K and that 15–30 minutes later it is converted into the mature receptor with apparent M_r ~160K. Since the receptor is a glycosylated protein, it would be reasonable to suppose that the increase in apparent molecular weight is related to the addition of oligosaccharide units to the primary translation product of receptor mRNA. As we saw in Chapter 3 (Section III), N-linked oligosaccharides are added cotranslationally to Asn residues of the growing peptide chain in the ER lumen and are then modified in the Golgi apparatus to "complex" or "high-mannose" oligosaccharides; O-linked oligosaccharides, on the other hand, are added posttranslationally to Ser/Thr residues, probably in the ER or the *cis* region of the Golgi apparatus.

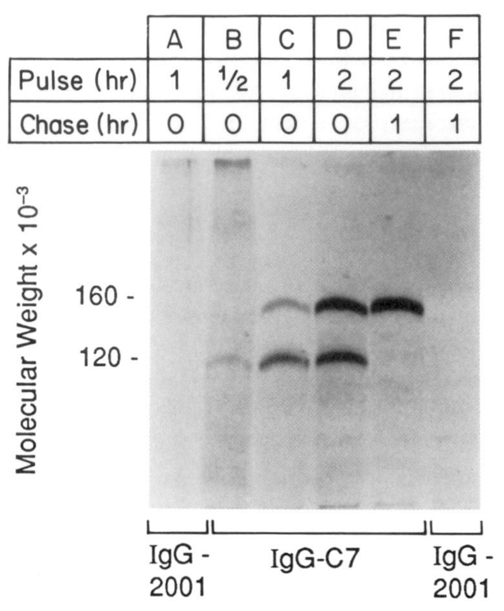

Figure 9.2. Autoradiograph of endogenously labeled precursor and mature LDL receptors separated by gel electrophoresis. Normal human fibroblasts in culture were pulse-labeled by incubation in the presence of [^{35}S]methionine for the indicated times and were then harvested either without a subsequent incubation in the presence of nonradioactive methionine (lanes A–D) or after a chase incubation for 1 hour (lanes E and F). After incubation, detergent-solubilized extracts of the cells were prepared and immunoprecipitates were formed with control antibody (IgG-2001) or with mouse monoclonal anti-LDL-receptor antibody IgG-C7. The labeled immunoprecipitates were subjected to SDS–polyacrylamide-gel electrophoresis followed by autoradiography. (From Tolleshaug *et al.*, 1982, with the permission of the authors.)

Cummings *et al.* (1983) have examined the time course of glycosylation of LDL receptors, and the effect of glycosylation on their apparent molecular weight, in human cells in culture. The precursor and mature forms of the LDL receptor were isolated from cells that had been incubated in the presence of radioactive sugars, and the radioactive oligosaccharide units synthesized during the incubation were characterized by chemical and enzymic methods. The 120K precursor was shown to contain one or two N-linked high-mannose units $(Asn)(GlcNAc)_2(Man)_{6-9}$ and up to 18 O-linked GalNAc residues (Ser/Thr)(GalNAc). The 160K mature receptor contained one or two asparagine-linked complex oligosaccharide units and several O-linked units containing galactose and sialic acid residues. When the incubations were carried out in the presence of tunicamycin, a potent inhibitor of N-glycosylation, the increase in apparent molecular weight from 120K to 160K took place normally. From these and other observations, Cummings *et al.* concluded that the apparent increase in molecular weight is an anomalous effect of the addition of galactose and sialic acid residues to O-linked GalNAc residues in the *trans* region of the Golgi stack. Since the precursor has O-linked GalNAc residues but does not contain N-linked complex oligosaccharides, the addition of O-linked GalNAc residues must occur before modification of the N-linked chains to complex oligosaccharides. Cell-fractionation studies have suggested that GalNAc transferase, the enzyme that catalyzes the addition of GalNAc to serine and threonine residues in O-linkage, is confined to the Golgi apparatus. However, this is not consistent with the behavior of receptors in class 2 mutant fibroblasts (see Section V below). In these cells, the precursor of the mature LDL receptor is synthesized in the rough ER but does not reach the Golgi apparatus. Nevertheless, these abnormal receptors contain core O-linked sugars. Pathak *et al.* (1988), using immunocytochemical methods, have shown that Class 2 mutant receptors are retained in the rough ER and in its tubulovesicular extensions. They suggest that the enzyme responsible for attaching the initial O-linked sugars to the precursor protein is located in the rough ER itself or in a transitional zone between the rough and smooth ER.

In the light of these observations and of information from other sources, the following sequence may be suggested for the biosynthesis of the LDL receptor. The receptor protein is synthesized in the rough ER, where N-linked high-mannose units are added. When this process is completed, the receptor moves to the smooth ER, where O-linked GalNAc residues are added. It then moves to the Golgi apparatus for modification of the N-linked units and the addition of galactose and sialic acid residues to the O-linked core sugars. As shown in Fig. 8.14, the fully assembled receptor is transported from the Golgi apparatus to the plasma membrane, possibly in small coated vesicles, where it moves into coated pits to begin the cycle of endocytosis.

C. Orientation

Schneider *et al.* (1983b) have shown that the LDL receptor on the surface of an intact cell is oriented with its N-terminus on the outside of the plasma membrane. To demonstrate this, they made use of the observation that antibodies raised against an isolated short segment of a protein will react specifically with the corresponding segment of the native protein (Walter *et al.*, 1980; Lerner *et al.*, 1981). An antibody was raised against a synthetic peptide corresponding to the first 16 N-terminal amino acids of the bovine LDL receptor. This antibody bound to the LDL receptor on normal human fibroblasts in culture but did not bind to FH homozygote cells. Thus, the N-terminus of the receptor must be accessible to antibodies outside the cell. The presence of potential ligand-binding sites within the N-terminal segment of the receptor (see next section) also indicates that this region is outside the plasma membrane.

Analysis of the fragments released by pronase digestion of LDL receptors has shown that the O-linked sugars are clustered mainly within a short segment of the protein outside the plasma membrane (Cummings *et al.*, 1983; Russell *et al.*, 1984). The deletion experiments of Davis *et al.* (1986a) show that this segment corresponds to the Ser/Thr-rich third domain (see Section II,D below).

As shown in the next section, the LDL receptor has a single membrane-spanning region 50 residues from its C-terminus, followed by a 50-residue C-terminal sequence. To determine the orientation of the C-terminal sequence, Russell *et al.* (1984) raised antibodies to synthetic peptides corresponding to short segments on the C-terminal and N-terminal sides of the membrane-spanning region of the bovine receptor; these site-specific antibodies were used to identify proteolytic fragments of the receptor. Sealed vesicles, rich in LDL receptors, were prepared from bovine adrenal cortex and were digested with a proteolytic enzyme, with or without detergent. In the absence of detergent the C-terminal segment was completely destroyed, but the N-terminal segment, detected by its reactivity with the antibody to the N-terminal peptide, was protected against digestion. When the digestion was carried out in the presence of detergent, all protein reacting with the two site-specific antibodies was destroyed. Russell *et al.* concluded that the N-terminus of the receptor was inside the vesicle and that the C-terminus was outside. Since it was already known that the N-terminus of the receptor on the surface of an intact cell is outside the plasma membrane, the vesicles must have been "inside-out." Therefore the C-terminal sequence of the receptor in an intact cell must project into the cytoplasm. The observation that the C-terminal sequence was completely digested in the absence of detergent indicates that it passes through the membrane only once. Repeated looping in and out of the membrane would have resulted in the release of multiple peptide fragments after proteolytic digestion.

D. Domain Structure and Function

Yamamoto *et al.* (1984) isolated a full-length cDNA for the human LDL receptor encoding a sequence of 860 amino acids. This includes a 21-amino-acid N-terminal signal sequence that is cleaved co-translationally, leaving a receptor protein with 839 amino acids. Examination of the amino acid sequence of the receptor, and of the exon–intron organization of its gene, shows that the mature receptor is divided into the five structural domains depicted in Fig. 9.3. Each domain is adapted to carrying out a separate component of the overall functional activity of the receptor, which includes the binding of ligands and participation in the endocytic cycle. The function of each domain has been deduced from its orientation and amino acid pattern, and these deductions have been tested by deletion analysis. Owing to the modular arrangement of the receptor, a single domain may be deleted by spontaneous or *in vitro* mutation without abolishing the functions of the remaining domains. From the effects of such deletions on

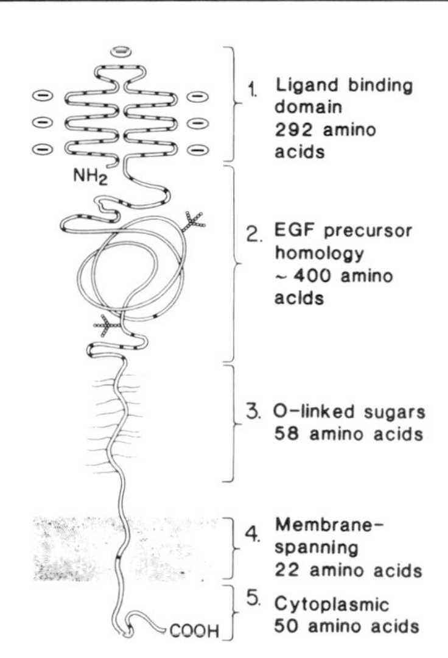

Figure 9.3. A representation of the mature LDL receptor, showing the five structural domains and their orientation in relation to the plasma membrane. The folding of the seven repeats in the first domain, in which the cysteine residues would form bridges *within* repeats, is arbitrary. The two N-linked complex oligosaccharides are shown in the second domain. The 18 O-linked oligosaccharides in the third domain are shown as horizontal lines. (From Brown and Goldstein, 1986, © The Nobel Foundation 1986, and with the permission of the authors.)

receptor function it has been possible to build up a picture of the relationship between structure and function of separate domains of the LDL receptor. Until recently, it was necessary to wait for the identification and analysis of very rare natural mutations at the receptor locus. However, developments in genetic engineering have now made it possible to produce virtually any desired modification to the receptor by the synthesis *in vitro* of genes with deletions, insertions, or point mutations. The mutant genes may then be expressed in cells *in vitro*. The only limitation to this approach is the theoretical possibility that the effect of a mutation on receptor function in the whole organism is not always revealed in cells *in vitro*. The use of transgenic animals should help to resolve this potential difficulty.

The five structural domains of the mature receptor are discussed as follows.

1. The first domain consists of the N-terminal 292 amino acids of the mature receptor (lacking the signal sequence). These are divided into seven imperfect repeat sequences each comprising about 40 amino acids, six of which are cysteines. Figure 9.4 shows the amino acid sequence of the first domain, with a consensus sequence of 19 amino acids. When the repeats are optimally aligned the cysteine residues occur at the same positions in all seven repeats. Labeling studies with iodoacetamide have shown that all the cysteines in this domain

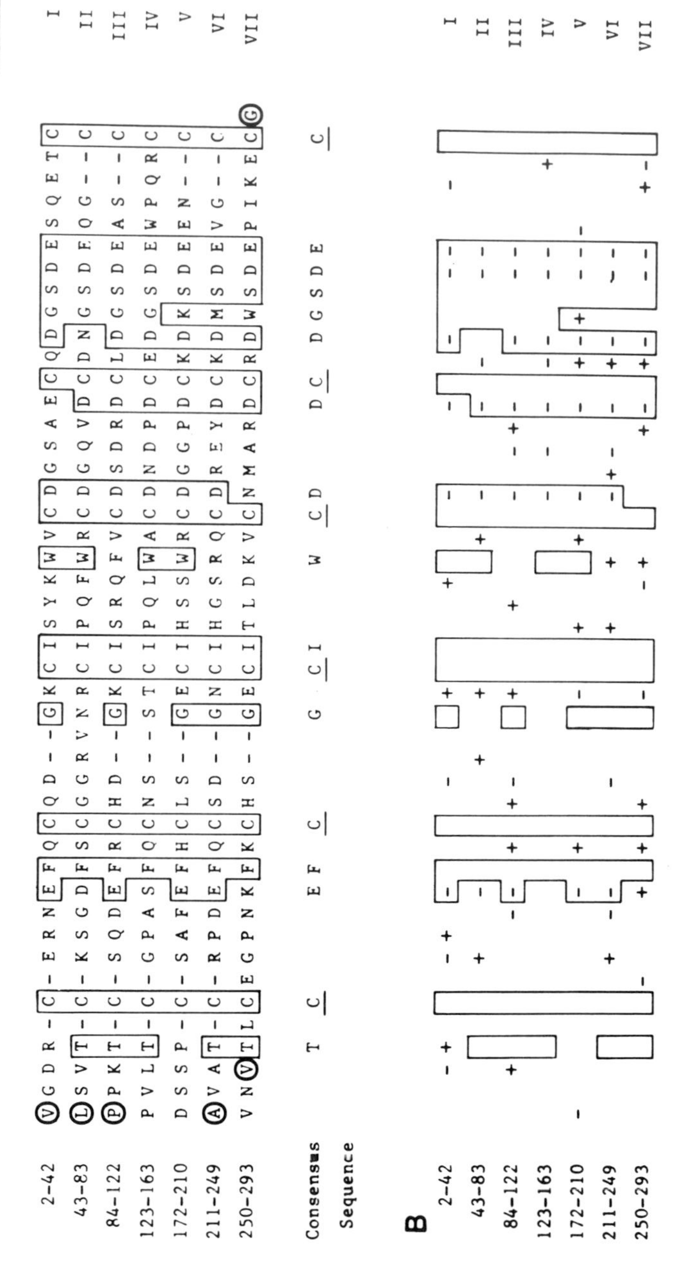
A
Residue No.
Repeat No.
Consensus
Sequence
B

Protein	Species	Residue	Amino Acid Sequence
LDL Receptor (A)	Human	297–331	C - - - L D N N G G C S H V C . (8) . C L C P D G F Q L V A Q - R R C
LDL Receptor (B)	Human	337–371	C - - - Q D P - D T C S Q L C . (8) . C Q C E E G F Q L D P H T K A C
LDL Receptor (C)	Human	646–690	C E R T T L S N G G C Q Y L C .(14) . C A C P D G M L L A R D M R S C
EGF-Precursor (1)	Mouse	366–401	C - - - A T Q N H G C T L G C . (8) . C T C P T G F V L L P D G K Q C
EGF-Precursor (2)	Mouse	407–442	C - - - P G N V S K C S H G C . (8) . C I C P A G S V L G R D G K T C
EGF-Precursor (3)	Mouse	444–482	C - - S S P D N G G C S Q I C . (9) . C D C F P G Y D L Q S D R K S C
EGF-Precursor (4)	Mouse	751–786	C - - - L Y R N G G C E H I C . (8) . C L C R E G F V K A W D G K M C
Factor X	Human	89–124	C - - - S L D N G D C D Q F C . (8) . C S C A R G Y T L A D N G K A C
Factor IX	Human	88–124	C - - - N I K N G R C E Q F C . (9) . C S C T E G Y R L A E N Q K S C
Protein C	Bovine	98–133	C - - - S A E N G G C A H Y C . (8) . C S C A P G Y R L E D D H Q L C
	Consensus		C - - - x x x N G G C x x x C . (8) . C x C x x G Y/F x L x x D x K x C

Figure 9.4. Amino acid sequence of the first domain, and of the growth-factor repeats in the second domain, of the human LDL receptor. In (A) the amino acids in the 7 repeat units have been aligned optimally by the computer programs ALIGN and RELATE, with modifications based on the positions of introns in the LDL-receptor gene (see Section III,D). Boxes show amino acids that are present in a given position in more than 50% of repeats. The positions at which introns interrupt the coding sequence of the receptor gene are denoted by encircled amino acids. A consensus sequence of 19 amino acids, including 6 cysteine residues, is shown below (A); the cysteine residues in the consensus sequence are underlined. In (B), the net charge in each of the amino acids in (A) is shown. Note that all conserved amino acids are negatively charged. Note also that repeats 4 and 5 are separated by 8 amino acids (164–171). In (C), the amino acids in the three growth-factor repeats of domain 2 of the human receptor are aligned to give maximum homology. Homologous regions from the mouse EGF precursor, Factor X, Factor IX, and protein C are shown for comparison. Note that the spacing of the cysteine residues in the growth-factor repeats differs from that in the binding-domain repeats (compare especially the intervals between the fourth and fifth and the fifth and sixth cysteines in the two sequence motifs). Note also that there is no cluster of negatively charged residues at the C-terminal ends of the growth-factor repeats. The amino acid code is as follows. A, Ala; C, Cys; D, Asp; E, Glu; F, Phe; G, Gly; H, His; I, Ile; K, Lys; L, Leu; M, Met; N, Asn; P, Pro; Q, Gln; R, Arg; S, Ser; T, Thr; V, Val; W, Trp; Y, Tyr. (From Südhof *et al.*, 1985a, *Science*, Vol. 228, pp. 815–822, with the permission of the authors. Copyright 1985 by the AAAS.)

participate in the formation of disulfide bonds (Lehrman *et al.*, 1987a). Hence, the first domain of the receptor must exist in a tightly folded state due to cross-linking either between or, more probably, within the repeats (see Section III,D,3 for discussion). Goldstein *et al.* (1985) suggest that extensive cross-linking stabilizes the receptor and thus enables it to withstand the strongly acidic environment within the endocytic vesicle without losing its ability to bind LDL when it returns to the plasma membrane. Each of the 7 repeats in the first domain of the LDL receptor is homologous to a single 40-residue sequence in human complement component C9 (Stanley *et al.*, 1985); 14 of the 19 amino acids in the consensus sequence of the LDL receptor repeats are present in the C9 sequence. The repeated sequence in the first domain has an even closer resemblance to one of the multiple cysteine-rich repeats in the human LDL-receptor-related protein (LRP) described by Herz *et al.* (1988) (see Chapter 8 for details). The cysteine-rich sequence motif present in domain 1 of the receptor and in complement components is designated class A by Stanley *et al.* (1985).

At the C-terminal end of each repeat in the first domain of the receptor there is a cluster of negatively charged amino acids whose consensus sequence is partially complementary to the positively charged receptor-recognition sequences in apoE and apoB-100 (Fig. 9.5). This suggests that the negatively charged clusters of amino acids in the first domain of the receptor constitute multiple binding sites for lipoprotein ligands. In agreement with an important functional role for these clusters of amino acids, there is a high degree of homology between the C-terminal residues in the seven repeats of domain 1 of the human receptor (Fig. 9.4) and between these sequences in the human, rabbit, and hamster receptors (Esser *et al.*, 1988). The conclusion that the first domain of the receptor contains the sequences required for ligand binding has been substantiated and made more precise by the study of natural mutations in the receptor gene and by analyzing the effects of synthetic mutations (mutational analysis).

Hobbs *et al.* (1986) have described a receptor from which the sixth repeat of the first domain has been deleted by natural mutation (patient FH 626, Section

	3357 3367
ApoB–100	Thr–Thr–Arg–Leu–Thr–Arg–Lys–Arg–Gly–Leu–Lys
Receptor	(Cys–Asp–X–X–X–Asp–Cys–X–Asp–Gly–Ser–Asp–Glu)$_7$
ApoE3 (0)	His–Leu–Arg–Lys–Leu–Arg–Lys–Arg–Leu–Leu–Arg
	140 150

Figure 9.5. Complementarity between amino acid sequences in the LDL receptor and in apoB-100 and apoE3. Residues 3357–3367 of apoB-100 and residues 140–150 of apoE3 are shown above and below the consensus for the cluster of negatively charged residues in the repeat sequences of the first domain of the LDL receptor (see Fig. 9.4). Charged amino acids are underlined. (Modified from Goldstein *et al.*, 1985.)

V). When observed in fibroblasts in culture, the mutant receptor is transported to the cell surface after undergoing normal glycosylation. At the cell surface it reacts with monoclonal antibody IgG-C7 in the presence of Ca^{2+} but is unable to bind LDL. However, β-VLDL is bound with normal affinity and is internalized and degraded at 37°C.

Van Driel *et al.* (1987b) have constructed two plasmids containing cDNA inserts encoding defective human LDL receptors. When these plasmids are introduced *in vitro* into receptor-negative hamster cells of the ldlA-7 line, the plasmid cDNA directs the synthesis of receptor protein. The first mutant receptor lacks only the first repeat of domain 1. This receptor reaches the cell surface but fails to bind antibody IgG-C7. However, it binds and internalizes LDL and β-VLDL and then recycles to the cell surface, all with normal efficiency. The second mutant contains the first repeat of domain 1 as essentially the only extracellular portion of the receptor. This receptor also reaches the cell surface. It binds antibody IgG-C7 in the presence of Ca^{2+}, but it does not bind LDL or β-VLDL. Van Driel *et al.* concluded that repeat 1 contains the epitope recognized by antibody IgG-C7 but that this repeat plays no part in ligand binding.

Esser *et al.* (1988) have extended these observations by analyzing the functional effects of a larger number of oligonucleotide-directed mutations in the human receptor gene. These mutations produced amino acid substitutions at highly conserved positions, or deletions of various combinations of complete repeats, in the first domain and in repeats A and B of the second domain (see Fig. 9.4). The mutant genes were expressed in monkey cells *in vitro*.

All receptors with mutations in domain 1 involving amino acid substitutions or deletions of whole repeats were converted into the mature form at the normal rate. This suggested that none of the mutations affected the folding of precursor protein into the conformation required for its transport to the Golgi apparatus. In confirmation of the findings of van Driel *et al.* (1987b), deletion of repeat 1 of the first domain had little or no effect on the surface binding of LDL or β-VLDL. Repeats 2+3 and 6+7 were required for maximal binding of LDL, but not of β-VLDL. Repeat 5 was required for binding of both ligands. Repeat A of domain 2 was required for binding of LDL, but not β-VLDL. In each case, defective binding was exhibited in the presence of saturating concentrations of ligand, indicating that the defect was due to a decrease in the number of functional binding sites on the cell surface rather than to a change in binding affinity.

On the basis of these findings, Esser *et al.* (1988) suggest a model for the arrangement of the seven repeats of domain 1 and repeat A of domain 2 (Fig. 9.6). Repeats 2–7 form a regular hexagon in which 2, 3, and 4 are separated from 5, 6, and 7 by a conserved sequence of eight amino acids (see Fig. 9.4) linking repeats 4 and 5. In this model, repeats 2+3 and 6+7 should be functionally equivalent, while repeats 4 and 5, which lie at the center of symmetry of the proposed binding structure, should be the most crucial for ligand binding.

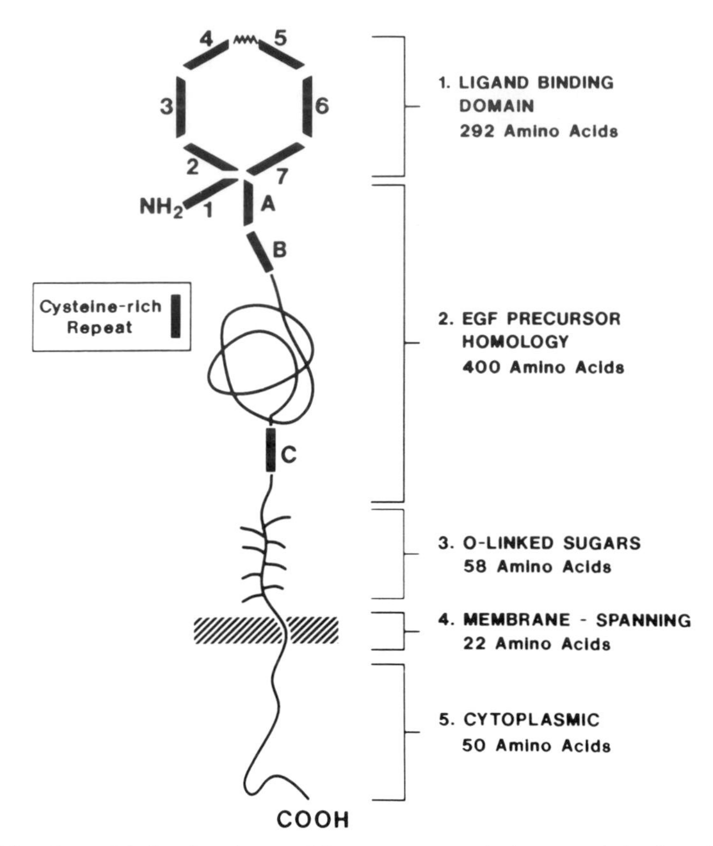

Figure 9.6. A model showing the possible arrangement of elements of the ligand-binding domain of the human LDL receptor. The essential feature is a hexagonal structure composed of repeats 2, 3, and 4 joined to repeats 5, 6, and 7 by a linker sequence of eight amino acids (zigzag). Repeat 1 plays no part in ligand binding. Repeat A of domain 2 is required for binding of LDL to the receptor but it does not interact with the ligand. (From Esser *et al.*, 1988, with the permission of the authors.)

Since repeat A of domain 2 lacks a cluster of negatively charged residues, it is unlikely to interact directly with ligands. Esser *et al.* suggest that this repeat in some way facilitates access of LDL particles to binding sites on receptors at the cell surface.

The more stringent structural requirement for the binding of LDL than for that of β-VLDL is difficult to explain and is likely to remain so until the question of the stoichiometry of lipoprotein binding to the receptor has been settled. Presumably, the difference is related to the fact that binding of a particle of LDL is mediated by a single molecule of apoB-100 with one binding site, whereas binding of a β-VLDL particle is mediated by several molecules of apoE. It is worth noting that the mutant LDL receptor from which the sixth repeat of domain 1 has been deleted is functionally similar to the LDL receptor on mouse macrophages investigated by Koo *et al.* (1986) (see Chapter 8, Section III,D). Both receptors bind β-VLDL with high affinity but exhibit little or no binding of LDL. As discussed in Section III,D below, the exon–intron organization of the LDL-receptor gene is such that single exons coding for segments of the first domain could be spliced out, resulting in the production of a receptor lacking one or more of the seven repeats. Hence, it is possible that LDL receptors with specific affinity for β-VLDL, similar to the mutant receptor produced in the fibroblasts of patient FH 626, are produced in some cells under physiological conditions.

2. The second domain, extending from residue 293 to residue 692, shows a considerable degree of homology with a region in the extracellular portion of the mouse EGF precursor, a protein that spans the plasma membrane (Yamamoto *et al.*, 1984; Südhof *et al.*, 1985a). Within this region of the LDL receptor there are three imperfect cysteine-rich repeats, each of ~40 amino acids, that are homologous to a sequence repeated 4 times in the mouse EGF precursor and 22 times in human LRP. [The mouse EGF precursor contains an additional six cysteine-rich repeats with lower degrees of homology with the three repeats in domain 2 of the LDL receptor (see Fig. 9.29).] The three repeats in the human receptor are denoted A (residues 293–332), B (333–375), and C (642–692) (Fig. 9.4). A single imperfect copy of the A-B-C repeat sequence in the LDL receptor is present in several proteins of the blood-clotting system (Factor IX, Factor X, proteins C and S, urokinase, and tissue plasminogen activator). Doolittle *et al.* (1984) call this sequence motif the growth-factor repeat. Stanley *et al.* (1986) refer to it as the class B repeat. Although growth-factor repeats and the repeats in domain 1 each have six cysteine residues, the two classes differ from each other in the spacing of the cysteine residues and in other ways (see Fig. 9.4), reflecting their different evolutionary origins (see Section VIII).

Two sites for N-linked glycosylation are present at residues 494 (Asn-Gly-Ser) and 636 (Asn-Leu-Thr). These are the most likely sites at which the two N-linked oligosaccharide units are attached to the receptor. Three other potential sites for N-linked glycosylation are present in the first domain of the receptor, but these are unlikely to be glycosylated in view of the unusual cross-linked structure of this region.

Davis *et al.* (1987b) have constructed a plasmid with a cDNA insert encoding the complete amino acid sequence of the human LDL receptor minus the sequence of 400 residues in the EGF-precursor homology region (domain 2). When introduced into hamster ldlA-7 cells, this plasmid directed the synthesis of a truncated receptor with the expected apparent molecular weight of 72K. The precursor was rapidly converted into a mature glycosylated receptor with apparent MW 125K. The mature receptor bound LDL and β-VLDL on nitrocellulose blots; in intact hamster cells it bound β-VLDL but failed to bind LDL. When expressed in ldlA-7 cells, the truncated mutant receptor mediated the binding, internalization, and degradation of β-VLDL with kinetics indistinguishable from the kinetics shown by normal receptors expressed in the same cells. However, the mutant receptor differed from normal receptors in that it failed to dissociate normally from its bound ligand at acid pH in endocytic vesicles. Hence, after internalization of receptor–ligand complexes, most of the receptors were delivered to lysosomes and degraded by lysosomal enzymes without recycling to the plasma membrane. Davis *et al.* (1987b) have prepared another DNA construct encoding a human LDL receptor lacking only the growth-factor repeats A and B (see Fig. 9.11). This mutant receptor also failed to dissociate normally from bound β-VLDL. When present in intact cells, it bound β-VLDL normally but failed to bind LDL.

Soutar *et al.* (1989) have described a natural mutation resulting in the substitution of leucine for proline at position 664 in growth-factor repeat C. Fibroblasts with this mutation exhibit impaired binding of LDL (see Section V for details). The effect of deleting growth-factor repeats A and B singly was mentioned above.

The above observations, together with those of Esser *et al.* (1988), show that selective deletion of the whole of domain 2 has no effect on the transport of newly synthesized LDL receptors to the Golgi apparatus or on the processing of their O-linked oligosaccharide residues. Nor does this deletion affect the ability of isolated receptors to bind lipoprotein ligands. However, it is evident that repeats A and C are required for normal binding of LDL by receptors on the cell surface, and that sequences within this domain are required for the release of bound ligand from the receptor at acid pH and for recycling of receptors internalized with bound ligand. Davis *et al.* (1987b) suggest that domain 2 normally undergoes a conformational change at acid pH which causes bound ligand to dissociate from the binding domain. The effects of mutations in domain 2 that lead to defective transport of receptors from the ER to the Golgi apparatus are discussed in Section V.

3. The third domain is just outside the membrane-spanning region. It extends for 58 amino acids (residues 693–750) and includes 18 Ser or Thr residues to which most of the O-linked oligosaccharide units are attached.

Goldstein *et al.* (1985) have pointed out that another cell-surface receptor, the IL-2 receptor on T-lymphocytes, also contains a cluster of O-linked oligosaccharides immediately outside the plasma membrane. They suggested that in both receptors the oligosaccharide chains act as struts to keep the ligand-binding domain at a distance from the surface of the cell. The chains might also help to anchor the receptor in the membrane. However, deletion analysis, involving short-term studies with mutated LDL receptors expressed in hamster cells *in vitro*, has failed to provide direct evidence as to the function of the third domain of the receptor. Davis *et al.* (1986a) constructed a mutant cDNA for the LDL receptor lacking the base sequence encoding amino acids 700–747. (This is the shortest amino acid sequence that contains all the 18 Ser/Thr residues of domain 3). Cells of the ldlA-7 line transfected with the mutant gene expressed truncated receptors without clustered O-linked sugars, thus providing formal proof that the cluster of O-linked sugars is indeed located in domain 3. Despite the deletion of this segment, the mutant receptors were transported to the cell surface, where they bound and internalized LDL normally. Moreover, the internalized receptors delivered their bound ligand to lysosomes, recycled to the plasma membrane at the normal rate, and exhibited a normal half-life.

It should be noted that about 20% of the O-linked sugars of the LDL receptor are not present as a cluster in domain 3 but are present as isolated chains elsewhere in the external portion of the receptor. Kingsley *et al.* (1986a) have isolated mutant lines of CHO cells in which the synthesis of N-linked and O-linked sugars on LDL receptors and other glycoproteins is markedly defective. LDL receptors are synthesized in these cells but are rapidly degraded without being processed (see Section V,D for details). These findings are consistent with the possibility that the isolated O-linked oligosaccharide chains on the receptor are essential for its stability *in vivo*.

4. The fourth domain (residues 768–789) is a membrane-spanning sequence of 22 hydrophobic amino acids, adapted to anchoring the receptor in the plasma membrane. The membrane-spanning domains in the human and bovine receptors differ at seven positions, but all the substituted amino acids are hydrophobic.

The function of the fourth domain of the receptor may be deduced by comparing the effects of two classes of natural mutation at the receptor locus. As discussed below, two mutations have been identified in FH homozygotes that lead to loss of the bulk of the cytoplasmic tail, while leaving the membrane-spanning region intact (patients FH 683 and 763; see Table 9.2 for references). The mutant receptors produced in the cells of these patients are anchored normally to the plasma membrane, though they fail to cluster in coated pits. Two other mutations have been described that lead to the formation of a truncated receptor from which both the transmembrane and cytoplasmic domains have been selectively deleted (patients FH 274 and 781; see Table 9.2 for references). The LDL receptors synthesized by these patients are not capable of anchoring themselves efficiently in a bilayer membrane. Consequently, most of those synthesized in the rough ER pass completely across the ER membrane into the lumen and are secreted into the external medium, while the bulk of those that do reach the cell surface dissociate from the external face of the plasma membrane (Lehrman *et al.*, 1987b). Thus, the function of the hydrophobic fourth domain is to anchor the newly synthesized receptor to the ER membrane in specific orientation (C-terminal on the cytoplasmic face and N-terminal on the opposite face) and to maintain this orientation during membrane-to-membrane transport of the receptor throughout repeated cycles to and from the plasma membrane.

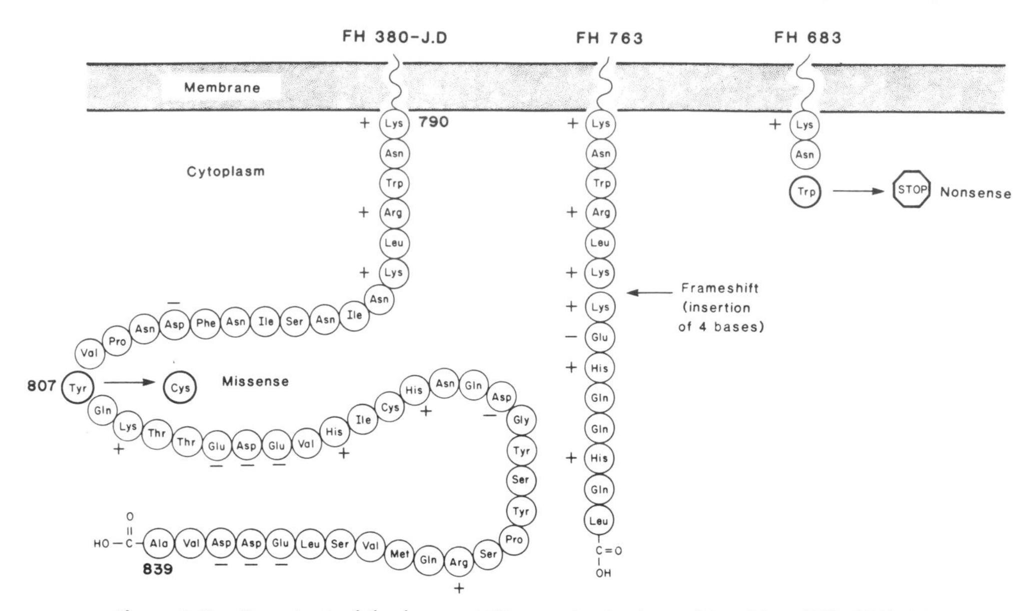

Figure 9.7. Domain 5 of the human LDL receptor (amino acid residues 790–839). Mutant forms of the receptor in three FH homozygotes (FH 380, FH 763, and FH 683) are also shown. The amino acids at some of the positions referred to in the text are indicated by residue numbers. (From Brown and Goldstein, 1986, © The Nobel Foundation 1986, and with the permission of the authors.)

5. The fifth domain is the C-terminal tail of 50 amino acids (residues 790–839) that projects into the cytoplasm as shown in Fig. 9.7. In keeping with its important functional role the amino acid sequence of the cytoplasmic tail of the LDL receptor is highly conserved between species. For example, the sequences in this region of the human and bovine receptors differ at only four positions, and at each substitution the charge on the amino acid is unchanged. (See also Davis *et al.*, 1987a). Bovine adrenal cortex contains a protein kinase that phosphorylates serine 833 in the cytoplasmic domain of the receptor (Kishimoto *et al.*, 1987). However, the functional significance of this enzyme is not clear. Kishimoto *et al.* were unable to demonstrate phosphorylation of LDL receptors in intact human fibroblasts *in vitro*. Moreover, Davis *et al.* (1987a) have shown that substitution of serine 833 by alanine has no effect on internalization of the receptor.

Three naturally occurring mutations in the receptor gene (Table 9.1, patients FH 380, 683, and 763) have helped to reveal the function of the cytoplasmic domain. Each of the three mutant genes encodes a receptor that reaches the cell surface and binds LDL with normal affinity but does not localize in a coated pit. Hence, the receptors and their bound ligands are not internalized at the normal rate. These are the internalization-defective alleles mentioned in Chapter 8 (Section II,C) and discussed in more detail in Section V below. In each case, the only effect of the mutation on receptor structure is to modify the amino acid sequence

Table 9.1
Classification of Mutations at the LDL-Receptor Locus That Give Rise to FH, Based on Structure/Function Changes in the Gene Product

Class of Mutation	Allele	Apparent M_r ($\times 10^{-3}$)		Receptor Location		LDL-binding to cells	References
		Precursor	"Mature"	Coated Pits	Non-Coated Surface		
No detectable receptor protein	*R-0*	None	None	—	—	None	1,2
Precursor not processed	*R-100*	100	100	—	—	None	2
	R-120	120	120	—	—	None	3
	R-135	135	135	—	—	None	2
Variant; precursor processed slowly and binds LDL poorly	*R-120* $\xrightarrow{slow}$ *R-160*	120	160	(+)	–	Reduced	4,9
Precursor processed normally. Mature receptor binds LDL poorly.	*R-140b*$^-$	100	140	+	–	Reduced	2
	R-160b$^-$	120	160	+	–	Reduced	2
	R-120b$^-$	170	210	+	–	Reduced	3
Precursor processed normally. Mature receptor binds LDL but does not enter coated pits.	*R-160i*$^-$	120	160	None	+	Normal binding, defective internalization	1,5
	R-155i$^-$	115	155	None	+		6
Variant; same as above, but >90% of mature receptors secreted from cell.	*R-155i*$^-$, *sec*	115	155	None	(+)	Normal binding, defective internalization	7,8

[From Goldstein *et al.* (1985) with additions.] The allele numbers refer to the apparent molecular weight (M_r) of the "mature" receptor protein; b^-, defective binding; i^-, defective internalization; *sec*, secreted; "mature" refers to the predominant form of the radioactive receptor protein present in cell extracts after a 2-h pulse followed by a 2-h chase. Apparent M_rs of Class 4 receptors are estimated from the number of amino acids deleted from the receptor protein. Plus signs in brackets denote minor populations. Numbers in brackets refer to mutant receptors mentioned in the text. Note that several of the mutant receptors included in the table are described in more than one reference. (1) Beisiegel *et al.*, 1982 (FH 380); (2) Tolleshaug *et al.*, 1983 (refers to several Class 1, 2 and 3 mutants, some of which are genetic compounds); (3) Tolleshaug *et al.*, 1982 (FH 295); (4) Schneider *et al.*, 1983*a* (WHHL rabbit, FH 692 and FH 429); (5) Davis *et al.*, 1986*b* (FH 380); (6) Lehrman *et al.*, 1985*a* (FH 683 and FH 763); (7) Lehrman *et al.*, 1985*b* (FH 274); (8) Lehrman *et al.*, 1987*b* (FH 781); (9) Knight *et al.*, 1989 (MM).

of the cytoplasmic tail, indicating that a major function of this domain is to facilitate the clustering of receptors in coated pits.

The effect of the mutation in J.D. (patient FH 380) is of particular interest. In J.D., a point mutation results in the substitution of a cysteine residue for tyrosine at position 807 in the cytoplasmic domain. Davis *et al.* (1986b) have shown conclusively that this substitution is sufficient to prevent receptors that have reached the plasma membrane from clustering in coated pits. They constructed a receptor cDNA in which the mutation was reproduced. When transfected into receptor-negative hamster cells, the mutant cDNA directed the synthesis of receptors that failed to enter coated pits. In the other two patients with internalization-defective receptors, the mutation leads to the formation of a truncated cytoplasmic domain with only the first two (FH 683) or the first six (FH 763) amino acids of the normal sequence.

Davis *et al.* (1987a) have carried out a more detailed analysis of the structural features of the cytoplasmic domain that are required for efficient clustering of LDL receptors in coated pits. Using oligonucleotide-directed mutagenesis, they generated 24 mutant receptors, each with an amino acid substitution or deletion in the C-terminal 50-residue segment. The modified receptors were expressed in ldlA-7 cells so that the efficiency with which they internalized bound LDL could be determined. The results of these experiments are summarized in Fig. 9.8.

Deletion of amino acids 812–839 had little effect on the internalization index of the receptor (see figure legend for explanation). Hence, receptors with only the first 22 amino acids of the cytoplasmic domain (residues 790–811) must be capable of clustering in coated pits. Within this sequence there was a stringent requirement for an aromatic amino acid at position 807 (the site of the substitution in the J.D. mutation). Replacement of the normal Tyr^{807} by a nonaromatic residue, whether charged or uncharged, markedly diminished the internalization index. On the other hand, replacement by phenylalanine or tryptophan had little effect. Thus, the functional effect of the mutation in J.D. is due, not to the introduction of a cysteine at 807, but to the loss of a tyrosine at this position. Termination at residue 792, leaving only two amino acids projecting into the cytoplasm (as in the FH 683 mutation), produced a receptor with a greatly diminished internalization index.

In view of the extent to which the amino acid sequence of the cytoplasmic domain of the LDL receptor is conserved between species, it is surprising that receptors lacking the C-terminal 28 amino acids are internalized normally in hamster cells *in vitro*. Davis *et al.* (1987a) suggest that this segment of the receptor is required for a regulatory function, perhaps involving the phosphorylation of serine 833, in some specialized cells such as hepatocytes or cells of the adrenal cortex.

The finding that an aromatic amino acid at position 807 is required for normal internalization raises the possibility that this residue is necessary for recognition

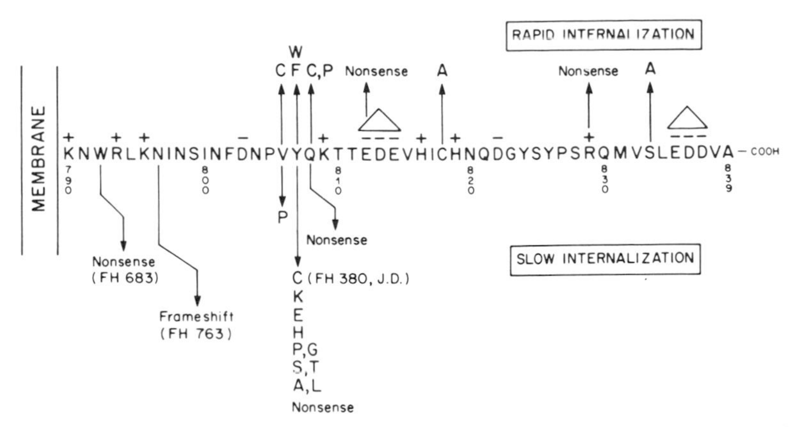

Figure 9.8. The effect of natural and synthetic mutations in the cytoplasmic domain of the human LDL receptor on the ability of receptors to internalize LDL via coated pits in receptor-negative Chinese hamster ovary cells in culture. The amino acid sequence of the cytoplasmic domain is shown in the single-letter code. Amino acid substitutions that diminish internalization are indicated below the normal sequence; those without effect on internalization are indicated above. Deleted amino acids are denoted by open triangles. Internalizing ability was determined as the "internalization index," defined as the sum of internalized and degraded [^{125}I]LDL divided by the amount of surface-bound [^{125}I]LDL in the steady state when cells were incubated in a medium containing [^{125}I]LDL. A, Ala; C, Cys; D, Asp; E, Glu; F, Phe; G, Gly; H, His; I, Ile; K, Lys; L, Leu; M, Met; N, Asn; P, Pro; Q, Gln; R, Arg; S, Ser; T, Thr; V, Val; W, Trp; Y, Tyr. (From Davis *et al.*, 1987a, with the permission of the authors.)

of the receptor by a coat protein in the coated pit. Alternatively, it is possible that clustering of LDL receptors in coated pits is dependent upon the association of two or more receptor molecules to form oligomers and that tyrosine 807 acts as a self-recognition signal for this process (see Davis *et al.*, 1987a, for discussion). The effects of selective modification of the cytoplasmic domain on the ability of LDL receptors to form oligomers and to cluster in coated pits provide only limited support for this view. Van Driel *et al.* (1987a) have shown that most of the normal LDL receptors in adrenal-cortex membranes, and a substantial proportion of those in intact fibroblasts, exist as noncovalently associated dimers. Deletion of the C-terminal 33 amino acids of the cytoplasmic domain (809–839) abolishes dimerization and the internalization of receptors in hamster cells, indicating that this domain is required for both processes. However, deletion of the C-terminal 28 residues (812–839) appears to abolish dimerization but, as noted above, has little effect on the clustering of receptors in coated pits and their subsequent internalization. As discussed in Chapter 8 (p. 266), a short conserved sequence of amino acids that includes a tyrosine residue is present in the cytoplasmic domain of the LDL receptor, LRP, and poly(Ig) receptors. This sequence may be part of a signal for the clustering of some transmembrane proteins in coated pits.

III. The LDL-Receptor Gene and Its Message

A. Chromosomal Localization

Analysis of the chromosomes in human–hamster hybrid cells expressing human LDL receptors, combined with hybridization *in situ* with a radioactive cDNA probe, has shown that the human receptor gene is on bands p13.1–p13.3 of the short arm of chromosome 19 (Francke *et al.*, 1984; Lindgren *et al.*, 1985). Assignment of the receptor gene to chromosome 19 may also be inferred from family studies showing that clinical expression of FH segregates with the third component of complement (Berg and Heiberg, 1978), the gene for which is known to be on chromosome 19 (Whitehead *et al.*, 1982). Analysis of the molecular basis of FH at the DNA level has shown that the mutation causing the disease is in the LDL-receptor gene itself rather than in some other gene whose product is required for normal receptor function (see Section V below). Hence, the LDL-receptor gene must be in the same chromosome as the gene for C3.

B. Cloning the Human Gene

Yamamoto *et al.* (1984) used a partial cDNA for the bovine LDL receptor (referred to in Section I) as a probe to isolate a cloned fragment near the 3′ end of the receptor gene from a human genomic library. An exon from this genomic clone was used in conjunction with synthetic oligonucleotides, corresponding to the N-terminal sequence of the bovine receptor, to isolate a 4.9-kb cDNA from a human fetal adrenal cDNA library. The 4.9-kb cDNA was extended by genetic engineering to produce a 5.3-kb cDNA in plasmid pLDLR2. The cDNA insert in this plasmid corresponded to the entire coding region, the entire 3′-untranslated region, and a portion of the 5′-untranslated region of the human LDL-receptor message. The coding region of the cDNA insert in pLDLR2 was used to deduce the complete amino acid sequence of the human receptor, as discussed in Section II. Plasmid pLDLR2 was also used for the preparation of radioactive probes for the isolation of a series of overlapping human genomic clones spanning most of the human LDL-receptor gene (Südhof *et al.*, 1985a).

C. The Receptor Message

Yamamoto *et al.* (1984) used probes prepared from pLDLR2 to investigate human LDL-receptor mRNA. Radioactive fragments of the insert from this plasmid were shown to hybridize with a 5.3-kb message on Northern blots of mRNA from human tissues.

Sequencing of a full-length cDNA showed that the 3′-untranslated region of mRNA for the human LDL receptor is 2.5 kb long. It is also unusual in that it contains the RNA complement of two and one-half repetitive Alu sequences, so called because they were first detected in human genomic DNA with the restriction endonuclease *Alu*I.

Alu sequences are partial repeats scattered throughout the human genome, usually in intergenic regions or in the introns of genes. Together, they account for up to 5% of the DNA in the whole human genome. As shown in Fig. 9.9A, the consensus Alu sequence is about 300 bases long and consists of two partial repeat units in tandem. The left-hand (5′) unit has ~130 bases; the right-hand unit is longer (~160 bases) owing to the addition of a sequence of about 30 bases. Alu sequences are often flanked by direct repeats of a few bases, suggesting that they originated as transposable elements. (For an explanation of why transposons are flanked by direct repeats, see Lewin, 1987.)

The arrangement of the Alu sequences in the receptor message is shown in Fig. 9.9B. The first sequence has 131 bases in the left-hand unit and 155 bases in the right-hand unit and is flanked by 15-base direct repeats. The second Alu sequence is a right-hand unit from which the 5′ 30 bases have been deleted. The third Alu sequence has left-hand and right-hand repeat units of normal length but is not flanked by direct repeats.

The function of Alu sequences in the human genome is not known. However, their presence within the LDL-receptor gene could increase the likelihood of misalignment in this region during meiosis and may, therefore, be partially responsible for the high frequency of deletion and insertion mutations in the human receptor gene, as discussed in Section V.

D. The Receptor Gene

1. Structure

The human LDL-receptor gene is about 45 kb long and is divided into 18 exons, the 17 introns together comprising about 40 kb (Südhof *et al.*, 1985a). Figure 9.12 shows the 5′ end and 5′-flanking region of the human receptor gene. S1-nuclease analysis and primer extension experiments have shown that transcription of the gene is initiated at sites between bases −79 and −93 (see Fig. 9.12 for numbering) and that there is no intron in the 5′-untranslated region. At a position 20–30 bases upstream from the sites for initiation of transcription there are two AT-rich sequences that may represent TATA boxes. Further upstream there are three imperfect direct repeats of 16 bases each and an imperfect inverted repeat of 14 bases (see Section IV,B for details). There are three polyadenylation sequences (AATAAA and AATTAA) in the 3′-untranslated region of the receptor gene. The presence of Alu sequences in several introns of the receptor gene is discussed in Section V below.

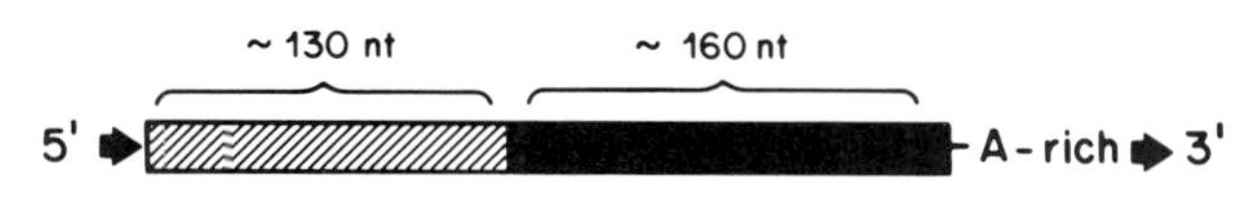

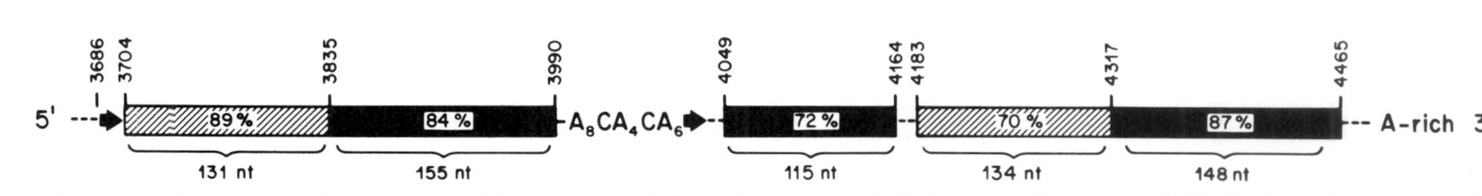

Figure 9.9. Organization of the consensus Alu sequence and of the Alu sequences in the human LDL receptor mRNA. (A) shows the consensus Alu sequence. The left (hatched) and right (solid) tandem repeat units are flanked by short direct repeats (arrows). (B) shows the Alu sequences in the 3′ untranslated region of the human receptor mRNA. The symbols are as in (A). Vertical numbers refer to the base numbers in receptor mRNA (1 is the first base in the initiator codon). Percentages within the repeat units refer to the percent homology with the sequence in the corresponding unit in the consensus sequence. (From Yamamoto *et al.*, 1984, with the permission of the authors.)

2. Exon–Intron Organization in Relation to Protein Domains

With a few exceptions, the exons in the LDL-receptor gene correspond closely to domains in the gene product. As shown in Fig. 9.10, most of the 17 introns interrupt the coding sequence for receptor mRNA at, or close to, boundaries between domains or between units within a domain. For example, the first intron interrupts the codon for the second amino acid in the mature receptor, so that exon 1 consists essentially of the code for the 5′-untranslated region and the signal sequence. Likewise, introns occur at the boundaries of the first three domains. The siting of introns in the base sequences encoding the ligand-binding and EGF-precursor-homology domains (1 and 2) is particularly informative from an evolutionary point of view (see Section VIII below).

In the receptor gene, introns are present just before the coding sequence for repeat I in the first domain and immediately after the sequences coding for repeats I, II, V, VI, and VII (see Fig. 9.4). Thus, the siting of introns in this region of the receptor gene provides strong support for the model shown in Fig. 9.3, in which the first domain of the receptor is divided into seven cysteine-rich repeat units each of ~40 amino acids. Südhof *et al.* (1985a) suggest that the seven repeats arose by successive duplications of an ancestral exon and that introns originally present between repeats III and IV and repeats IV and V have been lost during evolution. They also point out that exons in this region of the receptor gene could be spliced out to give mRNAs coding for receptors with different ligand-binding characteristics.

The coding sequence for the second domain of the receptor is bounded by introns 6 and 14, and each of the repeats denoted A, B, and C is encoded by a single exon (Fig. 9.11). Comparison between the homologous sequences in the human genes for the LDL receptor and the EGF precursor shows a striking similarity in the positions of their introns (Südhof *et al.*, 1985b). In both genes,

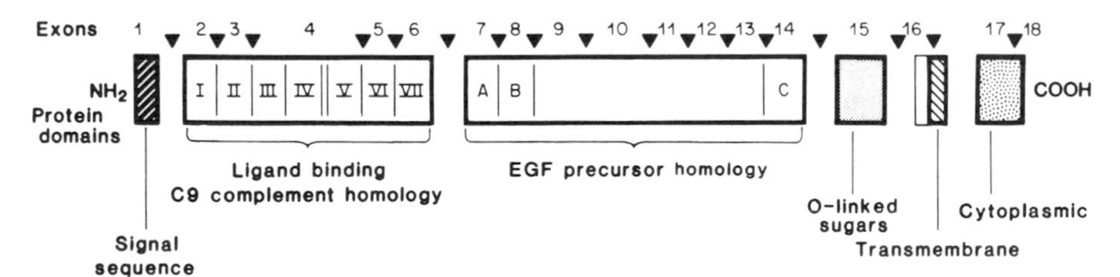

Figure 9.10. Exon–intron organization of the human LDL-receptor gene in relation to the protein domains. The signal sequence and the five domains of the mature receptor are delineated by thick lines. The seven repeats in the first domain are denoted I to VII; the eight amino acids separating repeats IV and V are shown by a double vertical line. The three repeats in the second domain are denoted A, B, and C. Exons are numbered 1–18 and the positions at which introns interrupt the gene are shown by triangles. (From Südhof *et al.*, 1985a, *Science*, Vol. 228, pp. 815–822, with the permission of the authors. Copyright 1985 by the AAAS.)

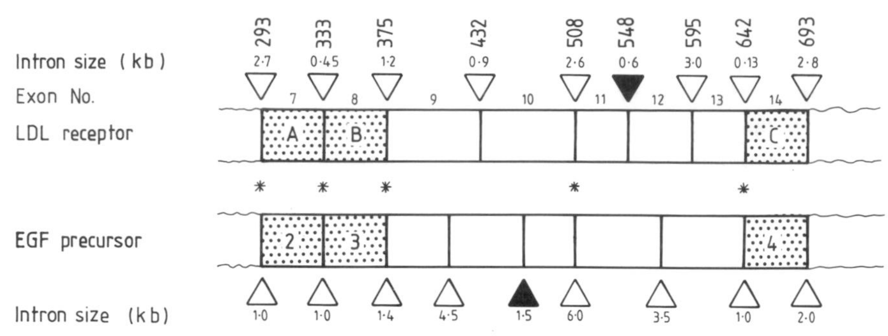

Figure 9.11. Positions of introns in the region of the human LDL-receptor gene encoding the second domain and in the homologous position of the human EGF precursor gene. Repeats A, B, and C in the receptor correspond to repeats 2, 3, and 4, respectively, in the EGF precursor. Positions of introns are denoted by triangles; black triangles denote unshared introns. Vertical numbers show the amino acid positions in the receptor gene that are interrupted by introns. Asterisks show the five introns that interrupt codons at identical positions in homologous regions of the receptor and EGF precursor genes. (From Südhof *et al.*, 1985b, *Science,* Vol. 228, pp. 893–895, with the permission of the authors. Copyright 1985 by the AAAS.)

the homologous sequence is interrupted by nine introns. When the two protein sequences are optimally aligned, five of the nine introns are seen to interrupt the coding sequence at the same amino-acid position in the two sequences. Inspection of Fig. 9.11 shows that the introns in three pairs (those at introns 9, 12, and 14 in the receptor gene) are imperfectly aligned and that each gene has an unpaired intron in this region. Südhof *et al.* (1985b) have discussed how these anomalies may have come about during evolution. Among other suggestions, they postulate that each of the unpaired introns originally had a partner in the homologous region of the other gene, and that the partner has been lost.

The 58 amino acids of domain 3 of the receptor (residues 693–750) are encoded in exon 15. Part of exon 16 encodes the remaining C-terminal segment of the extracellular portion of the receptor. The remainder of exon 16 and the 5′ end of exon 17 encode the 22 amino acids of the membrane-spanning domain. The remaining 3′ end of exon 17 and a part of exon 18 encode the 50-residue cytoplasmic tail. Exon 18 also encodes the whole of the 3′-untranslated region of the message. The relevance of these exon positions to deletion mutations in the receptor gene is discussed below.

The presence of repeat sequences in the LDL-receptor gene, the high degree of correspondence between exons and protein domains, and the sharing of homologous exons with other genes all have a bearing on the evolutionary origins of the receptor gene. These points are discussed in Section VIII of this chapter.

IV. Regulation of Expression of the Receptor Gene

A. General Remarks

As discussed in the previous chapter, synthesis of LDL receptors is suppressed by the cellular uptake of sterol from the medium and is stimulated when cells are incubated in the presence of LPDS. Russell *et al.* (1983) have shown that these alterations in rates of synthesis are accompanied by parallel changes in the intracellular concentration of LDL-receptor mRNA due, presumably, to changes in the rate of transcription of the receptor gene. Work carried out in the laboratory of Brown and Goldstein is beginning to show how sterols modulate expression of the receptor gene. The experimental approaches used in this work will be more readily understood if we first consider one or two general aspects of the expression of genes that function in eukaryotic cells.

Specific sequences in the 5′-flanking region of the gene are required for the binding of RNA polymerase to the DNA helix and the initiation of transcription. The region containing these sequences is the *promoter*, exemplified by the promoter of the thymidine kinase (*tk*) gene of herpes simplex virus, a gene that can be expressed in a wide variety of eukaryotic cells. McKnight and co-workers (McKnight *et al.*, 1981; McKnight, 1982) have shown that the *tk* promoter has at least three sequence elements within a segment of the 5′-flanking region extending 105 bp upstream from the startpoint for transcription. The proximal sequence (nearest to the startpoint) includes a hexanucleotide (TATTAA) homologous to the consensus sequence of the TATA box, present in many eukaryotic genes and required for initiation of transcription at the correct nucleotide (see Chapter 3, Section III). Upstream from the proximal sequence are two GC-rich sequences (GC boxes) called the first and second distal signals (see Fig. 9.13). Synthetic point mutations in either of the distal signals abolish transcription or markedly reduce its efficiency, showing that both signals are required for transcription. Point mutations in the stretches of DNA between the three sequence elements have no effect on the efficiency or accuracy of transcription of the *tk* gene. However, when the distance between adjacent sequences is extended by insertion of more than 50 bp of a random sequence, transcription is abolished.

As a general rule, the promoter for one gene will function efficiently when ligated to any other gene, provided that the sequence elements are in the correct orientation and in the correct positions relative to the startpoint for transcription of the new gene. Thus, the *tk* promoter, when correctly placed, will act as promoter for the chloramphenicol acetyltransferase (CAT) gene.

In addition to an intact promoter, transcription of eukaryotic genes requires the presence of protein *transcription factors* that bind to specific sequences in the promoter region. Transcription factors are thought to act by facilitating the binding of RNA polymerase to the promoter. An example of a transcription factor is the Sp1 protein, shown by Dynan and Tjian (1985) to bind to GC-box sequences present in several eukaryotic promoters and in the two distal signals of the herpes simplex *tk* promoter. An important aspect of transcription factors is their potential capacity for selective control of the expression of single genes or of groups of genes whose promoters share a specific recognition sequence for a transcription factor. If a specific transcription factor were required for expression of a gene, the gene would be expressed only when the transcription factor was present and would be suppressed when it was absent or was prevented by a repressor molecule from binding to the promoter.

The promoters of some eukaryotic and viral genes are activated by the presence of *cis*-acting sequences called *enhancers,* usually comprising several partial repeats each of ~10–20 bp. Enhancers cannot act as promoters for nearby genes. Moreover, they differ from promoters in two respects. An enhancer can act in either orientation and its position relative to the promoter can be varied within wide limits, either upstream or downstream, without effect on its ability to activate the promoter. An enhancer can activate any promoter near which it is placed, a property that may have played a part in the evolution of the regulation of gene networks in higher organisms (see Yamamoto, 1985).

Of potential interest in the present context are enhancers that are regulated by sterol-derived hormones. A well-studied example is the group of genes, including mammary tumor virus (MTV) DNA, whose expression is under the control of glucocorticoid (for details, see Yamamoto *et al.,* 1983; Chandler *et al.,* 1983; Yamamoto, 1985). Stimulation of the expression of MTV DNA by glucocorticoids is mediated by the presence of multiple sequences, called glucocorticoid response elements (GRE), located within a 340-bp stretch of DNA upstream from the startpoint for transcription. Glucocorticoids form a complex with intracellular glucocorticoid-receptor protein, which then binds to the response elements, leading to increased transcription of viral DNA. GRE behave like enhancers in that they can function in either orientation and when inserted at various distances from the promoter. When GRE are ligated upstream of the herpes simplex *tk* gene, including its intact promoter, expression of the *tk* gene becomes responsive to glucocorticoids *in vivo.* Thus, GRE may be regarded as hormone-regulated enhancers. Sequences homologous to the GRE near MTV DNA have been identified near other genes that are regulated by glucocorticoids. This could provide the basis for coordinate regulation of several genes by a single hormone.

While induction of many eukaryotic genes may be mediated by enhancers acting as positive regulatory elements, more complex mechanisms for gene regulation, involving both positive and negative elements, are beginning to be uncovered. One example is the human β-interferon gene; another is the LDL-receptor gene, discussed below.

Maniatis and co-workers (Goodbourn *et al.,* 1985, 1986; Zinn and Maniatis,

1986) have shown that induction of the β interferon gene by poly(I)–poly(C) is mediated by a regulated enhancer in the 5′-flanking region (the β-interferon gene regulatory element, or IRE). The IRE is active in either orientation and at distances up to 900 bp from the promoter, and is capable of stimulating nearby heterologous promoters. Deletion analysis shows that the IRE consists of two closely adjacent or overlapping sequences, a 5′ segment that acts as a positive transcription element and a 3′ segment that represses the activity of the 5′ region in the uninduced state.

Zinn and Maniatis (1986), using the method of DNase I footprinting,[1] have identified the DNA sequences near the β-interferon gene that are bound by proteins in the induced and uninduced states. Their findings suggest the following model for the regulation of this gene.

In the uninduced state, a repressor protein binds to the 3′ negative element of the IRE. This prevents an interferon transcription factor (ITF), possibly related to Sp1, from binding to the 5′ positive element. The presence of an inducer, such as poly(I)–poly(C), causes the repressor protein to dissociate from the 3′ element and thus allows ITF to bind to the 5′ element, leading to transcription of the gene. Goodbourn *et al.* (1986) have shown that the nucleotide sequence of the 5′ positive component of the IRE is strongly homologous to the sequences of several viral enhancers.

B. The Receptor Gene

Figure 9.12 shows the 5′ end and a part of the 5′-flanking region of the LDL-receptor gene. The major startpoint for transcription, 93 bp upstream from A of the ATG codon at which translation of the message is initiated, is designated nucleotide +1 (see figure legend). There are two closely adjacent TATA-like sequences between positions −8 and −23, the presence of two TATA boxes possibly accounting for the multiple startpoints for transcription of the receptor gene. Further upstream there are three imperfect direct repeats of 16 bp designated 1, 2, and 3. Repeats 2 and 3 (−37 to −68) are contiguous.

Brown and Goldstein and their co-workers have investigated the possible role of the repeat sequences in the regulation of the LDL-receptor gene by sterols. In their initial studies (Südhof *et al.*, 1987a) the *tk* promoter of herpes simplex virus fused to the *CAT* gene was used as a test system for regulatory activity. The normal *tk* promoter is not regulated by sterols. Hence, CAT mRNA was produced in cells transfected with plasmids containing the *tk–CAT* construct,

[1]DNase I footprinting is a method for determining the sequences in a stretch of DNA that are protected from DNase I cleavage by the binding of specific proteins. In principle, the radioactive fragments observed on a DNA ladder derived from the digestion of protected DNA by DNase I are compared with the fragments derived from a control sample incubated in the absence of binding protein, but similarly digested with DNase I.

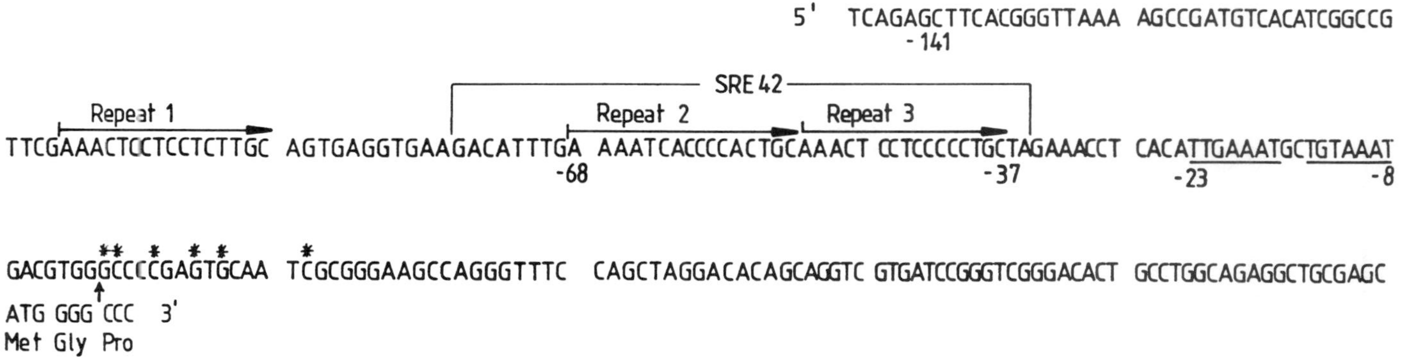

5' TCAGAGCTTCACGGGTTAAA AGCCGATGTCACATCGGCCG
-141

SRE 42

Repeat 1 Repeat 2 Repeat 3

TTCGAAACTCCTCCTCTTGC AGTGAGGTGAAGACATTTGA AAATCACCCCACTGCAAACT CCTCCCCCTGCTAGAAACCT CACATTGAAATGCTGTAAAT
-68 -37 -23 -8

GACGTGGGCCCCGAGTGCAA TCGCGGGAAGCCAGGGTTTC CAGCTAGGACACAGCAGGTC GTGATCCGGGTCGGGACACT GCCTGGCAGAGGCTGCGAGC

ATG GGG CCC 3'
Met Gly Pro

Figure 9.12 The nucleotide sequence (coding strand) of the 5′ end and 5′-flanking region of the human LDL-receptor gene, showing the positions of repeats 1, 2, and 3, the two TATA-like boxes (underlined), and the SRE 42 fragment. The vertical arrow shows the position of nucleotide 1, the major initiation site for transcription (93 bp upstream from A of the ATG codon used for initiation of translation). Some of the nucleotide positions referred to in the text are shown beneath the sequence. Transcription initiation sites are denoted by asterisks. (From Südhof *et al.*, 1987a, with modified numbering system and with the permission of the authors.)

whether or not sterols were present in the medium. When a 42-bp sequence containing repeats 2 and 3 (sterol regulatory element SRE 42) was inserted in either orientation into a 10-bp linker sequence between the TATA box and the first distal signal of the *tk* promoter (see Fig. 9.13), correctly initiated CAT mRNA was produced by transfected cells. However, in contrast to the behavior of the primary construct lacking the SRE 42 insert, transcription was suppressed when sterol was added to the medium. As noted in Section IV,A above, insertion of a *random* sequence of more than 50 bp between adjacent elements of the *tk* promoter abolishes transcription. Thus, SRE 42 acts both as a positive promoter element in association with the *tk* TATA box, and as a negative regulator of transcription. DNase I footprinting of sequences near the 5′ end of the receptor gene revealed the presence of two sequences, corresponding closely to repeats 1 and 3, that bound a protein present in a HeLa cell nuclear extract. The possible significance of this is mentioned below.

Extending these observations, Südhof *et al.* (1987b) investigated the sequences in the 5′-flanking region of the receptor gene that are required for sterol-regulated expression in the absence of a heterologous promoter. They showed that a 177-bp fragment extending upstream to nucleotide 141 (numbering as in Fig. 9.12) is capable of initiating transcription of the *CAT* gene at the correct startpoint and in a sterol-responsive manner. This fragment includes repeats 1, 2, and 3 and the two TATA-like sequences. The essential transcriptional elements within this fragment were identified by analyzing the ability of a series of mutant fragments to initiate transcription of the *CAT* gene. The results showed that all three repeats and the 5′ TATA-like sequence are required for normal transcription. Mutation involving only 3 bp in one repeat led to a 50% reduction in the amount of CAT mRNA produced.

Südhof *et al.* (1987b) noted that each of the three direct repeats includes a 10-bp sequence homologous to the "GC-box" consensus shown at the bottom of Fig. 9.14; homology with the GC-box consensus is stronger for repeats 1 and 3

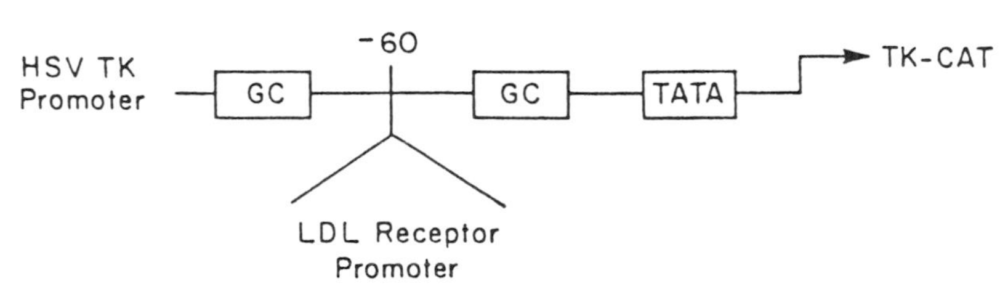

Figure 9.13. The herpes simplex virus (HSV) thymidine kinase (tk) promoter ligated to the 5′ end of the chloramphenicol acetyltransferase (CAT) gene, showing the site of insertion of synthetic oligonucleotides corresponding to repeats 2 or 3 of the human LDL receptor promoter. The two distal signals, denoted GC, are upstream to the TATA box of the *tk* promoter. (From Dawson *et al.*, 1988, with the permission of the authors.)

(8/10 and 9/10) than for repeat 2 (7/10). Kadonaga *et al.* (1986) have identified this consensus as the recognition sequence for transcription factor Sp1. (As mentioned above, Sp1 activates transcription of many genes by binding to specific GC-rich sequences in the promoter.) In view of these homologies, Südhof *et al.* (1987b) suggested that repeats 1 and 3 of the receptor gene are Sp1 binding sites and that the HeLa cell nuclear protein that binds to these repeats is, in fact, Sp1. In confirmation of this suggestion, Dawson *et al.* (1988) have shown by DNAase I footprinting that purified Sp1 binds *in vitro* to repeats 1 and 3 but not to repeat 2.

Dawson *et al.* (1988) have investigated the separate transcriptional and regulatory activities of the SRE 42 segment of the receptor promoter (see above) by inserting repeats 2 and 3 into the herpes simplex *tk* promoter ligated to the *CAT* gene, as in Fig. 9.13. The efficiency of transcription of the fusion gene and its ability to respond to the presence of sterol in the medium were tested in transfected hamster cells. By introducing synthetic mutations into the two repeats, Dawson *et al.* were able to determine separately the functions of each one.

The results of these experiments showed that repeat 3 is a constitutive positive transcriptional element that does not respond to sterol in the absence of repeat 2. In the presence of an intact repeat 2, transcription initiated by repeat 3 is suppressed by sterol. Changing the nucleotide sequence in repeat 3 so as to maximize its affinity for Sp1 has no effect on transcriptional efficiency but markedly reduces the response to sterol. Changing the nucleotides at five positions in repeat 3 abolishes both transcription and the binding of Sp1 by this repeat. The ability of repeat 2 to confer responsiveness to sterols upon repeat 3 is unaffected by reversing the orientation of repeat 2. However, suppressive activity of repeat 2 is abolished if mutations are introduced at 3 of the 4 positions at which the sequence in this repeat differs from the sequence in repeat 3 (see Fig. 9.14).

Putting all these experimental observations together, Dawson *et al.* (1988) suggest the following scheme for the sterol-regulated expression of the native LDL-receptor gene *in vivo*.

Each of the three 16-bp repeats is a weak positive transcriptional element and the coordinate action of all three is required for transcription of the gene. Repeats 1 and 3 are activated by Sp1; in the absence of sterol, repeat 2 is activated by a positive transcription factor other than Sp1. When sterols are present, a second (? sterol-binding) protein binds strongly to repeat 2 and inhibits the activity of repeat 3, possibly by preventing it from binding Sp1. Since the integrity of all three repeats is required for transcription, inhibition of repeats 2 and 3 would be sufficient to suppress expression of the gene.

On the basis of these proposals, Dawson *et al.* suggest that the promoter of the LDL-receptor gene has evolved from a single Sp1-driven positive element that was not regulated by sterol. The ancestral element then duplicated to give two contiguous repeats (2 and 3). This was followed by mutation at four positions in

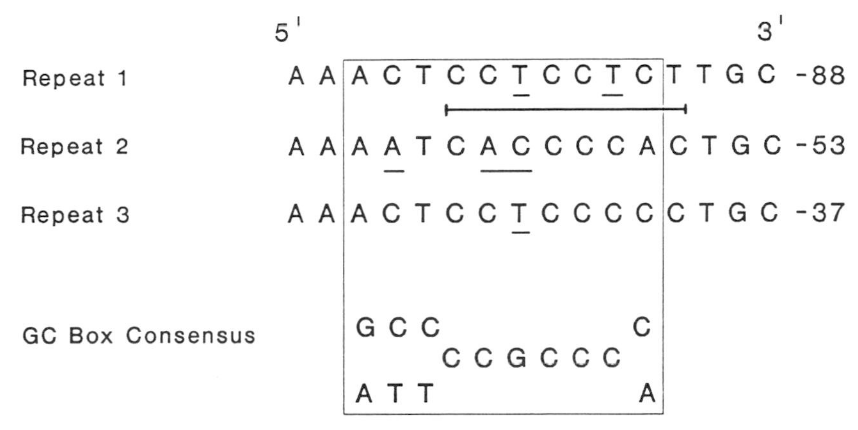

Figure 9.14. Repeats 1, 2, and 3 of the human LDL-receptor promoter (coding strands), aligned to show homology between the 10-bp core sequence in each repeat and the GC-box consensus identified by Kadonaga *et al.* (1986). The complement of the 8-bp sequence overlined in repeat 2 (nucleotides −56 to −63) has 7/8 homology with the sequence GTGCGGTG in the promoter region of the HMG-CoA reductase gene (see text). Nucleotides in repeats 1, 2, and 3 that differ from the nucleotides in the GC-box consensus are underlined. For numbering of nucleotides, see Fig. 9.12. (From Südhof *et al.*, 1987b, with the permission of the authors.)

repeat 2, giving it the ability (in the presence of sterol) to bind another protein that interferes with the binding of Sp1 to repeat 3. An additional event may have been mutation at two positions in repeat 3 that diminished its affinity for Sp1 and thus increased its susceptibility to negative regulation by repeat 2.

The above scheme for repression of the LDL-receptor gene by sterol has features in common with the scheme proposed earlier by Stuart *et al.* (1985) for induction of the mouse metallothionein-I (MT-I) gene by heavy metals. The promoter of the *MT-I* gene includes several 12-bp homologous sequences [metal-regulatory elements (MRE)] that are necessary for metal-regulated expression of the gene. Stuart *et al.* have described experiments which suggest that four of the MREs in the MT-I gene promoter contain the recognition sequence for a metal-dependent transcription factor. They suggest that binding of this factor to its recognition sequence in the four MREs enhances the binding of Sp1 to a nearby Sp1-recognition sequence and thus activates transcription of the gene. Since the MRE sequences are homologous to the consensus of the Sp1-recognition sequence, Stuart *et al.* suggest that the two recognition sequences have a common evolutionary origin. Thus, the LDL-receptor gene may not be the only eukaryotic gene to have adopted a strategy for regulation in which binding of a regulatory protein to one or more short repeats in the promoter region modifies the binding of Sp1 (or a similar positive transcription factor) to its recognition sequence.

As discussed in the previous section, coordinate induction of groups of genes by corticosteroids is thought to involve the binding of a corticosteroid-receptor protein to a specific recognition sequence in the vicinity of the promoter. At least three genes (those for HMG-CoA reductase, HMG-CoA synthase, and the LDL receptor) are repressed coordinately by sterol in animal cells. Preliminary evidence is beginning to suggest that coordinate repression by sterols is mediated by a mechanism analogous, in an opposite sense, to that responsible for induction of corticosteroid-sensitive genes. As we saw in Chapter 3, sterol-mediated repression of the reductase gene in intact cells requires the presence of a specific recognition sequence included within a 20-bp segment of the 5′-flanking region of the gene. DNase I footprinting experiments also suggest that this segment binds a sterol-dependent repressor protein. Osborne *et al.* (1988) have noted that the coding strand of the 20-bp segment of the reductase gene contains the 8-bp sequence GTGCGGTG (nucleotides −165 to −172 in the human gene) that has 7/8 homology with the sequence GTGGGGTG (nucleotides −56 to −63) in the noncoding strand of repeat 2 of the human LDL-receptor gene (see Fig. 9.14). In view of this homology, Osborne *et al.* suggest that the consensus $\mathrm{GTG}^{\mathrm{G}}_{\mathrm{C}}\mathrm{GGTG}$ is the recognition sequence for a specific protein that suppresses transcription of sterol-sensitive genes in the presence of sterol.

To substantiate these ideas, it will be necessary to identify the repressor protein presumed to bind to repeat 2 of the receptor-gene promoter in the presence of sterol. One possibility worth exploring is that the repressor is one of the sterol-binding proteins that bind oxygenated metabolites of cholesterol, as discussed in Chapter 3. How the effects of hormones and growth factors on receptor synthesis fit into this general scheme are questions for future research.

C. Failure to Compensate for Defective Alleles

The study of normal cells *in vitro* has shown beyond doubt that cholesterol, derived either from endogenous synthesis or from internalized lipoprotein, exerts a major regulatory influence upon the synthesis of LDL receptors. The results discussed above also show how the regulatory effect of intracellular cholesterol may be mediated indirectly by modulation of the expression of the LDL-receptor gene. However, observations on mutant cells point to the existence of additional mechanisms by which the expression of receptors is regulated.

Fibroblasts from FH heterozygotes have one normal and one defective allele at the receptor locus. In the presence of LPDS, heterozygote cells express half the number of LDL receptors expressed by normal cells. This is not surprising, since a single gene would be expected to direct the synthesis of half as much product as two genes under conditions of maximal induction. However, Goldstein *et al.* (1976) obtained unexpected results when they compared the regulation of receptor expression in heterozygote cells with that in normal cells. As the number of LDL receptors is suppressed progressively by adding increasing concentrations of cholesterol to the medium in which fully induced cells are growing, at any given concentration of cholesterol the number of receptors expressed by the heterozygote cells is half that expressed by the normal cells. This relationship holds even at cholesterol concentrations at which heterozygote cells have the capacity to produce as many receptors as normal cells. Thus, in a heterozygote cell, the one normal allele does not compensate for the nonfunctional allele by producing twice its normal amount of gene product. It is difficult to interpret this finding in terms of what we know about the regulation of LDL-receptor expression in normal cells. However, it does suggest that each allele responds independently to some signal other than the total functional activity of the receptors present on the cell surface.

This conclusion is supported by observations of Knight *et al.* (1987) on the synthesis of receptor protein in the cells of FH homozygote MM. As discussed in Section V,C, the cells of this patient produce LDL-receptor protein whose conversion into mature receptors is markedly delayed. In addition, the LDL-binding capacity of MM cells is only 20% of the normal. Knight *et al.* (1987) have shown that when receptor synthesis is induced in normal cells in culture by transferring them to LPDS, the synthesis of total LDL-receptor protein rises to a maximum after about 8 hours and then declines, presumably as a consequence of feedback repression. Surprisingly, MM cells behave in the same way. On transfer to LPDS, synthesis of receptor protein increases to a maximum and then declines, although the newly synthesized receptor protein does not appear on the cell surface as mature receptor. Thus, the synthesis of receptor protein in MM cells, as observed in these experiments, cannot be responding to the number of functional receptors on the cell surface. Possibly, the regulatory signal is the amount of receptor precursor present in the cell. In any case, the mode of regulation of receptor synthesis in MM cells is such that they do not compensate for their failure to produce a normal number of surface receptors by synthesizing precursors at a greater rate than normal.

V. Natural Mutations at the Receptor Locus

A. Clinical Classification

Mutations at the LDL-receptor locus that prevent the formation of functionally normal receptors give rise to the monogenic disorder, FH. The studies of Khachadurian (1964) on a Lebanese FH population showed that the FH mutation is inherited as a dominant trait, with more marked clinical expression in homozygotes than in heterozygotes (for details, see Chapter 10). The cells of an FH heterozygote, who has inherited a normal receptor gene from one parent and a mutant gene from the other, express half the normal number of functional receptors. In these patients the FCR of the plasma LDL is about half the normal, resulting in a two- to threefold increase in plasma LDL concentration. The cells of an FH homozygote, who has inherited a mutant gene from both parents, express little or no receptor activity. In FH homozygotes the plasma LDL concentration rises to 6 to 10 times the normal level, resulting in early death from coronary atherosclerosis.

B. Genetic Heterogeneity, Functional

1. Three Classes of Mutant Receptors

When receptor function began to be studied in cultured skin fibroblasts from large numbers of FH homozygotes it became apparent that there is more than one mutation at the receptor locus (Goldstein *et al.*, 1975; Brown and Goldstein, 1977). The cells from most homozygotes were found to express no surface receptors capable of binding LDL. These were called *receptor negative*. Cells from a smaller group of homozygotes were found to express a finite but reduced number of surface receptors that bound LDL (usually 5–10% of the normal). These were called *receptor defective*. The two groups could not be distinguished clinically. Subsequently, FH homozygote cells of a third class were identified. These are the *internalization-defective* cells, already mentioned in Section C of the previous chapter and in Section II,D above, and discussed in more detail below.

As we shall see, some patients with all the clinical features of homozygous FH are, in fact, genetic compounds; that is, they have two different mutant alleles on opposite chromosomes at the receptor locus. However, for clinical purposes these patients are classified as FH homozygotes because they cannot be distinguished clinically from true homozygotes, who have an identical mutant receptor gene on the two homologous chromosomes.

Two subclasses of receptor-negative cells have been identified by their response to antibodies to the LDL receptor. Beisiegel *et al.* (1982) and Tolleshaug *et al.* (1983) noted that some receptor-negative fibroblasts in culture bind detectable amounts of anti-receptor antibodies, i.e., they produce surface material that cross-reacts with antibody to the receptor (cross-reacting material or CRM). These cells were called CRM$^+$. Other receptor-negative cells were found to exhibit no surface binding of anti-receptor antibody. These were called CRM$^-$. As discussed below, some receptor-negative cells that exhibit no surface binding of anti-receptor antibody, and therefore have the "surface phenotype" CRM$^-$, produce detectable amounts of immunoreactive LDL-receptor protein that never reaches the cell surface. The term CRM$^-$ is now restricted to cells that produce no immunoreactive receptor protein detectable either on the cell surface or in solubilized cell extracts (Goldstein *et al.*, 1985). The mutant receptor genes carried in these cells are called *crm*$^-$ alleles (Hobbs *et al.*, 1988). Thus, a *crm*$^-$ allele fails to produce immunodetectable receptor protein.

2. The Internalization-Defective Mutation

Three examples of internalization-defective receptors have already been discussed in the section on the cytoplasmic domain. The first of these to be discovered was the mutation expressed in the J.D. family (see Brown and Goldstein, 1977, for review). The pedigree of this family is shown in Fig. 9.15. J.D., the index patient, was a 14-year-old boy with all the clinical features of homozygous FH. In keeping with this, both his parents had the clinical features of heterozygous FH. His brother, a first cousin, and his maternal uncle were also FH heterozygotes.

Skin fibroblasts from J.D. and from five of his relatives were examined in culture. At 4°C, J.D.'s cells bound about 70% of the amount of [^{125}I]LDL bound by the cells of his normal sister. However, when J.D.'s cells were warmed to 37°C, none of the surface-bound LDL was internalized or degraded. The cells of his mother and maternal uncle behaved like fibroblasts from the great majority of FH heterozygotes. They bound half the normal amount of LDL, but all the bound LDL was degraded; hence, the absolute rate of degradation was about 50% of normal. The behavior of his father's and brother's cells was strikingly different. They bound normal or greater than normal amounts of [^{125}I]LDL, but the rate of internalization and degradation was only about 50% of the normal. Kinetic studies showed that this anomaly was due to the presence of two populations of receptors on the cells of J.D.'s father and brother, one population that bound

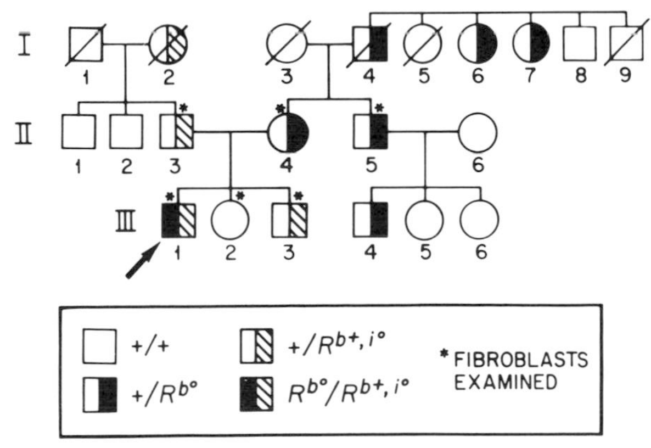

Figure 9.15. Pedigree of the J.D. family. J.D., the index patient (III-1), is shown by an arrow. He is designated FH 380 in Table 9.1. His mother (II-4) is FH 381 and his father (II-3) is FH 382 in Table 9.1. The symbols denoting the genotypes are enclosed in the box. The + allele encodes a normal receptor; R^{b^o} encodes a receptor that cannot bind LDL; R^{b+,i^o} encodes a receptor that binds LDL but cannot internalize it. The six subjects whose fibroblasts were examined are shown by asterisks. (From Goldstein *et al.*, 1977, with the permission of the authors.)

LDL and internalized it normally and another that bound LDL but could not internalize it.

Electron-microscopic studies of the binding of ferritin-labeled LDL by fibroblasts from J.D. and his father have revealed the ultrastructural basis of the abnormal behavior of these cells (Anderson *et al.*, 1977). As we saw in Chapter 8, at 4°C J.D.'s cells bound ferritin-LDL at random over the whole surface, with no clustering in coated pits. When warmed to 37°C, none of the surface-bound LDL was internalized, although coated pits invaginated and re-formed at the normal rate. Quantitative analysis showed that the total amount of ferritin-LDL bound at the cell surface was only slightly less than that bound by normal cells. When tested under the same conditions, the cells of J.D.'s father bound more than the normal amount of ferritin-LDL, but the amount clustered in coated pits was only half that in normal cells.

On the basis of these functional and ultrastructural observations, Brown and Goldstein (1977) drew the following conclusions. J.D. is a genetic compound who has inherited two different mutations in the receptor gene from his heterozygous parents, as shown in Fig. 9.16. His mother has one normal allele and one mutant allele that does not specify a receptor capable of binding LDL at the cell surface. His father, on the other hand, has one normal allele and a mutant (internalization-defective) allele specifying a receptor protein with a defective cytoplasmic domain such that the receptors reach the cell surface and bind LDL normally but do not cluster in coated pits. Hence, these receptors are unable to mediate the internalization and degradation of LDL. J.D. has inherited a receptor-negative allele from his mother and an internalization-defective allele from his father. Therefore J.D.'s cells have one allele that cannot produce a receptor capable of binding LDL on the cell surface and another allele specifying a receptor that binds LDL but cannot internalize it. Subsequent observations have shown that the mutant allele in J.D.'s mother is a null allele that produces no receptor protein (Lehrman *et al.*, 1986).

The presence of these two mutations in the J.D. family explains why there is a discrepancy between surface binding and internalization of LDL in the cells of J.D., his father, and his brother. However, it is a little surprising that the cells of J.D.'s father bind more LDL than do normal cells and that J.D.'s cells bind more than half as much LDL as normal cells (Anderson *et al.*, 1977). One possibility is that internalization-defective receptors accumulate on the cell surface in increased numbers because they are not exposed to intracellular catabolic enzymes and therefore have a prolonged lifespan. It is also possible that, owing to absence of steric hindrance, more LDL particles can be accommodated by a given number of receptors if they are distributed over the whole cell surface than if they are clustered in coated pits.

In parenthesis, it should be noted that the observations on J.D.'s father, taken

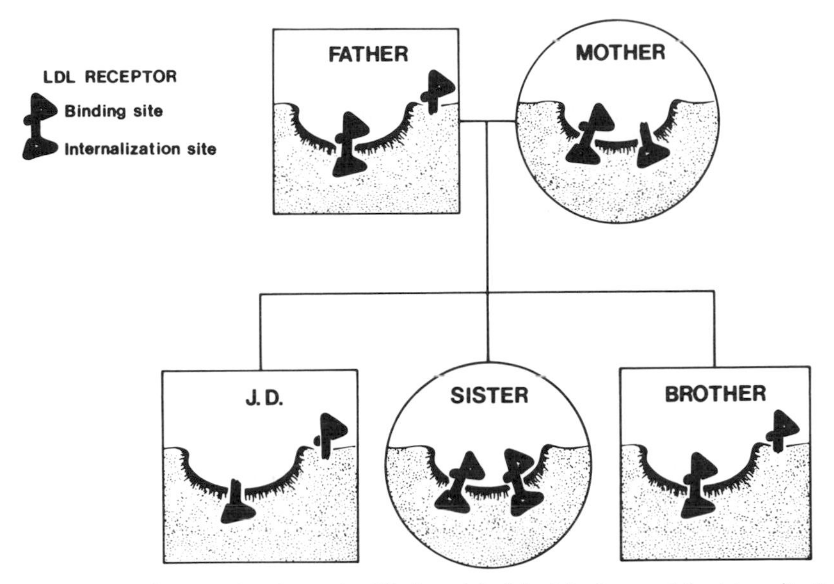

Figure 9.16. A diagram showing a simplified model of the inheritance of the internalization defect in the J.D. family. Each panel shows a coated pit and the products of the two alleles at the LDL-receptor locus. The normal receptor is shown with a binding domain projecting inside the cell, a transmembrane domain, and a cytoplasmic domain required for retention of the receptor in a coated pit. J.D.'s father has one normal receptor and one internalization-defective receptor. His mother has one normal receptor and one receptor lacking the binding domain (see text). J.D. has inherited the mutant allele carried by each parent and he therefore has one receptor lacking the binding domain and one internalization-defective receptor. J.D.'s sister has inherited the normal allele carried by each parent. J.D.'s brother has inherited a normal allele from his mother and his father's mutant allele. (From Brown and Goldstein, 1977, with the permission of the authors.)

by themselves, do not prove that the mutation leading to the failure of LDL receptors to cluster in coated pits is in the receptor gene. In theory, the mutation could be in another gene whose product is necessary for the movement of receptors into coated pits. However, Goldstein *et al.* (1977) concluded that the two mutations inherited by J.D. are both in the receptor gene itself because they do not complement each other to produce a normal receptor when present together in his cells (for explanation, see footnote on p. 28). This has since been proved conclusively by analysis of genomic DNA from J.D.'s mother (Lehrman *et al.*, 1986) and from J.D. himself (Davis *et al.*, 1986b). In both cases, the mutation was shown to lie within the receptor gene (for details, see Section V,D below). Other mutations giving rise to internalization-defective receptors are also mentioned in Section V,D.

C. Genetic Heterogeneity, Structural

Analysis of the immunoprecipitable receptor protein synthesized in fibroblasts from a large number of FH patients has shown that genetic heterogeneity in FH is even greater than that revealed at the level of receptor function (Tolleshaug *et al.*, 1982, 1983). These studies have also shown that, like J.D., many FH homozygotes are genetic compounds.

The genetic analysis of receptor proteins produced by cells of the FH 295 family (Tolleshaug *et al.*, 1982) provides a good illustration of these points. FH 295, the index patient, was a receptor-defective homozygote whose fibroblasts bound about 15% of the amount of LDL bound by normal cells. His fibroblasts were incubated in the presence of [^{35}S]methionine, and the LDL-receptor proteins synthesized during the incubation were identified by the methods described in Fig. 9.2. As shown in Fig. 9.17, FH 295's cells synthesized one receptor precursor with apparent MW 120K that was not converted into the normal 160K product, and another (170K) precursor that was converted into a 210K protein by the normal addition of 40K. When the intact cells were treated with pronase (a proteolytic enzyme) after the 2-hour chase period, all the radioactive 210K protein was destroyed but the radioactive 120K protein was unaffected (Tolleshaug *et al.*, 1983). Hence, the 210K protein was expressed on the cell surface but the 120K protein was retained within the cell.

Analysis of the cells of FH 295's parents showed that the father's fibroblasts synthesized a 120K precursor converted into the normal 160K receptor and a 170K protein converted into the 210K protein synthesized in his son's cells. The mother's cells synthesized two populations of 120K receptor proteins, one that was converted into the normal 160K mature receptor and another that persisted without change in molecular weight during the 2-hour chase incubation.

Thus, FH 295's father carried one normal LDL-receptor gene and one mutant gene specifying a precursor protein with apparent MW 170K that underwent the normal 40K increase, while his mother carried one normal gene and one mutant gene specifying an abnormal 120K protein that could not be processed to the mature receptor. FH 295 inherited the mutant gene from each parent and was therefore a genetic compound. In the notation proposed by Goldstein *et al.* (1985), the two mutant alleles in the FH 295 family are designated $R\text{-}210b^-$ (paternal) and *R-120* (maternal), and the normal allele is designated *R-160*. Hence, the genotype of FH 295 was $R\text{-}120/R\text{-}210b^-$.

Similar studies of fibroblasts from more than 100 FH homozygotes and their parents have revealed a remarkable degree of heterogeneity, expressed by differences in apparent molecular weight of the receptor proteins synthesized *in vitro* (Tolleshaug *et al.*, 1983; Goldstein *et al.*, 1985). On the basis of these structural differences and of the differences in receptor function discussed above, Goldstein *et al.* (1985) have divided mutations at the LDL-receptor locus into the

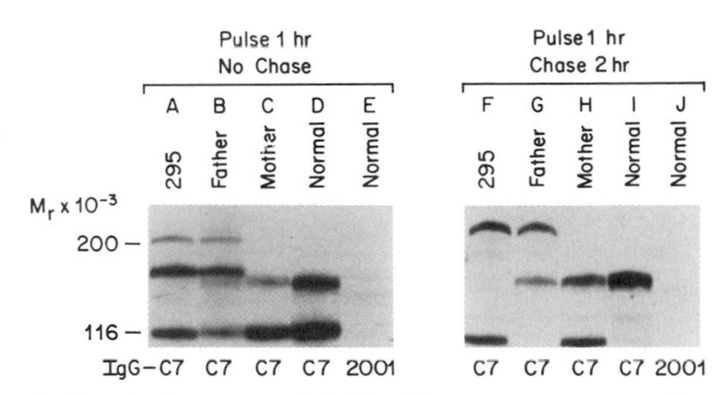

Figure 9.17. Radioactive immunoprecipitable LDL-receptor proteins synthesized by cultured fibroblasts obtained from FH 295 and his heterozygous parents. The cells were preincubated in medium containing LPDS to induce receptor synthesis. They were then incubated for 1 hour in the presence of [^{35}S]methionine and harvested immediately (lanes A to E) or after a further chase incubation for 2 hours in the presence of nonradioactive methionine (lanes F to J). Solubilized cell protein was precipitated with monoclonal anti-receptor antibody IgG-C7 (lanes A–D and F–I) or with the irrelevant antibody IgG-2001 (lanes E and J). The immunoprecipitates were submitted to SDS–gel electrophoresis and the radioactive bands located by autoradiography. The positions of molecular weight markers are shown at the left. Lane A shows the two major bands at 120K and 170K produced by FH 295's cells during the 1-hour pulse; after the 2-hour chase the 120K band remained unchanged but the 170K band was converted into a 210K band (lane F). Lane B shows the 120K and 170K bands produced by the father's cells during the 1-hour pulse; each band underwent a 40K increase during the 2-hour chase (lane G). Lane C shows the 120K band produced by the mother's cells during the 1-hour pulse; during the 2-hour chase half the 120K band was converted into a 160K band, the other half persisting without change in MW (lane H). Note that some radioactivity has begun to appear in the 210K and 160K bands during the 1-hour pulse in lanes A, B, and C. Note also that there is no radioactivity in lanes E and J (samples precipitated with IgG-2001). (From Tolleshaug *et al.*, 1982, with the permission of the authors.)

four classes shown in Fig. 9.18 and Table 9.1. The molecular basis of these mutations is discussed in Section V,D below.

Mutations in class 1, designated *R-O*, are null alleles that fail to produce receptor protein (either on the cell surface or in cell extracts) detectable with any of the available monoclonal or polyclonal antibodies to the LDL receptor. These are the *crm*$^-$ alleles referred to in the previous section. They probably account for at least 50% of all the mutations in the receptor gene that give rise to FH. Homozygotes whose cells produce no detectable receptor protein have two *crm*$^-$ alleles. Many are true homozygotes with two identical mutant alleles, but some have inherited two different *crm*$^-$ alleles from their parents and are therefore genetic compounds. As discussed in the next section, the molecular defects in alleles of class 1 include large deletions, as in the mutation inherited by J.D. from his mother (FH 381) (see Section V,D below). They may also include defects in the splicing of exons, and point mutations that give rise to stop codons near the 5′ end of the message.

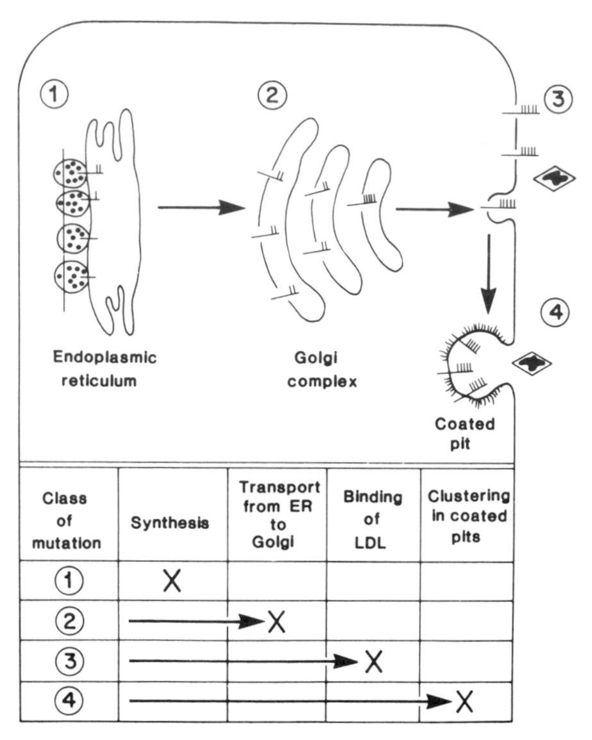

Class of mutation	Synthesis	Transport from ER to Golgi	Binding of LDL	Clustering in coated pits
1	X			
2	→	X		
3	→	→	X	
4	→	→	→	X

Figure 9.18. Diagrammatic representation of the four classes of mutations in the LDL receptor gene. Class 1 mutations prevent the formation of immunodetectable receptor protein in the ER. Class 2 mutations disrupt the transport of receptors to the Golgi apparatus. Class 3 mutations reach the plasma membrane but bind LDL with decreased efficiency. Class 4 mutations reach the cell surface and bind LDL normally but do not cluster in coated pits. (From Goldstein and Brown, 1984, with the permission of the authors.)

Class 2 mutations specify receptor proteins that fail to undergo the normal 40K increase in apparent molecular weight and do not reach the cell surface. Cells bearing two class 2 mutations are unable to bind LDL on their surfaces and therefore have the CRM$^-$ surface phenotype. Tolleshaug *et al.* (1983) have described three mutations of this class that specify proteins with apparent MW 100K, 120K, and 135K (alleles designated *R-100, R-120,* and *R-135*). The *R-120* mutant allele inherited by FH 295 from his mother is an example of a class 2 allele. Analysis of the oligosaccharide side-chains of the 120K mutant receptor protein from one family showed that in this abnormal precursor the N-linked sugars had not been converted into the mature endoglycosidase-H resistant form and that sialic acid had not been added to the O-linked chains. Moreover, the 120K protein was not destroyed by treatment of intact cells with pronase. Thus, it seems likely that this 120K protein failed to be transported from the endoplasmic reticulum to the Golgi apparatus. The abnormal molecular weights of the 100K and 135K proteins may be due to changes in the lengths of the amino acid chains.

Other mutations, comprising a subgroup of class 2, specify receptor proteins whose conversion into the mature form is incomplete or greatly delayed. These include the mutation responsible for the abnormal receptors in WHHL rabbits and FH 692 (Schneider *et al.*, 1983a), in FH 563 (Yamamoto *et al.*, 1986), in an Afrikaner family (Fourie *et al.*, 1988), in a black South African family (Leitersdorf *et al.*, 1988), and in FH homozygote MM (Knight *et al.*, 1989). In the WHHL rabbit, in addition to delayed processing of LDL-receptor protein, those receptors that do reach the cell surface exhibit defective binding of LDL. The mutant receptors in the black South African family, on the other hand, bind LDL with normal affinity once they have appeared on the cell surface. The reason for this difference is discussed in Section V,D below. Schneider *et al.* (1983a) have shown that in the abnormal receptors of WHHL rabbits the N-linked and O-linked core sugars are not processed, apparently because the newly formed precursor protein is not transported efficiently to the Golgi apparatus.

FH patient MM is a true homozygote (Knight *et al.*, 1989). His cells are receptor defective, with maximal LDL-binding capacity about 20% of normal. Cultured fibroblasts from this patient synthesize receptor protein of normal apparent molecular weight at the normal rate. However, the conversion of precursor into mature receptor is markedly delayed, as shown in Fig. 9.19 (Knight *et al.*, 1987). Despite the slow turnover of the precursor, time-course experiments with radioactive receptor protein (Fig. 9.20) show that almost all the precursor synthesized in MM cells is eventually converted into mature receptor. Additional experiments have shown that once mature receptors are formed in MM cells, they reach the surface and undergo normal endocytosis and recycling.

In view of the normal rate of synthesis of receptors and the nearly complete conversion of precursor into mature receptor in MM cells, their content of total receptor protein (precursor plus mature) should be almost equal to that in normal cells. The proportions of total receptor protein present in precursor and mature form in cells in the steady state may be deduced from the relative fractional rates of turnover of receptor proteins determined from time-course experiments such as the one shown in Fig. 9.20. Knight *et al.* (1989) conclude that in MM fibroblasts about half the total receptor protein is in precursor form, whereas in normal fibroblasts more than 95% is present as mature receptor. All these conclusions have been confirmed directly by quantitative immunoblotting of solubilized extracts of MM and normal fibroblasts. Knight *et al.* (1989) have also used radioimmunoassay to determine the total amounts of pronase-sensitive and pronase-resistant receptor protein in intact fibroblasts at 37°C. In MM cells about half the total receptor protein is removed by pretreatment with pronase. In normal cells virtually all receptor protein is destroyed by pronase. These findings

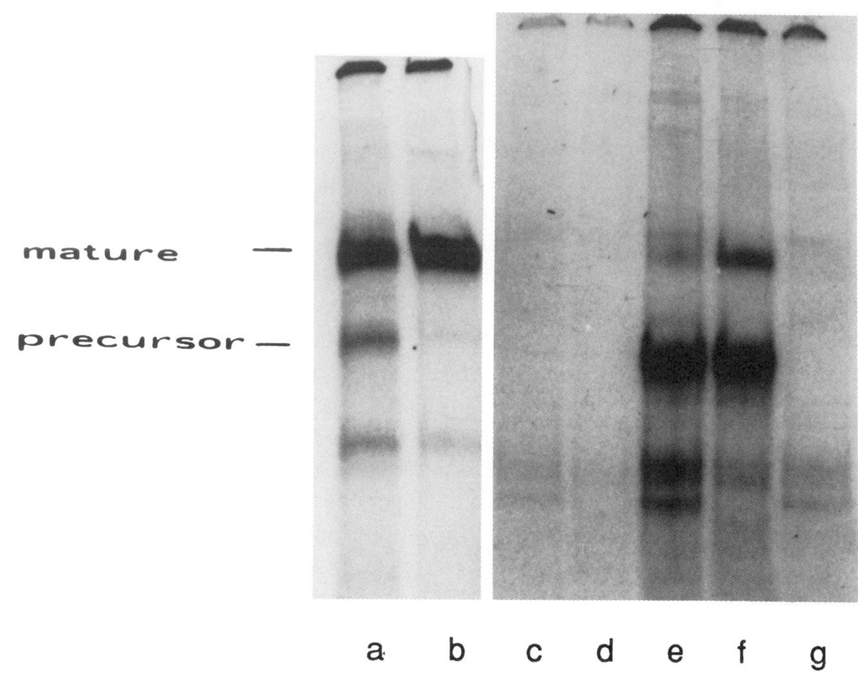

Figure 9.19. Autoradiograph of immunoprecipitates of [^{35}S]methionine-labeled LDL-receptor proteins from normal (lanes a and b), FH receptor-negative (lanes c and d), and MM (lanes e to g) fibroblasts in culture. The cells were preincubated for 14 hours in medium containing LPDS. They were then incubated for 2 hours with [^{35}S]methionine, either without a subsequent chase period (lanes a, c, and e) or with a further 2-hour incubation with 200 μM nonradioactive methionine (lanes b, d, and f). Cell extracts were immunoprecipitated with monoclonal anti-LDL-receptor antibody 4B3 (lanes a to f) or with an irrelevant antibody (lane g) and the precipitates were submitted to gel electrophoresis followed by autoradiography of the gels. Note that at the end of the 2-hour period all the precursor formed during the pulse incubation of normal cells was converted into mature receptor (lane b), whereas in MM cells most of the radioactivity was still present in the precursor (lane f). (From Knight *et al.*, 1987, with the permission of the authors.)

are consistent with the supposition that, under conditions in which LDL receptors are undergoing rapid endocytosis and recycling, the great majority of mature receptors are exposed to the outside of the cell and, hence, that there is little or no intracellular reservoir of mature receptors at 37°C. This confirms the earlier observations of Basu *et al.* (1978), based on a comparison between the LDL-binding capacity of intact cells and that of membranes prepared from solubilized cells.

The above results suggest that MM cells contain a nearly normal amount of receptor protein, half of which is present as mature receptors on the surface of the cell. Yet the LDL-binding capacity of MM cells is only about 20% of normal at 37°C. Knight *et al.* (1989) suggest that the anomalous reduction in binding

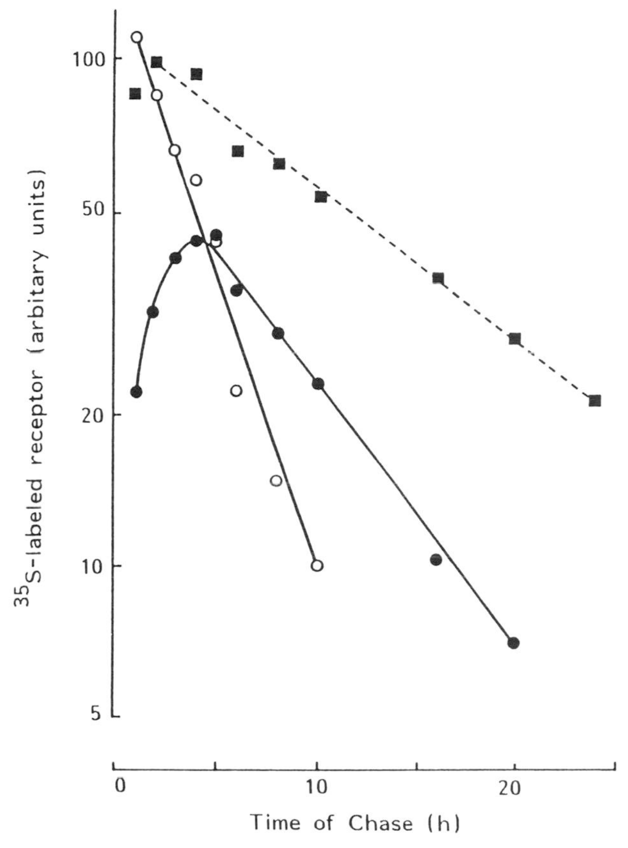

Figure 9.20. Time-course of radioactivity in precursor and mature receptor protein in MM fibroblasts after preincubation with [^{35}S]methionine. Cells were preincubated for 24 hours in medium containing LPDS followed by a 2-hour incubation with [^{35}S]methionine. Beginning at zero time on the horizontal axis, the cells were then incubated for the indicated times with 200 μM nonradioactive methionine. The solubilized cell extracts were immunoprecipitated with monoclonal antibody 4B3 and the radioactive receptor proteins were separated by gel electrophoresis followed by autoradiography of the gels, as in Fig. 9.19. The autoradiographs were scanned to determine the relative amounts of radioactivity in the bands corresponding to precursor and mature receptor. Cells from a normal subject were run in parallel with MM cells to determine the time course of radioactivity in mature receptor in normal cells. Symbols represent radioactivity in precursor (○) and (●) mature receptors of MM cells and in mature receptors (■) of normal cells. (From Knight *et al.*, 1989, with the permission of the authors.)

capacity of MM cells is due to dimerization of surface receptors at 37°C, so that half the receptors exposed at the surface are, in effect, nonfunctional. In support of this, they point to several similarities between the behavior of MM cells and that of normal cells in the presence of monoclonal anti-LDL-receptor antibody 10A2. This antibody has been shown to halve the maximal binding capacity of normal fibroblasts at 37°C by cross-linking receptor molecules on the cell surface through divalent interaction with antibody-binding sites, thus leading to masking of half the LDL-binding sites (Gavigan *et al.*, 1988).

Class 3 alleles specify receptor proteins that are normally processed in the Golgi apparatus and then reach the cell surface at the normal rate. However, the ability of the mature receptors to bind LDL is markedly defective. The majority of class 3 alleles specify precursors with normal apparent molecular weight (120K) that are converted into mature receptors with MW 160K. These alleles are designated *R-160b*$^-$. Tolleshaug *et al.* (1983) have described two other class 3 alleles that specify precursors with apparent MW 100K and 170K that undergo the normal 40K increase in molecular weight to produce a mature receptor with defective LDL-binding ability (*R-140b*$^-$ and *R-210b*$^-$).

Class 4 alleles are the internalization-defective mutations specifying receptors that bind LDL on the cell surface but do not internalize it because they cannot cluster in coated pits. A total of five different class 4 alleles has been identified in six FH homozygotes from North America, Saudi Arabia, Japan, and Europe. Internalization-defective receptors in three FH homozygotes (FH 380, 683, and 763) have already been discussed in relation to the function of the fifth domain of the LDL receptor. In each case, the effect of the mutation is to alter the amino acid sequence of the cytoplasmic 50-residue tail. Studies of synthetic receptor mutants expressed in cells *in vitro* showed that these alterations are sufficient to prevent LDL receptors from clustering in coated pits. As we saw in the previous section, FH 380 is a genetic compound who inherited an internalization-defective allele from his father and a null allele from his mother. FH 683, who is one of a pair of FH homozygote sibs (the offspring of a consanguineous marriage), is a true homozygote (Lehrman *et al.*, 1985a).

The internalization-defective receptors produced by FH 380, 683, and 763 have a normal membrane-spanning domain and are therefore anchored firmly in the ER membrane and, subsequently, in the plasma membrane. Hence, when produced in cultured fibroblasts from these patients, the mutant receptors are not secreted into the medium. However, as discussed in Section II,D above, two other internalization-defective alleles have been identified that specify receptors from which both the cytoplasmic tail and the membrane-spanning domain are deleted. One of these was identified in patient FH 274, a genetic compound who had inherited an internalization-defective allele from his mother and a null allele from his father (Lehrman *et al.*, 1985b). The other was discovered in a Japanese FH patient (FH 781 in Lehrman *et al.*, 1987b) shown by Miyake *et al.* (1981) to be homozygous for the internalization defect.

Lehrman *et al.* (1985b, 1987b) have shown that most of the receptors synthesized in cultured cells from both these patients are secreted into the medium. The small number of receptors that are not secreted remain anchored to the plasma membrane, where they bind LDL but cannot internalize it (Takaichi *et al.*, 1985;

Lehrman *et al.*, 1985b). The reason why some of the internalization-defective receptors produced by FH 274 and 781 remain attached to the plasma membrane is discussed in Section V,D below.

D. Molecular Basis

More than 20 different FH mutations have now been identified at the level of receptor protein (see Section V,C), of receptor mRNA, or of the gene itself. Many of these mutations have been analyzed by sequencing cloned fragments of the receptor gene isolated with cDNA or genomic probes, and in some cases the analysis is complete enough to permit deductions as to how the ancestral mutation arose. In some instances the probable site of the mutation within the gene could be inferred from the behavior of the mutant receptor and the domain structure of the normal receptor. Thus, in class 3 mutations, normal processing combined with failure to bind LDL pointed to the 5′ portion of the gene encoding domain 1. In class 4 mutations, on the other hand, attention was focused initially on the 3′ end of the gene encoding the membrane-spanning and cytoplasmic domains of the receptor. In class 1 mutations, in which no clues are provided by the behavior of the receptors because none are detectable, a more empirical approach has been adopted.

Table 9.2 summarizes what is known of the molecular basis of the mutations causing FH, including the mechanisms by which they arose and the effects they have on the phenotype.

1. Class 1 Mutations

Hobbs *et al.* (1988) examined the fibroblasts from 132 FH homozygotes. They identified 16 whose cells produced no immunoprecipitable receptor protein and were therefore carrying two class 1 (*crm*$^-$) alleles. They then estimated the minimum number of different FH mutations present in the 16 *crm*$^-$/*crm*$^-$ homozygotes by determining the total number of distinct haplotypes, each based on a set of 10 RFLPs at the receptor locus, in the 32 genes contributed by these patients. (The validity of this analysis depends upon the assumption that each FH mutation originated on a chromosome in which the RFLP haplotype at the receptor locus had already been established and that no recombination took place thereafter. If these assumptions are true, the probability that two identical rare FH mutations would be found on chromosomes with different 10-allele haplotypes must be very small.) Thirteen different haplotypes were found in this group of FH homozygotes, indicating the existence of at least 13 distinct *crm*$^-$ mutations. If we add to this list the 2 *crm*$^-$ alleles contributed by the unrelated FH heterozygotes 381 and T.D. (see Table 9.2) the total comes to 15.

Of particular interest in the present context are the French Canadian FH homozygotes (FH 49, 549, 808, and 859). All of the eight *crm*$^-$ alleles from

Table 9.2
Mutations in the LDL-Receptor Gene: Molecular Basis[a]

FH patient		Genotype	Change in DNA	Probable mechanism	Effect on phenotype	Reference[b]
Class 1	381 (T.D.'s M)	Hetero; *R-160/crm*$^-$ (*crm*$^-$ allele examined)	≈5 kb deleted from exon 13 to left arm of Alu B in intron 15 of *crm*$^-$ allele, with loss of 3′ end of exon 13 plus exon 14 and 15. (See Figure 9.26.)	Intrastrand double stem-loop between inverted repeats separated by ≈5 kb, with excision of loop.	5 kb mRNA. Should produce truncated receptor lacking part of D3 and whole of D4 and D5. ? Rapidly degraded.	1
	T.D.	Hetero; *R-160/crm*$^-$ (*crm*$^-$ allele examined)	4 kb deleted from intron 12 to intron 14 in *crm*$^-$ allele. Exon 12 joined to E15. Changes reading frame to give STOP at 16th codon in exon 15.	Misalignment of chromatids at meiosis due to base-pairing of direct Alu repeats in intron 12 and intron 14, followed by crossing over.	Loss of 100 residues from D2 and whole of D3, D4, and D5. ? Protein rapidly degraded.	2
	49, 549, 808, 859 (Canadian)	True homo; *crm*$^-$/*crm*$^-$ (by haplotype analysis)	Deletion of >10 kb, including promoter and exon 1.	Not known.	No mRNA. No receptor protein formed.	3
	26	Compound; *crm*$^-$(M)/*crm*$^-$(F) (from parental alleles)	*crm*$^-$(M) has no detectable deletion. *crm*$^-$(F) has 6-kb deletion including promoter and exon 1.	Not known. Not known.	No mRNA. ? Due to improper processing at transcription. No mRNA.	3
	651	True homo; *crm*$^-$/*crm*$^-$ (by haplotype analysis)	4 kb deleted, including E13 and E14. ? Same as T.D.	? Same as T.D.	5 kb mRNA. Predicts truncated protein, as with T.D.	3
	132	Compound; *crm*$^-$/*crm*$^-$ (by haplotype analysis)	One *crm*$^-$ (a) has 20-kb deletion including promoter and most of gene. The other *crm*$^-$ (b) has complex rearrangement.	Not known. Not known.	(a) No mRNA. (b) mRNA 6.2 kb ? with interrupted reading frame.	3

	431,[c] 485, 573, 664, 842, 61	? True homo; *crm⁻/crm⁻* (by haplotype analysis)	No detectable deletion.	Not known. ? STOP codon near 5′ end.	5.3 kb mRNA.	3
	551	True homo; *crm⁻/crm⁻* (by haplotype analysis)	No detectable deletion.	Not known.	8.4 kb mRNA.	3
	790	Compound; *crm⁻/crm⁻* (by haplotype analysis)	No detectable deletion.	Not known.	One normal-length mRNA (5.3 kb).	3
Class 2	264 (Lebanese)	True homo; *R-100/R-100*	Point mutation at codon 660 (TGC → TGA = STOP) in exon 14, generating a new *Hin*fI site.	Point mutation.	Deletes all residues from Cys 660 in repeat C of D2 to COOH terminus in D5. Truncated protein retained in ER.	4
	550, 786, 793	Lebanese homozygotes shown to be the same as FH 264 by Southern blotting after digestion with *Hin*fI.				
	WHHL rabbit	True homo; $R120 \xrightarrow{slow} R160b^-/$ $R120 \xrightarrow{slow} R160b^-$	Deletion of 12 bases (codons 115–118) in exon 4. S1 nuclease gives two protected fragments.	Not known.	Deletes 4 residues from repeat 3 in D1. ? Protein held in ER. Binds β-VLDL but not LDL (by ligand blotting).	5,6
	563	True homo; $R120 \xrightarrow{slow} R160b^-/$ $R120 \xrightarrow{slow} R160b^-$	Deletion of 3 bases in exon 4.	Not known.	Deletes a glycine residue from repeat 5 in D1. ? Protein retained in ER.	5,15
	429	True homo; *R-120/R-120*	Point mutation at codon 544 in exon 11, abolishing an *Sfa*NI site.	Point mutation.	5.3 kb mRNA. Changes Gly 544 to Val in D2. Protein retained in ER?	16

(*continued*)

Table 9.2 (*Continued*)

FH patient		Genotype	Change in DNA	Probable mechanism	Effect on phenotype	Reference[b]
	MM	True homo	Point mutation at codon 664 in exon 14, generating a *Pst*I site.	Point mutation.	5.3 kb mRNA. Changes Pro 664 to Leu in repeat C of D2. Delayed processing; defective binding.	17
	TT	True homo *R-120*/*R-120*	Deletion of 6 bases in exon 2, generating a *Pst*I site.	Not known.	Deletes Asp 26 and Gly 27 from repeat 1 of D1. Delayed processing but normal LDL binding. Fails to bind IgG-C7.	14
Class 3	295	Compound; *R-120*(M)/ *R-210b*$^-$(F) (F allele examined)	14-kb insert due to duplication of E2 to E8 joined 3′ to E8.	Misalignment at meiosis between direct Alu repeats in intron 1 and intron 8 followed by crossing over.	Duplicates 7 repeats of binding domain plus repeats A and B of D2. Receptor normally processed but decreased synthesis.	7
	626	Compound; *R-140b*$^-$(F)/ *R-120*(M) (F allele examined)	Deletion of exon 5.	Misalignment of chromatids at meiosis due to base-pairing of direct Alu repeats in intron 4 and intron 5, followed by crossing over.	Deletion of repeat 6 in D1. All other domains preserved. Binds β-VLDL but not LDL.	8
	359 454 (sibs)	True homo	4 kb deleted, removing exon 7 and exon 8. Exon 6 spliced to exon 9 with retention of reading frame.	Not known.	Receptor lacks repeats A and B of D2. Reaches cell surface and binds β-VLDL but not LDL.	9
Class 4	682, 683	True homo; *R-i*$^-$/ *R-i*$^-$	TGG → STOP in codon for third residue in D5.	Point mutation.	Deletes all but 2 residues of cytoplasmic tail. Internalization-defective receptor.	10

763	Not known. (Parent not examined)	Frameshift.	4-base insertion	Deletes all but 6 residues of cytoplasmic tail. Internalization-defective receptor.	10
380 (J.D.)	Compound; *R-i*⁻(F)/*crm*⁻(M) (F allele examined)	TAT → TGT in codon for residue 807 in D5.	Point mutation.	Changes Tyr 807 to Cys. Internalization-defective receptor.	11
274	Compound; *R-i*⁻,*sec*(M)/*crm*⁻(F)	≃5.5 kb deleted from left arm of *Alu*E in intron 15 to right arm of *Alu*F in exon 18, with loss of exon 16, exon 17, and 5′ end of exon 18. (See Fig. 9.25.)	Intrastrand double stem-loop between inverted repeats separated by ≃5.5 kb, with excision of loop. D	7.7 kb mRNA due to abnormal splicing. Deletion of D4 and D5, with addition of 55 abnormal terminal residues. Internalization-defective receptor, mostly secreted.	12
781	True homo; *R-i*⁻,*sec*/*R-i*⁻,*sec*	7.8 kb deleted from left arm of *Alu*C in intron 15 to *Alu*H in exon 18, with loss of exon 16, exon 17, and 5′ end of exon 18. (See Fig. 9.25.)	Misalignment of chromatids at meiosis due to base-pairing of direct Alu repeats in intron 15 and exon 18, followed by crossing over.	6.1 kb mRNA due to abnormal splicing. Deletion of D4 and D5, with addition of 55 abnormal terminal residues. Internalization-defective receptor, mostly secreted.	13

[a]Abbreviations: Compound, genetic compound (or compound heterozygote); D, LDL-receptor domain; ER, endoplasmic reticulum; F, allele inherited from father; hetero, heterozygote; homo, homozygote; M, allele inherited from mother. b^- denotes defective LDL binding; i^- denotes defective internalization; *sec* denotes secreted from the cell; crm^- denotes absence of cross reacting material; *R-100* to *R-210* refer to apparent MW of the receptors observed after a 2-hour pulse followed by a 2-hour chase.

It should be noted that the designation of mutant alleles responsible for FH (and, for that matter, for any other monogenic disorder) has evolved as knowledge of the underlying mechanism has advanced. What was originally referred to as the FH gene was first subdivided into three classes based on the functional properties of fibroblasts in culture (e.g., the i^- class of internalization-defective alleles). The three classes were then subdivided on the basis of the behavior and apparent MW of the product of the mutant allele (e.g., the crm^- and *R-120* alleles). Most, if not all, of the subclasses include several different mutations, many of which have already been identified at the DNA level. Ultimately, it should be possible to devise a system of nomenclature based solely on the underlying molecular abnormalities in the LDL-receptor gene, analogous to the nomenclature used for mutations affecting hemoglobin. In the meantime, current nomenclature can only reflect the present level of our knowledge of FH mutations. (Note that this table was constructed before several mutations detected by screening of FH heterozygotes had been reported. Any attempt to keep this table up to date would be futile.)

[b]References: (1) Lehrman *et al.*, 1986; (2) Horsthemke *et al.*, 1987a; (3) Hobbs *et al.*, 1988; (4) Lehrman *et al.*, 1987c; (5) Yamamoto *et al.*, 1986; (6) Schneider *et al.*, 1983a; (7) Lehrman *et al.*, 1987a; (8) Hobbs *et al.*, 1986; (9) Russell *et al.*, 1986; (10) Lehrman *et al.*, 1985a; (11) Davis *et al.*, 1986b; (12) Lehrman *et al.*, 1985b; (13) Lehrman *et al.*, 1987b; (14) Leitersdorf *et al.*, 1988; (15) Russell *et al.*, 1989; (16) Esser and Russell, 1988; (17) Soutar *et al.*, 1989.

[c]FH 431, 485, and 664, though unrelated, were homozygous for the same haplotype and may therefore have the same receptor mutation.

these four homozygotes were carried on chromosomes with the same RFLP haplotype. This is consistent with the supposition that most of the FH homozygotes in the French Canadian population (in which the frequency of FH is unusually high) are the descendants of a single individual who emigrated to Canada from France in the early 1700s and who happened to carry this rare mutation at the receptor locus. If the French Canadian population then increased while remaining genetically isolated, the gene would be expected to attain a high frequency in the descendants of the small number of ''founders.'' Additional evidence for this is discussed below.

The conditions required for a ''founder effect'' have been discussed by Harris (1977). There are many cases where a mutation giving rise to a severe clinical disorder is relatively common in a small, genetically isolated population but is very rare in the rest of the world. These cases are thought to be due to a founder effect. Examples are hereditary tyrosinosis in Northern Quebec and porphyria variegata in the white population of South Africa. Mutations at the LDL-receptor locus itself provide other examples, including the mutations responsible for the high frequency of FH in Lebanon (see p. 403) and among the Afrikaners of South Africa (Brink *et al.*, 1987). A substantial proportion of carriers of an FH allele in the Finnish population may also be descended from a common ancester (Aalto-Setälä *et al.*, 1988).

As shown in Table 9.2, some class 1 alleles fail to produce detectable receptor mRNA (*mRNA*$^-$ alleles), whereas others produce mRNA which may be of normal or abnormal length. The *mRNA*$^-$ and *mRNA*$^+$ alleles will be considered in turn.

In one of the *mRNA*$^-$ alleles identified by Hobbs *et al.* (1988) (one of the FH 26 alleles) no deletion was detectable in the gene. In other *mRNA*$^-$ alleles, including that responsible for FH in the four French Canadian homozygotes, there was an extensive deletion. Hobbs *et al.* (1987) had previously shown that the deletion in the French Canadian mutation removes a segment of DNA at least 10 kb long, including the promoter and exon 1, and that it gives rise to an abnormal 15-kb fragment in *Xba*I/*Kpn*I digests of genomic DNA. Using the 15-kb fragment as a marker for this mutant gene, Hobbs *et al.* (1987) screened a small group of French Canadian FH patients. The mutation was present in 9/14 chromosomes from seven unrelated homozygotes and in 53/84 heterozygotes. Thus, the ≥10-kb deletion accounts for nearly two-thirds of all the FH alleles in French Canada. This supports the view that most people with FH in French Canada are descended from a common ancestor.

Several class 1 alleles examined by Hobbs *et al.* (1988) produced mRNA of normal length. None of these had detectable alterations in gene structure. The abnormality in this group of alleles is probably a point mutation giving rise to a stop codon near the 5′ end of the message.

Among the *crm*$^-$ alleles that produce mRNA of abnormal length, two (FH

381 and 651) have deletions and one of them (FH 132 allele *b*) has an extensive rearrangement of DNA sequences. The mutant allele in T.D., which has a 4-kb deletion, may also be included in this group of mutations. The molecular basis of the deletions in the mutant alleles of FH 381 and T.D., both of whom were FH heterozygotes, has been worked out in considerable detail.

The abnormality in the mutant FH 381 allele was elucidated by Lehrman *et al.* (1986). They began by isolating an abnormal *Bam*HI fragment from the patient's DNA. This fragment was shown to contain a 5-kb deletion that removes a segment of the normal gene extending from exon 13 to an Alu repeat sequence in intron 15. Analysis of the nucleotide sequence on either side of the deletion joint revealed a sequence of 11 bases in exon 13, just upstream from the joint, that has 10/11 complementarity with the reverse of a sequence 5 kb downstream on the same DNA strand in the left arm of Alu repeat B in intron 15 (Fig. 9.21). Immediately 5′ to the deletion joint there is another inverted repeat of 8 bases.

Lehrman *et al.* (1986) suggest that the mutation that occurred in an ancestor of FH 381 was initiated by base-pairing between the two 11-base sequences to form the stem of a 5-kb loop, together with the formation of a small loop by pairing of the shorter 8-base sequences. The formation of this double stem-loop system was then followed by excision of the 5-kb loop and religation of the cut ends. The postulated base-pairing would have occurred at some stage during DNA replication, when segments of the two strands of the double helix were separated from each other and were therefore free to pair with complementary single strands. The mutant allele resulting from this deletion should produce a shortened message encoding a truncated protein lacking part of domain 3 and the whole of the membrane-spanning and cytoplasmic domains. As noted in Table 9.2, the mutant allele from FH 381 produces a 5-kb message (normal length 5.3 kb). However, Lehrman *et al.* (1986) could not detect any truncated receptor in cultures of fibroblasts from FH 381, either in the cells or in the medium. Presumably, any

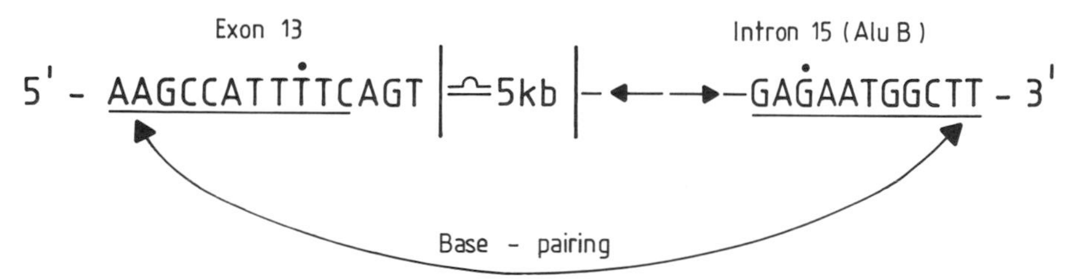

Figure 9.21. Nucleotide sequences at the deletion joint in the coding strand of the mutant allele from FH 381. The two 11-base sequences that fold back on each other to form the stem of a 5-kb loop are underlined. The 8-base inverted repeat that forms the small loop is shown by the two arrows. The two ends of the deletion are shown by vertical lines. Dots are placed over the two noncomplementary bases in the 11-base repeats. For designation of the Alu repeat sequences in the receptor gene, see Fig. 9.26. (From Lehrman *et al.*, 1986).

receptor protein synthesized under the direction of this allele is rapidly degraded within the cell.

Horsthemke *et al.* (1987a) have shown that the mutant gene in T.D. has a 4-kb deletion. Analysis of the nucleotide sequence across the deletion joint in an abnormal *Xba*I fragment, isolated from the patient's DNA, showed that an Alu repeat sequence in intron 12 is joined to another Alu repeat, in the same orientation, in intron 14 (see Fig. 9.22). Thus, the deletion removes the whole of exons 13 and 14. The splice junctions at introns 12 and 14 are such that fusion of the transcripts of exons 12 and 15 during the formation of the message would change the reading frame downstream from the first nucleotide transcribed from exon 15. This would result in a premature stop codon after only 15 amino acids. The abnormal mRNA would encode a truncated protein shorter than the normal receptor by 230 amino acids. If such a protein is formed it is not detectable in cultured cells from T.D. (Horsthemke *et al.*, 1987a), perhaps because it is rapidly broken down.

Horsthemke *et al.* suggest that the T.D. mutation arose during meiosis by base-pairing between the Alu repeats in introns 12 and 14 in homologous chromatids, followed by unequal crossing over, as in Fig. 9.22. This would have resulted in the formation of the T.D. allele lacking exons 13 and 14, plus another allele containing duplicates of exons 13 and 14; the latter may or may not have survived in the population.

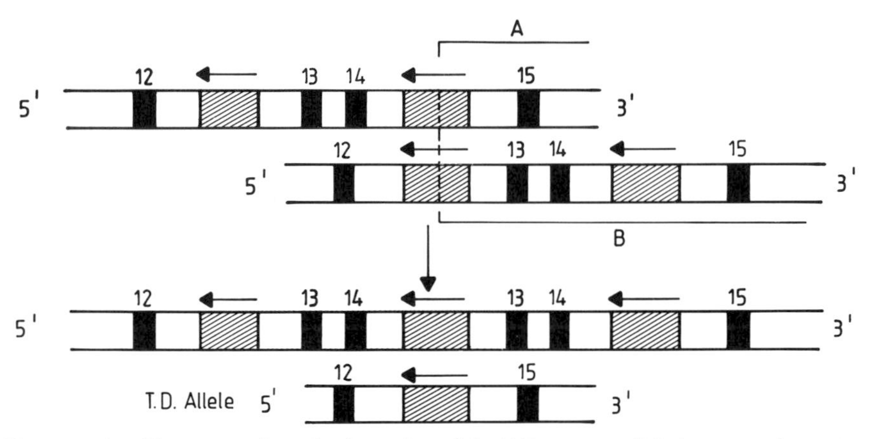

Figure 9.22. Diagram to show the formation of the T.D. mutant allele by unequal crossing over at meiosis. Exons are shown by solid bars, Alu repeats shown by hatched bars. Horizontal arrows show the orientation of the Alu repeats. Base sequences in intron 14 pair with sequences in intron 12, followed by exchange of segment A for segment B, producing the T.D. mutation and an allele with duplicates of exons 13 and 14. The position at which crossing over occurs within the two aligned introns, shown by vertical broken line, is not known exactly. (Modified from Horsthemke *et al.*, 1987a.)

2. Class 2 Mutations

The mutation in the Lebanese homozygote FH 264 has been analyzed by Lehrman *et al.* (1987c). This patient's cells synthesize only a 100K precursor of the receptor that is not processed by the normal addition of 40K and does not reach the cell surface. Biochemical analysis of the precursor formed in cultured cells from FH 264 showed that it contains little or no O-linked sugar and that the N-linked sugars are in the unprocessed "high-mannose" form. Despite the reduced length of the receptor protein, the mRNA transcribed from the mutant allele is of normal length. Lehrman *et al.* have shown that the molecular basis of the mutation is a single-base substitution producing a stop codon at the base triplet encoding residue Cys 660 in repeat C of domain 2 (the "EGF precursor homology" domain). The protein encoded by this gene lacks the C-terminal portion of domain 2, the whole of domain 3 (which includes the region containing the clustered O-linked sugars), and the whole of domains 4 and 5.

The point mutation in the FH 264 gene creates a new *Hin*fI restriction site, giving rise to a characteristic pattern on Southern blots of the patient's DNA. This gave Lehrman *et al.* the opportunity to test for the presence of the FH 264 allele in FH patients in and around Lebanon, where FH is exceptionally frequent. Four out of five unrelated FH homozygotes (four from Lebanon and one from Syria) were found to have two copies of the "Lebanese" allele. Thus, it is likely that further testing will show that the high frequency of FH in Lebanon is due to a founder effect.

Leitersdorf *et al.* (1988) have identified the molecular lesion responsible for the class 2 mutation in the black South African FH family mentioned in Section V,C above. Cells from the homozygous index patient (TT) synthesized a precursor protein of normal molecular weight, but conversion of the precursor into mature LDL receptors was markedly delayed and the number of receptors expressed on the cell surface was only about 30% of normal. A clue to the site of the mutation in the receptor gene was provided by the observation that the LDL receptors on TT's cells bound LDL with normal affinity but failed to bind monoclonal antibody IgG-C7. These are the properties exhibited by the mutant receptor constructed by van Driel *et al.* (1987b) (see Section II,D), in which the whole of the first repeat of domain 1 (encoded in exon 2) was deleted. Taking their cue from this observation, Leitersdorf *et al.* (1988) amplified exon 2 of TT's receptor gene by the polymerase chain reaction[2] so as to obtain enough

[2]The polymerase chain reaction (PCR) is an ingenious procedure devised by Mullis and Faloona (see Mullis and Faloona, 1987) for amplifying a specific genomic segment of DNA whose base sequence is known. Chemically synthesized oligonucleotides, complementary to short sequences flanking the two ends of the segment, are used as primers for the repetitive enzymic synthesis of complementary copies of each strand of the segment. After completion of the first polymerase reaction, the new strands of DNA are separated from their templates by heating and the process is repeated. At each repetition the amount of DNA corresponding to the specific segment is doubled.

DNA from this segment for analysis. Sequencing of the amplified segment revealed an in-frame deletion of 6 bp from the 3′ half of exon 2, resulting in the elimination of Asp 26 and Gly 27 from the first repeat of domain 1 of the receptor.

The mutations producing transport-deficient receptor protein have now been identified in four other naturally occurring class 2 alleles.

The defect in the FH 563 allele (see Table 9.2) is a deletion of three bases in the segment of exon 4 encoding the fifth repeat of the binding domain, resulting in elimination of a glycine residue (Russell *et al.*, 1989). The mutation in the WHHL rabbit is a deletion of 12 bases in exon 4 of the receptor gene (Yamamoto *et al.*, 1986). This exon encodes repeats 3, 4, and 5 of the binding domain. Since neither of these deletions alters the reading frame, the mutant genes encode nearly complete receptors lacking only one (FH 563) or four (WHHL rabbits) amino acid residues.

The cells of FH homozygote 429 synthesize a 120K receptor that is not processed to the mature form and does not reach the cell surface (Tolleshaug *et al.*, 1983). Esser and Russell (1988) have shown that the mutation in this patient is a single-base change in exon 11 leading to substitution of valine for glycine at position 544. This amino acid position is in the cysteine-poor sequence of ~280 residues between growth-factor repeats B and C of domain 2 (Fig. 9.6). The mutation in FH homozygote MM (see p. 357) is a single-base change resulting in the substitution of leucine for proline at position 664 in growth-factor repeat C (Soutar *et al.*, 1989). The FH 429 mutation abolishes an *Sfa*NI restriction site, while the MM mutation creates a new *Pst*I site. Both mutations may therefore be detected by Southern blotting of digests of the patient's DNA.

3. The Cause of Defective Transport in Class 2 Mutations

In normal cells, LDL receptors synthesized in the rough ER are transported to the Golgi apparatus, where their carbohydrates are extensively modified. This results in an increase in apparent molecular weight from ~120K to ~160K and loss of sensitivity of N-linked sugars to endoglycosidase H (see Section II above). In class 2 mutant cells, newly synthesized receptor protein is retained in the rough ER (Pathak *et al.*, 1988) and the modifications that normally occur in the Golgi are abolished or markedly delayed. How do class 2 mutations interrupt the transport of LDL-receptor protein from the rough ER to the Golgi apparatus?

A question of current interest to cell biologists is the nature of the signals and recognition systems that determine whether a protein synthesized in the rough ER is retained in the ER or is transported to the Golgi. Proteins destined for processing by Golgi enzymes are thought to cluster in regions of the ER that bud off to form transport vesicles (see Lodish *et al.*, 1986). Such a mechanism would require specific positive or negative signals in proteins synthesized in the ER and "recognition" molecules in potential budding regions of the ER membrane.

Progress in the identification of these elements has been slow. However, some of the factors that lead to retention of abnormal proteins in the ER are beginning to be understood.

Proteins synthesized in the ER are usually folded into a stable conformation while they are still in the lumen of the ER. The conformation adopted is determined by the primary amino acid sequence and by covalent modifications, including the formation of intrapeptide S-S bonds between cysteine residues. Evidence derived from the study of proteins modified by mutation or other means indicates that unless a newly formed protein undergoes normal folding it cannot leave the rough ER (see Lodish, 1988, for examples). It has been suggested that improperly folded proteins, or proteins containing free cysteine residues that normally form disulfide bonds, are trapped within the ER, partially or completely, by specific ''gatekeeper'' proteins (see discussion in Yamamoto *et al.*, 1986).

Five of the class 2 mutations that have been identified are in segments of the receptor gene encoding cysteine-rich repeats. They are therefore likely to interfere with the formation of S-S bonds between cysteine residues, either by changing the spacing between adjacent cysteines (FH 563, FH TT, FH MM, and the WHHL rabbit) or by deleting one or more cysteine residues that are normally paired by S-S bonding (FH 264). In this regard, the mutation in FH 264 is especially significant. Inspection of Fig. 9.4C shows that the cysteine at position 660 (the site of the stop codon in the mutant gene) is the third cysteine residue in repeat C. It is likely that in growth-factor repeats, S-S bonds are formed between the first and third cysteines, the second and fourth, and the fifth and sixth (see Esser and Russell, 1988). If this is the pattern in repeat C of the LDL receptor, deletion of the third cysteine would leave the first two as free cysteine residues. This might seem to support the view that delayed or absent processing of receptors in class 2 mutant cells is due to failure to form S-S bonds between cysteines.

However, while interference with S-S bond formation may explain the defective transport of LDL receptors in some mutant cells, this is unlikely to be the cause of the abnormal behavior of receptors in FH 429, because the mutation in this patient does not affect a cysteine-rich region of the receptor. Moreover, Esser and Russell (1988) have shown that the presence of free cysteine residues in repeat C of the FH 264 receptor is not essential for the defect in intracellular transport. To demonstrate this, Esser and Russell constructed a mutant cDNA encoding a receptor identical to the FH 264 receptor, except that the first two cysteines in repeat C were replaced by serines. When this gene was expressed in CHO cells, the mutant LDL receptors failed to undergo normal processing in the Golgi apparatus.

Esser and Russell (1988) suggest that specific amino acid sequences in certain regions of the receptor protein are essential for normal folding of the polypeptide backbone and that some class 2 mutations act by altering these sequences. They note that the glycine residue at position 544 is conserved between species and that glycine residues are involved in the folding of at least one other protein. The mutation in MM leads to substitution of leucine for proline in a proline-rich region of the receptor that is also highly conserved. Soutar *et al.* (1989) point out that substitution of leucine for proline has been shown to alter the stability of several other proteins. Thus, it seems likely that some class 2 mutations, including some of those that affect cysteine-rich segments of the receptor, act by altering sequences required for normal folding of the peptide chain rather than by generating free cysteine residues.

The effect of class 2 mutations on the ligand-binding properties of receptors that reach the cell surface depends upon the functional role of the segment in which the mutation occurs. Binding of LDL is unaffected by a mutation in the first repeat of domain 1 (the TT mutation), as in the synthetic mutants studied by van Driel *et al.* (1987b) and Esser *et al.* (1988). On the other hand, receptors produced by the cells of the WHHL rabbit and of FH homozygote MM bind LDL with reduced affinity. The mutation in the WHHL rabbit affects the fifth repeat of domain 1, shown by Esser *et al.* (1988) to be essential for high-affinity binding of LDL by the human receptor. The impaired binding of LDL by MM receptors suggests that disruption of the normal folding of repeat C in some way causes the binding domain to adopt an unfavorable conformation.

4. Class 3 Mutations

Three class 3 mutations have been investigated by Brown and Goldstein and their co-workers. In each case, the mutant receptor is normally processed and reaches the cell surface at the normal rate, but it binds LDL with markedly reduced efficiency.

The mutation in the paternal allele inherited by FH 295 (*R-210b*$^{-}$) results from an insertion of 14 kb, in effect between exons 8 and 9 (Lehrman *et al.*, 1987a). The insertion duplicates an uninterrupted sequence of the gene including exon 2 to exon 8, as in Fig. 9.23. Nucleotide sequencing across the deletion joint showed that the isolated left arm of an Alu repeat in intron 1 is joined to the left arm of another Alu repeat in intron 8. Since all the splice sites are in the same reading frame (see Südhof *et al.*, 1985a), the mutant gene would give rise to a message encoding a receptor of extended length in which the seven cysteine-rich repeats in the binding domain and the A and B repeats in domain 3 are duplicated. Lehrman *et al.* (1987a) suggest that the ancestral mutation arose by misalignment of chromatids at meiosis due to base-pairing of Alu repeats in intron 1 and intron 8, followed by unequal crossing over (Fig. 9.23). This event should gave given rise to a complementary allele lacking exons 2 to 8.

The normal processing of the precursor of the 210K receptor shows that it is transported from the ER to the Golgi apparatus. This suggests that the nine additional repeats are normally folded, since incorrect folding of a nascent pro-

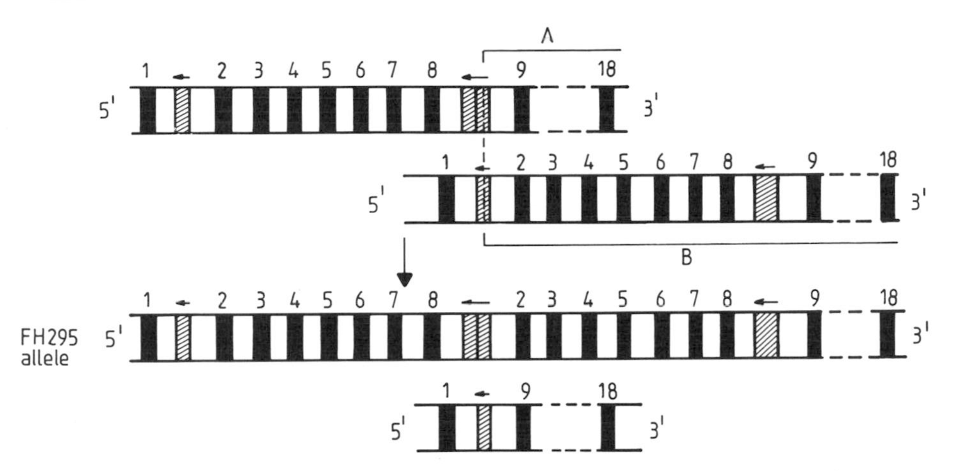

Figure 9.23. Diagram to show the formation of the *R-210b*⁻ allele inherited by FH 295 (intervals not to scale). Exons are shown by solid bars; Alu repeats are shown by hatched bars. Horizontal arrows show the orientation of Alu repeats in relation to the consensus Alu repeat. Vertical broken line shows the position of the recombination joint. DNA segment A exchanges with segment B during meiosis. (Constructed from Lehrman *et al.*, 1987a.)

tein is thought to impede its movement out of the ER. Hence, it seems likely that each complete cysteine-rich repeat folds independently by the formation of intra-repeat disulfide bonds, as postulated by Lehrman *et al.* (1987c).

The abnormality in the mutant allele inherited by FH 626 from his father (see Table 9.2) is a deletion of 0.8 kb (Hobbs *et al.*, 1986). The deletion extends from the left arm of an Alu repeat sequence in intron 4 to the left arm of another Alu repeat in the same orientation in intron 5, resulting in the removal of exon 5. Since the splice sites at introns 4 and 5 are in the same reading frame, splicing of the transcript of exon 4 to that of exon 6 produces a message encoding a complete receptor protein except for the sixth repeat of the binding domain (encoded by exon 5). The deletion probably arose by unequal crossing over of chromatids that were incorrectly aligned at meiosis by complementary base-pairing between Alu repeats in introns 4 and 5 (Fig. 9.24). As noted in Section II,D above, the receptor encoded by this allele binds β-VLDL with normal affinity but does not bind LDL.

Russell *et al.* (1986) have reported two homozygous FH siblings (FH 359 and 454) who carry a mutant receptor gene from which exons 7 and 8 have been removed. The message transcribed from this gene has a complete reading frame encoding a shortened receptor protein lacking only repeats A and B of domain 2. The abnormal receptor is processed normally and is transported to the cell surface. It fails to bind LDL but retains an ability to bind β-VLDL.

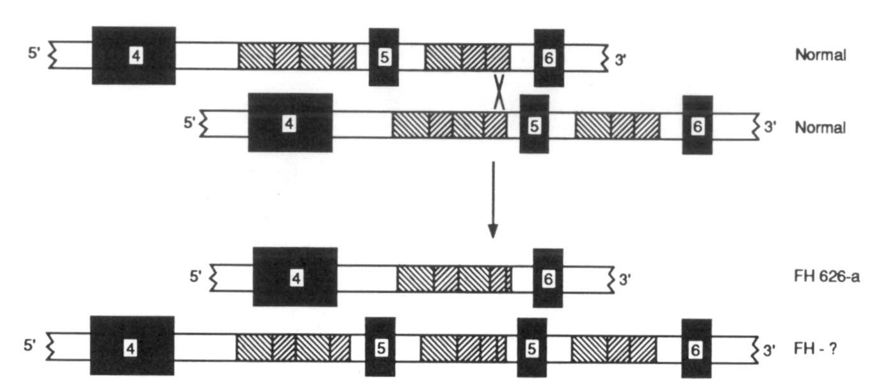

Figure 9.24. Diagram to show the formation of the paternal mutant allele in FH 626. Exons are shown by solid bars; Alu repeats are shown by hatched (right arms) or stippled (left arms) bars. The probable position of the crossover at meiosis is shown by a cross. (From Hobbs *et al.*, 1986, with the permission of the authors.)

5. Class 4 Mutations

The functional and structural characteristics of the internalization-defective receptors in three FH homozygotes [FH 380 (J.D.), 683, and 763] have already been discussed in Sections II,D and V,B above. The molecular basis of each mutation is shown in Fig. 9.7.

The internalization-defective allele carried by FH 380 has a point mutation (TAT → TGT) in the codon for residue 807 in the cytoplasmic tail of the receptor, resulting in a change from tyrosine to cysteine. The FH 683 allele has a point mutation (TGG → TGA) resulting in a stop codon after the codon for the second amino acid (residue 791) in the cytoplasmic tail. The message produced by this allele encodes a receptor lacking all but the first two amino acids of domain 5. The FH 763 allele has a duplication of four nucleotides immediately after the codon for residue 795. This leads to a change in the reading frame, producing a message that encodes a receptor with a normal amino acid sequence as far as residue 795, followed by a short sequence of eight abnormal amino acids. Thus, the FH 763 receptor has only the first six residues of the normal sequence of the cytoplasmic tail.

Two internalization-defective alleles encoding receptors lacking both the membrane-spanning and cytoplasmic domains (FH 274 and 781) have already been mentioned in Sections II,D and V,C. The molecular basis of both mutations is of sufficient general interest to warrant discussion in some detail.

FH 274, as noted earlier in this chapter, inherited an internalization-defective allele from his mother. Lehrman *et al.* (1985b) have analyzed the mutation in this allele. An abnormal *Xba*I fragment common to the genomic DNA of FH 274 and his mother was identified by Southern blotting. Analysis of this fragment showed

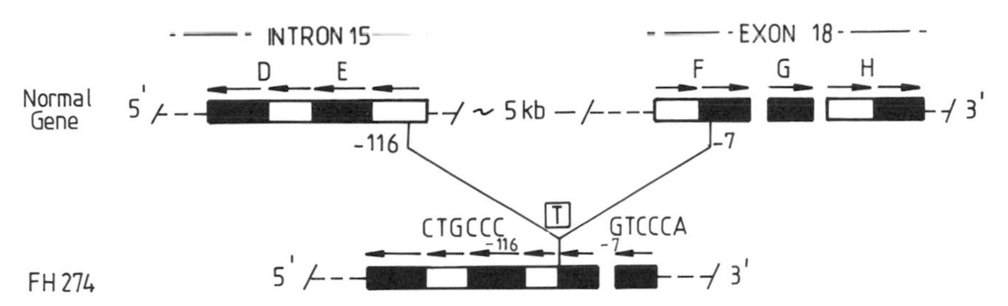

Figure 9.25. The deletion joint in the internalization-defective FH 274 allele. The deletion removes all sequences between nucleotide −116 in the left arm of Alu repeat E in intron 15 and nucleotide −7 in the right arm of Alu repeat F in exon 18 (see Fig. 9.26 for notation of Alu sequences in the receptor gene). Six nucleotides are shown on either side of the joint, plus the additional T. The sequence

−CTGCCC T GTCCCA−
−116 −7

is shown in the stem of the stem-loop system in Fig. 9.27 inset. The left arm of each repeat is shown as an open rectangle; the right arm is shown as a black rectangle. Arrows show the orientation of the Alu sequences. Note that in Alu sequences with orientation opposite to that of the consensus Alu sequence (D and E) the left arm is to the right. By convention, the nucleotides in each Alu repeat sequence are numbered separately; position +1 refers to the C of AGCT, the *Alu*I restriction site. (Modified from Lehrman *et al.*, 1985b.)

that it corresponds to the 3′ end of the receptor gene from which a segment ~5 kb long has been deleted (subsequently revised to 5.5 kb by Lehrman *et al.*, 1987b).

Nucleotide sequencing of the region containing the deletion joint showed that the 3′ end of the left arm of Alu repeat E in intron 15 is joined to the right arm of an oppositely oriented Alu repeat (F) in exon 18 (Fig. 9.25, and see Fig. 9.26 for designation of Alu repeats). Alignment of the sequences upstream and downstream from the deletion joint with the corresponding sequences in the normal gene showed that the deletion may have been preceded by the formation of an intrastrand stem-loop, analogous to that responsible for the generation of the FH 381 mutation. Lehrman *et al.* (1985b) postulate that two sequences, complementary in reverse on the same DNA strand in intron 15 and exon 18, formed the stem of a 5.5-kb loop (Fig. 9.27A). The loop was then excised at staggered breaks at nucleotide −116 in Alu repeat E and −6 in Alu repeat F (Fig. 9.27C). This was followed by folding back of the single 106-base strand resulting from the staggered break to form a second loop (Fig. 9.27D). This strand was then ligated to the 5′ end of Alu repeat F, with the insertion of an additional T to match an unpaired A (Fig. 9.27, inset). The new intact sequence then unfolded and underwent normal duplication. For details of this unusually complicated and very rare mutational event (only one family with the FH 274 allele has been discovered), see the text of the paper by Lehrman *et al.* (1985b).

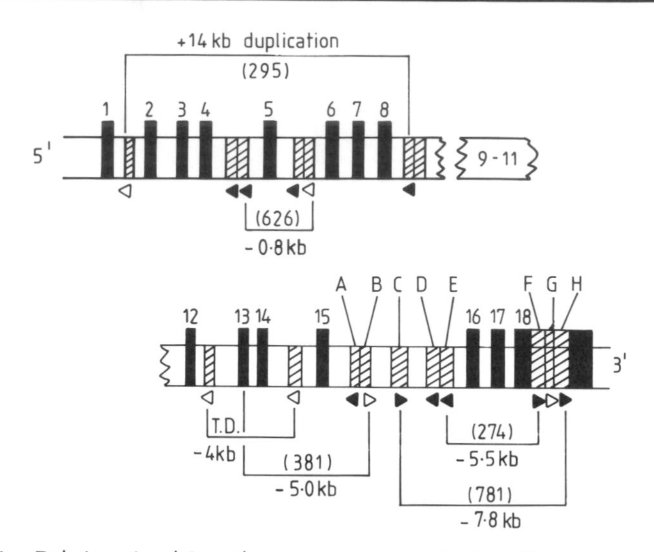

Figure 9.26. Deletions involving Alu repeat sequences in the LDL-receptor gene. Exons, introns, and Alu repeats are shown as solid, open, and striped segments respectively. Alu repeats in intron 15 and exon 18 are denoted by capital letters. The orientation of each Alu repeat in relation to the orientation of the consensus Alu repeat is indicated by an arrowhead [closed for complete repeat (left and right arms), open for half repeat]. The deletions discussed in the text are shown as horizontal lines. See Fig. 9.9 for a diagram of the consensus Alu repeat sequence. (Modified from Lehrman *et al.*, 1987a.)

Excision of the loop removed exons 16 and 17 and the 5′ end of exon 18, together with all the normal splice-acceptor sites downstream from intron 15. Lehrman *et al.* (1987b) have shown that this results in the production of a 7.7-kb message in which transcription has been continued from exon 15 through the remaining portions of intron 15 and exon 18. This message encodes a normal sequence of 749 amino acids as far as the 3′ end of exon 15, followed by a sequence of 55 abnormal residues before the first stop codon is reached. Lehrman *et al.* noted that the additional sequence contains a stretch of 17 uncharged amino acids that could act as a membrane-spanning domain. Thus, the mutant receptor lacks the normal membrane-spanning and cytoplasmic domains, but it contains a new segment at its C-terminal end that could anchor it to the plasma membrane, although with reduced efficiency. The presence of the anomalous hydrophobic sequence explains why a small proportion of the receptors produced in FH 274 cells remain attached to the plasma membrane and are not secreted into the external medium. Lacking a normal cytoplasmic domain, they are, of course, unable to cluster in coated pits.

The mutation in FH 781 is a 7.8-kb deletion that includes exons 16 and 17 and the 5′ portion of exon 18. Lehrman *et al.* (1987b) have shown that the deletion

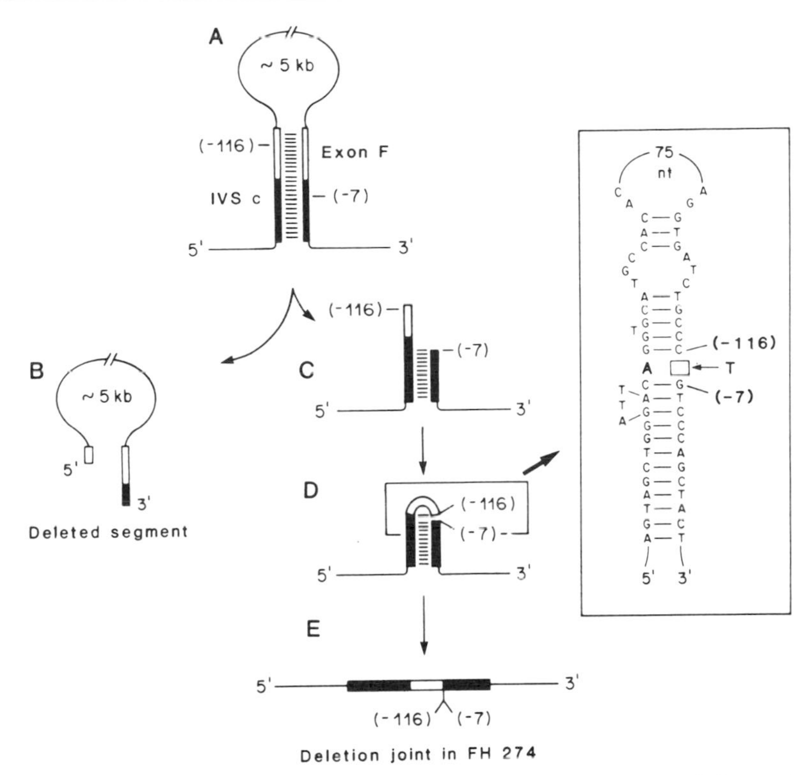

Figure 9.27. The probable mechanism for the gene deletion in the FH 274 allele. (A) shows the formation of a loop by base-pairing between Alu sequences in intron 15 (IVSc) and exon 18 (Exon F) on a single strand of DNA. (B) and (C) show the deletion of a segment extending from nucleotide −116 in the left arm of the 5′ Alu sequence to nucleotide −7 in the right arm of the 3′ Alu. (D) shows the formation of a second loop by folding of the unpaired strand left by the staggered break. The inset shows the base-pairing at (D), including the insertion of a T to complement the unpaired A in the stem of the second loop. The left arm of an Alu sequence is shown as an open bar; the right arm is shown as a closed bar. For numbering of Alu sequences, see Fig. 9.22. (From Lehrman *et al.*, 1985b, *Science*, Vol. 227, pp. 140–146, with the permission of the authors. Copyright 1985 by the AAAS.)

joins the left arm of Alu repeat C in intron 15 to the left arm of Alu repeat H in exon 18. The two Alu repeats at the deletion joint are oriented in the same direction. The deletion presumably resulted from the misalignment of chromatids during meiosis due to base-pairing between homologous sequences in Alu repeats C and H (see Fig. 9.26), followed by unequal crossing over. This produced a mutant allele encoding a truncated receptor lacking the membrane-spanning and cytoplasmic domains. Deletion of the 7.8-kb segment from the normal receptor gene removed all the splice-acceptor sites downstream from intron 15. As with the FH 274 mutant allele, this resulted in the production of a message encoding a protein with an abnormal C-terminal sequence of 55 amino acids containing a membrane-spanning sequence of 17 amino acids. The presence of this sequence explains why some of the receptors produced by the cells of patient FH 781 are retained by the plasma membrane and are able to bind LDL, though they fail to cluster in coated pits.

6. Summary of Known Mutations

Figure 9.28 summarizes the mutations in the LDL receptor gene that have been shown to be responsible for FH. In addition to the mutations discussed above, in which the molecular basis has been analyzed (numbers 1–14 and 25–27), 10 different partial deletions have been detected in 11 individuals by Southern blotting of restriction fragments of the genomic DNA from a total of 339 unrelated FH heterozygotes. These additional mutations have not yet been analyzed at the molecular level, although their approximate positions have been deduced by restriction mapping with cDNA probes complementary to limited segments of

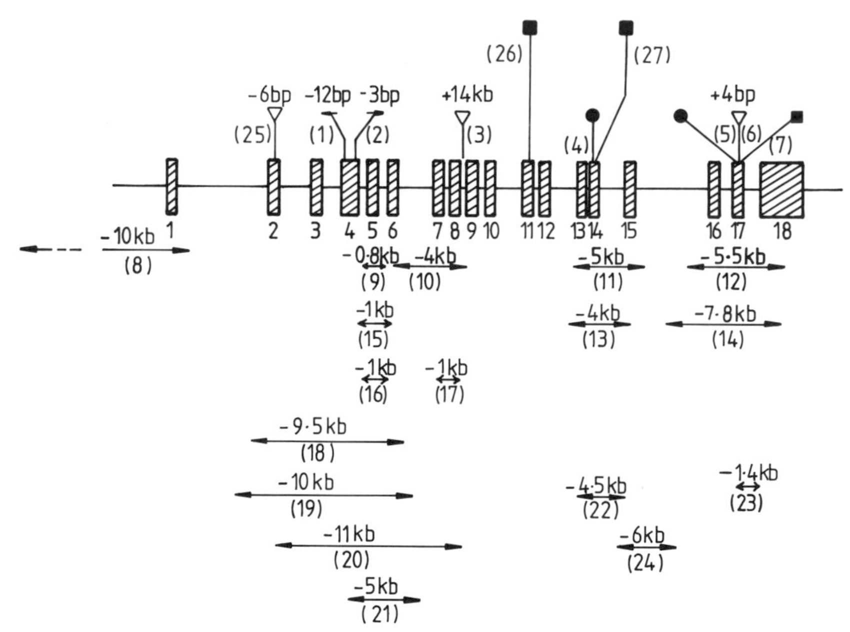

Figure 9.28. Mutations in the LDL receptor gene responsible for FH. Hatched boxes are exons. Introns are shown by horizontal lines (not drawn to scale). (1) WHHL rabbit; (2) FH 563; (3) FH 295; (4) FH 264, 550, 786, and 793; (5) FH 683; (6) FH 763; (7) FH 380; (8) FH 49, 549, 808, and 859; (9) FH 626; (10) FH 359 and 459; (11) FH 381; (12) FH 274; (13) T.D. and FH 651; (14) FH 781; (15–17) from Horsthemke *et al.*, 1987b; (18–23) from Langlois *et al.* (1988); (24) from Kajinami *et al.* (1988); (25) FH TT; (26) FH 429; (27) FH MM. Symbols: ↔, deletion; ▽, insertion; ●, nonsense; ■, missense. (Based on Russell *et al.*, 1986.)

the gene. Three of the 10 deletions were detected in a sample of 70 heterozygotes from the UK (Horsthemke *et al.*, 1987b) (numbers 15–17), 6 were found in a sample of 234 patients from British Columbia (Langlois *et al.*, 1988) (numbers 18–23), and 1 (number 24) was found in two unrelated heterozygotes in a sample of 35 patients from the Tonami region of Japan (Kajinami *et al.*, 1988). In several cases, confirmation that the deletion in the receptor gene was the cause of the FH was obtained by segregation analysis of the patient's family.

The FH mutations identified to date include 17 major deletions involving at least one exon, 2 minor deletions within exon 4 and 1 within exon 2, a major insertion of seven exons, a minor insertion of four nucleotides, three missense mutations, and two nonsense mutations.

It is clear, from the mutations listed in Fig. 9.28 and from the haplotype analysis of crm^- mutations carried out by Hobbs *et al.* (1988), that FH can be caused by a large number of different mutations in the receptor gene. It is equally clear that many more FH mutations will be uncovered when screening of FH patients by Southern-blot hybridization is extended to other parts of the world. However, it should be noted that this approach can only reveal large rearrangements in the genome. Point mutations and small deletions or insertions of a few base pairs do not give rise to detectable alterations in the pattern of restriction fragments, except in the rare instances where the mutation happens to abolish an existing restriction site or create a new one, such as the *Hin*fI site in the FH 264 mutation. These limitations explain why the mutations in less than 3% of the above 339 FH heterozygotes were identified. Other methods of screening will have to be used for the identification of mutations that cannot be detected by Southern blotting.

It is a striking fact that many of the FH mutations detected in unrelated heterozygotes, other than in those from populations where there is a founder effect, have been restricted to one or two individuals. This marked degree of genetic heterogeneity accounts for the high frequency of genetic compounds among the homozygous offspring of unrelated parents (see Table 9.2). The possible relationship between genetic heterogeneity of FH and the unusual organization of the receptor gene is discussed below.

7. The Significance of Alu Repeat Sequences

As shown in Fig. 9.26, eleven different Alu repeat sequences have participated in the generation of six receptor-gene mutations involving large deletions or insertions. In two of these (the FH 274 and 381 mutations) the initial event was a recombination between sequences oriented in opposite directions. In the other four mutations, unequal crossing over took place after misalignment between Alu sequences oriented in the same direction (FH 295, 626, 781, and T.D.). Lehrman *et al.* (1987b) have pointed out that in six out of the seven breakpoints they were able to analyze, the break occurs in the *left* arm of an Alu sequence between the A and B sequences that are homologous to the internal promoters for genes transcribed by RNA polymerase III (Paolella *et al.*, 1983). They also point out that Alu sequences are involved in four deletions of the γ,δ,β-globin genes, suggesting that a limited region in the left arm of an Alu repeat predisposes to recombination, perhaps when the DNA strands unwind during transcription by polymerase. Thus, the unusually large number of Alu repeats in the human receptor gene may help to explain why FH mutations are so common in the human population. In this regard, Langlois *et al.* (1988) have drawn attention to the fact that all the known major rearrangements in the gene occur in the two regions in which Alu repeats are clustered (the regions including exons 1–8 and 13–18).

VI. Somatic-Cell Mutations

We have already seen how naturally occurring and synthetic mutations in the receptor gene have been used to determine the function of separate domains in the receptor. Additional information on receptor function has also been obtained from studies of mutations induced in cells in culture by chemical mutagens or by X-irradiation. Mutations produced in this way are random in the sense that it is not possible to predetermine the sites within the genome at which they occur. However, it may be possible to select for mutated cells expressing a particular functional defect. An example is the ldlA class of Chinese hamster ovary cell mutants selected on the basis of their failure to produce LDL receptors that are capable of internalizing LDL from the medium. These cells have been invaluable as vehicles for testing the functional properties of synthetic mutants of the human receptor. As noted above (Section II,D), ldlA-7 cells were used in several studies of synthetic constructs of the receptor gene designed to reveal the functions of four of the five receptor domains.

Another advantage of somatic-cell mutants with defective LDL-receptor function is that they may uncover the existence of genes, other than the receptor gene, whose products are needed for normal receptor function. An example is the cell line in which a mutation has occurred in the gene encoding an enzyme required for glycosylation of the normal precursor of the mature LDL receptor (see below for details).

One of the strategies used for isolating somatic-cell mutants with abnormal receptor function is based on a method described by Krieger *et al.* (1979) for preparing LDL particles in which nonpolar lipids in the particle have been removed by heptane extraction and have then been replaced with other lipids. These reconstituted particles can be used to deliver abnormal lipids, including those that are toxic, to the cell interior. Krieger *et al.* (1981) used this approach to isolate CHO cell lines with mutations affecting the LDL-receptor pathway.

Cells growing in a lipoprotein-free medium were treated with the mutagen ethyl methanesulfonate. LDL in which the normal esterified cholesterol had been replaced by esterified 25-hydroxycholesterol was then added to the medium and growth of the cells was allowed to continue. Under these conditions, cells with a normal LDL-receptor pathway and normal sensitivity to suppression of HMG-CoA reductase by 25-hydroxycholesterol fail to survive because they cannot obtain cholesterol by uptake from the medium or by endogenous synthesis. Cells that survive are either resistant to 25-hydroxycholesterol but have a normal LDL-receptor pathway, or are lacking a normal receptor pathway. The latter are selected from the survivors of the incubation with toxic LDL by their failure to fluoresce when incubated with LDL containing fluorescent cholesteryl ester.

An alternative method for isolating receptor-pathway mutants is to grow the cells for short periods in a medium containing compactin and LDL. Cells lacking the LDL receptor pathway lose cholesterol from their plasma membranes and may be identified by their relative resistance to amphotericin B, a cytotoxic antibiotic that binds to cholesterol (Krieger *et al.*, 1983).

All the mutants isolated so far by these procedures have turned out to be phenotypically similar to the mutant cells from patients with the more usual form of familial hypercholesterolemia, in which high-affinity surface binding of LDL is defective. None has exhibited the internalization-defective phenotype.

Kingsley and Krieger (1984) have identified four distinct complementation groups among CHO somatic-cell mutants with defective receptor function. These were designated ldlA, ldlB, ldlC, and ldlD. Receptor function was normalized when cells from one group were fused pairwise with those from another group, but was unaffected when cells from the same group were used. Fibroblasts from a receptor-negative homozygous FH patient restored receptor activity to normal when fused with ldlB or ldlC CHO cells, but had no effect on ldlA cells. Kingsley and Krieger concluded that the mutation in ldlA cells is in the structural gene for the LDL receptor and that the mutations in cells of the other groups are in genes required for the normal processing or functioning of receptors.

In several of the mutant cell lines isolated by Krieger *et al.* (1981), LDL-receptor activity was less than 5% of that in the wild-type parental cells. Thus, these cells are phenotypically similar to cells from a receptor-negative FH homozygote, suggesting that the CHO mutants have no normal allele at the locus at which the mutation has occurred. This could be because both genes have mutated or because CHO cells are hemizygous at this locus. (In the cells of some established lines, one of the two alleles at certain loci is suppressed, a condition known as hemizygosity. Only one "hit" by a mutagenic agent at a hemizygous locus is required to produce a functionally homozygous mutant cell.) The unexpectedly high frequency of somatic-cell mutants with the receptor-negative phenotype led Krieger *et al.* (1981) to suggest that wild-type CHO cells are hemizygous at the locus of the LDL-receptor gene. However, Kingsley and Krieger

(1984) have isolated heterozygous revertants derived from receptor-negative ldlA mutants that are apparently homozygous. The heterozygous cells exhibited 50% of the receptor activity determined in wild-type CHO cells. The existence of these heterozygotes shows that CHO cells have two alleles at the receptor locus. In agreement with the conclusion that CHO cells are diploid at this locus, some ldlA cells have been shown to be genetic compounds, i.e., they carry two different mutant receptor genes (see below).

Kozarsky *et al.* (1986) have investigated the structure and processing of the LDL receptors synthesized in cells of the ldlA complementation group. The labeled proteins synthesized in the presence of [^{35}S]methionine were precipitated with an antibody to the receptor and were then size-separated by polyacrylamide-gel electrophoresis, as in the studies of Tolleshaug *et al.* (1982) discussed in Section II,B above. Three classes of mutation at the ldlA locus were identified, corresponding to class 1, 2, and 3 mutations in the human receptor gene. Cells with two class 1 alleles produce no detectable receptor protein. Class 2 alleles encode a receptor protein that fails to undergo normal processing to the mature form and is rapidly degraded within the cell. (The ldlA-7 cells used in the deletion experiments described earlier in this chapter were class 2 mutants.) Class 3 alleles encode a receptor that is normally processed but cannot bind LDL. Cells of one mutant line (ldlA-5) synthesized two different forms of defective receptor and were therefore genetic compounds.

Sege *et al.* (1986) have examined the mutant genes from several ldlA lines at the DNA level. Genomic DNA was analyzed by Southern blotting after digestion with restriction enzymes. Three ldlA mutants induced by X-irradiation of wild-type CHO cells had deletions in one of the two homologous alleles. The remaining allele in one of these mutants had a disruption involving exons 13 to 17. The message transcribed from this allele encodes a protein which lacks its C-terminal portion and in which the oligosaccharide chains are not processed because the precursor is not transported to the Golgi region.

Mutants of complementation groups ldlB, ldlC, and ldlD have been examined by Kingsley *et al.* (1986a,b). Cells from all three groups produce a precursor of normal molecular weight that is not processed to the high-molecular-weight form by the usual glycosylation reactions. The incompletely glycosylated receptors are rapidly degraded within the mutant cells. In these mutants, failure to synthesize normal oligosaccharide chains on LDL receptors is accompanied by defective glycosylation of certain other cell constituents. On the other hand, many glycosylation mutants derived from the parental CHO cell line by mutagenesis produce normally processed receptors. Kingsley *et al.* (1986b) concluded that the mutations in ldlB, ldlC, and ldlD cells affect enzymes concerned in a limited number of glycosylation reactions, including one or more of those required for the processing of LDL receptors.

The molecular basis of defective glycosylation in ldlB and ldlC mutants has not been elucidated. However, Kingsley *et al.* (1986a) have shown that the mutation in ldlD cells affects specifically the activity of what appears to be a single enzyme that catalyzes the formation of UDP-galactose and UDP-*N*-acetyl-galactosamine (UDP-GalNAc) from their corresponding glucose precursors. In the absence of this enzyme (UDP-galactose/UDP-GalNAc 4-epimerase), cells grown in the usual glucose-containing medium cannot carry out glycosylation reactions involving the addition of galactose or GalNAc. If ldlD cells are grown in a medium containing galactose and GalNAc, the two UDP sugars can be synthesized via salvage pathways. Under these conditions, the cells synthesize mature LDL receptors at the normal rate. The ability of galactose and GalNAc to reverse the effect of the mutant gene in ldlD cells explains an earlier observation that the genetic defect can be corrected by growing the cells in close proximity to human fibroblasts (Krieger, 1983). Hobbie *et al.* (1987) have shown that cells with normal epimerase activity supply the mutant cells with the missing substrates for glycosylation and that transfer of sugar molecules takes place through junctions between cells in direct contact with each other.

It is worth noting that mutations in enzymes required for glycosylation of LDL receptors have not been identified in any FH patients. As noted in Chapter 10, in every FH family examined so far, the mutation has turned out to be in the receptor gene or (rarely) the apo-B gene. Possibly, a mutation affecting a glycosylation reaction would have general effects incompatible with fetal development.

VII. Polymorphism in the LDL-Receptor Gene

The LDL-receptor mutations discussed above give rise either to a distinct clinical abnormality expressed in every individual who carries at least one mutant allele (Section V) or to a defect in the handling of LDL detectable in cells in culture (Section VI). However, in addition to rare mutations that disrupt receptor function, there must be many other receptor-gene mutations in the general population that, by themselves, have no obvious effect on the phenotype. As discussed in relation to the apoB gene (Chapter 6), a mutation has no effect on the structure of the gene product if it occurs in a noncoding region of the gene or if it affects a single base in a coding triplet without changing the amino acid. A mutation causing a change in amino acid sequence may still have little effect on the function of the gene product if the sequence involved is functionally "silent." Mutations that have no effect, or only a very small deleterious effect, on the gene product are free to spread through the population and may eventually give rise to a balanced polymorphism (for a definition of genetic polymorphism, see p. 143).

The observed frequency of the alleles at the polymorphic locus will depend, among other factors, on the evolutionary age of the mutation(s) and the extent to which the population has mixed genetically with other populations.

The potential importance of polymorphism at the LDL-receptor locus is two-fold. In the first place, polymorphism affecting the amino acid sequence of the receptor, or the rate of transcription of the receptor gene, could contribute to the genetic variability in plasma LDL concentration observed in normal populations. This possibility is now being investigated in several laboratories by examining samples of the normal population for associations between plasma LDL concentration and the frequency of an allele at a polymorphic site in the receptor gene detected with a restriction endonuclease. A significant association would suggest that the allele in question is causally related to the plasma LDL level or that it is in linkage disequilibrium with another allele that is so related. As shown in Table 9.3, 13 RFLPs have already been detected in the human LDL-receptor gene with 11 different restriction enzymes. In view of the length of the receptor

Table 9.3
Restriction-Fragment-Length Polymorphism in the Human LDL-Receptor Gene[a]

Restriction endonuclease	Frequency of rare allele (%)	Position of site
*Pvu*II	24	≃19 kb upstream
*Rsa*I	25	≃39 kb upstream
*Bsm*I	15	5′ flanking region
*SpH*I	35	Intron 6
*Stu*I	9	Exon 8
*Hinc*II	45	Exon 12
*Ava*II	44	Exon 13
*Spe*I	10	Intron 15
*Apa*LI	42	Intron 15
*Pvu*II	24	Intron 15
*Nco*I	34	Exon 18
*Pst*I	40	3′ flanking region
*Apa*LI	39	3′ flanking region

[a]All the polymorphic sites listed, except that detected with *Hinc*II and the two upstream sites, are mentioned in Hobbs *et al.* (1988). The *Hinc*II site is reported in Leitersdorf and Hobbs (1988); the two upstream sites are reported in Hegele *et al.* (1988). The positions of the sites are taken from various sources. Note that two different RFLPs are detected with the restriction enzyme *Apa*LI. Note also that the frequencies of some alleles were determined in small genetically heterogeneous populations and may not be applicable to other populations. See also Leitersdorf *et al.* (1989) for a restriction map of the polymorphic sites and a list of the population frequencies of 31 haplotypes constructed from 10 of the RFLPs included in this table.

gene, it is certain that a search with other restriction enzymes will reveal more RFLPs in this gene. Taylor *et al.* (1988) could find no differences between mean plasma total cholesterol levels in a group of normal men and women with different RFLP genotypes detected with four restriction enzymes. With two of these enzymes (*Stu*I and *Nco*I) the polymorphic sites detected were in exons of the receptor gene, though none of the sites was in the region of the gene encoding the LDL-binding domain.

As noted in Section V,D above, most point mutations or small rearrangements in the genome do not generate RFLPs. Hence it is probable that screening by Southern blotting of restriction fragments reveals only a small proportion of all the polymorphic sites present in the receptor gene in the general population. A more fruitful approach to the problem of detecting clinically relevant polymorphic sites in the receptor gene will be to amplify selected short segments of the gene known to encode functionally important regions of the receptor (see footnote, p. 369). Amplified segments of DNA from small groups of individuals could then be sequenced and the base sequences examined for differences between individuals. A positive finding could be the starting point for a more detailed investigation into the relation between plasma LDL concentration and alleles at polymorphic loci in the receptor gene.

The second reason why polymorphism in the receptor gene is important is that alleles at polymorphic loci may be used as linkage markers in segregation analysis of FH families. A promising example is the study of Humphries *et al.* (1985). In each of two informative families, the chromosome carrying the FH gene in a heterozygous parent was identified by showing that the disease segregated with one of the two alleles at a polymorphic site in the receptor gene detected with *Pvu*II. Early detection of an FH gene in the offspring of an affected family would be useful in confirming a diagnosis based on plasma LDL concentration. In principle, it should also be possible to diagnose FH antenatally by Southern blotting of fetal DNA in families informative for an RFLP detected with a restriction enzyme. In families in which both parents are FH heterozygotes and both are heterozygous for an RFLP in the receptor gene, it should be possible, in some cases, to tell whether an affected fetus is heterozygous or homozygous for FH. This information would be of great help in giving the parents genetic advice.

VIII. Evolutionary Aspects of the Receptor Gene

A comparison between the nucleotide sequences and exon–intron organizations of homologous genes suggests how many eukaryotic genes have evolved. In Chapter 6 (Section VI,G) we saw that the genes of the apoprotein multigene family appear to have arisen from a common ancestor by one or more complete duplications, each duplication providing an opportunity for increasing divergence by a succession of mutations. The mutations responsible for this divergence include single-base substitutions, short deletions and insertions, and repeated duplication of segments within genes.

Gilbert (1978, 1985) has pointed out that the existence of introns has also enabled eukaryotic genes to increase their functional complexity by acquiring complete exons or blocks of exons from elsewhere in the genome, the exons often encoding functional units of a protein such as ligand-binding domains. (See Gilbert, 1986, for a discussion of the mechanisms by which exon shuffling may have been achieved.) The newly acquired exons may then undergo further mutation to produce the gene in its present form. In this way, multifunctional proteins could evolve very rapidly by the addition of functional "modules" that have already been improved by long periods of natural selection. Modular units that are rearranged between proteins as a result of exon shuffling are usually short amino acid sequences (40–100 residues) with several disulfide bonds. Examples mentioned in earlier chapters and in Section II of this chapter are the kringle repeats common to apo(a) and plasminogen, and the cysteine-rich repeats common to the LDL receptor, LRP, and complement component C9. The "growth-factor" repeats present in domain 2 of the LDL receptor and in the EGF precursor are very widely distributed and must have appeared at an early stage of evolution. Rothberg *et al.* (1988) list a total of 22 proteins in which the growth-factor motif is present, including several insect proteins and a protein from the nematode *Caenorhabditis elegans*.

Doolittle (1985) has suggested the term *mosaic protein* for a protein that contains segments derived from, or common to, several other proteins. The gene encoding a mosaic protein may thus be related by sequence homology to genes belonging to more than one multigene family.

The LDL receptor is a typical mosaic protein. Fig. 9.29 shows the structural relationships between the receptor and six other vertebrate proteins with which it shares one or both of its two classes of cysteine-rich repeats—the repeat sequence in domain 1 (class A) and the growth-factor repeat in domain 2 (class B). The class A motif occurs 7 times in the LDL receptor, once in complement component C9, and 31 times in LRP. The class B motif occurs 3 times in the LDL receptor, once in C9, 10 times in the EGF precursor, and 22 times in LRP. The class B motif is also present as a single copy in Factor X, tissue plasminogen activator (TPA), and urokinase. In addition, TPA and urokinase each have one copy of the kringle repeat sequence (D in Fig. 9.29), while TPA shares another cysteine-rich repeat (C) with fibronectin (not shown).

There are clearly a great many possible routes by which the modular structures shown in Fig. 9.29 could have evolved. However, traces of the evolutionary history of the LDL receptor, LRP, and the EGF precursor are still discernible in segments of these proteins. The sequence of ~400 amino acids comprising domain 2 of the LDL receptor is homologous to similar segments in LRP and the

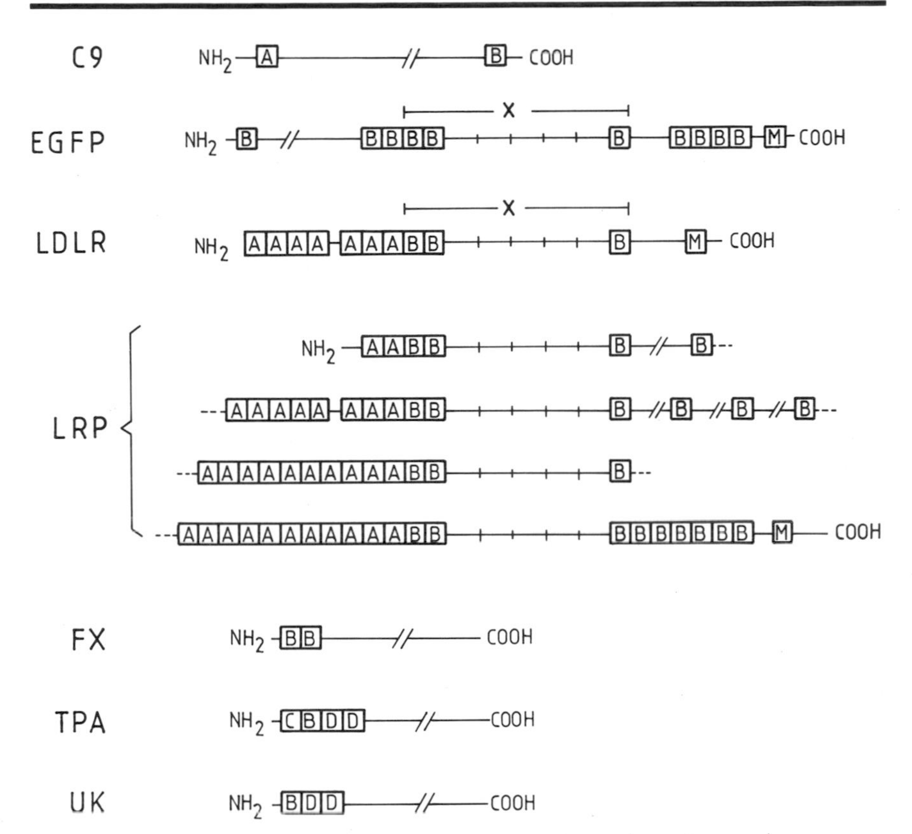

Figure 9.29. Diagram showing the modular structure of several vertebrate mosaic proteins related by sequence homology. Three of the proteins are membrane bound and four are secreted. C9, complement component 9; EGFP, epidermal growth factor precursor; LDLR, LDL receptor; LRP, LDL-receptor-related protein; FX, clotting Factor X; TPA, tissue plasminogen activator; UK, urokinase. The cysteine-rich modules are A, repeats present in domain 1 of the LDL receptor (class A motif of Stanley *et al.*, 1986); B, EGF precursor repeats (class B motif of Stanley *et al.*, 1986); C, the fibronectin type I 45-residue repeat; D, the 80-residue kringle; M, membrane-spanning segment. Vertical lines show Tyr-Trp-Thr-Asp repeats separated by about 50 amino acids. Note that the LRP is arranged in four sections to show that it consists essentially of four copies of the LDL receptor, with different numbers of A and B repeats in each copy. The segments labeled x in the EGF precursor and the LDL receptor have almost identical exon–intron organizations (see Fig. 9.11). (Modified and rearranged from Doolittle, 1985, and Herz *et al.*, 1988.)

EGF precursor, and in all these proteins the homologous segment is divided into a block of two class B repeats separated from a third class B repeat by a sequence of about 280 residues containing the sequence Tyr-Trp-Thr-Asp at intervals of about 50 residues. Moreover, as shown in Fig. 9.11, the positions of the introns in the segments of the receptor gene and the EGF-precursor gene encoding this region are almost identical. These similarities suggest that the progenitors of the present-day genes for the three related proteins arose by two sequential duplications of an ancestral gene in which the exons encoding a domain-2-like structure had already been assembled. Subsequent divergence to produce the LDL receptor and LRP may have involved the addition of class A motifs by exon shuffling, either as blocks of several exons or as single exons which then multiplied by internal duplication. At some stage in this evolutionary process, a complete multifunctional unit containing domains corresponding to the five domains of the LDL receptor may have undergone successive internal duplications to produce an LRP-like protein. This would then have been modified by further internal duplications of class A and B motifs, together with other mutational events.

Once a primitive receptor had appeared, the interaction between the receptor and its two apoprotein ligands could have evolved toward greater efficiency by selection of favorable mutations in both receptor and ligands. (A possible analogy to what may have occurred is the parallel evolution, toward increasing mutual adaptation, of some flowering plants and the insects that pollinate them.) It is not necessary to assume that the earliest LDL receptor had a binding domain with at least six class A repeat sequences. The experiments of Esser *et al.* (1988) suggest that, although the most efficient binding of LDL is achieved with six repeats, binding domains with fewer repeats could have provided the organism with some selective advantage. Thus, it is possible that a stepwise addition of repeat units to the binding domain contributed to the evolution of the LDL receptor.

There is very little information on which to base speculation about the evolutionary age of the LDL receptor. The observations of Peacock *et al.* (1988) show that the oocytes of the clawed toad (*Xenopus laevis*) contain the specific signals that recognize elements of the human LDL receptor, including the sequence containing tyrosine 807 in the cytoplasmic tail. This suggests that the *Xenopus* oocyte expresses an LDL receptor with at least some components homologous to those of the human receptor. If this is so, the LDL receptor may have existed more than 300 million years ago. George *et al.* (1987) have isolated a 95K membrane protein from chicken oocytes that has several properties in common with the mammalian LDL-receptor. The oocyte receptor binds native chicken LDL with high affinity and in a saturable manner but does not bind reductively methylated chicken LDL. It also binds human LDL, though with lower affinity. Polyclonal antibodies to the bovine LDL receptor cross-react specifically with the oocyte receptor. These preliminary observations suggest that an LDL receptor with a binding domain homologous to the binding domain of the mammalian receptor was present in the ancestors common to reptiles, birds, and mammals. Any speculation about the early evolutionary history of the LDL receptor must, of course, take into account the parallel evolution of its ligands.

As regards the more recent history of the receptor gene, the observations of Hobbs *et al.* (1985) suggest that the three Alu repeats in exon 18 of the human gene (see Fig. 9.26) are present in gorillas and chimpanzees but are absent from baboons and other lower species. The genetic events giving rise to the insertion of Alu repeats into exon 18 may, therefore, have occurred less than 33 million years ago. Comparison between the nucleotide sequences in the 3′-untranslated regions of the bovine and human receptor mRNAs indicates that the three Alu sequences were inserted into exon 18 in two separate events (Hobbs *et al.*, 1985).

As discussed in Section V, the presence of repetitive sequences in the receptor gene may have contributed to the high frequency of mutations involving deletions and insertions of this gene. A striking example is the FH 626 mutation. This appears to have arisen as a result of misalignment of the two chromatids at meiosis, due to incorrect pairing between Alu sequences in introns 4 and 5, followed by unequal crossing over (see Fig. 9.24). This event would have produced a mutant allele from which the whole of exon 5 (encoding repeat 6 of the binding domain) was deleted without changing the reading frame.

In the homozygous state the FH 626 allele gives rise to a severe and life-shortening disease. However, as shown in Fig. 9.24, the exon that was lost from one chromatid would have been inserted into its homolog to produce an allele with two copies of exon 5, encoding a receptor with an extra repeat 6 in the binding domain. We do not know what effect this would have on the binding properties of an LDL receptor in which the present-day complement of seven repeats was already in place. However, it does suggest a mechanism, involving repeated sequences, for increasing the numbers of class A repeats in a primordial LDL receptor. If the allele with the additional exon were selectively advantageous it would spread through the population, thus benefiting the species at the expense of a few individuals. If a recombination event of the kind that is thought to have produced the FH 626 allele occurred between segments of two different genes, this could lead to the transfer of exons, singly or in blocks, from one gene to another. Hence, unequal crossing over due to the presence of repetitive sequences in introns may have played a part in the shuffling of exons that almost certainly contributed to the assembly of the LDL-receptor gene in evolutionary time.

References

Aalto-Setälä, K., Gylling, H., Miettinen, T., and Kontula, K. (1988). Identification of a deletion in the LDL receptor gene. A Finnish type of mutation. *FEBS Lett.* **230,** 31–34.

Anderson, R. G. W., Goldstein, J. L., and Brown, M. S. (1977). A mutation that impairs the ability of lipoprotein receptors to localise in coated pits on the cell surface of human fibroblasts. *Nature (London)* **270,** 695–699.

Basu, S. K., Goldstein, J. L., and Brown, M. S. (1978). Characterization of the low density lipoprotein receptor in membranes prepared from human fibroblasts. *J. Biol. Chem.* **253,** 3852–3856.

Beisiegel, U., Kita, T., Anderson, R. G. W., Schneider, W. J., Brown, M. S., and Goldstein, J. L. (1981a). Immunologic cross-reactivity of the low density lipoprotein receptor from bovine adrenal cortex, human fibroblasts, canine liver and adrenal gland, and rat liver. *J. Biol. Chem.* **256,** 4071–4078.

Beisiegel, U., Schneider, W. J., Goldstein, J. L., Anderson, R. G. W., and Brown, M. S. (1981b). Monoclonal antibodies to the low density lipoprotein receptor as probes for study of receptor-mediated endocytosis and the genetics of familial hypercholesterolemia. *J. Biol. Chem.* **256,** 11923–11931.

Beisiegel, U., Schneider, W. J., Brown, M. S., and Goldstein, J. L. (1982). Immunoblot analysis of low density lipoprotein receptors in fibroblasts from subjects with familial hypercholesterolemia. *J. Biol. Chem.* **257,** 13150–13156.

Berg, K., and Heiberg, A. (1978). Linkage between familial hypercholesterolemia with xanthomatosis and the C3 polymorphism confirmed. *Cytogenet. Cell Genet.* **22,** 621–623.

Brink, P. A., Steyn, L. T., Coetzee, G. A., and Van der Westhuyzen, D. R. (1987). Familial hypercholesterolemia in South African Afrikaners. *Pvu*II and *Stu*I DNA polymorphisms in the LDL-receptor gene consistent with a predominating founder effect. *Hum. Genet.* **77,** 32–35.

Brown, M. S., and Goldstein, J. L. (1977). Familial hypercholesterolemia: model for genetic receptor disease. *Harvey Lect. Ser.* **73,** 163–201.

Brown, M. S., and Goldstein, J. L. (1986). A receptor-mediated pathway for cholesterol homeostasis. *Science* **232,** 34–47.

Chandler, V. L., Maler, B. A., and Yamamoto, K. R. (1983). DNA sequences bound specifically by glucocorticoid receptor *in vitro* render a heterologous promoter hormone responsive *in vivo.* *Cell* **33,** 489–499.

Cummings, R. D., Kornfeld, S., Schneider, W. J., Hobgood, K. K., Tolleshaug, H., Brown, M. S., and Goldstein, J. L. (1983). Biosynthesis of N- and O-linked oligosaccharides of the low density lipoprotein receptor. *J. Biol. Chem.* **258,** 15261–15273.

Daniel, T. O., Schneider, W. J., Goldstein, J. L., and Brown, M. S. (1983). Visualization of lipoprotein receptors by ligand blotting. *J. Biol. Chem.* **258,** 4605–4611.

Davis, C. G., Elhammer, A., Russell, D. W., Schneider, W. J., Kornfeld, S., Brown, M. S., and Goldstein, J. L. (1986a). Deletion of clustered O-linked carbohydrates does not impair function of low density lipoprotein receptor in transfected fibroblasts. *J. Biol. Chem.* **261,** 2828–2838.

Davis, C. G., Lehrman, M. A., Russell, D. W., Anderson, R. G. W., Brown, M. S., and Goldstein, J. L. (1986b). The J.D. mutation in familial hypercholesterolemia: amino acid substitution in cytoplasmic domain impedes internalization of LDL receptors. *Cell* **45,** 15–24.

Davis, C. G., van Driel, I. R., Russell, D. W., Brown, M. S., and Goldstein, J. L. (1987a). The low density lipoprotein receptor. Identification of amino acids in cytoplasmic domain required for rapid endocytosis. *J. Biol. Chem.* **262,** 4075–4082.

Davis, C. G., Goldstein, J. L., Südhof, T. C., Anderson, R. G. W., Russell, D. W., and Brown, M. S. (1987b). Acid-dependent ligand dissociation and recycling of LDL receptor by growth factor homology region. *Nature (London)* **326,** 760–765.

Dawson, P. A., Hofmann, S. L., Van der Westhuyzen, D. R., Südhof, T. C., Brown, M. S., and Goldstein, J. L. (1988). Sterol-dependent repression of low density lipoprotein receptor promoter mediated by 16-base pair sequence adjacent to binding site for transcription factor Sp1. *J. Biol. Chem.* **263,** 3372–3379.

Doolittle, R. F. (1985). The genealogy of some recently evolved vertebrate proteins. *Trends Biochem. Sci.* **10,** 233–237.

Doolittle, R. F., Feng, D. F., and Johnson, M. S. (1984). Computer-based characterization of epidermal growth factor precursor. *Nature (London)* **307,** 558–560.

Dynan, W. S., and Tjian, R. (1985). Control of eukaryotic messenger RNA synthesis by sequence-specific DNA-binding proteins. *Nature (London)* **316,** 774–778.

Esser, V., and Russell, D. W. (1988). Transport-deficient mutations in the low density lipoprotein receptor. Alterations in the cysteine-rich and cysteine-poor regions of the protein block intracellular transport. *J. Biol. Chem.* **263,** 13276–13281.

Esser, V., Limbird, L. E., Brown, M. S., Goldstein, J. L., and Russell, D. W. (1988). Mutational analysis of the ligand binding domain of the low density lipoprotein receptor. *J. Biol. Chem.* **263,** 13282–13290.

Fourie, A. M., Coetzee, G. A., Gevers, W., and Van der Westhuyzen, D. R. (1988). Two mutant low density lipoprotein receptors in Afrikaners slowly processed to surface forms exhibiting rapid degradation or functional heterogeneity. *Biochem. J.*, **255,** 411–415..

Francke, U., Brown, M. S., and Goldstein, J. L. (1984). Assignment of the human gene for the low density lipoprotein receptor to chromosome 19: Synteny of a receptor, a ligand and a genetic disease. *Proc. Natl. Acad. Sci. U.S.A.* **81,** 2826–2830.

Gavigan, S. J. P., Patel, D. D., Soutar, A. K., and Knight, B. L. (1988). An antibody to the low-density lipoprotein (LDL) receptor that partially inhibits the binding of LDL to cultured human fibroblasts. *Eur. J. Biochem.* **171,** 355–361.

George, R., Barber, D. L., and Schneider, W. J. (1987). Characterization of the chicken oocyte receptor for low and very low density lipoproteins. *J. Biol. Chem.* **262,** 16838–16847.

Gilbert, W. (1978). Why genes in pieces? *Nature (London)* **271,** 501.

Gilbert, W. (1985). Genes-in-pieces revisited. *Science* **228,** 823–824.

Gilbert, W. (1986). The RNA world. *Nature (London)* **319,** 618.

Goldstein, J. L., and Brown, M. S. (1984). Progress in understanding the LDL receptor and HMG-CoA reductase, two membrane proteins that regulate the plasma cholesterol. *J. Lipid Res.* **25,** 1450–1461.

Goldstein, J. L., Dana, S. E., Brunschede, G. Y., and Brown, M. S. (1975). Genetic heterogeneity in familial hypercholesterolemia: evidence for two different mutations affecting functions of low-density lipoprotein receptors. *Proc. Natl. Acad. Sci. U.S.A.* **72,** 1092–1096.

Goldstein, J. L., Sobhani, M. K., Faust, J. R., and Brown, M. S. (1976). Heterozygous familial hypercholesterolemia: failure of normal allele to compensate for mutant allele at a regulated gene locus. *Cell* **9,** 195–203.

Goldstein, J. L., Brown, M. S., and Stone, N. J. (1977). Genetics of the LDL receptor: evidence that the mutations affecting binding and internalization are allelic. *Cell* **12,** 629–641.

Goldstein, J. L., Brown, M. S., Anderson, R. G. W., Russell, D. W., and Schneider, W. J. (1985). Receptor-mediated endocytosis: concepts emerging from the LDL receptor system. *Ann. Rev. Cell Biol.* **1,** 1–39.

Goodbourn, S., Zinn, K., and Maniatis, T. (1985). Human β-interferon gene expression is regulated by an inducible enhancer element. *Cell* **41,** 509–520.

Goodbourn, S., Burstein, H., and Maniatis, T. (1986). The human β-interferon gene enhancer is under negative control. *Cell* **45,** 601–610.

Harris, H. (1977). "The Principles of Human Biochemical Genetics," 3rd Ed. Elsevier, Amsterdam.

Hegele, R. A., Emi, M., Nakamura, Y., Lalouel, J-M., and White, R. (1988). RFLPs upstream of the low-density lipoprotein receptor (LDLR) gene. *Nucleic Acids Res.* **16,** 7214.

Herz, J., Hamann, U., Rogne, S., Myklebost, O., Gausepohl, H., and Stanley, K. K. (1988). Surface location and high affinity for calcium of a 500 kDa liver membrane protein closely related to the LDL-receptor suggest a physiological role as lipoprotein receptor. *EMBO J.* **7,** 4119–4127.

Hobbie, L., Kingsley, D. M., Kozarsky, K. F., Jackman, R. W., and Krieger, M. (1987). Restoration of LDL receptor activity in mutant cells by intercellular junctional communication. *Science* **235,** 69–73.

Hobbs, H. H., Lehrman, M. A., Yamamoto, T., and Russell, D. W. (1985). Polymorphism and evolution of *Alu* sequences in the human low density lipoprotein receptor gene. *Proc. Natl. Acad. Sci. U.S.A.* **82,** 7651–7655.

Hobbs, H. H., Brown, M. S., Goldstein, J. L., and Russell, D. W. (1986). Deletion of exon encoding cysteine-rich repeat of low density lipoprotein receptor alters its binding specificity in a subject with familial hypercholesterolemia. *J. Biol. Chem.* **261,** 13114–13120.

Hobbs, H. H., Brown, M. S., Russell, D. W., Davignon, J., and Goldstein, J. L. (1987). Deletion in the gene for the low-density-lipoprotein receptor in a majority of French Canadians with familial hypercholesterolemia. *N. Engl. J. Med.* **317,** 734–737.

Hobbs, H. H., Leitersdorf, E., Goldstein, J. L., Brown, M. S., and Russell, D. W. (1988). Multiple crm$^-$ mutations in familial hypercholesterolemia. Evidence for 13 alleles, including four deletions. *J. Clin. Invest.* **81,** 909–917.

Horsthemke, B., Beisiegel, U., Dunning, A., Havinga, J. R., Williamson, R., and Humphries, S. (1987a). Unequal crossing-over between two Alu-repetitive DNA sequences in the low-density-lipoprotein-receptor gene. A possible mechanism for the defect in a patient with familial hypercholesterolaemia. *Eur. J. Biochem.* **164,** 77–81.

Horsthemke, B., Dunning, A., and Humphries, S. (1987b). Identification of deletions in the human low density lipoprotein receptor gene. *J. Med. Genet.* **24,** 144–147.

Humphries, S. E., Kessling, A. M., Horsthemke, B., Donald, J. A., Seed, M., Jowett, N., Holm, M., Galton, D. J., Wynn, V., and Williamson, R. (1985). A common DNA polymorphism of the low-density lipoprotein (LDL) receptor gene and its use in diagnosis. *Lancet* **1,** 1003–1005.

Kadonaga, J. T., Jones, K. A., and Tjian, R. (1986). Promoter-specific activation of RNA polymerase II transcription by Sp1. *Trends Biochem. Sci.* **11,** 20–23.

Kajinami, K., Mabuchi, H., Itoh, H., Michishita, I., Takeda, M., Wakasugi, T., Koizumi, J., and Takeda, R. (1988). New variant of low density lipoprotein receptor gene, FH-Tonami. *Arteriosclerosis* **8,** 187–192.

Khachadurian, A. K. (1964). The inheritance of essential familial hypercholesterolemia. *Am. J. Med.* **37,** 402–407.

Kingsley, D. M., and Krieger, M. (1984). Receptor-mediated endocytosis of low density lipoprotein: somatic cell mutants define multiple genes required for expression of surface-receptor activity. *Proc. Natl. Acad. Sci. U.S.A.* **81,** 5454–5458.

Kingsley, D. M., Kozarsky, K. F., Hobbie, L., and Krieger, M. (1986a). Reversible defects in O-linked glycosylation and LDL receptor expression in a UDP-Gal/UDL-GalNAc 4-epimerase deficient mutant. *Cell* **44,** 749–759.

Kingsley, D. M., Kozarsky, K. F., Segal, M., and Krieger, M. (1986b). Three types of low density lipoprotein receptor-deficient mutant have pleiotropic defects in the synthesis of N-linked, O-linked, and lipid-linked carbohydrate chains. *J. Cell Biol.* **102,** 1576–1585.

Kishimoto, A., Brown, M. S., Slaughter, I. C. A., and Goldstein, J. L. (1987). Phosphorylation of serine 833 in cytoplasmic domains of low density lipoprotein receptor by a high molecular weight enzyme resembling casein kinase II. *J. Biol. Chem.* **262,** 1344–1351.

Knight, B. L., Patel, D. D., and Soutar, A. K. (1987). Regulation of synthesis and cell content of the low-density-lipoprotein receptor protein in cultured fibroblasts from normal and familial hypercholesterolaemic subjects. *Eur. J. Biochem.* **163,** 189–196.

Knight, B. L., Gavigan, S. J. G., Soutar, A. K., and Patel, D. D. (1989). Defective processing and binding of low-density lipoprotein receptors in fibroblasts from a familial hypercholesterolaemic subject. *Eur. J. Biochem.* **179,** 693–698.

Koo, C., Wernette-Hammond, M. E., and Innerarity, T. L. (1986). Uptake of canine β-very low density lipoprotein by mouse peritoneal macrophages is mediated by a low density lipoprotein receptor. *J. Biol. Chem.* **261,** 11194–11201.

Kozarsky, K. F., Brush, H. A., and Krieger, M. (1986). Unusual form of low density lipoprotein receptors in hamster cell mutants with defects in the receptor structural gene. *J. Cell Biol.* **102,** 1567–1575.

Krieger, M. (1983). Complementation of mutations in the LDL pathway of receptor-mediated endocytosis by cocultivation of LDL receptor-defective hamster cell mutants. *Cell* **33,** 413–422.

Krieger, M., McPhaul, M. J., Goldstein, J. L., and Brown, M. S. (1979). Replacement of neutral lipids of low density lipoprotein with esters of long chain unsaturated fatty acids. *J. Biol. Chem.* **254,** 3845–3853.

Krieger, M., Brown, M. S., and Goldstein, J. L. (1981). Isolation of Chinese hamster cell mutants defective in the receptor-mediated endocytosis of LDL. *J. Mol. Biol.* **150,** 167–184.

Krieger, M., Martin, J., Segal, M., and Kingsley, D. (1983). Amphotericin B selection of mutant Chinese hamster cells with defects in the receptor-mediated endocytosis of low density lipoprotein and cholesterol biosynthesis. *Proc. Natl. Acad. Sci. U.S.A.* **80,** 5607–5611.

Langlois, S., Kastelein, J. J. P., and Hayden, M. R. (1988). Characterization of six partial deletions in the low-density-lipoprotein (LDL) receptor gene causing familial hypercholesterolemia (FH). *Am. J. Hum. Genet.* **43,** 60–68.

Lehrman, M. A., Goldstein, J. L., Brown, M. S., Russell, D. W., and Schneider, W. J. (1985a). Internalization-defective-LDL receptors produced by genes with nonsense and frameshift mutations that truncate the cytoplasmic domain. *Cell* **41,** 735–743.

Lehrman, M. A., Schneider, W. J., Südhof, T. C., Brown, M. S., Goldstein, J. L., and Russell, D. W. (1985b). Mutation in LDL receptor: Alu-Alu recombination deletes exons encoding transmembrane and cytoplasmic domains. *Science* **227,** 140–146.

Lehrman, M. A., Russell, D. W., Goldstein, J. L., and Brown, M. S. (1986). Exon-*Alu* recombination deletes 5 kilobases from the low density lipoprotein receptor gene, producing a null phenotype in familial hypercholesterolemia. *Proc. Natl. Acad. Sci. U.S.A.* **83,** 3679–3683.

Lehrman, M. A., Goldstein, J. L., Russell, D. W., and Brown, M. S. (1987a). Duplication of seven exons in LDL receptor gene caused by Alu-Alu recombination in a subject with familial hypercholesterolemia. *Cell* **48,** 827–835.

Lehrman, M. A., Russell, D. W., Goldstein, J. L., and Brown, M. S. (1987b). Alu-Alu recombination deletes splice acceptor sites and produces secreted low density lipoprotein receptor in a subject with familial hypercholesterolemia. *J. Biol. Chem.* **262,** 3354–3361.

Lehrman, M. A., Schneider, W. J., Brown, M. S., Davis, C. G., Elhammer, A., Russell, D. W., and Goldstein, J. L. (1987c). The Lebanese allele at the low density lipoprotein receptor locus. Nonsense mutation that produces truncated receptor that is retained in endoplasmic reticulum. *J. Biol. Chem.* **262,** 401–410.

Leitersdorf, E., and Hobbs, H. H. (1988). Human LDL receptor gene: *Hinc*II polymorphism detected by gene amplification. *Nucleic Acids Res.* **16,** 7215.

Leitersdorf, E., Hobbs, H. H., Fourie, A. M., Jacobs, M., and Van der Westhuyzen, D. R. (1988). Deletion in first cysteine-rich repeat of low density lipoprotein receptor impairs its transport but not lipoprotein binding in subjects with familial hypercholesterolemia. *Proc. Natl. Acad. Sci. U.S.A.*, **85,** 7912–7916..

Leitersdorf, E., Chakravarti, A., and Hobbs, H. H. (1989). Polymorphic DNA haplotypes at the LDL receptor locus. *Am. J. Hum. Genet.* **44,** 409–421.

Lerner, R. A., Sutcliffe, J. G., and Shinnick, T. M. (1981). Antibodies to chemically synthesized peptides predicted from DNA sequences as probes of gene expression. *Cell* **23,** 309–310.

Lewin, B. (1987). "Genes III," 3rd Ed. Wiley, New York.

Lindgren, V., Luskey, K. L., Russell, D. W., and Francke, U. (1985). Human genes involved in cholesterol metabolism: chromosomal mapping of the loci for the low density lipoprotein receptor and 3-hydroxy-3-methylglutaryl-coenzyme A reductase with cDNA probes. *Proc. Natl. Acad. Sci. U.S.A.* **82,** 8567–8571.

Lodish, H. F. (1988). Transport of secretory and membrane glycoproteins from the rough endoplasmic reticulum to the Golgi. A rate-limiting step in protein maturation and secretion. *J. Biol. Chem.* **263,** 2107–2110.

Lodish, H., Darnell, J., and Baltimore, D. (1986). "Molecular Cell Biology." Scientific American Books, New York.

McKnight, S. L. (1982). Functional relationships between transcriptional control signals of the thymidine kinase gene of herpes simplex virus. *Cell* **31,** 355–365.

McKnight, S. L., Gavis, E. R., Kingsbury, R., and Azel, R. (1981). Analysis of transcriptional regulatory signals of the HSV thymidine kinase gene: identification of an upstream control region. *Cell* **25,** 385–398.

Miyake, Y., Tajima, S., Yamamura, T., and Yamamoto, A. (1981). Homozygous familial hypercholesterolemia mutant with a defect in internalization of low density lipoprotein. *Proc. Natl. Acad. Sci. U.S.A. 78,* 5151–5155.

Mullis, K. B., and Faloona, F. A. (1987). Specific synthesis of DNA *in vitro* via a polymerase-catalyzed chain reaction. *Methods Enzymol.* **155,** 335–350.

Osborne, T. F., Gil, G., Goldstein, J. L., and Brown, M. S. (1988). Operator constitutive mutation of 3-hydroxy-3-methylglutaryl coenzyme A reductase promoter abolishes protein binding to sterol regulatory element. *J. Biol. Chem.* **263,** 3380–3387.

Paolella, G., Lucero, M. A., Murphy, M. H., and Baralle, F. E. (1983). The Alu family repeat promoter has tRNA-like bipartite structure. *EMBO J.* **2,** 691–696.

Pathak, R. K., Merkle, R. K., Cummings, R. D., Goldstein, J. L., Brown, M. S., and Anderson, R. G. W. (1988). Immunocytochemical localization of mutant low density lipoprotein receptors that fail to reach the Golgi complex. *J. Cell Biol.* **106,** 1831–1841.

Peacock, S. L., Bates, M. P., Russell, D. W., Brown, M. S., and Goldstein, J. L. (1988). Human low density lipoprotein receptor expressed in *Xenopus* oocytes. Conserved signals for O-linked glycosylation and receptor-mediated endocytosis. *J. Biol. Chem.* **263,** 7838–7845.

Rothberg, J. M., Hartley, D. A., Walther, Z., and Artavanis-Tsakonas, S. (1988). Slit: an EGF-homologous locus of *D. melanogaster* involved in the development of the embryonic central nervous system. *Cell* **55,** 1047–1059.

Russell, D. W., Yamamoto, T., Schneider, W. J., Slaughter, C. J., Brown, M. S., and Goldstein, J. L. (1983). cDNA cloning of the bovine low density lipoprotein receptor: Feedback regulation of receptor mRNA. *Proc. Natl. Acad. Sci. U.S.A.* **80,** 7501–7505.

Russell, D. W., Schneider, W. J., Yamamoto, T., Luskey, K. L., Brown, M. S., and Goldstein, J. L. (1984). Domain map of the LDL receptor: sequence homology with the epidermal growth factor precursor. *Cell* **37,** 577–585.

Russell, D. W., Lehrman, M. A., Südhof, T. C., Yamamoto, T., Davis, C. G., Hobbs, H. H., Brown, M. S., and Goldstein, J. L. (1986). The LDL receptor in familial hypercholesterolemia: use of human mutations to dissect a membrane protein. *Cold Spring Harbor Symp. Quant. Biol.* **51,** 811–819.

Russell, D. W., Esser, V., and Hobbs, H. H. (1989). Molecular basis of familial hypercholesterolemia. *Arteriosclerosis* **9,** I8–I13.

Schneider, W. J., Goldstein, J. L., and Brown, M. S. (1980). Partial purification and characterization of the low density lipoprotein receptor from bovine adrenal cortex. *J. Biol. Chem.* **255,** 11442–11447.

Schneider, W. J., Beisiegel, U., Goldstein, J. L., and Brown, M. S. (1982). Purification of the low density lipoprotein receptor, an acidic glycoprotein of 164,000 molecular weight. *J. Biol. Chem.* **257,** 2664–2673.

Schneider, W. J., Brown, M. S., and Goldstein, J. L. (1983a). Kinetic defects in the processing of low density lipoprotein receptor in fibroblasts from WHHL rabbits and a family with familial hypercholesterolemia. *Mol. Biol. Med.* **1,** 353–367.

Schneider, W. J., Slaughter, C. J., Goldstein, J. L., Anderson, R. G. W., Capra, D. J., and Brown, M. S. (1983b). Use of anti-peptide antibodies to demonstrate external orientation of the NH_2-terminus of the low density lipoprotein receptor in the plasma membrane of fibroblasts. *J. Cell Biol.* **97,** 1635–1640.

Sege, R. D., Kozarsky, K. F., and Krieger, M. (1986). Characterization of a family of gamma-ray-induced CHO mutants demonstrates that the ldlA locus is diploid and encodes the low-density lipoprotein receptor. *Mol. Cell Biol.* **6,** 3268–3277.

Soutar, A. K., Knight, B. L., and Patel, D. D. (1989). Identification of a point mutation in growth factor repeat C of the low density lipoprotein-receptor gene in a patient with homozygous familial hypercholesterolemia that affects ligand binding and intracellular movement of receptors. *Proc. Natl. Acad. Sci. U.S.A.* **86,** 4166–4170.

Stanley, K. K., Kocher, H.-P., Luzio, J. P., Jackson, P., and Tschopp, J. (1985). The sequence and topology of human complement component C9. *EMBO J.* **4,** 375–382.

Stanley, K. K., Page, M., Campbell, A. K., and Luzio, J. P. (1986). A mechanism for the insertion of complement component C9 into target membranes. *Molec. Immunol.* **23,** 451–458.

Stuart, G. W., Searle, P. F., and Palmiter, R. D. (1985). Identification of multiple metal regulatory elements in mouse metallothionein-I promoter by assaying synthetic sequences. *Nature (London)* **317,** 828–831.

Südhof, T. C., Goldstein, J. L., Brown, M. S., and Russell, D. W. (1985a). The LDL receptor gene: a mosaic of exons shared with different proteins. *Science* **228,** 815–822.

Südhof, T. C., Russell, D. W., Goldstein, J. L., Brown, M. S., Sanchez-Pescador, R., and Bell, G. I. (1985b). Cassette of eight exons shared by genes for LDL receptor and EGF precursor. *Science* **228,** 893–895.

Südhof, T. C., Russell, D. W., Brown, M. S., and Goldstein, J. L. (1987a). 42 bp element from LDL receptor gene confers end-product repression by sterols when inserted into viral TK promoter. *Cell* **48,** 1061–1069.

Südhof, T. C., Van der Westhuyzen, D. R., Goldstein, J. L., Brown, M. S., and Russell, D. W. (1987b). Three direct repeats and a TATA-like sequence are required for regulated expression of the human low density lipoprotein receptor gene. *J. Biol. Chem.* **262,** 10773–10779.

Takaichi, S., Tajima, S., Miyake, Y., and Yamamoto, A. (1985). Histochemical characterization of low density lipoprotein receptors in internalization-defective familial hypercholesterolemia. *Arteriosclerosis* **5,** 238–243.

Taylor, R., Jeenah, M., Seed, M., and Humphries, S. (1988). Four DNA polymorphisms in the LDL receptor gene: their genetic relationship and use in the study of variation at the LDL receptor locus. *J. Med. Genet.* **25,** 653–659.

Tolleshaug, H., Goldstein, J. L., Schneider, W. J., and Brown, M. S. (1982). Posttranslational processing of the LDL receptor and its genetic disruption in familial hypercholesterolemia. *Cell* **30,** 715–724.

Tolleshaug, H., Hobgood, K. K., Brown, M. S., and Goldstein, J. L. (1983). The LDL receptor locus in familial hypercholesterolemia: multiple mutations disrupt transport and processing of a membrane receptor. *Cell* **32,** 941–951.

van Driel, I. R., Davis, C. G., Goldstein, J. L., and Brown, M. S. (1987a). Self-association of the low density lipoprotein receptor mediated by the cytoplasmic domain. *J. Biol. Chem.* **262,** 16127–16134.

van Driel, I. R., Goldstein, J. L., Südhof, T. C., and Brown, M. S. (1987b). First cysteine-rich repeat in ligand-binding domain of low density lipoprotein receptor binds Ca^{2+} and monoclonal antibodies but not lipoproteins. *J. Biol. Chem.* **262,** 17443–17449.

Wade, D. P., Knight, B. L., and Soutar, A. K. (1985). Detection of the low-density-lipoprotein receptor with biotin-low-density lipoprotein. A rapid new method for ligand blotting. *Biochem. J.* **229,** 785–790.

Walter, G., Scheidtman, K. H., Carbone, A., Laudano, A. P., and Doolittle, R. F. (1980). Antibodies specific for the carboxy- and amino-terminal regions of Simian virus 40 large tumor antigen. *Proc. Natl. Acad. Sci. U.S.A.* **75,** 5197–5200.

Whitehead, A. S., Solomon, E., Chambers, S., Bodmer, W. F., Dovey, S., and Fey, G. (1982). Assignment of the structural gene for the third component of human complement to chromosome 19. *Proc. Natl. Acad. Sci. U.S.A.* **79,** 5021–5025.

Yamamoto, K. R. (1985). Steroid receptor regulated transcription of specific genes and gene networks. *Ann. Rev. Genet.* **19,** 209–252.

Yamamoto, K. R., Payvar, F., Firestone, G. L., Maler, B. A., Wrange, O., Carlstedt-Duke, J., Gustafsson, J-A., and Chandler, Y. L. (1983). Biological activity of cloned mammary tumor virus DNA fragments that bind purified glucocorticoid receptor protein *in vivo*. *Cold Spring Harbor Symp. Quant. Biol.* **47,** 977–984.

Yamamoto, T., Davis, C. G., Brown, M. S., Schneider, W. J., Casey, M. L., Goldstein, J. L., and Russell, D. W. (1984). The human LDL receptor: a cysteine-rich protein with multiple Alu sequences in its mRNA. *Cell* **39,** 27–38.

Yamamoto, T., Bishop, R. W., Brown, M. S., Goldstein, J. L., and Russell, D. W. (1986). Deletion in cysteine-rich region of LDL receptor impedes transport to cell surface in WHHL rabbit. *Science* **232,** 1230–1237.

Zinn, K., and Maniatis, T. (1986). Detection of factors that interact with the human β-interferon regulatory region *in vivo* by DNAase I footprinting. *Cell* **45,** 611–618.

Chapter 10

Normal and Defective LDL-Receptor Function *in Vivo*

I. Introduction

The experiments described in Chapter 7 (Section II) show beyond doubt that the LDL receptor functions in the whole body under physiological conditions. They also provide quantitative information on the contributions of individual tissues to total catabolism of LDL via the LDL-receptor pathway *in vivo*. In particular, it has been shown that the liver makes by far the largest single contribution to LDL catabolism in most species and that hepatic catabolism of LDL is mediated predominantly by LDL receptors. However, there remains the question as to how far the behavior of LDL receptors *in vivo* resembles their behavior in cells *in vitro,* especially in skin fibroblasts in culture, the system from which so much of our knowledge is derived.

The LDL-binding affinity of LDL receptors on fibroblasts in culture is such that half-saturation is reached at about 10 μg of LDL protein/ml of extracellular medium at 37°C. Is this consistent with a significant contribution by LDL receptors *in vivo* to the catabolism of LDL in the extravascular fluids? As discussed in Chapters 3 and 8, and elsewhere in this book, sterol synthesis and LDL-receptor expression in cells *in vitro* are regulated coordinately so as to supply the cell with just the amount of cholesterol it needs for optimal growth, maintenance, and specialized function. Can a similar regulation of LDL-receptor expression be demonstrated in cells in the intact organism? These and other related questions are considered in this chapter. To some extent, this will mean bringing together points that have already been mentioned in a different context in earlier chapters.

It will be clear from the evidence assembled in Chapters 7, 8, and 9 that much of our understanding of the normal LDL receptor is based on the study of lipoprotein metabolism in isolated cells or whole bodies of human subjects and rabbits with genetically defective receptors. Indeed, it would be no exaggeration to say that the discovery of the LDL receptor was made possible by the existence of FH, and that at every stage in the investigation of the structure and function of LDL receptors the study of FH has yielded conclusive information. Accordingly, I shall end this chapter by focusing directly on the biochemical and pathological changes that occur in FH patients and WHHL rabbits as a consequence of LDL-receptor deficiency. I shall also discuss approaches to the diagnosis and treatment of FH that have arisen from studies of the LDL receptor and its gene. For a more detailed account of the clinical aspects of FH, see Myant (1981) and Goldstein and Brown (1983).

II. Comparison between LDL-Receptor Function *in Vitro* and *in Vivo*

A. Uptake as a Function of LDL Concentration

As we saw in Chapter 8, LDL receptors on fibroblasts in culture in the fully induced state are saturated when the LDL concentration in the medium reaches about 50 μg of LDL protein/ml. Moreover, when incubated for 48 hours in the presence of LDL at this concentration receptor activity on fibroblasts is suppressed. Hence, receptors on fibroblasts maintained indefinitely in the presence of human plasma (in which the LDL concentration is normally about 1000 μg/ml) would be expected to make little or no contribution to the total amount of LDL catabolized by the cells. In agreement with this, human circulating monocytes and lymphocytes express only a negligible level of high-affinity binding and catabolism of LDL when freshly isolated. However, numerous observations on intact animals have shown that extravascular extrahepatic tissues, including adrenal cortex, kidney, and intestine, catabolize LDL by the LDL-receptor pathway *in vivo* (see Chapter 7, Section II). This raises the question of the concentration of LDL in the fluid at the surfaces of cells in these tissues; the liver is a special case and is discussed below. Reichl *et al.* (1977) showed that the concentration of LDL in lymph draining the human foot is about 100 μg of protein/ml. If this is representative of the LDL concentration in interstitial fluid, LDL receptors on the cells of extravascular tissues would be repressed *in vivo* if their response to LDL in the extracellular medium is similar to that of fibroblasts in culture.

The fact that many extravascular tissues in a wide variety of animal species express functionally active LDL receptors *in vivo* could be due to a combination of several factors. The concentration of LDL in interstitial fluid may be much lower than that in peripheral lymph; the regulation of LDL-receptor activity in cells that have been through many passages in culture may differ from that in cells in the living body; the presence of growth factors and of hormones such as insulin may enable cells *in vivo* to express LDL receptors in the presence of LDL at relatively high concentrations; the ability to express receptors in the presence of LDL in the external medium may differ intrinsically from one cell type to another, exemplified by the difference between monocyte-macrophages and skin fibroblasts *in vitro* (see Chapter 8, Section III,D). In the present context it is worth noting that if the concentration of LDL in interstitial fluid is, in fact, below that required to saturate receptors on cell surfaces, a rise in LDL concentration would result in an increase in the rate of uptake and catabolism of LDL by the LDL-receptor pathway, as in the experiment *in vitro* shown in Fig. 8.4. At LDL concentrations above the level required to saturate LDL receptors, any further increase in LDL-receptor-mediated catabolism could only be achieved by increasing the number of receptors on the surfaces of cells.

The high level of LDL-receptor activity in the liver, demonstrated unequivocally *in vivo,* poses a special problem because of the high concentration of LDL that must exist in the space of Disse. This has already been mentioned in Chapter 8, Section IV. Several factors could contribute to the expression of LDL receptors on liver cells *in vivo* in the presence of LDL concentrations at which receptors on fibroblasts in culture would be repressed. As discussed in Chapter 8, LDL receptors are expressed by Hep G2 cells *in vitro* in the presence of LDL at high concentration (see Fig. 8.17). Moreover, Wade *et al.* (1988) have shown that insulin induces LDL receptors in Hep G2 cells incubated in the presence of LDL. If hepatocytes *in vivo* behave in these respects like Hep G2 cells *in vitro,* this would go some way toward explaining why the liver makes such a substantial contribution to LDL-receptor-dependent catabolism of LDL in the intact organism.

Another possibility that has a bearing on the behavior of hepatic LDL receptors *in vivo* is suggested by the experiments of Dietschy and Spady (1985) on the uptake of plasma LDL by the liver of the intact rat. Inspection of the curves shown in Fig. 7.5 (panel B) suggests that high-affinity uptake of native LDL does not approach saturation until the plasma LDL concentration rises to about 150 mg of LDL cholesterol/100 ml. If the LDL concentration in the space of Disse is only a little below that in plasma, this would indicate that the LDL-binding affinity of hepatic LDL receptors *in vivo* is much lower than that of receptors on fibroblasts *in vitro.* If this were so, hepatic LDL receptors would be operating below saturation over the normal range of plasma LDL concentration. In this case, clearance of LDL by the LDL-receptor pathway could greatly exceed that by nonsaturable pathways, as in Fig. 7.5 (panel A).

B. Ligand-Binding Specificity

As discussed in Chapter 8, LDL receptors, from whatever tissue they have been obtained, have been shown to have dual binding specificity when examined *in vitro.* They bind lipoproteins containing either apoB-100 or apoE in favorable conformation and their binding affinity for lipoproteins containing several molecules of apoE is many times higher than their affinity for LDL. These properties are exhibited by LDL receptors on the surfaces of intact cells or on solubilized subcellular membranes and are probably a consequence of the multiple ligand-binding sites on the native receptor molecule. LDL receptors exhibit these binding specificities *in vivo*. Many observations on intact animals have shown that LDL and apoE-containing lipoproteins, including HDL_c and VLDL remnants, are cleared selectively from the circulation by the LDL-receptor pathway (see Chapters 7 and 8 for details). These observations have also shown that the clearance of lipoproteins in which apoE is the receptor-binding ligand is higher than that of LDL. The dual specificity of LDL receptors *in vivo* is also reflected in the selective accumulation of LDL, VLDL remnants, and apoE-enriched HDL in the plasma of FH patients and WHHL rabbits (see below).

C. Coordinate Regulation

As a general rule, LDL-receptor activity and cholesterol synthesis in cells *in vitro* change in parallel in a variety of different conditions. A good example is the decrease in both cholesterol synthesis and receptor activity that occurs when cells are grown in a medium containing LPDS and are then switched to one containing LDL. Other examples are the parallel increase in receptor activity and cholesterol synthesis in cells that are dividing rapidly in response to a mitogenic stimulus, and the induction of HMG-CoA reductase and of LDL receptors when LDL is replaced by LPDS in the medium in which cells are growing. These and many other examples of coordinate regulation give biological meaning to the presence of a specific, sterol-sensitive, regulatory sequence common to the promoter regions of the genes for reductase and the LDL receptor.

Parallel changes in the rate of synthesis of cholesterol and in LDL-receptor number or activity have also been demonstrated in the tissues of intact animals or human subjects (see Chapter 7 for details). Examples are the increase in cholesterol synthesis and receptor number in the livers of young dogs given colestipol (Kovanen *et al.*, 1981), and the suppression of sterol synthesis and of LDL-receptor activity, determined by the continuous infusion procedure, in the livers of male hamsters fed cholesterol plus saturated fat (Spady and Dietschy, 1985). Another probable example of coordinate regulation of LDL-receptor activity *in vivo* is the stimulation of LDL-receptor-mediated catabolism of LDL by thyroid hormone in myxoedematous human subjects (Thompson *et al.*, 1981). There is ample evidence that thyroid hormone stimulates cholesterol synthesis in the liver by inducing HMG-CoA reductase (see Gibbons *et al.*, 1982).

In each of these cases, the regulation of LDL-receptor expression resembles that observed in cells *in vitro* and is best interpreted as a component of a coordinated response to a change in the cell's requirement for cholesterol. However, Dietschy and his co-workers have shown that in intact animals the regulation of LDL-receptor expression may, in some instances, be dissociated from that of cholesterol synthesis. Thus, the rate of synthesis of cholesterol in the small intestine (Stange and Dietschy, 1983) and liver (Spady *et al.*, 1985) of the intact rat can be varied experimentally within wide limits by cholestyramine treatment or cholesterol feeding with little or no effect on LDL-receptor activity. Likewise, the age-related decrease in the rate of synthesis of cholesterol in the liver and other tissues of adult rats is not accompanied by a significant change in hepatic uptake of plasma LDL by the LDL-receptor pathway (Stange and Dietschy, 1984). Spady and Dietschy (1985) point out that the rate of hepatic synthesis of cholesterol in rats is much higher than in most other species. They suggest that this enables the rat hepatocyte to vary its supply of cholesterol over a wide range by varying reductase activity without changing the rate of uptake of LDL from the external medium.

D. Regulation by Receptor-Independent Uptake of LDL

In most cells growing in culture, lipoprotein cholesterol internalized by receptor-independent pathways does not suppress HMG-CoA reductase activity. This was noted by Brown and Goldstein in their initial studies of human skin fibroblasts. In these studies it was shown that LDL does not suppress reductase in FH homozygote cells *in vitro,* even at LDL concentrations high enough to permit uptake and intracellular degradation of LDL by nonsaturable pathways at a substantial rate (see Fig. 8.4, panel E). This observation, later confirmed in other types of cells, has never been satisfactorily explained (see Chapter 8, Section II,B for discussion). As we shall see below, the rate of synthesis of cholesterol is not above normal in the whole bodies of FH homozygotes or in individual tissues of WHHL rabbits. Presumably, in the long term, the regulatory pool of cholesterol in the cells of FH homozygotes and WHHL rabbits is maintained at the normal level by LDL cholesterol taken up by nonsaturable pathways and by cholesterol synthesized within the cells. Whatever the mechanism by which a normal level of reductase activity is achieved *in vivo* in the absence of LDL receptors, it is clear that observations made over a limited period on cholesterol homeostasis in cells *in vitro* are not always applicable to the behavior of cells in the living body. It should also be noted that reductase activity in human monocyte-macrophages in culture is suppressed by LDL taken up by LDL-receptor-independent pathways (Knight *et al.*, 1983, and see Chapter 8, Section III,D).

III. Familial Hypercholesterolemia: A Consequence of LDL-Receptor Deficiency

A. Historical Background

By the 1950s FH was recognized as a familial disorder characterized by hypercholesterolemia, xanthomatosis, and premature heart disease (see Thannhauser,

1940, 1950). Once a clear distinction had been made between FH and other lipoprotein disorders, the way was open for investigation of its mode of inheritance and of the underlying metabolic abnormality. These investigations led to the demonstration that FH is due to mutation at a single gene locus (Khachadurian, 1964) and that the plasma abnormality responsible for the hypercholesterolemia is a selective increase in LDL concentration (Gofman *et al.*, 1954; Fredrickson *et al.*, 1967). This, in turn, led on to the discovery of the LDL-receptor and the identification of FH mutations at the level of the genome (see Brown and Goldstein, 1986).

B. Definition

Familial hypercholesterolemia is usually defined in terms of the major biochemical abnormality as *a monogenically inherited increase in plasma LDL concentration.* As noted below, other plasma lipoprotein abnormalities may also occur as a result of defective LDL-receptor function. However, these abnormalities are small in relation to the change in plasma LDL level and are not essential for a diagnosis of FH. In every instance in which a patient with FH in the homozygous form has been investigated at the molecular level, the primary abnormality has turned out to be defective LDL-receptor function due to mutation in the receptor gene. Hence, there is a case for defining FH in terms of the underlying molecular lesion as a *familial absence or deficiency* of LDL receptors. This would be consistent with the designation *familial defective apoB-100* (FDB) for the condition in which the molecular lesion underlying a raised plasma LDL concentration is a specific amino acid substitution in apoB-100 (Innerarity *et al.*, 1987, and see Chapter 6). The range of clinical expression of FDB cannot yet be assessed. However, it may well turn out that some individuals with the FDB mutation are clinically indistinguishable from "classical" FH patients with defective LDL-receptor function. Moreover, other mutations in the apoB gene nearer the region encoding the receptor-binding domain may, in the future, be found to produce true clinical copies of FH. In this case, it would be logical to regard FH as a clinical syndrome due either to defective receptor function (a receptor disease) or, more rarely, to an abnormal apoB-100 (a ligand disease). Whether or not the definition of FH will change, laboratory tests for the presence of known mutations in the apoB gene (see legend to Fig. 6.17) must now become routine in every lipid clinic.

Extending the term "FH homozygote" to include patients who are genetic compounds, as well as those who are truly homozygous at the receptor locus, is clinically useful and should be continued. Accordingly, FH homozygotes are defined here as *individuals with the severe form of the disorder who have inherited an FH allele from both parents, whether or not the two alleles are identical.* However, it is worth noting that it is sometimes necessary to identify each of the two FH alleles in a clinically homozygous individual, as in the analysis of haplotypes within families and also in population studies, e.g., in the investigation of a founder effect in isolated communities.

C. Evidence that FH is Monogenic

It is not always easy to work out the genetics of an inherited disorder that is detected as a change in a continuous variable, such as plasma cholesterol concentration, related only indirectly to the product of the mutant gene. However, Khachadurian (1964) was able to carry out definitive studies of the genetics of FH by taking advantage of conditions in Lebanon that were especially favorable for his investigations. This work, a landmark in the development of our understanding of FH, is worth placing on record.

The mean plasma cholesterol level in the population in Lebanon was low by Western standards at the time of Khachadurian's investigations. Hence, there was virtually no overlap between plasma cholesterol levels in carriers of the FH gene and the levels in the normal population. Secondly, the frequency of FH in Lebanon is exceptionally high. This is due to the presence of the Lebanese allele (see Chapter 9), which has achieved a high local frequency owing to a founder effect combined with a rate of consanguineous marriages equal to about 20%. Using plasma total cholesterol concentration as a marker for FH, Khachadurian (1964) examined the first-degree relatives of nearly 50 severely affected index patients. His observations showed that FH is a monogenic disorder inherited as an autosomal dominant trait, with more marked expression in homozygotes than in heterozygotes. Pedigree analysis was consistent with the conclusion that all his index patients were homozygotes. In the whole group of first-degree relatives the plasma cholesterol concentrations were distributed in three modes: normal levels, corresponding to individuals with two normal alleles at the FH locus; levels about twice as high as the normal, corresponding to heterozygotes; and levels about four times the normal, corresponding to homozygotes with two mutant alleles.

In several of Khachadurian's larger families, the trimodal distribution of plasma cholesterol levels may be seen within a single sibship, as in Fig. 10.1. All the features of the inheritance of FH are shown in this one family, in which the parents were first cousins. Three of the nine offspring were normal, four were heterozygotes [II (2), (3), (6), and (8)], and two were homozygotes [II (4) and (7)]. In keeping with the autosomal inheritance of FH, three of the affected offspring were female and three were male. The ratio of homozygous normal : heterozygotes : homozygous FH was 3 : 4 : 2. With a sibship of nine, this ratio could not be closer to the expected Mendelian ratio of 1 : 2 : 1 for the offspring of two heterozygotes. Both parents, who were obligate heterozygotes by virtue of the presence of homozygotes among their offspring, had plasma cholesterol levels close to the age-adjusted levels of their heterozygous children.

Additional evidence for monogenic inheritance of FH was obtained subse-

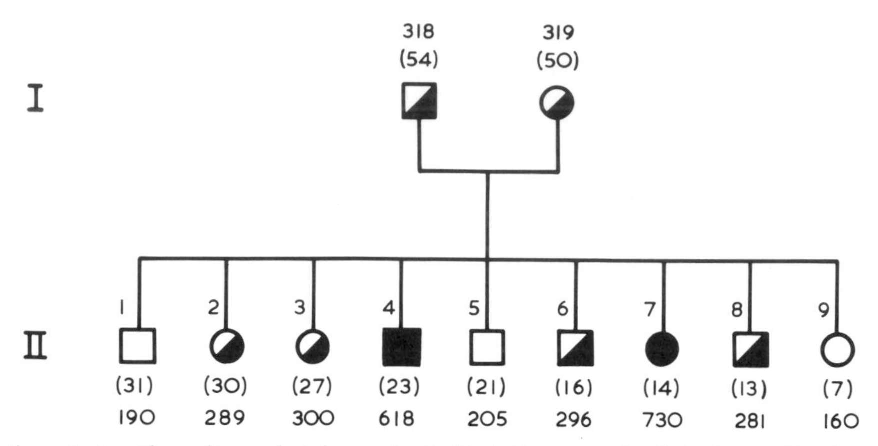

Figure 10.1. The pedigree of a Lebanese family illustrating the mode of inheritance of familial hypercholesterolemia. The parents were first cousins and both are presumed to have FH in the heterozygous form. II (1), (5), and (9) are normal; II (2), (3), (6), and (8) are heterozygotes; II (4) and (7) are homozygotes. Patients II (4) and (7) had extensive planar and tuberous skin xanthomas; the father had xanthelasmas. Numbers above parents and below offspring are ages (in parentheses) and plasma total cholesterol concentrations. (Modified from Khachadurian, 1964.)

quently by Schrott *et al.* (1972), who observed a bimodal distribution of plasma total cholesterol levels in a four-generation Alaskan FH family comprising 92 individuals. The results of a segregation analysis of matings between normal (N) parents and between presumed heterozygous (H) and normal parents were also consistent with monogenic inheritance of an autosomal dominant trait. N × N matings gave only normal offspring and N × H gave the expected 1 : 1 ratio for affected : unaffected offspring.

Taken together, the observations of Khachadurian (1964) and of Schrott *et al.* (1972) showed beyond doubt that FH is a monogenically inherited disorder and, hence, that the underlying molecular lesion is a mutation at a single gene locus. If FH were polygenic, the distribution of plasma cholesterol levels in the first-degree relatives of index patients would be unimodal, rather than bimodal, with a mean value higher than the normal mean.

D. Distribution and Frequency

FH is widely distributed throughout the world. It has been reported in most European countries and in Africa, America, Australia, Japan, the Middle East, and the Republic of China. The frequency of heterozygotes in Britain, estimated from the probable number of homozygotes in the whole population and on the assumption that the FH gene is in Hardy-Weinberg equilibrium in the British population, is about 1 : 500 (Myant and Slack, 1976). Goldstein *et al.* (1973), by analyzing a large sample of hypercholesterolemic survivors of myocardial infarction, arrived at a similar value for the frequency of FH heterozygotes in the white population of the United States. The observation of Mabuchi *et al.* (1978) suggest that the frequency of FH in Japan as a whole is comparable with that in Britain and the United States. The frequency of FH homozygotes in most of the world is probably less than 1 in 10^6.

In Lebanon and among the Afrikaners of Transvaal Province of South Africa, FH is much more common than in most other parts of the world. The frequency of heterozygotes in Lebanon is probably as high as 1 : 80 (Myant and Slack, 1976) and is about 1 : 100 in Transvaal Province (Seftel *et al.*, 1980). In both cases, genetic analysis of carriers of the FH mutation has shown that the unusually high prevalence of FH is due to a founder effect (see Chapter 9, Section V, for references). A founder effect has also been shown to have contributed to the prevalence of FH in two other genetically isolated populations—the Finns (Aalto-Setälä *et al.*, 1988) and the French-speaking population of Quebec Province in Canada (Hobbs *et al.*, 1987).

Even in populations where there has been no genetic isolation, as in Britain and throughout most of the United States, FH mutations are more common than the majority of other mutations that give rise to serious clinical disorders. The relatively high global frequency of FH may be due to a high mutation rate at the LDL-receptor locus, possibly related to the large number of Alu repeat sequences within the gene, coupled with the fact that the majority of female heterozygotes survive beyond the reproductive period.

E. Clinical, Biochemical, and Pathological Features

Thc FH mutation, in single or double dose, is expressed at birth as an increase in the plasma concentrations of total and LDL cholesterol. The subsequent course of the disease differs strikingly between heterozygotes and homozygotes. The two genotypes will therefore be dealt with separately.

1. Homozygotes

In homozygotes the plasma total cholesterol concentration is increased to more than four times the normal level, as shown in Table 10.1. The increase is due almost entirely to an increased plasma LDL-cholesterol concentration, which usually exceeds 600 mg/100 ml and is similar in males and females. The rise in plasma LDL concentration is accompanied by changes in the composition of LDL (Table 10.2). Expressed in terms of LDL protein, there is an increase in free and esterified cholesterol and a decrease in triglyceride, with no change in total phospholipid. The plasma concentrations of IDL (Soutar *et al.*, 1982) and of the apoE-rich fraction of HDL (Schmitz and Assmann, 1982) are also in-

Table 10.1
Lipid Concentrations in Fasting Plasma in FH, with Normal Values Shown for Comparison

		Total cholesterol (mg/100 ml)	Lipoprotein cholesterol (mg/100 ml)		
			LDL	VLDL	HDL
Normal[a] (30–40 years)	Male	201	135	19	43
	Female	190	116	14	60
Heterozygous[b] (30–40 years)	Male	369	299	29	40
	Female	365	269	24	48
	All ages, both sexes	340 (250–500)			
Homozygous[c] (1–19 years)		750 (600–1100)	625	19	34

[a]From various sources.
[b]Kwiterovich *et al.*, 1974.
[c]Fredrickson *et al.*, 1978.

creased, though the contribution of these changes to the increase in total cholesterol concentration is negligible. The increase in IDL and apoE-rich HDL concentrations gives rise to a twofold increase in plasma apoE concentration in FH heterozygotes (Blum *et al.*, 1980). Plasma VLDL concentration does not differ from that in people who are not carriers of an FH mutation. In homozygous FH, the plasma total HDL-cholesterol concentration is decreased in most, but not all,

Table 10.2
The Lipid Composition of LDL in Familial Hypercholesterolemia[a]

Lipid fraction		Normal	Heterozygote	Homozygote
Total cholesterol / protein	(W/W)	1.37	1.60	1.91
Phospholipid / protein	(W/W)	1.00	1.03 (NS)[b]	1.13 (NS)
Triglyceride / protein	(W/W)	0.28	—	0.10
Free cholesterol / protein	(Molar)	0.77	1.03	1.08
Sphingomyelin / protein	(W/W)	0.26	0.31	0.38
Lecithin / protein	(W/W)	0.80	0.69	0.62
Lecithin / sphingomyelin	(Molar)	2.7	2.1	1.4

[a]Values from Slack and Mills (1970), Shattil *et al.* (1977), and Jadhav and Thompson (1979).
[b]NS, not significantly different from normal. All other values for FH are significantly different from normal.

patients and is similar in males and females. Abnormal accumulation of chylomicrons and their remnants in the plasma after ingestion of a fatty meal has not been reported in homozygous FH.

The severe hypercholesterolemia present throughout life results in the accumulation of LDL-derived cholesterol in skin, tendons, and arterial walls, giving rise to cutaneous and tendon xanthomas, corneal arcus, and atheroma of the aorta, the coronaries, and other medium-sized arteries. The cutaneous xanthomas may be flat (planar) yellowish lesions or tuberous outgrowths. Planar xanthomas have a very characteristic distribution (see Myant, 1981, and Fig. 10.2 legend) and are seen in FH heterozygotes rarely, if at all.

The cardiovascular lesions of greatest clinical significance are xanthoma-like thickenings of the intima involving especially the root of the ascending aorta and the aortic valve cusps, causing narrowing of the openings of the coronary arteries and aortic stenosis. The intimal thickening at the base of the aorta gives rise to a characteristic appearance in aortograms (Fig. 10.3). Atherosclerosis of the coronary arteries tends to be most marked at their proximal ends, as in Fig. 10.3. These changes give rise to left ventricular hypertrophy and coronary occlusion, leading to early death from myocardial infarction. The clinical course of the disease is rather variable. In the homozygotes studied by Khachadurian (1972), planar xanthomas usually appeared in infancy; the median age at first appearance was about 4 years. Tendon xanthomas and arcus corneae appeared later, but usually before the end of the first decade. Age at death from heart disease is more variable and may occur at any age from infancy to the 30s or 40s. In Khachadurian's series the mean age at death from heart disease was 21 years. Variation in the severity of clinical signs and in the age at death in FH is a question of considerable interest and is discussed below.

The skin and tendon xanthomas consist largely of accumulations of foam cells surrounded by fibrous tissue containing extracellular deposits of free and esterified cholesterol. The foam cells of xanthomas are macrophage-like cells filled with droplets of lipid that are not surrounded by a bilayer membrane (Fig. 10.4). The major lipid in the foam-cell droplets is cholesterol esterified predominantly with oleic acid. The atheromatous lesions in the arteries of homozygous FH patients are similar in histological appearance to those seen in the general population, except for the predominance of foam cells in the unusual lesions affecting the aortic valves and the root of the aorta in FH.

Foam cells in atherosclerotic lesions were long thought to be derived from smooth-muscle cells that had migrated into the intima from the media (see Wissler, 1968, and Geer and Haust, 1972). However, more recent observations, including histochemical studies with monoclonal antibodies that react specifically with monocytes (Aqel *et al.*, 1984) or macrophages (Gown *et al.*, 1986) have shown that a substantial proportion of the foam cells in atherosclerotic lesions in human subjects and fat-fed animals is derived from circulating blood monocytes

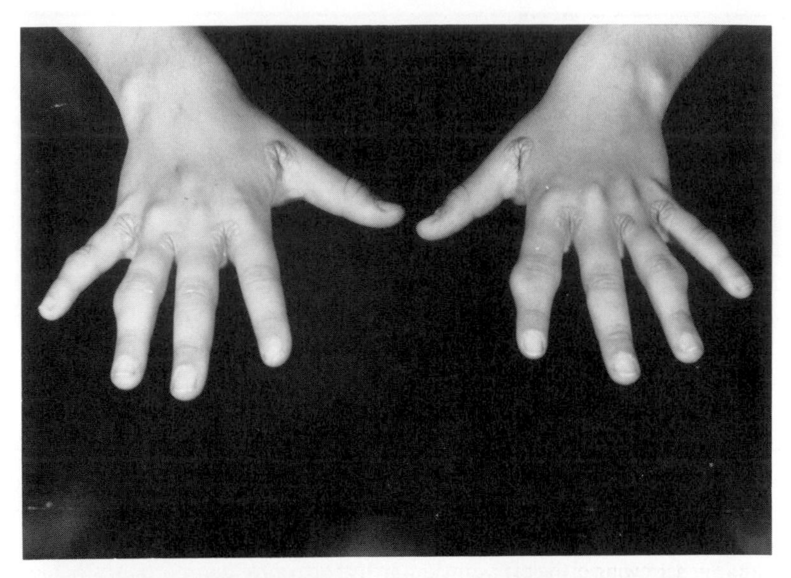

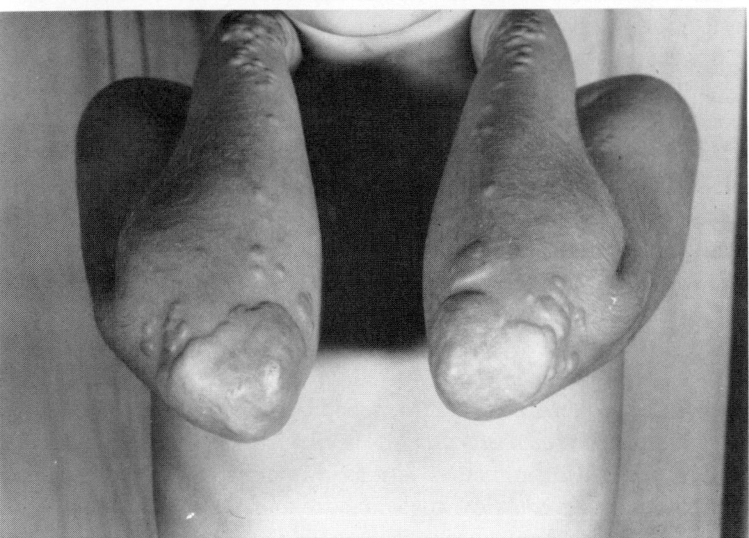

Figure 10.2. Planar skin xanthomas in two FH homozygotes. Upper panel: Xanthomas in the webs of the fingers of a 12-year-old receptor-defective boy; plasma total cholesterol concentration, 600–800 mg/100 ml before treatment. This patient died from cardiac failure at age 19. The parents were unrelated. Both had raised plasma cholesterol levels. Lower panel: Xanthomas over the elbows and forearms of a 10-year-old girl (genotype not established); plasma total cholesterol concentration, 900–1000 mg/100 ml before treatment. Angina was present at age 9 and death from coronary thrombosis occurred at age 11. The parents were first cousins and both had hypercholesterolemia. (From Myant, 1981.)

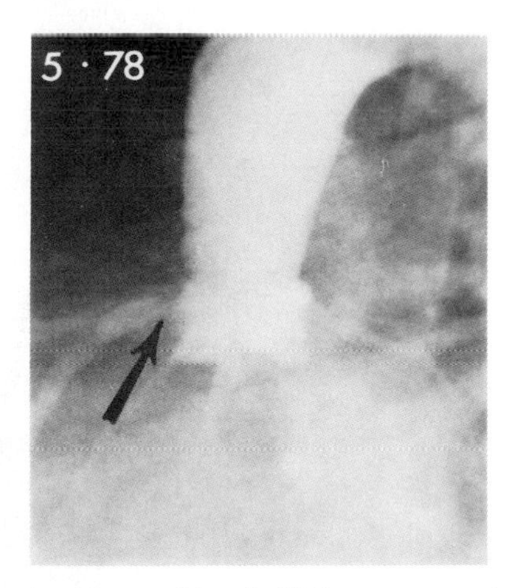

Figure 10.3. Aortogram of a 25-year-old male FH homozygote, showing the characteristic narrowing and irregularity of the lumen of the aortic root due to infiltration with xanthomatous lipid. Note the narrowing of the right coronary ostium (shown by the arrow). (From Thompson *et al.*, 1980.)

(see Ross, 1986, for discussion). This suggests that many of the foam cells in atherosclerotic lesions in FH are derived from monocytes that have crossed the arterial endothelium from the blood circulation into the subendothelial space. Presumably, the foam cells in skin and tendon xanthomas (in which there are no smooth-muscle cells) are derived entirely from cells of the macrophage class, either resident or originating from blood monocytes.

The massive accumulation of LDL-derived cholesterol within macrophage-like cells in the arterial wall and xanthomas in FH homozygotes raises the question as to whether or not cholesterol accumulates in other cells of the RE system in these patients. The evidence on this point is inconclusive. Buja *et al.* (1979) have summarized the findings from a dozen or so reports on FH homozygotes whose extravascular tissues were examined *post mortem* by microscopy. In some patients the lesions were confined to the cardiovascular system and the xanthomas in skin and tendons. In others, foam cells were present in additional tissues, including renal medullary papillae, lymph glands, spleen, liver, and bone marrow. It is noteworthy that in two patients, aged 11 and 22 years at death, lipid deposits in Kupffer cells were described as ''rare.''

2. Heterozygotes

Representative values for plasma lipid and lipoprotein concentrations in FH heterozygotes are shown in Table 10.1. Plasma total- and LDL-cholesterol concentrations are about twice the normal levels, but the range of values is considerable. Mean values for LDL cholesterol are similar in males and females in the

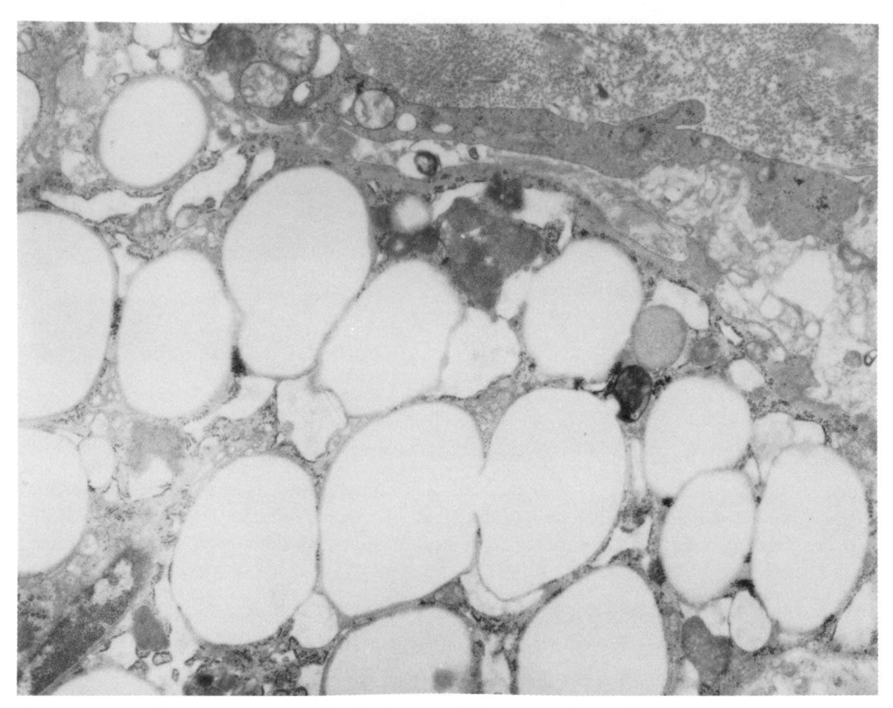

Figure 10.4. Electron micrograph of a foam cell from an atherosclerotic plaque taken from a patient with homozygous FH. Note that the lipid droplets are not surrounded by a bilaminar membrane (×15,110). (From Buja *et al.*, 1979, with the permission of the authors.)

age range within which women in the general population have lower plasma LDL levels than men. HDL-cholesterol concentration is slightly below normal in both sexes, but the sex difference observed in the general population (HDL-cholesterol level higher in women than in men) is present in FH heterozygotes. Changes in the composition of LDL, similar to those that occur in FH homozygotes but less marked, are seen in heterozygotes (see Table 10.2). Teng *et al.* (1986) have analyzed the "light" and "heavy" subfractions of LDL from normal subjects and FH heterozygotes, using the method of ultracentrifugation in a discontinuous density gradient. As mentioned in Chapter 5, this procedure separates LDL into two major subfractions differing in hydrated density. In FH heterozygotes, the proportion of the total LDL-apoB in the light fraction and the ratio of cholesterol to apoB in this fraction are both significantly higher than in normal subjects.

The severity of clinical signs and the age at which they appear are more variable in heterozygotes than in homozygotes. The natural history of heterozygous FH has been studied in single large families (Harlan *et al.*, 1966; Schrott *et al.*, 1972), in which all affected members within each family have inherited an identical FH allele, and in large numbers of patients from many different families (Slack, 1969; Stone *et al.*, 1974; Heiberg, 1975). Corneal arcus and xanthomas begin to appear toward the end of the second decade and are present in about 50% of all heterozygotes by age 30. In a small proportion of heterozygotes, tendon xanthomas are not detectable at the time of death from heart disease. Within the single large family studied by Schrott *et al.* (1972) the age at which tendon xanthomas first appeared ranged from the second decade to the end of the fourth decade.

Clinical signs of ischemic heart disease make their first appearance at a later age in heterozygotes than in homozygotes. Nevertheless, the prevalence of overt coronary artery disease in heterozygotes is many times greater than in the general population. For example, Jensen *et al.* (1967), in a study of 11 Danish families, found that ischemic heart disease was about 25 times as common in heterozygotes as in their unaffected relatives. Coronary artery disease develops at an earlier age in heterozygous men than in heterozygous women, despite their similar mean plasma LDL concentrations. Slack (1969), in a study of 104 heterozygotes, found that the mean age at which clinical signs of heart disease first appeared was 43 years in men and 53 years in women. A similar difference in mean age at death in heterozygous men (55 years) and women (64 years) was noted by Heiberg (1975).

F. Metabolic Changes

1. Cholesterol Metabolism

The rate of synthesis of cholesterol in the whole bodies of homozygous FH adults (Grundy and Ahrens, 1969) and children (Martin and Nestel, 1979) is not significantly higher than in age-matched normal subjects. Limited observations on sterol synthesis in freshly isolated tissues are also consistent with the conclusion that cholesterol synthesis *in vivo* is essentially normal in homozygous FH (see Myant, 1981). In agreement with this, comparable observations on WHHL rabbits (see below) have shown that HMG-CoA reductase activity is normal in most tissues of these animals. As discussed in Section II,D above, and in the section on lymphocytes in Chapter 8, normal suppression of cholesterol synthesis *in vivo* in the presence of defective LDL-receptor function is probably mediated by nonsaturable uptake of LDL present in the extracellular fluids at very high concentration, together with feedback suppression by cholesterol synthesized *in situ*.

In FH homozygote fibroblasts grown in the presence of LDL, ACAT is inactive and the cholesteryl-ester content of the cells is only a small fraction of that in normal cells grown under the same conditions (Brown *et al.*, 1975). There is no evidence to suggest that there is a similar deficiency in the synthesis or cell content of cholesteryl esters in FH patients *in vivo*. Presumably, cholesterol delivered to cells by nonsaturable uptake of LDL is capable of activating ACAT under conditions *in vivo*. The massive accumulation of esterified cholesterol in cells of the macrophage class in FH patients calls for a different explanation and is discussed below.

2. LDL Catabolism

Langer *et al.* (1972) provided the first clear indication that the underlying metabolic lesion in FH is defective catabolism of LDL, due to a defect in the catabolic mechanism itself. They showed that the FCR of ^{125}I-labeled autologous LDL is decreased in FH heterozygotes (Fig. 10.5) and that normal LDL and LDL from FH heterozygotes are catabolized at the same reduced rates when injected intravenously into an FH heterozygote. Subsequent observations by other workers (Reichl *et al.*, 1974; Simons *et al.*, 1975; Bilheimer *et al.*, 1975) showed that there is an even greater reduction in the FCR of LDL in homozygotes and that in these patients there is also a marked increase in the rate of production of LDL. Langer *et al.* (1972) had found normal rates of production of LDL in their heterozygous patients. Others, however, have since reported moderate increases in LDL-production rate in heterozygotes (Packard *et al.*, 1976, Bilheimer *et al.*, 1979).

Reichl *et al.* (1974), in their study of a receptor-negative FH homozygote, prepared radioiodine-labeled LDL from the patient and from a normal subject. Both LDLs were then injected simultaneously into the patient and the normal subject. As shown in Fig. 10.6, the two LDLs were catabolized at identical rates in both recipients (normal in the normal subject and slow in the homozygote). Thus, the reduced FCR of autologous LDL in the patient was due entirely to a defect in the catabolic mechanism. It may also be concluded that the abnormal composition of the patient's LDL (see Table 10.2) was without effect on its rate of catabolism *in vivo*.

Representative values for the fractional and absolute rates of catabolism of LDL-apoB in normal subjects and FH patients are shown in Table 10.3, together with other values mentioned below. Since all measurements were made in the steady state, the rates of production of LDL are in each case equal to the absolute catabolic rates (ACR). As discussed in Chapter 7, estimates of the FCR and ACR of LDL-apoB in human subjects vary quite widely from one laboratory to another, probably owing to differences in the methods used and in the management of the subjects while the observations were made. In the studies of Simons *et al.* (1975) and Bilheimer *et al.* (1979), for example, mean values for the total FCR of LDL-apoB in normal subjects were 0.45 to 0.50 pools/day and the FCRs in FH heterozygotes and homozygotes were correspondingly higher than those shown in Table 10.3. However, in all published reports the FCR in normal subjects is 1.5 to 2.0 times that in FH heterozygotes and about three times that in homozygotes.

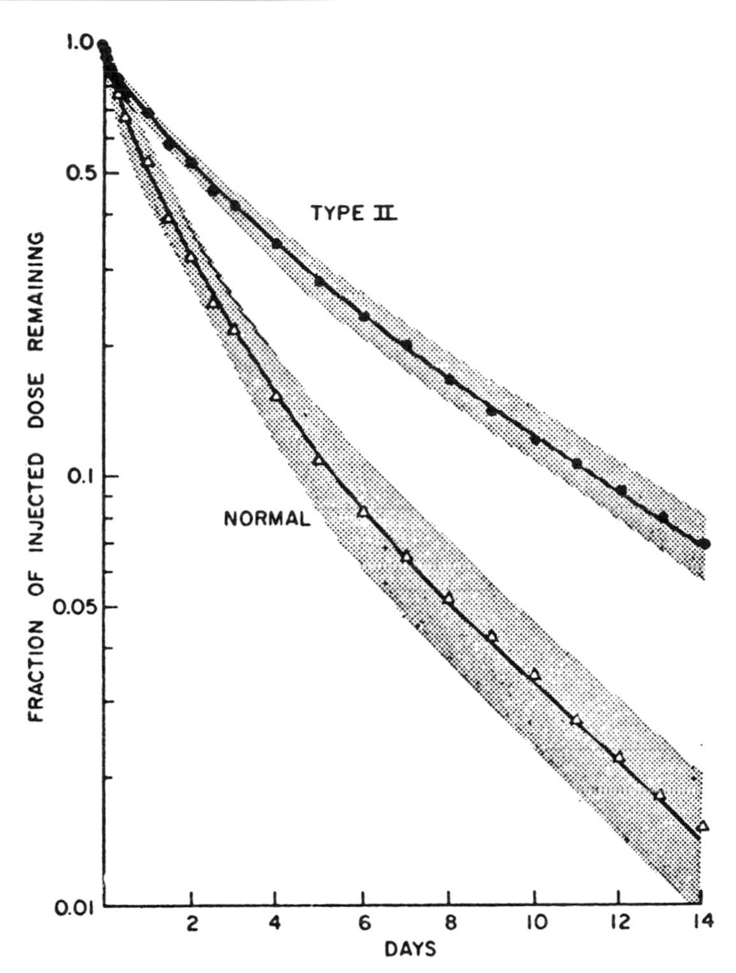

Figure 10.5. Plasma radioactivity after intravenous injection of ^{125}I-labeled autologous LDL into normal subjects (△) and patients with heterozygous FH (●) (designated type II). Each curve is the mean of curves from 10 subjects. Shaded areas indicate 1 standard deviation on either side of the mean. (From Langer *et al.*, 1972. Reproduced from the *Journal of Clinical Investigation*, 1972, Vol. 51, pp. 1528–1536 by copyright permission of the American Society for Clinical Investigation.)

Thompson *et al.* (1977) have examined the possibility that the increased production of LDL in FH contributes to the reduced FCR of LDL by saturating a catabolic mechanism that is not saturated at normal plasma LDL concentrations. They made serial measurements of the FCR of autologous LDL in FH homozygotes while the plasma LDL concentration was rising rapidly from near-normal levels after routine plasmapheresis to lower the plasma LDL level. Figure

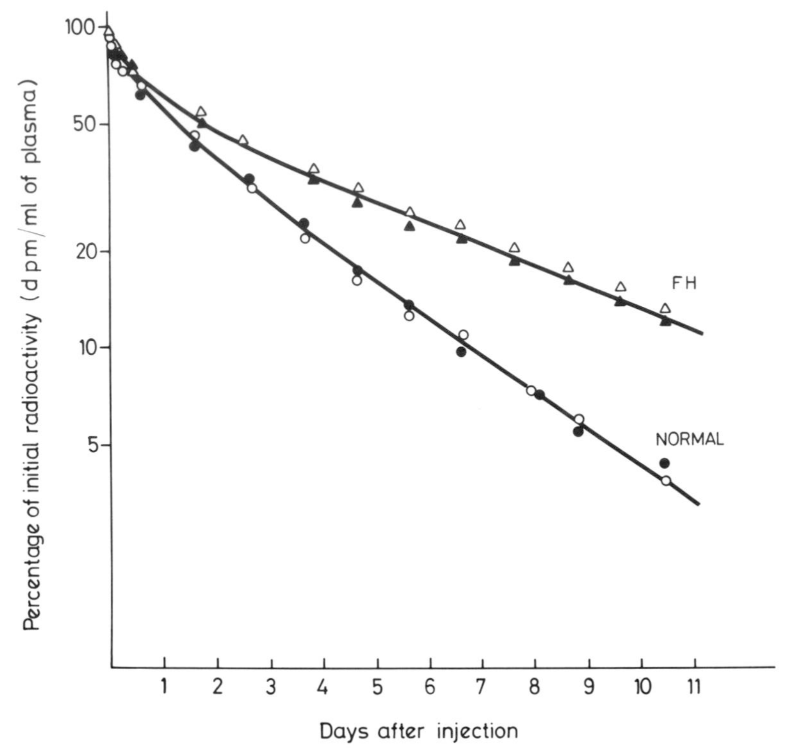

Figure 10.6. Turnover of LDL in a patient with homozygous FH (▲,△) and in a normal recipient (●,○). LDL from the patient (▲,●) was labeled with ^{125}I; LDL from the normal donor (△,○) was labeled with ^{131}I. The two labeled samples were injected simultaneously at zero time into the patient and the normal subject. (From Reichl *et al.*, 1974).

10.7 shows a representative series of observations on a female homozygote. When the plasma LDL concentration rose almost twofold, the FCR remained essentially constant at about 0.12 pools/day. These observations indicate that in the absence of the LDL-receptor pathway LDL is catabolized by mechanisms that are not saturated at the high plasma LDL levels seen in homozygous FH.

3. The Mechanism of Increased Production of LDL

The mechanisms responsible for the increased rate of production of LDL in FH have not been completely elucidated. Soutar *et al.* (1977) determined the rates of turnover (mg/kg/day) of VLDL-apoB and LDL-apoB in three FH homozygotes, using the single-pool model for VLDL-apoB and the Matthews model for LDL-apoB (see Chapter 7, Section I,C). In all three patients the rate of turnover of LDL-apoB was about twice that of VLDL-apoB, indicating that about half the LDL produced was not derived from VLDL. As discussed in Chapter 7, esti-

Table 10.3
Fractional and Absolute Catabolic Rates of LDL-ApoB via the LDL-Receptor-Dependent and LDL-Receptor-Independent Pathways in Normal Human Subjects, in FH, and in Myxoedema

Condition	Plasma LDL cholesterol (mg/100 ml)	FCR[a] (pools/day)			ACR[b] (mg/kg/day)			Reference
		Total	Receptor-dependent	Receptor-independent	Total	Receptor-dependent	Receptor-independent	
Normal	102	0.33	0.11	0.22	9.5	3.0	6.5	Shepherd *et al.* (1979)
FH heterozygote	342	0.19	0.03	0.16	16.1	2.8	13.3	Shepherd *et al.* (1979)
FH heterozygote on cholestyramine	267	0.24	0.07	0.17	16.5	4.8	11.7	Shepherd *et al.* (1980)
FH homozygote	510	0.08	0	0.08	15.5	0	15.5	Thompson *et al.* (1981)
Normal off cholestyramine	98	0.33	0.06	0.26	10.1	1.8	8.3	Thompson *et al.* (1981)
Normal on cholestyramine	66	0.35	0.14	0.21	7.1	2.5	4.6	Thompson *et al.* (1981)
Myxoedema	319	0.11	0.04	0.07	11.6	2.5	9.1	Thompson *et al.* (1981)
Myxoedema on T4[c]	137	0.29	0.14	0.15	20.1	9.7	10.4	Thompson *et al.* (1981)

[a]FCR, fractional catabolic rate.
[b]ACR, absolute catabolic rate.
[c]T4, thyroxine

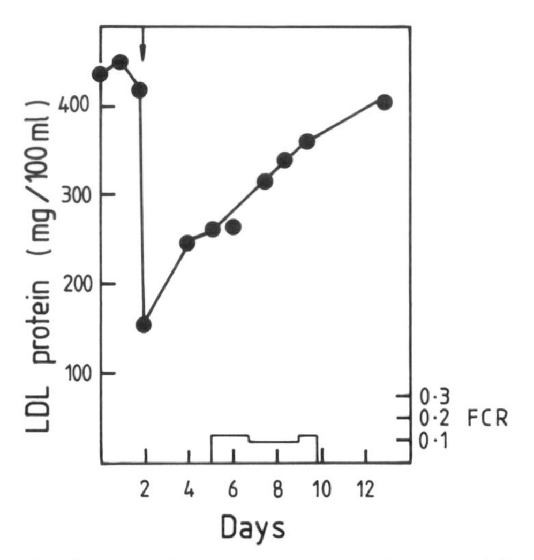

Figure 10.7. Relationship between the concentration in plasma and the fractional catabolic rate of LDL-apoB in an FH homozygote. At the time shown by the arrow, the patient was given a 3-liter plasma exchange. This caused a fall in plasma LDL concentration from 413 to 170 mg of protein/100 ml. During the subsequent rise in plasma LDL level the fractional catabolic rate of LDL, estimated by the U/P method (see Thompson *et al.*, 1977), did not change significantly. (From Thompson *et al.*, 1977.)

mates of VLDL-apoB turnover, based on the assumption that VLDL in the plasma is metabolically homogeneous, are subject to error. Nevertheless, additional evidence for VLDL-independent production of LDL in FH has been obtained by Janus *et al.* (1980), using a different experimental approach. In a group of FH heterozygotes, Janus *et al.* determined the rate of conversion of VLDL-apoB into LDL (mg/kg/day) by deconvolution analysis, and the net rate of production of LDL-apoB by the Matthews method. In the six patients examined, an average of 42% of the total LDL-apoB produced was not derived from VLDL. Thus, despite the uncertainties involved in kinetic analysis of lipoprotein metabolism (see Chapter 7 for discussion), it seems clear that VLDL-independent production of LDL occurs in both heterozygous and homozygous FH.

Soutar *et al.* (1979, 1982) tried to determine the nature of the particles responsible for VLDL-independent formation of LDL in FH by examining the precursor–product relationships between VLDL, IDL, and LDL in heterozygous and homozygous FH patients. Each patient was given an intravenous injection of radioiodine-labeled VLDL, followed by serial measurements of the specific radioactivity of apoB in plasma VLDL, IDL, and LDL. In both heterozygotes and homozygotes the VLDL and IDL curves intersected at or near the maximum of the IDL curve, indicating that all or most of the IDL produced in the plasma in FH is derived from VLDL. However, in each of the four homozygotes examined, the LDL curve reached its maximum before crossing the IDL curve, as in the example shown in Fig. 10.8, showing that the radioactive LDL-apoB was diluted with unlabeled LDL-apoB not derived from IDL. Soutar *et al.* concluded that a proportion of the LDL in the plasma of these patients was secreted directly into the circulation, presumably by the liver, as particles with the density of LDL. A rough calculation (Soutar *et al.*, 1982) suggested that only about 10% of the LDL produced in the FH homozygote shown in Fig. 10.8 was derived from IDL.

Teng *et al.* (1986) carried out a more detailed analysis of the metabolism of the plasma apoB in FH heterozygotes, using deconvolution analysis to estimate the proportion of light LDL converted into heavy LDL and the Matthews method to estimate total LDL production rate. Their findings showed that IDL-independent production of LDL occurs in these patients and that in some there is direct secretion of heavy LDL into the circulation. As discussed below, the rate of production of LDL is markedly increased in WHHL rabbits. Much, if not all, of this increase is attributable to reduced LDL-receptor-mediated uptake of IDL by the liver. In FH, the clearance rate of IDL from the plasma is decreased (Soutar *et al.*, 1982), suggesting that a mechanism similar to that in WHHL rabbits contributes to increased production of LDL in this disorder.

4. The LDL-Receptor Pathway in FH

The rates of catabolism of LDL-apoB via the LDL-receptor-dependent and LDL-receptor-independent pathways in FH patients have been estimated by the CHD method in several laboratories (see Chapter 7). Figure 10.9 shows the catabolism of native and CHD-modified radioactive LDL in a normal subject and a patient with homozygous FH. In the normal subject the FCRs of native and CHD-modified LDL were 0.353 and 0.212 pools/day, respectively. Hence, the FCR via the LDL-receptor pathway, estimated as the difference between the two FCRs, was 0.141 pools/day. In the FH homozygote the FCRs of native and modified LDL were almost identical, indicating that in this patient there was no catabolism of LDL via the LDL-receptor route.

Table 10.3 shows representative values for the fractional and absolute rates of catabolism of LDL in various conditions. For reasons discussed in Chapter 7, the contribution of the LDL-receptor pathway to total catabolism of LDL is probably underestimated by the CHD method. It was also pointed out above that estimates of the catabolic rates of LDL in human subjects differ between laboratories. Despite these limitations, the values shown in Table 10.3 illustrate the quantitative changes that occur in LDL catabolism *in vivo* when LDL-receptor activity is changed genetically or nongenetically.

In heterozygous FH the two- or threefold rise in plasma LDL concentration is accompanied by a roughly equivalent fall in FCR via the LDL-receptor pathway, with little or no change in FCR via LDL-receptor-independent pathways. The net

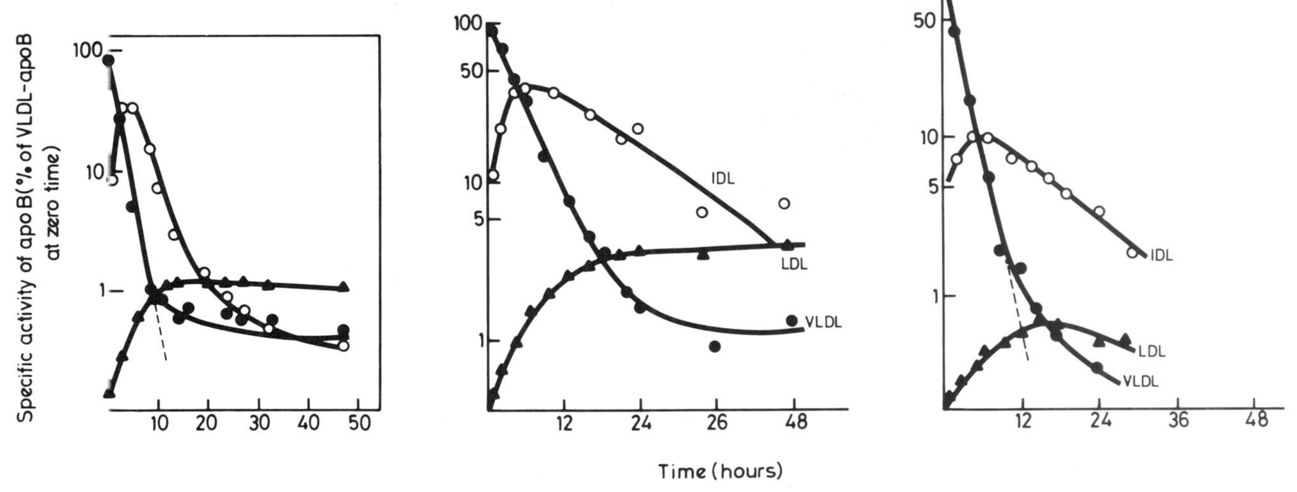

Figure 10.8. Specific radioactivity of apoB in VLDL (●), IDL (○), and LDL (▲) in the plasma of a normal subject (left), an FH heterozygote (center), and a homozygote (right) after intravenous injection of radioiodine-labeled VLDL. Autologous VLDL was labeled with ^{125}I before injection into the subject. Lipoproteins were isolated from serial samples of plasma by sequential ultracentrifugation at appropriate densities. Specific radioactivity was measured in apoB isolated from each sample of VLDL or IDL by column chromatography with Sephadex G150. Specific radioactivity of LDL apoB was determined from the radioactivity per milligram of total LDL protein in each sample. (From Soutar *et al.*, 1979, 1982.)

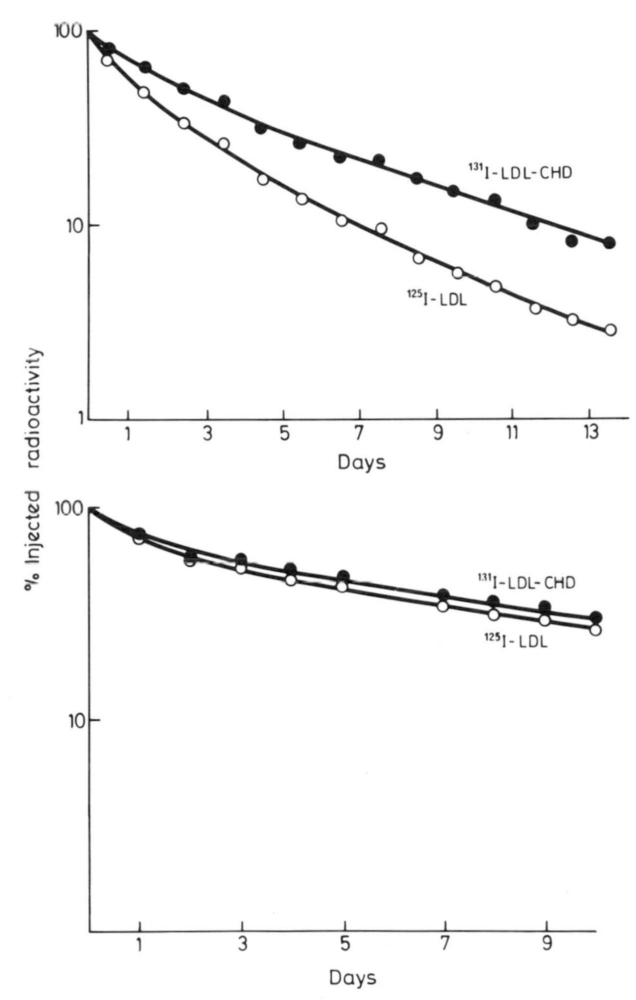

Figure 10.9. The effect of blocking arginine residues in apoB on the catabolism of the plasma LDL. Upper panel shows the plasma radioactivity in a normal subject after simultaneous intravenous injection of ^{125}I-labeled autologous LDL (○) and cyclohexanedione-treated ^{131}I-labeled autologous LDL (●) ([^{131}I]LDL-CHD). Lower panel shows the same experiment in a patient with receptor-defective homozygous FH. (From Shepherd *et al.*, 1979, and Thompson *et al.*, 1981, with the permission of the authors.)

result is that the absolute catabolic rate of LDL (FCR × plasma LDL concentration) by the LDL-receptor pathway remains essentially unchanged, while that by LDL-receptor-independent pathways increases markedly. This results in an increase in the total catabolic rate of LDL (from 9.5 to 16.1 mg/kg/day in the study of Shepherd *et al.*, 1979). In the FH homozygote shown in Table 10.3, no LDL was catabolized via the LDL-receptor route and the FCR via the LDL-receptor-independent route was less than half that in the normal subjects studied by Shepherd *et al.* (1979). However, the fivefold increase in plasma LDL concentration more than compensated for the fall in FCR, resulting in a net increase in absolute catabolic rate by all pathways.

As noted above, FH heterozygotes catabolize LDL via LDL receptors at the same rate as normal subjects, but at a two- to threefold higher plasma LDL concentration. This would be expected if the behavior of LDL receptors on cells *in vivo* resembles their behavior on fibroblasts *in vitro* and if the concentration of LDL in the fluid surrounding cells in normal subjects and FH heterozygotes is below that required to saturate LDL receptors. Figure 10.10 shows the rate of catabolism of LDL by normal and FH heterozygote fibroblasts in culture, as a function of LDL concentration in the medium. At LDL concentrations *above* the level required to saturate receptors, the heterozygote cells (being able to express only half the normal number of LDL receptors) catabolize LDL at half the normal rate. However, at LDL concentrations *below* saturation, heterozygote cells in the presence of LDL at a given concentration catabolize LDL at the same rate as normal cells in the presence of half that concentration (see broken line in Fig. 10.10). Thus, the normal absolute rate of catabolism of LDL by the LDL-receptor pathway in FH heterozygotes is consistent with the conclusion that in these patients the bulk of the cells responsible for LDL-receptor-mediated cata-

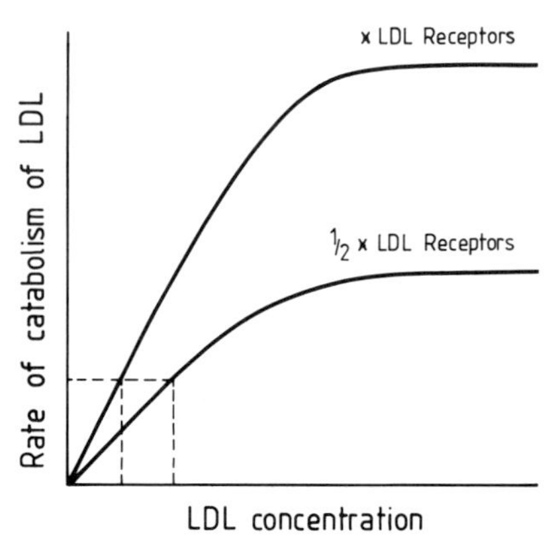

Figure 10.10. Idealized curves showing the rate of catabolism of LDL as a function of LDL concentration by normal cells (upper curve) and cells from an FH heterozygote (lower curve). In the presence of saturating concentrations of LDL, the rate of catabolism of LDL by the normal cells is twice that by the FH heterozygote cells. At concentrations below saturation, the normal and heterozygote cells catabolize LDL at the same rate when the concentration of LDL in the medium surrounding the heterozygote cells is twice that in the medium surrounding normal cells (see broken lines). (Modified from Brown and Goldstein, 1976.)

bolism are surrounded by fluid in which the LDL concentration is below that needed to saturate receptors.

The FCR of LDL in the homozygote shown in Table 10.3, who expressed no LDL-receptor activity *in vivo,* was lower than the LDL-receptor-independent FCR in normal subjects and FH heterozygotes. This might suggest that catabolism of LDL by independent routes includes a component that is saturated at very high plasma LDL concentrations. However, the FCR in this patient (0.08 pools/day) may not be typical of homozygotes in general. Much higher FCRs in homozygotes, ranging from 0.15 to 0.20 pools/day, have been reported by Simons *et al.* (1975) and by Bilheimer *et al.* (1979). Moreover, Bilheimer *et al.* (1975) and Thompson *et al.* (1977) (see Fig. 10.7) have shown that the FCR of LDL in FH homozygotes does not increase when the plasma LDL level is lowered acutely to within, or close to, the normal range by a portacaval shunt or plasmapheresis. The possibility that LDL-receptor-independent catabolism of LDL in man *in vivo* includes an adsorptive process was discussed in Chapter 7 (Section II,E).

The effects of cholestyramine on LDL catabolism are of particular interest in relation to the treatment of FH. As we saw in Chapter 7, cholestyramine lowers the plasma LDL concentration in FH heterozygotes by selectively increasing the FCR of LDL via the LDL-receptor route. This increase is probably a consequence of induction of LDL receptors on liver cells in response to depletion of hepatic cholesterol, itself a result of increased conversion of cholesterol into bile acids. (The one normal receptor gene in heterozygote cells can be stimulated to produce more receptors in conditions under which gene expression is submaximal.) As shown in Table 10.3, cholestyramine also stimulates LDL catabolism in normal human subjects. However, cholestyramine does not lower the plasma LDL concentration in FH homozygotes, even at a dose level high enough to cause a maximal increase in bile-acid synthesis (Moutafis *et al.*, 1977). This is readily explained by the absence of normal LDL-receptor genes in these patients.

The effect of cholestyramine on the absolute rates of catabolism of LDL by the two pathways is shown in Table 10.3. In the FH heterozygotes and the normal subject, the increase in FCR by the LDL-receptor pathway led to an increase in the absolute rate of catabolism of LDL by this pathway, despite the fall in plasma LDL concentration. However, the fall in LDL level caused a significant decrease in the absolute rate of catabolism via the LDL-receptor-independent route. Hence, an important consequence of the selective stimulation of LDL receptors is to divert LDL from non-LDL-receptor routes to the receptor pathway. In Chapter 8, it was suggested that foam cells present in atherosclerotic lesions are derived in part from macrophages that have taken up modified LDL by a pathway not involving LDL receptors. If this is so, cholestyramine, in addition to reducing the plasma LDL concentration, may divert LDL from a catabolic pathway that is potentially atherogenic.

The response of a myxoedematous patient to treatment with thyroid hormone was mentioned in Chapter 8, Section IV (and see Fig. 8.22). Table 10.3 shows the fractional and absolute catabolic rates of LDL in this patient before and after treatment. Note that the absolute rate of catabolism almost doubled after treatment and that the increase was due to an increase in the rate of catabolism of LDL via the LDL-receptor pathway.

G. Explanation of Metabolic and Clinical Features in Terms of Receptor Deficiency

1. Metabolic

The increase in plasma LDL concentration in FH is readily explained by partial or complete loss of a pathway normally responsible for 50% or more of total LDL catabolism, combined with increased LDL production. FH cells *in vitro* exhibit a gene-dosage effect; cells from heterozygotes express half the normal number of LDL receptors, while those from homozygotes express none or a very small number. The relative numbers of LDL receptors expressed in the whole bodies of FH patients and normal subjects cannot be determined *in vivo*. However, a gene-dosage effect may be discerned in FH patients insofar as the abnormalities in plasma LDL concentration, FCR of LDL, and LDL production rate are all greater in homozygotes than in heterozygotes. As noted above, the absolute catabolic rate of LDL by the receptor pathway in heterozygotes, whose plasma LDL concentration is about twice the normal level, is roughly equal to that in normal subjects. This is similar to what is observed in cells *in vitro* (see Fig. 10.10) and is consistent with the conclusion that the total number of LDL receptors expressed *in vivo* by heterozygotes is half that expressed by normal subjects.

The increase in plasma concentrations of LDL and apoE-enriched HDL is presumably a consequence primarily of the reduced number of LDL receptors expressed by hepatocytes.

The altered composition of LDL, more marked in homozygotes than in heterozygotes, is probably a consequence of increased residence time of LDL particles in the circulation, itself a consequence of reduced FCR. In keeping with this, the compositional changes are reversible. Thus, Jadhav and Thompson (1979) have shown that the composition of LDL in the plasma of FH homozygotes changes toward the normal when the plasma LDL concentration is rising rapidly after a plasma exchange. Under these conditions, the proportion of newly produced to "old" LDL particles in the plasma is higher than in the steady state. There is also some evidence to suggest that LDL receptors are required for the normal conversion of light (cholesterol-rich) LDL into heavy LDL (Teng *et al.*, 1986). Delay in this step would accentuate the effect of LDL-receptor deficiency on the composition of total LDL in the circulation. The mechanism by which prolonged residence in the plasma results in the changes in LDL shown in Table 10.2 is not completely understood. However, it is reasonable to suppose that a long residence time would favor the transfer of esterified cholesterol from HDL to LDL by cholesteryl-ester transfer protein. This would also help to explain why plasma HDL-cholesterol concentration is usually reduced in FH.

As we saw in Section III,F above, the increased rate of production of LDL that occurs in FH probably results from direct secretion of LDL into the circulation, combined with decreased LDL-receptor-mediated uptake of IDL by the liver. In normal human subjects, all the LDL entering the plasma is derived from circulating IDL, and direct secretion of LDL does not take place (see Chapter 7). So why does deficiency of LDL receptors lead to secretion of preformed LDL particles? One possibility is that LDL receptors on hepatocytes in some way inhibit secretion of LDL particles and that this mechanism for controlling LDL secretion breaks down when the liver fails to express a normal number of receptors. There is no experimental evidence to support this speculation.

2. Clinical

The atherosclerotic lesions in heterozygous FH are similar in their morphology and distribution to the lesions that occur in the population as a whole. The higher incidence, greater severity, and earlier onset of CAD in carriers of the FH gene must be due largely to the life-long elevation of plasma LDL concentration. However, it is possible that the rise in IDL concentration contributes to the development of atherosclerosis in FH heterozygotes. The unusual lesions in the aortic valves and ascending aorta seen in homozygotes are best interpreted as a consequence of the very high plasma LDL level present from birth in these patients, although, again, the rise in plasma IDL level may be a contributory factor. Aortic intimal lesions, similar to those in FH homozygotes, develop in rats fed high-fat diets for long enough to produce marked hypercholesterolemia (Záhoř and Czabanová, 1980).

The development of xanthomas in skin and tendons may be explained by the accumulation of LDL-derived cholesterol in the macrophages of these tissues. In keeping with this, xanthomas in FH patients decrease in size when the plasma LDL level is reduced by cholestyramine treatment or plasma exchange. Tendon xanthomas, which develop sooner or later in most FH patients, are rare in patients with other forms of hyperlipidemia of comparable degree. (In cerebrotendinous xanthomatosis, a condition in which xanthomas develop in the tendons and central nervous system, plasma lipoprotein concentrations are normal.) The extent to which tendon xanthomas develop may depend partly on the physical properties, including size, of the lipoprotein particles that accumulate in the plasma in different hyperlipidemias. No satisfactory explanation can be offered for the very characteristic distribution of skin xanthomas in FH. In this disease, skin xanthomas tend to occur in areas subjected to injury or pressure. However, this does not explain the common occurrence of planar xanthomas in restricted areas, such as the webs of the fingers of homozygotes (see Fig. 10.2 for a striking example).

In Chapter 8 (Section III,D) I discussed at length the probable mechanism by which LDL-derived cholesterol accumulates in foam cells. In brief, it was suggested that LDL in the vicinity of vascular endothelial and smooth-muscle cells is converted into an electronegative particle that is taken up by cells of the RE system in which acetyl-LDL receptors are expressed. Because the acetyl-LDL-receptor pathway is not regulated, this results in the accumulation of cholesterol within the cells. This mechanism would explain the formation of foam cells in xanthomas, all of which appear to be derived from cells of the macrophage class, and it would also explain the formation of foam cells derived from monocyte-macrophages in atherosclerotic lesions. However, the formation of foam cells from smooth-muscle cells in atherosclerotic lesions poses a problem, in that these cells do not normally express acetyl-LDL receptors, nor do smooth-muscle cells in FH homozygotes express LDL receptors. Hence, if smooth-muscle cells are, indeed, the precursors of some foam cells in the arterial wall, they must develop an unregulated pathway by which they acquire LDL cholesterol. One possibility is that, after migrating into the intima, they develop the capacity to express acetyl-LDL receptors.

H. Variable Expression

Variability in plasma LDL concentration and in age at onset of clinical signs in FH was mentioned in Section E above.

1. Homozygotes

The variability observed within the homozygote population is less striking than that in heterozygotes, suggesting that a double dose of the FH gene tends to override the effects of environmental factors, and of other genes, on the plasma LDL concentration. In homozygotes, gender has no effect on plasma LDL concentration. Nor does it have any effect on age at death from coronary artery disease, possibly because the sex difference in plasma HDL concentration in the general population is not apparent in homozygous FH (Seftel *et al.*, 1980). Moreover, the mean plasma cholesterol concentrations are closely similar in FH homozygotes from different countries in which plasma cholesterol levels in the general population differ widely, largely as a result of differences in habitual diet. Nevertheless, homozygotes do exhibit considerable person-to-person variation in plasma LDL level (see Table 10.1), in age at onset of skin and tendon xanthomas, and in age at death.

One factor that may help to determine the severity of the disease in homozygotes is the nature of the mutation(s) at the receptor locus. Goldstein and Brown (1982) compared the clinical course of FH in receptor-negative and receptor-defective homozygotes. In the receptor-negative group, signs of ischemic heart disease appeared at an earlier average age and the coronary death rate was higher than in the receptor-defective patients. In agreement with this observation Sprecher *et al.* (1985) found a positive correlation between age at onset of cardiovascular symptoms and residual LDL-receptor activity, determined in cells *in vitro,* in a group of FH homozygotes. These observations suggest that some FH mutations have a greater effect on the phenotype than others. However, there must be additional factors that influence the severity of the disorder in homozygous FH, because there is wide variation in the expression of the disease in homozygotes within families and within genetically isolated populations in which most homozygotes carry an identical receptor-gene mutation, as in the French Canadian and Lebanese populations. The point is well illustrated by the French Canadian null mutation. This produces no receptor message and, therefore, no receptor protein. Yet Hobbs *et al.* (1987) noted a range in age at death from 3 to 33 years among homozygotes with this mutation. It is of particular interest that the wide variation in clinical expression in the French Canadian homozygotes studied by Hobbs *et al.* was not due to differences in plasma LDL concentration between different individuals.

As discussed below, there is reason to believe that polymorphism at other gene loci whose products influence proneness to coronary artery disease may affect the severity of heterozygous FH. It is reasonable to suppose that polymorphism at loci other than the receptor locus may also influence the phenotypic expression of FH mutations in homozygotes.

2. Heterozygotes

Variation in expression of the FH gene is very marked in heterozygotes. The plasma LDL concentration ranges in different patients from values close to the upper limit of the normal for the local population to values comparable with those seen in homozygotes. Variation in clinical expression is equally marked. In some heterozygous men, the disease is severe enough to cause death from coronary artery disease in the early 30s, while many heterozygous women survive into their 60s with minimal clinical signs; in the group of heterozygous men and women studied by Heiberg and Slack (1977), age at death ranged from 31 to 93 years.

The effect of sex on the clinical expression of the FH gene in heterozygotes is not mediated by a difference in plasma LDL concentration, but it may be due partly to the higher plasma HDL level in heterozygous women than in heterozygous men (see Table 10.1).

The wide variation in age at onset of heart disease in FH heterozygotes of the same sex cannot be explained by interindividual differences in plasma total- or

LDL-cholesterol concentration. This rather surprising conclusion follows from the detailed observations of Gagné *et al.* (1979) on 575 heterozygotes. Although males and females with tendon xanthomas had higher mean plasma LDL concentrations than those without, plasma LDL levels were similar in heterozygotes of the same sex whether or not ischemic heart disease was present.

The severity and time of onset of the clinical manifestations of heterozygous FH are likely to be determined by the background of environmental and polygenic factors against which the FH gene is expressed. The environmental determinants presumably include all those factors, such as diet and smoking habit, that affect proneness to coronary disease in the general population. The genetic contribution to variability of clinical expression is likely to include heterogeneity of the FH allele itself, together with polymorphic variation in the normal receptor allele and in other genes whose products are causally related to the development of heart disease. This view is supported by the observations of Heiberg and Slack (1977). In a combined study of Norwegian and British FH families, they found a significant sib–sib correlation for age at death from ischemic heart disease. Since most of the sib pairs had lived apart for at least 30 years before death, Heiberg and Slack concluded that the correlation between sibs was largely genetic in origin. They suggested that variability in age at death in their FH heterozygotes could be explained by the existence of two or more different FH genes, each having its own major effect on age at death but modified by environmental and polygenic influences.

Among the polymorphic genes that could contribute to variability in the prognosis of heterozygous FH are the genes for apoB, apoE, and the Lp(a) antigen. However, it should be noted that neither age at onset of ischemic heart disease (Gagné *et al.*, 1979) nor age at death (Heiberg and Slack, 1977) has been shown to be correlated with the plasma total cholesterol or LDL-cholesterol level in FH heterozygotes. Hence, it is likely that genetic variability in the development of heart disease in heterozygous FH is determined partly by polymorphic genes other than those that influence the plasma cholesterol concentration.

I. Diagnosis

1. Measurement of Receptor Function as an Aid to Diagnosis

Homozygous FH can be diagnosed without ambiguity from the clinical signs, plasma lipoprotein pattern, and family history, whatever the age of the patient. LDL-receptor function determined in skin fibroblasts or other cells *in vitro* is always grossly defective in homozygous FH and shows no overlap with the values obtained from normal cells. Indeed, an FH homozygote has been correctly diagnosed *in utero* by measuring LDL-receptor activity in amniotic cells grown in culture (Brown *et al.*, 1978). Heterozygous FH, on the other hand, may be difficult to diagnose at the clinical level when the plasma LDL concentration is only moderately raised or when clinical signs are minimal. The presence of tendon xanthomas in a patient with a raised plasma LDL level is almost diagnostic of FH, but tendon xanthomas in heterozygous FH are never present in childhood, are uncommon before the third decade, and, in a few patients, are not detectable at any age. It is particularly important to diagnose FH in childhood so that early preventive treatment can be instituted. Moreover, effective genetic counselling cannot be given to the family if the diagnosis is uncertain.

An improvement in diagnostic accuracy in doubtful cases can be achieved by measuring receptor activity in fully induced skin fibroblasts in culture or in lymphocytes *in vitro* with the improved methods discussed in Chapter 8 (Section III). With these methods it is possible to make a correct diagnosis in heterozygotes from families already known to be carrying an FH gene. However, it has yet to be shown that any biochemical test of receptor activity in cells *in vitro* is capable of picking out heterozygotes from the general population with 100% accuracy.

2. Diagnosis at the DNA Level

Detection of a deleterious mutation in the patient's receptor gene would obviate the need for tests of LDL-receptor function when the clinical diagnosis is equivocal. However, it is surprisingly difficult to devise a general strategy for detecting the presence of a mutant receptor gene in a potential FH patient unless the nature of the expected mutation is already known. As discussed in Chapter 9, in most populations a search for an abnormal restriction-fragment-length pattern would detect a mutant receptor gene in only a few % of carriers of an FH gene, because most mutations do not affect the fragment pattern.

In populations in which there is a founder effect the situation is more favorable. If the "founder" mutation gives rise to an abnormal restriction fragment pattern, as in the Lebanese and French Canadian mutations, a high proportion of FH carriers within the population may be diagnosed unequivocally by Southern blotting. If the founder mutation is a single-base substitution that does not change the restriction-fragment pattern it may still be possible to detect individuals carrying the mutant gene. Provided that the mutation has been identified, the patient's DNA may be examined by the "oligonucleotide melting" procedure of Conner *et al.* (1983). Two oligonucleotides are synthesized, one complementary to a short segment of the mutant gene containing the base substitution and the other complementary to the corresponding segment of the normal receptor gene. The two oligonucleotides are then tested for hybridization with the relevant segment of the patient's DNA that has been amplified. Under appropriate conditions, each probe hybridizes only with its own complementary DNA sequence.

When the patient comes from a genetically heterogeneous population, two complicating factors must be taken into account. In addition to the presence of many different LDL-receptor mutations in the population, there is also the fact that the effects of mutations in the apoB gene may mimic FH. The optimum strategy will depend upon the availability of affected and unaffected members of the patient's family.

If relatives can be examined, the first step would be to see whether or not a raised plasma LDL level segregates within the family with a haplotype derived from RFLPs in the receptor gene. A positive finding would indicate that the mutation responsible for the observed phenotype is in the LDL-receptor gene and, hence, that the patient has FH. A negative finding would show that the causal mutation is not in the receptor gene. The next step would then be to test for the presence of a deleterious mutation in the apoB gene, as discussed below.

If a segregation analysis cannot be carried out, or if the results are inconclusive, a possible approach would be to begin by testing for the presence of an apoB mutation as the cause of the patient's hyperlipidaemia. For this, a rapid and reliable method for assessing high-affinity binding of LDL by normal LDL receptors would be required. At present, the most promising are methods involving freshly-prepared blood lymphocytes, such as the procedure of Benhamamouch *et al.* (1988) or that of Frostegärd *et al.* (1990). If the patient's LDL is defective, his apoB gene could be tested for the presence of the apoB ($Arg_{3500} \rightarrow$ Gln) mutation (and of any other functionally similar apoB-gene mutations identified in the future), using the oligonucleotide melting procedure. If the LDL is normal, the next step would be to search for a causal mutation in the receptor gene. Segments of DNA extending over the whole coding region of the gene would be amplified by PCR and the amplified segments tested for the presence of mutations, using the method of mismatch analysis (see p. 157). A diagnosis of FH would be established by the finding of a nonsense mutation, an extensive deletion or insertion, a short deletion or insertion that changed the reading frame of the gene, or a mutation already known to cause FH. A point mutation involving an amino-acid change not corresponding to any known polymorphism in the receptor gene would be strongly suggestive of FH. [For complete proof, it would be necessary to construct a receptor gene containing the point mutation and to show that receptors synthesized under the direction of the mutant gene, placed in a receptor-negative cell line, were functionally defective. This could only be done in a laboratory specializing in molecular genetics.]

The salient points of the strategy outlined above are shown in Scheme 10.1. With relatively simple laboratory equipment, it should be possible to carry out Southern blotting, oligonucleotide melting and mismatch analysis with DNA amplified from cells present in a mouth wash or a single drop of blood (Lench *et al.*, 1988).

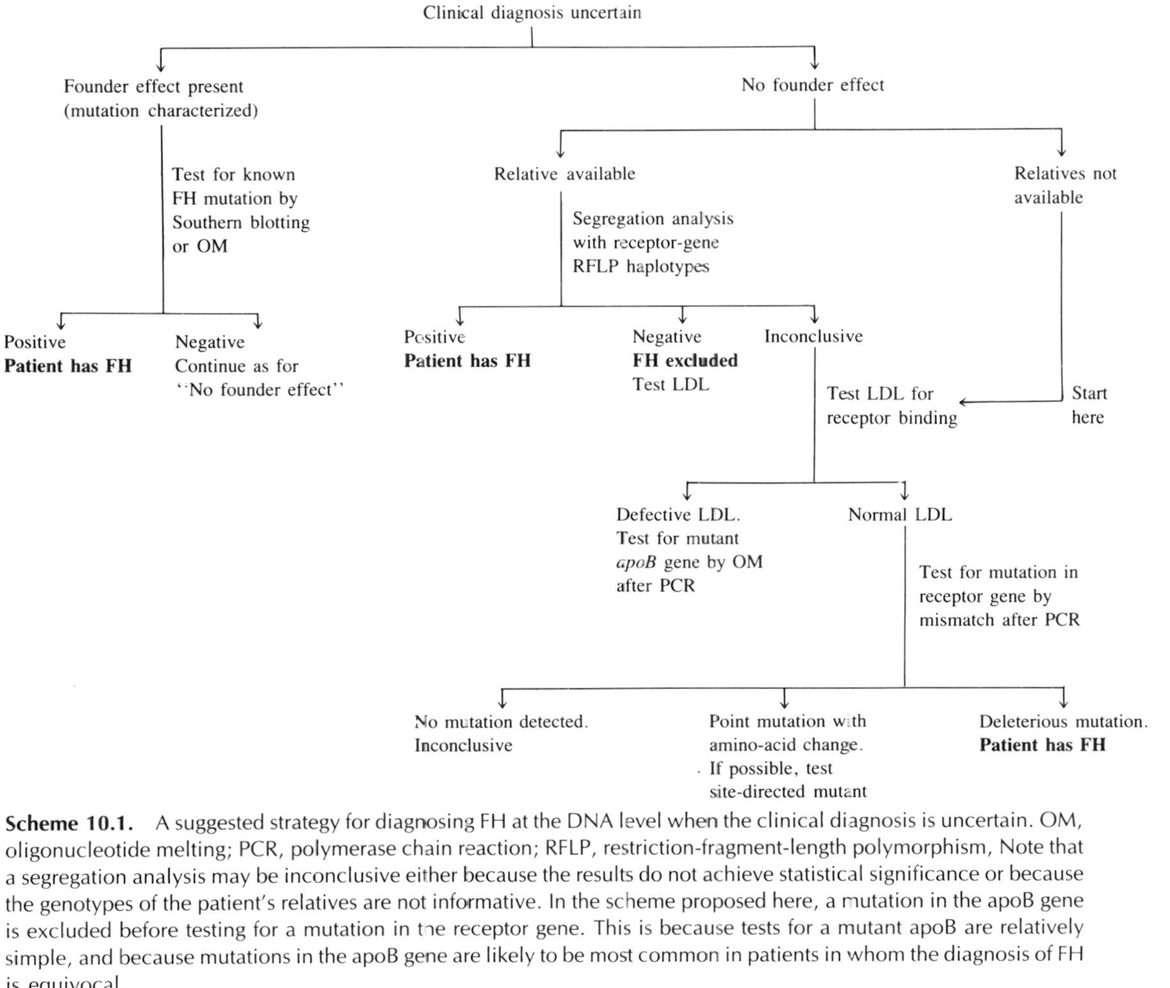

Scheme 10.1. A suggested strategy for diagnosing FH at the DNA level when the clinical diagnosis is uncertain. OM, oligonucleotide melting; PCR, polymerase chain reaction; RFLP, restriction-fragment-length polymorphism, Note that a segregation analysis may be inconclusive either because the results do not achieve statistical significance or because the genotypes of the patient's relatives are not informative. In the scheme proposed here, a mutation in the apoB gene is excluded before testing for a mutation in the receptor gene. This is because tests for a mutant apoB are relatively simple, and because mutations in the apoB gene are likely to be most common in patients in whom the diagnosis of FH is equivocal.

J. Current Approaches to Treatment

1. Heterozygotes

Until the importance of the LDL-receptor pathway in cholesterol metabolism was understood, treatment of FH was largely empirical. The objective was to lower the plasma cholesterol level by any method (diet, drugs, or ileal bypass) that seemed to be effective, without regard to the mechanism by which it worked. Once it was recognized that LDL receptors in the liver and other tissues make a major contribution to LDL catabolism in the whole body, a more rational approach could be adopted. Drug treatment for FH heterozygotes may now be designed to increase the requirement of the patient's cells for cholesterol, and thus to stimulate the one normal receptor gene to produce more LDL receptors. This, as we saw in Chapter 7 and in Section III,F above, is the mechanism by which cholestyramine is now known to lower the plasma LDL concentration in normal human subjects and in FH patients, although the drug had been in use for more than a decade before this was understood.

The effectiveness of cholestyramine as a means of inducing LDL-receptor activity in the liver is limited by the fact that increased conversion of cholesterol into bile acids results in a compensatory increase in hepatic synthesis of cholesterol. [In some species, such as the rat, the increased requirement for cholesterol as precursor for bile-acid formation is met by increased cholesterol synthesis without any increase in hepatic LDL-receptor activity (see Section II,C above).] Failure to achieve a satisfactory reduction in the plasma LDL level in many FH heterozygotes given maximal doses of cholestyramine could well be due to this effect. An obvious way to overcome this limitation would be to combine cholestyramine with another drug that blocks the compensatory increase in cholesterol synthesis. With this in mind, Mabuchi *et al.* (1983) treated a group of FH heterozygotes with a bile-acid-binding resin in combination with compactin (the competitive inhibitor of HMG-CoA reductase discussed in Chapter 3). The effect of the two drugs together on the plasma cholesterol level was greater than that of either alone, and in most patients the plasma LDL level fell to within the normal range. Following up these observations, Bilheimer *et al.* (1983) examined the effect of the bile-acid sequestrant colestipol combined with mevinolin (an analog of compactin shown in Fig. 3.9) on LDL metabolism in a patient with heterozygous FH. Using the glucosylated-LDL procedure for determining LDL-receptor-mediated catabolism of LDL, they showed that the two drugs together produced a twofold increase in the FCR of LDL, due entirely to stimulation of the LDL-receptor pathway.

These initial observations led to the setting up of clinical trials in several research centers to test the long-term safety and efficacy of the above combined therapy for FH heterozygotes who were refractory to treatment with single drugs. The results reported in 1988 were very encouraging (see Illingworth, 1988, and

Illingworth and Bacon, 1989, for reviews). Combinations of a bile-acid sequestrant with compactin (mevastatin) or mevinolin (lovastatin) bring about a 50% reduction in plasma LDL concentration in many patients. Over periods of up to 5 years, side-effects have been uncommon and have seldom been severe enough to outweigh the potential benefits of treatment. Despite the marked inhibition of cholesterol synthesis by mevinolin, Thompson *et al.* (1986) could find no evidence that treatment of FH patients with mevinolin plus cholestyramine interferes with the response of their adrenals to ACTH.

A logical extension of the two-drug approach would be to introduce a third drug whose mechanism of action on LDL metabolism differs from that of the other two. Malloy *et al.* (1987) treated a group of FH heterozygotes with a combination of colestipol, lovastatin, and nicotinic acid (a drug which decreases hepatic secretion of VLDL, the precursor of LDL). The effect of the three drugs given together was to reduce the mean plasma LDL concentration by 67%.

Clinical experience with HMG-CoA reductase inhibitors is now promising enough to suggest that when these drugs become generally available they will be used as the first choice, in combination with a bile-acid sequestrant, in the treatment of selected adult patients with heterozygous FH. Longer-term studies will be needed to assess the suitability of reductase inhibitors in the treatment of FH in children and in women within the reproductive period. A discussion of the use of other drugs in the treatment of FH will be found in the review by Illingworth (1988).

2. Homozygotes

In principle, it should be possible to stimulate the two abnormal genes in receptor-defective homozygotes to produce a greater number of functionally inadequate receptors, and thus to stimulate LDL catabolism. In practice, cholestyramine has no effect on the plasma LDL concentration in FH homozygotes, irrespective of their genotype, despite a normal increase in bile-acid synthesis (Moutafis *et al.*, 1977). Combined therapy with cholestyramine and nicotinic acid may cause the skin xanthomas to regress in some homozygotes in the absence of a fall in plasma LDL level (Moutafis *et al.*, 1971), but this is clearly not an adequate response to treatment. The two alternatives at present available for treating FH homozygotes are life-long repeated plasma exchange and liver transplantation.

Plasma exchange was first used for the treatment of homozygous FH at Hammersmith Hospital in 1974 (Thompson *et al.*, 1975). Since then, FH homozygotes and a few severely affected heterozygotes have been treated by this procedure at several other centers. The results obtained at all centers over a 12-year period have been reviewed by Thompson *et al.* (1989). Removal of 3–4 liters of the patient's plasma by a continuous-flow blood-cell separator, with simultaneous replacement by 4.5% human albumin solution, results in a rapid and profound fall in plasma LDL concentration, followed by a return to the baseline value over the following 2–3 weeks (as shown in Fig. 10.7). There is also a fall in plasma HDL level, but the return to baseline takes only a few days. When 3- to 4-liter exchanges are carried out at fortnightly intervals the mean plasma LDL level between exchanges falls to about half the preexchange value. A greater fall in mean LDL level may be achieved by combining plasma exchange with the administration of cholestyramine, nicotinic acid, or mevinolin. In the Hammersmith study, plasma exchange repeated at fortnightly intervals for 5–10 years brought about regression of skin and tendon xanthomas and arrested or delayed the progression of atherosclerotic lesions in the aorta and coronary arteries in some patients. A comparison between the current ages of homozygotes treated by plasma exchange with the age at death of their untreated siblings also showed that this procedure significantly prolongs life.

A major disadvantage of plasma exchange with substitution by albumin solution is that many normal constituents of the plasma, including HDL, are not replaced when the patient's plasma is removed. To overcome this objection, several methods have been developed for the selective removal of LDL by including an affinity-chromatography step in the procedure. In the most recent of these modifications, the patient's plasma is perfused through an affinity column, consisting of dextran sulfate coupled to cellulose, before return to the blood circulation (Yokoyama *et al.*, 1985). The column removes LDL efficiently but does not bind HDL. Preliminary results of a study of the value of this procedure when combined with the administration of compactin are encouraging (Yokoyama *et al.*, 1986).

In view of the predominant role of the liver in LDL-receptor-mediated catabolism of LDL, a liver transplant should prove a very effective means of providing an FH homozygote with an adequate supply of regulated LDL receptors. This is borne out by the successful treatment of a receptor-negative genetic compound by Bilheimer *et al.* (1984). The patient (FH 728) was a 6-year-old girl who had inherited a null allele from her mother and a class 2 allele from her father. She had a history of myocardial infarction and angina. Her plasma cholesterol concentration was above 1000 mg/100 ml, the FCR of her plasma LDL was 0.12 pools/day, and the rate of production of LDL-apoB was double the normal rate. After a combined heart–liver transplant, the plasma cholesterol level fell to 302 mg/100 ml, the FCR of LDL increased to 0.31 pools/day, and the rate of production of LDL-apoB fell to within the normal range. All these changes may be explained by the restoration of near-normal LDL-receptor activity in the patient's whole body. They also provide striking confirmation of the conclusion, emphasized in earlier chapters of this book, that the liver is the major site of LDL catabolism in man.

Since the above report was published, two other successful liver transplants have been carried out in homozygous FH patients, one in whom only the liver was replaced (Hoeg *et al.*, 1987) and the other in whom both liver and heart were replaced (Valdivielso *et al.*, 1988). In the light of these three reports, liver transplantation should now be considered in the management of receptor-negative FH patients in whom the development of coronary heart disease cannot be controlled by conventional treatment. However, patients with liver transplants receive life-long immunosuppressive treatment.

In Chapter 9 I mentioned the possible use of segregation analysis in the diagnosis of homozygous FH before birth, with polymorphic restriction sites in the receptor gene as linkage markers. As more RFLPs are identified in the receptor gene, the proportion of FH families to which this method is applicable will increase. In cases where a diagnosis of homozygous FH in the fetus can be established unequivocally, early termination of pregnancy may be offered to the parents. However, segregation analysis of FH families will always be limited to those in which DNA can be obtained from normal relatives and from relatives already known to be carrying an FH gene. Moreover, some homozygotes will continue to be born to parents not previously known to have FH. This raises the question of gene therapy.

K. Gene Therapy

1. FH as a Candidate for Gene Therapy

It would obviously be unwise to try to predict when the first FH homozygote will be treated by incorporating normal receptor genes into the genome. However, the spectacular progress of molecular genetics over the past decade suggests that it may be sooner than many are inclined to think.

Homozygous FH has several features in its favor as a potential candidate for gene therapy. We have a wealth of information about the receptor gene and about the flanking sequences required for its regulated expression in transfected[1] cells. Homozygous FH is invariably fatal at an early age, so that some degree of risk to the patient would be acceptable. Finally, and most important of all, the Watanabe rabbit provides the perfect system for testing the effectiveness *in vivo* of methods for delivering normal receptor genes to their target cells and also for studying the effects of such procedures on the whole organism under varying experimental conditions. It can hardly be doubted that successful long-term incorporation of receptor genes into the genome of a Watanabe rabbit would be followed rapidly by the first clinical trial in an FH homozygote. As discussed below, a difficulty that may have to be faced in perfecting gene therapy for FH is the probable need to place LDL-receptor genes in hepatocytes in their normal position in relation to the biliary system. This is the kind of problem that can be studied in the WHHL rabbit.

[1]*Transfection* refers to the introduction of foreign DNA into a cell, followed by integration of the DNA into the cell's genome.

2. Objectives in Receptor-Gene Therapy

Although gene therapy for human inborn errors is still at an exploratory stage it is already possible to see what the objectives should be for successful receptor-gene replacement and to discern in broad outline the steps by which these objectives will be achieved.

From the point of view of providing the body of an FH homozygote with an adequate number of functional LDL receptors, the ideal procedure would consist in the intravenous injection of a delivery system carrying a recombinant vector that contains the nucleotide sequence of the receptor gene, ligated to the sequences needed for its sterol-regulated expression. The delivery system would be capable of bringing about the integration of normal genes into the genome in a high proportion of all the hepatocytes in the liver, say in at least 50%. Selective targeting to the liver might be achieved by labeling the vehicle carrying the receptor gene with a ligand for a specific hepatic receptor (e.g., the asialoglycoprotein receptor). With current DNA technology, exogenous DNA that has entered the nucleus of a eukaryotic cell is integrated into genomic DNA, if at all, at random sites. Random integration could interfere with the functioning of normal genes in the cell and could produce new mutations. However, it might eventually be possible to construct a recombinant that would replace the mutant gene at the receptor locus by a normal gene (site-specific recombination). The normal gene would then be in its correct position in relation to its promoter and regulatory sequences and to neighboring genes.

3. Delivery Systems: Vectors Derived from Retroviruses

In principle, there is no difficulty in incorporating LDL-receptor genes into the genomic DNA of mammalian cells *in vitro*. Chapter 9 included several examples of the introduction of mutant human receptor genes into receptor-negative cells in culture, resulting in the production of stable cell lines capable of expressing the mutant gene. However, in these experiments a fragment of recombinant DNA was introduced into the cells by chemical methods, usually as a component of a plasmid expression vector. Such methods would not be suitable for gene therapy in human diseases, because foreign DNA introduced into cells in this way is integrated into the genome in only a very small proportion of treated cells. Moreover, with chemical methods of transfection it would not be possible to achieve selective introduction of DNA into cells of specific tissues.

The most promising delivery systems at present under investigation involve the use of vectors based on retroviruses. Since retrovirus-derived vectors are likely to play an important role in the development of gene therapy for human diseases, including FH, the principles underlying their construction are worth a brief mention at this point. (For a discussion of retroviral vectors in relation to gene therapy, see Anderson, 1984.)

The genome of a retrovirus consists of two identical single-stranded RNA molecules, each 7–10 kb long. This genomic RNA, together with an enzyme called *reverse transcriptase,* is enclosed in a protein coat that enables the complete virus particle (the virion) to bind to its target cell and to cross the plasma membrane into the cytoplasm. The viral genomic RNA contains sequences encoding reverse transcriptase and the coat proteins. Once the virion has entered the cell, one of the two RNA strands is transcribed by reverse transcriptase into a double strand of DNA called the *provirus,* which is rapidly and efficiently integrated into the host cell's genomic DNA (Fig. 10.11). At each end of the provirus there is a direct repeat called the long terminal repeat (LTR). (For details, see Watson *et al.,* 1987). The LTR contains all the sequences required for integration into the genome and for initiating transcription of the integrated provirus into RNA. Just inside the LTR is a short sequence, called the *packaging sequence,* that is transcribed into the viral RNA and is necessary for its packaging by coat proteins. As soon as the provirus is integrated into the host's genome, it is transcribed by the cell's enzymes into RNA copies of viral genome and into mRNA's for reverse transcriptase and the viral coat proteins. Viral genomic RNA, together with reverse transcriptase, is packaged by the coat proteins into virions, which leave the cell by budding from the plasma membrane, thus completing the life cycle of the virus.

Modified retroviruses containing foreign genes have been constructed by exploiting the complex life cycle of the virus in highly ingenious ways (see, for example, Mann *et al.*, 1983, and Cepko *et al.*, 1984). The first step is to develop a line of cells (producer cells) in which a modified provirus (the helper virus) has been integrated into the cell's genome. The helper virus contains LTRs and all the sequences encoding reverse transcriptase and viral coat proteins, but it has no packaging sequence. Hence, it produces viral proteins and viral RNA, but the RNA cannot be packaged to form virions. The second step consists in the construction of a recombinant provirus in which the DNA sequences encoding viral proteins have been replaced by a cDNA for the gene of interest (the LDL-receptor gene in the present context). Cells of the producer line are transfected with this recombinant to generate a modified line that expresses the helper virus and the reconstituted virus containing the foreign gene. The recombinant is transcribed by the cell's RNA polymerase into RNAs containing the sequence complementary to the cDNA for the foreign gene, linked to LTRs and a packaging sequence in their normal positions. These RNAs are packaged into retroviral particles in which the coat proteins and reverse transcriptase are supplied by the helper virus (see Fig. 10.11).

The modified retroviruses, containing RNA sequences encoding the foreign gene, are secreted into the culture medium and are used as delivery systems for introducing the foreign gene into animal cells. Having a normal viral coat, modified retroviruses enter their target cells with a high degree of efficiency.

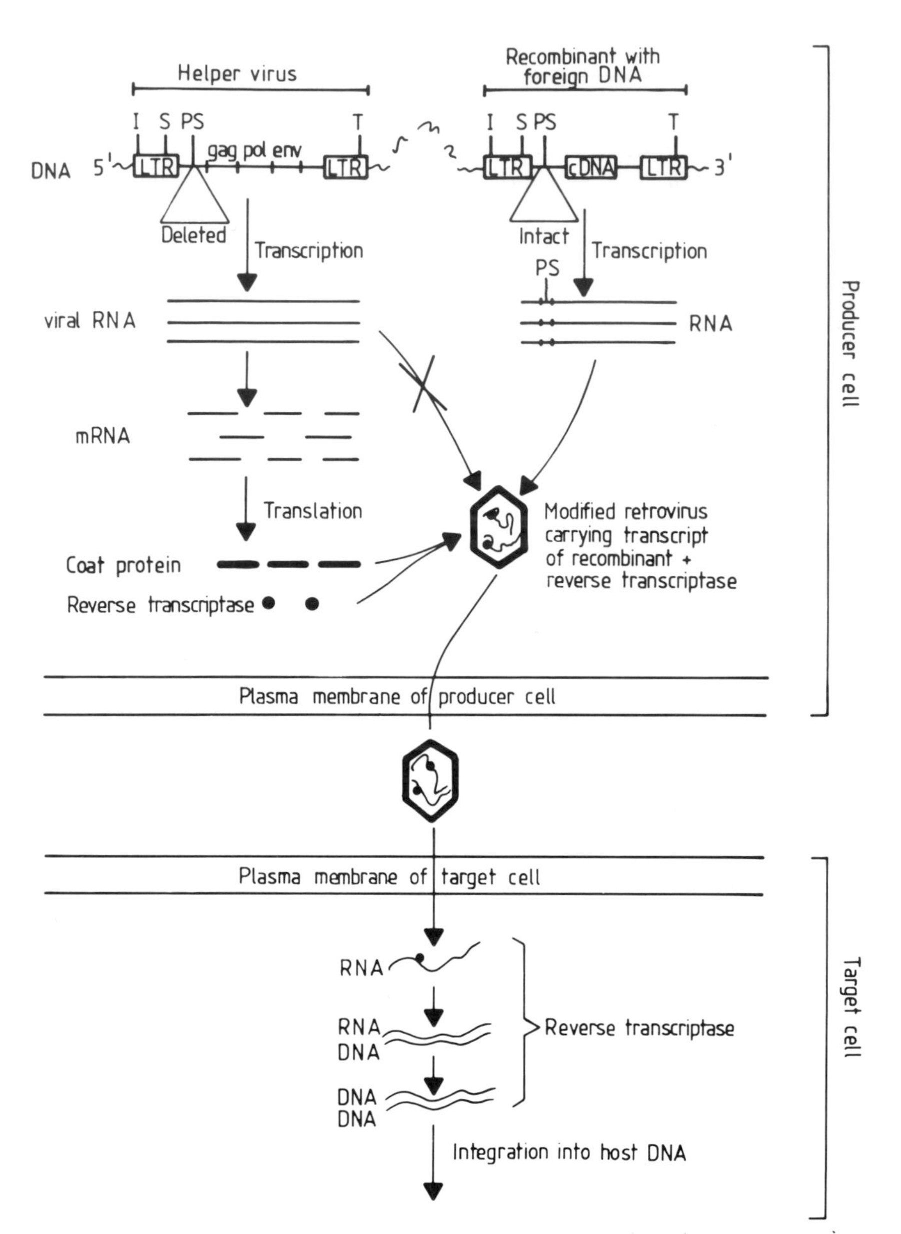

Figure 10.11. Diagram showing the major steps in the construction of a modified retrovirus particle carrying a foreign gene (the separate elements are not to scale). Cell DNA is shown by wavy lines.

A producer cell line is generated by transfecting cells in culture with proviral DNA from which the packaging sequence has been deleted (upper left) and with a recombinant provirus in which the genes encoding viral proteins have been replaced by a cDNA for the foreign gene (upper right). Note that proviral DNA contains sequences at each end that are not present in the viral RNA strand from which it was transcribed. These additional segments include promoter

Inside the cell, the viral RNA is transcribed into proviral DNA by reverse transcriptase carried in the particle, and the proviral DNA is efficiently integrated into the cell's genome by virtue of the presence of LTRs in the provirus. The modified retrovirus is unable to complete its life cycle by generating infective particles because it lacks the sequences encoding viral proteins.

4. What Progress Has Been Made?

In most investigations of the feasibility of gene therapy, bone marrow has been used as the test system because this has always seemed to be the one most likely to yield positive results. The usual strategy has been to remove marrow cells from an animal and then to introduce foreign genes into hematopoietic stem cells *in vitro*, followed by reinjection of the genetically altered cells into the donor. Transfection *in vitro* of marrow cells from mice (Williams *et al.*, 1984), monkeys (Anderson *et al.*, 1986), and human subjects (Miller *et al.*, 1986) has been achieved with foreign genes carried in retroviral vectors, usually derived from a mouse leukemia virus. In at least two studies with experimental animals, treated marrow cells carrying the foreign gene were shown to persist in the donor animals after reinjection (Williams *et al.*, 1984; Anderson *et al.*, 1986). The efficiency of transfection in these experiments was seldom greater than 20%. Nevertheless, these early results have provided a basis for attempts to improve efficiency of transfection and for long-term assessment of the safety of introducing cells transfected by retrovirus-derived vectors into living animals.

The introduction of LDL-receptor genes into bone-marrow cells *in vitro* and their reinjection into the donor would not be a worthwhile approach to gene therapy for FH. As noted above, therapeutic success will probably require the

---

and enhancer sequences (not shown) and the signal for integration into the host's genome. (See Watson *et al.*, 1987, for an explanation of how reverse transcriptase brings about this elongation.) Proviral DNA of the helper virus is transcribed by the cell's RNA polymerase II into viral RNA lacking the packaging sequence. Splicing of some viral RNA strands generates mRNAs for coat proteins and reverse transcriptase.

The recombinant provirus is transcribed into RNA strands containing the packaging sequence. These are packaged into retroviral particles by the coat proteins derived from the helper virus. (The structure of the coat has been simplified in the diagram.) The modified retroviral particles leave the producer cell by budding and infect their target cells. Inside the target cell, a single strand of RNA from the modified virus is transcribed into double-stranded DNA with normal LTRs containing the sequences necessary for integration into the target cell's DNA and for initiation of transcription. The double-stranded DNA is integrated into the host DNA, where it is transcribed into mRNAs for the foreign gene.

I, signal for integration into host genome (present at both ends of provirus); LTR, long terminal repeat; PS, packaging sequence; S, startpoint for transcription; T, termination of transcription; *gag* and *env* are genes encoding the coat proteins; *pol* (polymerase) is the gene for reverse transcriptase.

transfection of a substantial proportion of all the hepatocytes in the liver with normal receptor genes. This follows from the fact that the liver is the major site for the catabolism of LDL in the mammalian body. The close functional and anatomical relationship between the liver and the biliary system (the sole route for excretion of cholesterol delivered to the liver by LDL) also suggests that receptor genes will have to be placed in liver cells in their normal position in the body. This is why the intravenous route for transfection seems, in theory, a better ultimate objective than the alternative of introducing receptor genes into the patient's liver cells *in vitro* and then implanting the altered cells in the peritoneal cavity.

Although the intravenous route may in the end turn out to be the only effective one for gene therapy for FH, the development of methods for the efficient transfection of hepatocytes with receptor genes, and for studying the behavior of transfected hepatocytes in intact animals, is likely to be based, for some time to come, on an *in vitro–in vivo* strategy. In this regard, Wilson *et al.* (1988) have already succeeded in transfecting WHHL rabbit liver cells in culture with a retroviral vector containing normal rabbit LDL-receptor genes.

Successful treatment of homozygous FH by supplying the liver with normal receptor genes would leave the patient with a continuing absence of LDL-receptor function in extrahepatic tissues, including the adrenal cortex. Would these tissues be able to function normally, in the long term, in the presence of a normal plasma LDL concentration? The favorable response of FH homozygotes to liver transplantation and the apparently normal adrenal cortical function in abetalipoproteinemia (in which no LDL-receptor-mediated uptake of LDL can occur) suggest that they would.

Another problem that arises with liver transplantation and with selective introduction of receptor genes into the livers of FH homozygotes is that all the progeny from matings between treated patients and normal spouses will be heterozygotes. This will require a carefully considered approach to genetic counseling.

IV. The WHHL Rabbit

A. Origin and Uses

Watanabe and his colleagues (Watanabe, 1980) have developed an inbred strain of genetically hyperlipidemic rabbits [now called the Watanabe Heritable Hyperlipidemic (WHHL) strain] from a mutant identified in 1977. Initial studies in Watanabe's laboratory showed that hyperlipidemia develops in WHHL animals raised on a normal diet and that it is inherited as an autosomal recessive trait with slight expression in the heterozygous state. Homozygotes exhibit marked hypercholesterolemia, hypertriglyceridemia, a reduced plasma HDL concentration, and the presence of β-migrating lipoproteins in the plasma. Watanabe (1980) also noted that homozygous animals develop atherosclerosis of the aorta and coronary arteries, often with xanthomas in the digital tendons. (Throughout this chapter "WHHL" refers to the homoygous form of the disorder.)

In view of the plasma lipoprotein pattern and the increased frequency of atherosclerosis, Watanabe (1980) suggested that the genetic lesion in WHHL rabbits is analogous to that responsible for human familial type III hyperlipoproteinemia (broad β disease). However, Tanzawa *et al.* (1980) showed that the hypercholesterolemia is due to a selective increase in plasma LDL concentration, accompanied by a marked reduction in the FCR of LDL *in vivo,* and that high-affinity binding and catabolism of normal rabbit LDL *in vitro* by skin fibroblasts from WHHL rabbits is less than 5% of normal. In agreement with the latter finding, LDL-receptor deficiency was demonstrated in hepatic and adrenal-gland membranes (Kita *et al.*, 1981) and in cultured hepatocytes (Attie *et al.*, 1981) from WHHL rabbits. These observations showed that the lesion underlying the hyperlipidemia in WHHL rabbits is an inherited deficiency of LDL receptors.

A strain of experimental animals in which LDL-receptor function is deleted opened the way to the study of aspects of LDL metabolism *in vivo* that it had not been possible to investigate in FH patients. Colonies of WHHL rabbits have now been established in many laboratories throughout the world; homozygotes of both sexes are usually obtained by crossing a homozygous male with a heterozygous female. In a sense, the artificial maintenance of genetically isolated colonies of WHHL rabbits in different parts of the world provides a parallel to a naturally occurring founder effect in the human population.

Several examples of the use of WHHL rabbits in investigations of lipoprotein metabolism in the whole body were mentioned in Chapter 8. These include studies concerned with (1) the genetic relationship between LDL receptors and chylomicron-remnant receptors in the liver (Kita *et al.*, 1982b), (2) the contributions of LDL-receptor-mediated and LDL-receptor-independent uptake to total catabolism of LDL in the whole body and in particular tissues (Pittman *et al.*, 1982), and (3) the mechanism underlying the increased rate of production of LDL in LDL-receptor deficiency (Kita *et al.*, 1982a).

When Tanzawa *et al.* (1980) showed that the mutation in the WHHL rabbit affects primarily the function of LDL receptors, it was generally assumed that the mutation is at the same locus as the locus bearing the FH mutation. This assumption was later proved to be true. As discussed in Chapter 9, the LDL receptors produced in cells of WHHL rabbits resemble the defective receptors produced in the fibroblasts of FH patients with a class 2 receptor mutation (Schneider *et al.*, 1983). Moreover, in WHHL rabbits the receptor gene itself has a 12-base deletion in an exon encoding part of the binding domain of the receptor (Yamamoto *et al.*, 1986). Thus, the WHHL rabbit is an appropriate model for use in the development of gene replacement in the treatment of FH.

B. Plasma Lipoproteins

As noted above, plasma total cholesterol and triglyceride concentrations are increased in WHHL rabbits. Total cholesterol concentration is usually about 10 times that in normal rabbits, due largely to a 20- to 50-fold rise in plasma LDL level. VLDL and IDL concentrations are also markedly elevated, the raised concentrations of these lipoproteins accounting for the 2- to 4-fold increase in triglyceride level. The increased concentration of IDL explains the presence of lipoproteins with β mobility in the plasma of WHHL rabbits (Watanabe, 1980). The plasma HDL level is decreased to about one-third of the normal value. Plasma total apoE concentration in WHHL rabbits is 4 to 5 times the normal level. Representative values for normal New Zealand and homozygous WHHL rabbits are shown in Table 10.4.

The above changes in plasma LDL, IDL, and HDL concentrations are qualitatively similar to those seen in FH, though the rise in LDL is much greater in WHHL rabbits. A rise in plasma VLDL concentration is not characteristic of FH. The more marked changes in plasma lipoprotein concentrations in WHHL rabbits than in FH homozygotes, and the presence of hypertriglyceridemia in WHHL rabbits, are probably due to differences in the kinetics of lipoprotein metabolism in the two species (see Section IV,C below).

C. Metabolic Changes

1. Cholesterol Metabolism

Dietschy *et al.* (1983) have determined rates of sterol synthesis in the whole bodies and specific tissues of WHHL rabbits *in vivo*. The rate of whole-body

Table 10.4
Plasma Lipid and Lipoprotein Concentrations in Normal and Homozygous WHHL Rabbits, with Values for ApoE[a]

	Cholesterol						
	Total	VLDL	IDL	LDL	HDL	Triglyceride	ApoE
Control (NZW)	58.5	17.0	7.8	11.5	22.3	213.3	2.6
WHHL	487.5	110.3	116.3	218.0	8.3	435.0	10.3

[a] All values are mg/100 ml. Values are taken from Havel *et al.* (1982). Control animals were New Zealand White rabbits. All animals were males aged 4 to 5 months and were in the fasting state.

synthesis was similar to, or below, that in normal rabbits at all ages from prenatal to adult life. Rates of sterol synthesis per gram of tissue were normal in all tissues examined except liver and adrenal glands. In adult WHHL rabbits hepatic synthesis was slightly reduced, while synthesis in adrenal glands was increased five- to six-fold. Thus, with the exception of the adrenals, cholesterol synthesis in all tissues of WHHL rabbits is suppressed to the normal extent, despite the absence of LDL receptors. The reasons for this have been discussed by Dietschy *et al.* They point out that many tissues, including muscle, skin, and fat, satisfy most of their requirement for cholesterol *in vivo* by endogenous synthesis. In these tissues, the rate of sterol synthesis should not be affected by deficiency of LDL receptors. Other tissues, such as lung and kidney, normally obtain a substantial proportion of the cholesterol they require by LDL-receptor-mediated uptake of LDL from the external medium. In the absence of LDL receptors these tissues may take up LDL, present in the external medium at high concentration, by LDL-receptor-independent pathways. As discussed in Section II,D above, this might suppress cholesterol synthesis *in vivo* in the long term. The adrenal gland of the normal rabbit satisfies the bulk of its requirement for cholesterol via the LDL-receptor pathway (Pittman *et al.*, 1982). It is not surprising, therefore, that in the absence of LDL receptors the rate of synthesis of sterol in the adrenals is markedly increased.

In the tissues of WHHL rabbits examined by Dietschy *et al.* (1983), the contents of free and esterified cholesterol were normal or slightly raised. Thus, there is no evidence to suggest that the cells of these animals lack the ability to esterify free cholesterol *in vivo*. In this respect, their behavior differs from that of FH homozygote fibroblasts in culture, in which the ACAT reaction is completely inactive (see Fig. 8.4).

2. Lipoproteins

As noted above, hypercholesterolemia in WHHL rabbits is due to a marked deficiency of LDL-receptor function. Bilheimer *et al.* (1982) have shown that the FCR of LDL in WHHL rabbits is about 0.5 pools/day, compared with a value of about 1.6 pools/day in normal rabbits, and that this difference is due entirely to a difference in FCR via the LDL-receptor pathway. Bilheimer *et al.* (1982) have also shown that the low FCR in WHHL rabbits is accompanied by a six-fold increase in the absolute rate of production of LDL. The consequence of decreased catabolism combined with increased production is the massive increase in plasma LDL concentration shown in Table 10.4.

In contrast to the abnormal metabolism of LDL, when chylomicrons containing [^{3}H]cholesteryl esters are injected intravenously into rabbits, ^{3}H disappears from the plasma and is taken up by the liver at rates similar to those in normal rabbits (Kita *et al.*, 1982b). This indicates that in the presence of a nearly complete deficiency of hepatic LDL receptors, chylomicron remnants are removed from the plasma by the liver at the normal rate. Indeed, this is one of the strongest pieces of evidence for the view that the chylomicron-remnant receptor is genetically distinct from the LDL receptor (see Chapter 8, Section III,B).

Unlike the behavior of chylomicron remnants, the metabolism of VLDL and its remnants (present mainly in IDL) is markedly impaired in WHHL rabbits, as demonstrated by the observations of Kita *et al.* (1982a) shown in Fig. 10.12. When radioactive VLDL is injected intravenously into normal rabbits, radioactivity in VLDL-apoB disappears rapidly from the circulation. Radioactivity in IDL-apoB rises to a peak within minutes of the injection and then declines rapidly. Concomitantly, radioactivity appears in LDL-apoB, rising to a maximum and then declining with a half-life of about 12 hours. This sequential transfer of apoB from VLDL to IDL to LDL in normal rabbits is similar to that

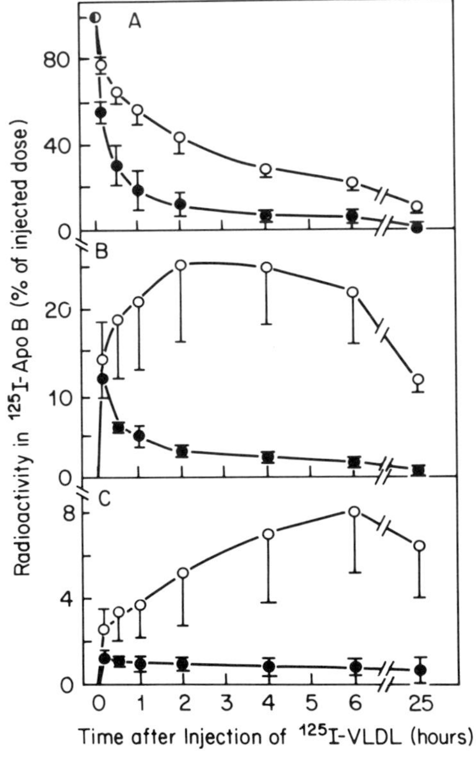

Figure 10.12. The metabolism of ^{125}I-labeled apoB in VLDL, IDL, and LDL after intravenous injection of labeled VLDL into normal (●) and WHHL (○) rabbits. The [^{125}I]apoB content of the three lipoproteins was determined after isopropanol precipitation of each lipoprotein fraction isolated by ultracentrifugation at appropriate densities. Each point is the mean of three determinations ± SEM: A, VLDL; B, IDL; C, LDL. (From Kita *et al.*, 1982a, with the permission of the authors.)

observed in normal human subjects (Fig. 10.8), though the fractional rates of transfer are much greater in rabbits. In WHHL rabbits, the disappearance of radioactivity in VLDL-apoB from the plasma is delayed. The time course of radioactivity in IDL-apoB also differs from that in normal rabbits. In this fraction, radioactivity rises progressively for about 2 hours and then falls slowly (Fig. 10.12, panel B). Radioactivity in LDL-apoB increases for at least 6 hours before falling slowly. Quantitative analysis of the two sets of curves shows that the fractions of the injected radioactive apoB present in IDL and LDL at 6 hours after the injection are several times greater in WHHL than in normal rabbits.

The interpretation of these results proposed by Kita *et al.* (1982a) is that the rabbit's liver normally removes a substantial proportion of the circulating VLDL and IDL by the LDL-receptor pathway. They suggest that when LDL-receptor function is defective, decreased hepatic uptake of VLDL and IDL results in increased conversion of VLDL into LDL via IDL. Kita *et al.* could not exclude the possibility that direct secretion of LDL by the liver also contributes to the increased production of LDL observed in WHHL rabbits. However, as we saw in Chapter 7 (Section I,C), at least 90% of the VLDL-apoB secreted into the rabbit's circulation is removed by the liver as VLDL or IDL without conversion into LDL. A severalfold increase in the rate of production of LDL could, therefore, be brought about by decreased hepatic uptake of VLDL-apoB alone. In this regard, it should be noted that Hornick *et al.* (1983) could find no evidence for direct secretion of LDL by perfused livers of WHHL rabbits.

D. Pathology

In his original observations on the WHHL rabbit, Watanabe (1980) noted that early atherosclerotic lesions were present in the thoracic aorta by 3 months and that by 1 year most animals had aortic atheromatous plaques. Watanabe (1980) also described xanthomas in the tendons of the paws of 15-month-old WHHL rabbits (Fig. 10.13).

Subsequently, Buja *et al.* (1983) made a detailed examination of the tissues of WHHL rabbits from birth to age 15 months. By 4 months, fatty streaks had developed in the intima of the thoracic aorta. By 15 months raised atheromatous plaques were seen, involving mainly the aorta but also present at the origin of the left main coronary artery. The raised plaques were histologically similar to those seen in FH patients and in members of the general population with advanced atherosclerosis. The intimal lesions contained foam cells, some of which had the morphological characteristics of smooth-muscle cells, while others appeared to be derived from macrophages. The presence of foam cells derived from smooth-muscle cells and from macrophages in the atherosclerotic lesions of WHHL rabbits has since been confirmed by immunocytochemical analysis with monoclonal antibodies (Tsukada *et al.*, 1986). Tendon xanthomas observed in the hind

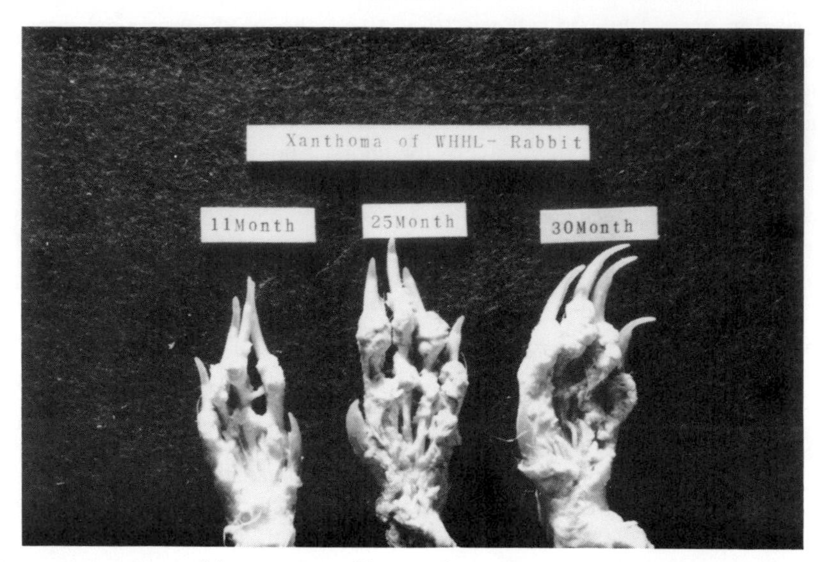

Figure 10.13. Paws of three WHHL rabbits aged 11 months, 25 months, and 30 months to show tendon xanthomas. (From Watanabe, 1980, with the permission of the author.)

paws of a 15-month-old rabbit consisted mainly of histiocytic foam cells containing esterified cholesterol, similar in appearance to the foam cells seen in the xanthomas of FH patients. Lesions elsewhere in the body were inconspicuous and were confined to lipid deposits in some hepatocytes and in a few macrophages of liver, spleen, and lymph nodes. This contrasts with the lesions in cholesterol-fed normal rabbits. In these animals, the arterial lesions consist mainly of intimal accumulations of foam cells and there is marked deposition of lipid in macrophages of liver, spleen, lymph glands, and bone marrow.

The similarity in appearance and distribution of the lesions in WHHL rabbits and FH patients is probably a consequence of the fact that in both cases the predominant lipoprotein abnormality is an increase in plasma LDL concentration. In cholesterol-fed rabbits, on the other hand, the hyperlipidemia is due largely to the presence of β-VLDL in the plasma. Whatever the reasons for the similarity between WHHL rabbits and FH patients, the WHHL rabbit is an "experiment" of nature from which we may expect to learn much about the pathogenesis of atherosclerosis in man and about the efficacy of procedures for preventing or postponing its appearance.

References

Aalto-Setälä, K., Gylling, H., Miettinen, T., and Kontula, K. (1988). Identification of a deletion in the LDL receptor gene. A Finnish type of mutation. *FEBS Lett.* **230,** 31–34.

Anderson, W. F. (1984). Prospects for human gene therapy. *Science* **226,** 401–409.

Anderson, W. F., Kantoff, P., Eglitis, M., McLachlin, J., Karson, E., Zwiebel, J., Nienhuis, A., Karlsson, S., Blaese, R. M., Kohn, D., Gilboa, E., Armentano, D., Zanzani, E. D., Flake, A., Harrison, M. R., Gillio, A., Bordignon, C., and O'Reilly, R. (1986). Gene transfer and expression in nonhuman primates using retroviral vectors. *Cold Spring Harbor Symp. Quant. Biol.* **51,** 1073–1081.

Aqel, N. M., Ball, R. Y., Waldmann, H., and Mitchinson, M. J. (1984). Monocytic origin of foam cells in human atherosclerotic plaques. *Atherosclerosis* **53,** 265–271.

Attie, A. D., Pittman, R. C., Watanabe, Y., and Steinberg, D. (1981). Low density lipoprotein receptor deficiency in cultured hepatocytes of the WHHL rabbit. *J. Biol. Chem.* **256,** 9789–9792.

Benhamamouch, S., Kuznierz, J-P., Agnani, G., Marzin, D., Lecerf, J-M., Fruchart, J-C., and Clavey, V. (1988). Determination of the LDL receptor binding capacity of human lymphocytes by immunocytofluorometric assay. *Biochim. Biophys. Acta* **1002,** 45–53.

Bilheimer, D. W., Goldstein, J. L., Grundy, S. M., and Brown, M. S. (1975). Reduction in cholesterol and low density lipoprotein synthesis after portacaval shunt surgery in a patient with homozygous familial hypercholesterolemia. *J. Clin. Invest.* **56,** 1420–1430.

Bilheimer, D. W., Stone, N. J., and Grundy, S. M. (1979). Metabolic studies in familial hypercholesterolemia. Evidence for a gene-dosage effect *in vivo*. *J. Clin. Invest.* **64,** 524–533.

Bilheimer, D. W., Watanabe, Y., and Kita, T. (1982). Impaired receptor-mediated catabolism of low density lipoprotein in the WHHL rabbit, an animal model of familial hypercholesterolemia. *Proc. Natl. Acad. Sci. U.S.A.* **79,** 3305–3309.

Bilheimer, D. W., Grundy, S. M., Brown, M. S., and Goldstein, J. L. (1983). Mevinolin and colestipol stimulate receptor-mediated clearance of low density lipoprotein from plasma in familial hypercholesterolemia heterozygotes. *Proc. Natl. Acad. Sci. U.S.A.* **80,** 4124–4128.

Bilheimer, D. W., Goldstein, J. L., Grundy, S. M., Starzl, T. E., and Brown, M. S. (1984). Liver transplantation to provide low-density-lipoprotein receptors and lower plasma cholesterol in a child with homozygous familial hypercholesterolemia. *N. Engl. J. Med.* **311,** 1658–1664.

Blum, C. B., Aron, L., and Sciacca, R. (1980). Radioimmunoassay studies of human apolipoprotein E. *J. Clin. Invest.* **66,** 1240–1250.

Brown, M. S., and Goldstein, J. L. (1976). New directions in human biochemical genetics: understanding the manifestations of receptor deficiency states. *Prog. Med. Genet. (New Ser.)* **1,** 103–119.

Brown, M. S., and Goldstein, J. L. (1986). A receptor-mediated pathway for cholesterol homeostasis. *Science* **232,** 34–47.

Brown, M. S., Faust, J. R., and Goldstein, J. L. (1975). Role of the low density lipoprotein receptor in regulating the content of free and esterified cholesterol in human fibroblasts. *J. Clin. Invest.* **55,** 783–793.

Brown, M. S., Kovanen, P. T., Goldstein, J. L., Eeckels, R., Vandenberghe, K., Van den Bergh, H., Fryns, J. P., and Cassiman, J. J. (1978). Prenatal diagnosis of homozygous familial hypercholesterolaemia: Expression of a genetic receptor disease *in utero*. *Lancet* **1,** 526–529.

Buja, L. M., Kovanen, P. T., and Bilheimer, D. W. (1979). Cellular pathology of homozygous familial hypercholesterolemia. *Am. J. Pathol.* **97,** 327–357.

Buja, L. M., Kita, T., Goldstein, J. L., Watanabe, Y., and Brown, M. S. (1983). Cellular pathology of progressive atherosclerosis in the WHHL rabbit. An animal model of familial hypercholesterolemia. *Arteriosclerosis* **3,** 87–101.

Cepko, C. L., Roberts, B. E., and Mulligan, R. C. (1984). Construction and applications of a highly transmissible murine retrovirus shuttle vector. *Cell* **37,** 1053–1062.

Conner, B. J., Reyes, A. A., Morin, C., Itakura, K., Teplitz, R. L., and Wallace, R. B. (1983). Detection of sickle cell β^s globin allele by hybridization with synthetic oligonucleotides. *Proc. Natl. Acad. Sci. U.S.A.* **80,** 278–282.

Dietschy, J. M., and Spady, D. K. (1985). Regulation of low density lipoprotein levels in animals and man with particular emphasis on the role of the liver. *In* "Receptor-mediated Uptake in the Liver" (H. Greten, E. Windler, and U. Beisiegel, eds.), pp. 56–65. Springer-Verlag, Berlin.

Dietschy, J. M., Kita, T., Suckling, K. E., Goldstein, J. L., and Brown, M. S. (1983). Cholesterol synthesis *in vivo* and *in vitro* in the WHHL rabbit, an animal with defective low density lipoprotein receptors. *J. Lipid Res.* **24,** 469–480.

Fredrickson, D. S., Levy, R. I., and Lees, R. S. (1967). Fat transport in lipoproteins—an integrated approach to mechanisms and disorders. *N. Engl. J. Med.* **276,** 34–44, 94–103, 148–156, 215–225, 273–281.

Fredrickson, D. S., Goldstein, J. L., and Brown, M. S. (1978). The familial hyperlipoproteinemias. *In* "The Metabolic Basis of Inherited Disease" (J. B. Stanbury, J. B. Wyngaarden, and D. S. Fredrickson, eds.), 3rd ed., pp. 604–655. McGraw-Hill, New York.

Frostegard, Jl., Hausten, A., Gidlund, M., and Nilsson, J. (1990). Low density lipoprotein-induced growth of U937 cells—a novel method to determine the receptor binding of low density lipoprotein. *J. Lipid Res.* **31,** in press.

Gagné, C., Moorjani, S., Brun, D., Toussaint, M., and Lupien, P-J. (1979). Heterozygous familial hypercholesterolemia. Relationship between plasma lipids, lipoproteins, clinical manifestations and ischaemic heart disease in men and women. *Atherosclerosis* **34,** 13–24.

Geer, J. C., and Haust, M. D. (1972). Smooth muscle cells in atherosclerosis. *In* "Monographs on Atherosclerosis," Vol. 2. Karger, Basel.

Gibbons, G. F., Mitropoulos, K. A., and Myant, N. B. (1982). "Biochemistry of Cholesterol." Elsevier, Amsterdam.

Gofman, J. W., Rubin, L., McGinley, J. P., and Jones, H. B. (1954). Hyperlipoproteinemia. *Am. J. Med.* **17,** 514–520.

Goldstein, J. L., and Brown, M. S. (1982). The LDL receptor defect in familial hypercholesterolemia: implications for pathogenesis and therapy. *Med. Clin. N. Amer.* **66,** 335–362.

Goldstein, J. L., and Brown, M. S. (1983). Familial hypercholesterolemia. *In* "The Metabolic Basis of Inherited Disease" (J. B. Stanbury, J. B. Wyngaarden, D. S. Fredrickson, J. L. Goldstein, and M. S. Brown, eds.), 5th ed., pp. 672–712. McGraw-Hill, New York.

Goldstein, J. L., Hazzard, W. R., Schrott, H. G., Bierman, E. L., and Motulsky, A. (1973). Hyperlipidemia in coronary heart disease. I. Lipid levels in 500 survivors of myocardial infarction. *J. Clin. Invest.* **52,** 1533–1543.

Gown, A. M., Tsukada, T., and Ross, R. (1986). Human atherosclerosis. II. Immunocytochemical analysis of the cellular composition of human atherosclerotic lesions. *Am. J. Pathol.* **125,** 191–207.

Grundy, S. M., and Ahrens, E. H., Jr. (1969). Measurements of cholesterol turnover, synthesis, and absorption in man, carried out by isotope kinetic and sterol balance methods. *J. Lipid Res.* **10,** 91–107.

Harlan, W. R., Jr., Graham, J. B., and Estes, H. E. (1966). Familial hypercholesterolemia: genetic and metabolic study. *Medicine* **45,** 77–110.

Havel, R. J., Kita, T., Kotite, L., Kane, J. P., Hamilton, R. L., Goldstein, J. L., and Brown, M. S. (1982). Concentration and composition of lipoproteins in blood plasma of the WHHL rabbit. *Arteriosclerosis* **2,** 467–474.

Heiberg, A. (1975). The risk of atherosclerotic vascular disease in subjects with xanthomatosis. *Acta Med. Scand.* **198,** 249–261.

Heiberg, A., and Slack, J. (1977). Family similarities in the age at coronary death in familial hypercholesterolaemia. *Br. Med. J.* **3,** 493–495.

Hobbs, H. H., Brown, M. S., Russell, D. W., Davignon, J., and Goldstein, J. L. (1987). Deletion in the gene for the low-density-lipoprotein receptor in a majority of French Canadians with familial hypercholesterolemia. *N. Engl. J. Med.* **317,** 734–737.

Hoeg, J. M., Starzl, T. E., and Brewer, H. B. (1987). Liver transplantation for treatment of cardiovascular disease: comparison with medication and plasma exchange in homozygous familial hypercholesterolemia. *Am. J. Cardiol.* **59,** 705–707.

Hornick, C. A., Kita, T., Hamilton, R. L., Kane, J. P., and Havel, R. J. (1983). Secretion of lipoproteins from the liver of normal and Watanabe heritable hyperlipidemic rabbits. *Proc. Natl. Acad. Sci. U.S.A.* **80,** 6096–6100.

Illingworth, D. R. (1988). Drug therapy of hypercholesterolemia. *Clin. Chem.* **34,** 123–132.

Illingworth, D. R., and Bacon, S. P. (1989). Treatment of heterozygous familial hypercholesterolemia with lipid-lowering drugs. *Artereosclerosis Supplement 1,* **9,** *T*-121–134.

Innerarity, T. L., Weisgraber, K. H., Arnold, K. S., Mahley, R. W., Krauss, R. M., Vega, G. L., and Grundy, S. M. (1987). Familial defective apolipoprotein B-100: low density lipoproteins with abnormal receptor binding. *Proc. Natl. Acad. Sci. U.S.A.* **84,** 6919–6923.

Jadhav, A. V., and Thompson, G. R. (1979). Reversible abnormalities of low density lipoprotein composition in familial hypercholesterolaemia. *Eur. J. Clin. Invest.* **9,** 63–67.

Janus, E. D., Nicoll, A., Wootton, R., Turner, P. R., Magill, P. J., and Lewis, B. (1980). Quantitative studies of very low density lipoprotein: conversion to low density lipoprotein in normal controls and primary hyperlipidaemic states and the role of direct secretion of low density lipoprotein in heterozygous familial hypercholesterolaemia. *Eur. J. Clin. Invest.* **10,** 149–159.

Jensen, J., Blankenhorn, D. H., and Kornerup, V. (1967). Coronary artery disease in familial hypercholesterolemia. *Circulation* **36,** 77–82.

Khachadurian, A. K. (1964). The inheritance of essential familial hypercholesterolemia. *Am. J. Med.* **37,** 402–407.

Khachadurian, A. K. (1972). A general review of clinical and laboratory features of familial hypercholesterolemia (Type II hyperbetalipoproteinemia). *In* "Protides of the Biological Fluids," *Proc. Colloq., 19th, Bruges, 1971* (H. Peeters, ed.), pp. 315–318. Pergamon, Oxford.

Kita, T., Brown, M. S., Watanabe, Y., and Goldstein, J. L. (1981). Deficiency of low density lipoprotein receptors in liver and adrenal gland of the WHHL rabbit, an animal model of familial hypercholesterolemia. *Proc. Natl. Acad. Sci. U.S.A.* **78,** 2268–2272.

Kita, T., Brown, M. S., Bilheimer, D. W., and Goldstein, J. L. (1982a). Delayed clearance of very low and intermediate density lipoproteins with enhanced conversion to low density lipoprotein in WHHL rabbits. *Proc. Natl. Acad. Sci. U.S.A.* **79,** 5693–5697.

Kita, T., Goldstein, J. L., Brown, M. S., Watanabe, Y., Hornick, C. A., and Havel, R. J. (1982b). Hepatic uptake of chylomicron remnants in WHHL rabbits: A mechanism genetically distinct from the low density lipoprotein receptor. *Proc. Natl. Acad. Sci. U.S.A.* **79,** 3623–3627.

Knight, B. L., Patel, D. D., and Soutar, A. K. (1983). The regulation of 3-hydroxy-3-methylglutaryl-CoA reductase activity, cholesterol esterification and low-density-lipoprotein receptors in cultured monocyte-derived macrophages. *Biochem. J.* **210,** 523–532.

Kovanen, P. T., Bilheimer, D. W., Goldstein, J. L., Jaramillo, J. J., and Brown, M. S. (1981). Regulatory role for hepatic low density lipoprotein receptors *in vivo* in the dog. *Proc. Natl. Acad. Sci. U.S.A.* **78,** 1194–1198.

Kwiterovich, P. O., Fredrickson, D. S., and Levy, R. I. (1974). Familial hypercholesterolemia (one form of familial type II hyperlipoproteinemia). *J. Clin. Invest.* **53,** 1237–1249.

Langer, T., Strober, W., and Levy, R. I. (1972). The metabolism of low density lipoprotein in familial type II hyperlipoproteinemia. *J. Clin. Invest.* **51,** 1528–1536.

Lench, N., Stainer, P., and Williamson, R. (1988). Simple non-invasive method to obtain DNA for gene analysis. *Lancet* **1,** 1356–1358.

Mabuchi, H., Tayami, R., Haba, T., Veda, K., Veda, R., Kametani, T., Itoh, S., Koizumi, J., Oota, M., Myamoto, S., Takeda, R., and Takeshita, H. (1978). Homozygous familial hypercholesterolemia in Japan. *Am. J. Med.* **65,** 290–297.

Mabuchi, H., Sakai, T., Sakai, Y., Yoshimura, A., Watanabe, A., Wakasugi, T., Koizumi, J., and Takeda, R. (1983). Reduction of serum cholesterol in heterozygous patients with familial hypercholesterolemia. Additive effects of compactin. *N. Engl. J. Med.* **308,** 609–613.

Malloy, M. J., Kane, J. P., Kunitaka, S. T., and Tun, P. (1987). Complementarity of colestipol, niacin and lovastatin in treatment of severe familial hypercholesterolemia. *Ann. Intern. Med.* **107,** 616–623.

Mann, R., Mulligan, R. C., and Baltimore, D. (1983). Construction of a retrovirus packaging mutant and its use to produce helper-free defective retrovirus. *Cell* **33,** 153–159.

Martin, G. M., and Nestel, P. (1979). Changes in cholesterol metabolism with dietary cholesterol in children with familial hypercholesterolaemia. *Clin. Sci.* **56,** 377–380.

Miller, A. D., Palmer, T. D., and Hock, R. A. (1986). Transfer of genes into human somatic cells using retrovirus vectors. *Cold Spring Harbor Symp. Quant. Biol.* **51,** 1013–1019.

Moutafis, C. D., Myant, N. B., Mancini, M., and Oriente, P. (1971). Cholestyramine and nicotinic acid in the treatment of familial hyperbetalipoproteinaemia in the homozygous form. *Atherosclerosis* **14,** 247–258.

Moutafis, C. D., Simons, L. A., Myant, N. B., Adams, P. W., and Wynn, V. (1977). The effect of cholestyramine on the faecal excretion of bile acids and neutral steroids in familial hypercholesterolaemia. *Atherosclerosis* **26,** 329–334.

Myant, N. B. (1981). "The Biology of Cholesterol and Related Steroids." Heinemann, London.

Myant, N. B., and Slack, J. (1976). Type II-hyperlipoproteinemia. *In* "Handbuch der inneren Medizin: Fettstoffwechsel" (G. Schettler, H. Greten, G. Schlierf, and D. Seidel, eds.), pp. 275–300. Springer-Verlag, Berlin.

Packard, C. J., Shepherd, J., Third, J. L. H. C., Lorimer, R., Morgan, H. G., and Lawrie, T. D. V. (1976). Low-density-lipoprotein metabolism in type II hyperlipoproteinaemia. *Biochem. Soc. Trans.* **4,** 105–107.

Pittman, R. C., Carew, T. E., Attie, A. D., Witztum, J. L., Watanabe, Y., and Steinberg, D. (1982). Receptor-dependent and receptor-independent degradation of low density lipoprotein in normal rabbits and in receptor-deficient mutant rabbits. *J. Biol. Chem.* **257,** 7994–8000.

Reichl, D., Simons, L. A., and Myant, N. B. (1974). The metabolism of low-density lipoprotein in a patient with familial hyperlipoproteinaemia. *Clin. Sci. Molec. Med.* **47,** 635–638.

Reichl, D., Myant, N. B., and Pflug, J. J. (1977). Concentration of lipoproteins containing apolipoprotein B in human peripheral lymph. *Biochim. Biophys. Acta* **489,** 98–105.

Ross, R. (1986). The pathogenesis of atherosclerosis: An update. *N. Engl. J. Med.* **314,** 488–500.

Schmitz, G., and Assmann, G. (1982). Isolation of human serum HDL by zonal ultracentrifugation. *J. Lipid Res.* **23,** 903–910.

Schneider, W. J., Brown, M. S., and Goldstein, J. L. (1983). Kinetic defects in the processing of low density lipoprotein receptor in fibroblasts from WHHL rabbits and a family with familial hypercholesterolemia. *Mol. Biol. Med.* **1,** 353–367.

Schrott, H. G., Goldstein, J. L., Hazzard, W. R., McGoodwin, M. M., and Motulsky, A. G. (1972). Familial hypercholesterolemia in a large kindred. Evidence for a monogenic mechanism. *Ann. Intern. Med.* **76,** 711–720.

Seftel, H. C., Baker, S. G., and Sandler, M. P. (1980). A host of hypercholesterolaemic homozygotes in South Africa. *Br. Med. J.* **281,** 633–636.

Shattil, S. J., Bennett, J. S., Colman, R. W., and Cooper, R. A. (1977). Abnormalities of cholesterol-phospholipid composition in platelets and low-density lipoproteins of human hyperbetalipoproteinemia. *J. Lab. Clin. Med.* **89,** 341–353.

Shepherd, J., Bicker, S., Lorimer, A. R., and Packard, C. J. (1979). Receptor-mediated low density lipoprotein catabolism in man. *J. Lipid Res.* **20,** 999–1006.

Shepherd, J., Packard, C. J., Bicker, S., Lawrie, T. D. V., and Morgan, H. G. (1980). Cholestyramine promotes receptor-mediated low-density-lipoprotein catabolism. *N. Engl. J. Med.* **302,** 1219–1222.

Simons, L. A., Reichl, D., Myant, N. B., and Mancini, M. (1975). The metabolism of the apoprotein of plasma low density lipoprotein in familial hyperbetalipoproteinaemia in the homozygous form. *Atherosclerosis* **21,** 283–298.

Slack, J. (1969). Risks of ischaemic heart-disease in familial hyperlipoproteinaemic states. *Lancet* **2,** 1380–1382.

Slack, J., and Mills, G. L. (1970). Anomalous low density lipoproteins in familial hyperbetalipoproteinaemia. *Clin. Chim. Acta* **29,** 15–25.

Soutar, A. K., Myant, N. B., and Thompson, G. R. (1977). Simultaneous measurement of apolipoprotein B turnover in very-low- and low-density-lipoproteins in familial hypercholesterolaemia. *Atherosclerosis* **28,** 247–256.

Soutar, A. K., Myant, N. B., and Thompson, G. R. (1979). Metabolism of apolipoprotein B-containing lipoproteins in familial hypercholesterolaemia. Effects of plasma exchange. *Atherosclerosis* **32,** 315–325.

Soutar, A. K., Myant, N. B., and Thompson, G. R. (1982). The metabolism of very-low-density and intermediate-density lipoproteins in patients with familial hypercholesterolaemia. *Atherosclerosis* **43,** 217–231.

Spady, D. K., and Dietschy, J. M. (1985). Dietary saturated triglycerides suppress hepatic low density lipoprotein receptors in the hamster. *Proc. Natl. Acad. Sci. U.S.A.* **82,** 4526–4530.

Spady, D. K., Turley, S. D., and Dietschy, J. M. (1985). Rates of low density lipoprotein uptake and cholesterol synthesis are regulated independently in the liver. *J. Lipid Res.* **26,** 465–472.

Sprecher, D. L., Hoeg, J. M., Schaefer, E. J., Zech, L. A., Gregg, R. E., Lakatos, E., and Brewer, H. B. (1985). The association of LDL receptor activity, LDL cholesterol level, and clinical course in homozygous familial hypercholesterolemia. *Metabolism* **34,** 294–299.

Stange, E. F., and Dietschy, J. M. (1983). Cholesterol synthesis and low density lipoprotein uptake are regulated independently in rat small intestinal epithelium. *Proc. Natl. Acad. Sci. U.S.A.* **80,** 5739–5743.

Stange, E. F., and Dietschy, J. M. (1984). Age-related decreases in tissue sterol acquisition are mediated by changes in cholesterol synthesis and not low density lipoprotein uptake in the rat. *J. Lipid Res.* **25,** 703–713.

Stone, N. J., Levy, R. I., Fredrickson, D. S., and Verter, J. (1974). Coronary artery disease in 116 kindred with familial type II hyperlipoproteinemia. *Circulation* **49,** 476–488.

Tanzawa, K., Shimada, Y., Kuroda, M., Tsujita, Y., Arai, M., and Watanabe, Y. (1980). WHHL-rabbit: a low density lipoprotein receptor-deficient animal model for familial hypercholesterolemia. *FEBS Lett.* **118,** 81–84.

Teng, B., Sniderman, A. D., Soutar, A. K., and Thompson, G. R. (1986). Metabolic basis of hyperapobetalipoproteinemia. Turnover of apolipoprotein B in low density lipoprotein and its precursors and subfractions compared with normal and familial hypercholesterolemia. *J. Clin. Invest.* **77,** 663–672.

Thannhauser, S. J. (1940). "Lipidoses," 1st ed. Oxford University Press, New York.

Thannhauser, S. J. (1950). "Lipidoses, Diseases of the Cellular Lipid Metabolism," 2nd ed. Oxford University Press, New York.

Thompson, G. R., Lowenthal, R., and Myant, N. B. (1975). Plasma exchange in the management of homozygous familial hypercholesterolaemia. *Lancet* **1,** 1208–1211.

Thompson, G. R., Spinks, T., Ranicar, A., and Myant, N. B. (1977). Non-steady-state studies of low-density lipoprotein turnover in familial hypercholesterolaemia. *Clin. Sci. Molec. Med.* **52,** 361–369.

Thompson, G. R., Myant, N. B., Kilpatrick, D., Oakley, C. M., Raphael, M. J., and Steiner, R. E. (1980). Assessment of long-term plasma exchange for familial hypercholesterolaemia. *Br. Heart J.* **43,** 680–688.

Thompson, G. R., Soutar, A. K., Spengel, F. A., Jadhav, A., Gavigan, S. J. P., and Myant, N. B. (1981). Defects of receptor-mediated low density lipoprotein catabolism in homozygous familial hypercholesterolemia and hypothyroidism *in vivo*. *Proc. Natl. Acad. Sci. U.S.A.* **78,** 2591–2595.

Thompson, G. R., Ford, J., Jenkinson, M., and Trayner, I. (1986). Efficacy of mevinolin as adjuvant therapy for refractory familial hypercholesterolaemia. *Quart. J. Med.* **60,** 803–811.

Thompson, G. R., Barbir, M., Okabayashi, K., Trayner, I., and Larkin, S. (1989). Plasmapheresis in familial hypercholesterolaemia. *Arteriosclerosis Supplement 1,* **9,** *T*-152–157.

Tsukada, T., Rosenfeld, M., Ross, R., and Gown, A. M. (1986). Immunocytochemical analysis of cellular components in atherosclerotic lesions. Use of monoclonal antibodies with the Watanabe and fat-fed rabbit. *Arteriosclerosis* **6,** 601–613.

Valdivielso, P., Escolar, J. L., Cuervas-Mons, V., Pulpón, L. A., Chaparro, M. A. S., and González-Santos, P. (1988). Lipids and lipoprotein changes after heart and liver transplantation in a patient with homozygous familial hypercholesterolemia. *Ann. Intern. Med.* **108,** 204–206.

Wade, D. P., Knight, B. L., and Soutar, A. K. (1988). Hormonal regulation of low-density lipoprotein (LDL) receptor activity in human hepatoma Hep G2 cells. Insulin increases LDL receptor activity and diminishes its suppression by exogenous LDL. *Eur. J. Biochem.* **174,** 213–218.

Watanabe, Y. (1980). Serial inbreeding of rabbits with hereditary hyperlipidemia (WHHL-rabbit). *Atherosclerosis* **36,** 261–268.

Watson, J. D., Hopkins, N. H., Roberts, J. W., Steitz, J. A., and Weiner, A. M. (1987). "Molecular Biology of the Gene," 4th ed., Vol. II, pp. 936–943. Benjamin-Cummings, Minlo Park, California.

Williams, D. A., Lemischka, I. R., Nathan, D. G., and Mulligan, R. C. (1984). Introduction of new genetic material into pluripotent haematopoietic stem cells of the mouse. *Nature (London)* **310,** 476–480.

Wilson, J. M., Johnston, D. E., Jefferson, D. M., and Mulligan, R. C. (1988). Correction of the genetic defect in hepatocytes from the Watanabe heritable hyperlipidemic rabbit. *Proc. Natl. Acad. Sci. U.S.A.* **85,** 4421–4425.

Wissler, R. W. (1968). The arterial medial cell, smooth muscle cell or multifunctional mesenchyme. *J. Atheroscler. Res.* **8,** 201–213.

Yamamoto, T., Bishop, R. W., Brown, M. S., Goldstein, J. L., and Russell, D. W. (1986). Deletion in cysteine-rich region of LDL receptor impedes transport to cell surface in WHHL rabbit. *Science* **232,** 1230–1237.

Yokoyama, S., Hayashi, R., Satani, M., and Yamamoto, A. (1985). Selective removal of low density lipoprotein by plasmapheresis in familial hypercholesterolemia. *Arteriosclerosis* **5,** 613–622.

Yokoyama, S., Kikkawa, T., Hayashi, R., Satani, M., Kishino, B., and Yamamoto, A. (1986). Conventional LDL-apheresis techniques to attain regression of atheromatous vascular lesions. *Circulation* **74,** Suppl. II, 335.

Záhŏr, Z., and Czabanová, V. (1980). Changes in experimental atherosclerotic lesions of rat heart valves following long-term regression. *Atherosclerosis* **35,** 425–432.

Glossary

For the reader's convenience, abbreviations defined in the text, together with a few useful definitions, are brought together in the following list.

ABL: Abetalipoproteinemia

ACAT: Acyl-CoA : cholesterol acyltransferase

ACR: Absolute catabolic rate

ACTH: Adrenocorticotrophic hormone

AMP-PK: AMP-activated protein kinase

4-APP: 4-Aminopyrazolopyrimidine

bp: base pairs

CAD: Coronary artery disease

CAT: Chloramphenicol acetyltransferase

CHD: Cyclohexanedione

CHO: Chinese hamster ovary

Cholestyramine: A bile-acid-binding resin

Codon: A sequence of 3 nucleotides encoding an amino acid or a signal for terminating translation of mRNA

Compactin: A fungal inhibitor of HMG-CoA reductase

Compound heterozygote: Also called "genetic compound". An individual with two different mutant alleles at a given locus.

Consensus sequence: An average sequence in which each nucleotide is the most frequent at that position in a set of homologous sequences

CRM: Cross-reacting material

CURL: Compartment of uncoupling of receptor and ligand

DH: DNase-hypersensitive

Direct repeats: Identical or homologous sequences present in two or more copies oriented in the same direction in a molecule of DNA

cDNA: DNA complementary to mRNA

Downstream: In the 3′ direction with respect to the DNA coding strand

EC-modified: Modified by incubation in the presence of endothelial cells

EGF: Epidermal growth factor

ER: Endoplasmic reticulum

Exon: A segment of an interrupted gene that is present in the final mRNA

FCR: Fractional catabolic rate

FDB: Familial defective apoB-100

FH: Familial hypercholesterolaemia

FPP: Farnesyl pyrophosphate

GalNAc: *N*-Acetylgalactosamine

Genetic compound: *See* Compound heterozygote

GRE: Glucocorticoid response element

Haplotype: A set of two or more alleles on a single chromosome. (Extended in this book to include a set of restriction sites on a single chromosome)

HBL: Hypobetalipoproteinemia

HDL: High-density lipoprotein

HDLc: HDL enriched with cholesterol and apoE

HMG: Hydroxymethylglutar-(yl) (ate)

IDL: Intermediate-density lipoprotein

Intron: An intervening sequence between two exons that is removed from the primary transcript by splicing together the exons on each side of it

IRE: β-Interferon gene regulatory element

kb: Kilobases

kDa: Kilodaltons

K: Molecular weight multiple of one thousand (45 K = 45,000)

LC: Light chains of clathrin

LDL: Low-density lipoprotein

Linkage disequilibrium: The association, in unrelated individuals, of two linked alleles more frequently than would be expected by chance

Lovastatin: The same as mevinolin

Lp(a): Lipoprotein (a)

LPDS: Lipoprotein-deficient serum

LRP: LDL-receptor-related protein

LTR: Long terminal repeat

Man: Mannose

Mevastatin: The same as compactin

Mevinolin: An analogue of compactin

MH: Micrococcal nuclease-hypersensitive

Missense mutation: A mutation that changes the amino acid encoded by a codon to a different amino acid

M6P: Mannose-6-phosphate

MRE: Metal regulatory element

MTI: Mouse metallothionein-I

Nonsense mutation: A mutation that changes a codon for an amino acid to a stop codon

nt: Nucleotide

Null allele: A mutant allele that expresses no detectable protein product

PCR: Polymerase chain reaction

PDGF: Platelet-derived growth factor

Plasmid: A closed circular extramitrochrondrial DNA duplex capable of independent replication within a cell

Probucol: A drug that lowers the plasma cholesterol concentration

Restriction site: A short segment of double-stranded DNA recognized by a restriction enzyme

RFLP: Restriction-fragment-length polymorphism

mRNA: Messenger RNA

tRNA: Transfer RNA

T_c: Phase transition temperature

Term: Stop codon

tk: Thymidine kinase

Transgenic animals: A line of animals produced by injecting DNA into the nucleus of the egg

Transposon (Transposable element): A sequence of genomic DNA that can replicate and insert itself at random elsewhere in the genome

UDP: Uridine diphosphate

Upstream: In the 5′ direction with respect to the coding strand

VLDL: Very-low-density lipoprotein

WHHL: Watanabe Heritable Hyperlipidemic

Appendix A: Abbreviations Used for Amino Acids

Amino acid	Three-letter word	One-letter symbol
Alanine	Ala	A
Arginine	Arg	R
Asparagine	Asn	N
Aspartic acid	Asp	D
Cysteine	Cys	C
Glutamic acid	Glu	E
Glutamine	Gln	Q
Glycine	Gly	G
Histidine	His	H
Isoleucine	Ile	I
Leucine	Leu	L
Lysine	Lys	K
Methionine	Met	M
Phenylalanine	Phe	F
Proline	Pro	P
Serine	Ser	S
Threonine	Thr	T
Tryptophan	Trp	W
Tyrosine	Tyr	Y
Valine	Val	V

Appendix B: Abbreviations for Nucleotide Bases

One-letter symbol	Base
A	Adenine (purine)
C	Cytosine (pyrimidine)
G	Guanine (purine)
T	Thymine (pyrimidine)
U	Uracil (pyrimidine)

Index